MXenes: Next-Generation 2D Materials

MXenes: Next-Generation 2D Materials

Fundamentals and Applications

Edited by

Jay Singh
Banaras Hindu University, Varanasi, Uttar Pradesh, India

Kshitij RB Singh
Banaras Hindu University, Varanasi, Uttar Pradesh, India

Ravindra Pratap Singh
Indira Gandhi National Tribal University, Amarkantak, Madhya Pradesh, India

Charles Oluwaseun Adetunji
Edo University, Iyamho, Nigeria

WILEY

This edition first published 2024.

Registered Offices
John Wiley & Sons, Inc., 111 River Street, Hoboken, NJ 07030, USA
John Wiley & Sons Ltd, The Atrium, Southern Gate, Chichester, West Sussex, PO19 8SQ, UK

For details of our global editorial offices, customer services, and more information about Wiley products visit us at www.wiley.com.

Wiley also publishes its books in a variety of electronic formats and by print-on-demand. Some content that appears in standard print versions of this book may not be available in other formats.

Library of Congress Cataloging-in-Publication Data Applied for:
Hardback ISBN: 9781119873990

Cover Design: Wiley
Cover Image: © Sergey Nivens/Shutterstock

Set in 9.5/12.5pt STIXTwoText by Straive, Chennai, India
Printed and bound by CPI Group (UK) Ltd, Croydon, CR0 4YY

C9781119873990_150324

*In loving memory of the **Late Professor Ravindra Pratap Singh**, whose insightful guidance and unwavering support shaped this book into its finest form. Your dedication to excellence and commitment to nurturing the written word will forever inspire us. Your legacy lives on through these pages. Sorrow fills our hearts, for we miss you dearly.*

Contents

Editor Biographies

Dr. Jay Singh

Dr. Jay has been working as an assistant professor at the Department of Chemistry, Institute of Sciences, Banaras Hindu University, Varanasi, Uttar Pradesh, since 2017. He received his PhD degree in polymer science from Motilal Nehru National Institute of Technology in 2010 and his MSc and BSc from Allahabad University, Uttar Pradesh, India. He was a postdoctoral fellow at National Physical Laboratory, New Delhi; Chonbuk National University, South Korea; and Delhi Technological University, Delhi. Dr. Jay has received many prestigious fellowships like CSIR (RA), DST-Young Scientist fellowship, and DST-INSPIRE faculty award. He is actively engaged in the development of nanomaterials (CeO_2, NiO, rare-earth metal oxide, Ni, $Nife_2O_4$, Cu_2O, graphene, RGO, etc.) based nanobiocomposite, conducting polymer and self-assembled monolayers-based clinically important biosensors for estimation of bioanalaytes such as cholesterol, xanthine, glucose, pathogens, and pesticides/toxins using DNA and antibodies. Dr. Jay has published more than 105 international research papers with total citations of more than 4300 and an h-index of 38. He has completed/running various research projects for different funding agencies. He has edited more than 8 books and has authored more than 40 book chapters for internationally reputed press publications, namely, Elsevier, Springer Nature, IOP, Wiley, and CRC. He is actively engaged in fabricating metal oxide-based biosensors for clinical diagnosis, food packaging applications, drug delivery, and tissue engineering applications. His research has contributed significantly toward the fundamental understanding of interfacial charge transfer processes and sensing aspects of metal nanoparticles.

Mr. Kshitij RB Singh

Mr. Singh is a postgraduate in biotechnology from Indira Gandhi National Tribal University, Amarkantak, Madhya Pradesh, India. He is currently working in the laboratory of Dr. Jay Singh, Department of Chemistry, Institute of Science, Banaras Hindu University, Varanasi, Uttar Pradesh, India. He has more than 60 peer-reviewed publications to his credit, has edited 12 books, and has authored more than 50 book chapters published in internationally reputed press, namely, Elsevier, IOP Publishing, Springer Nature, Wiley, and CRC Press. He is currently also involved in editing books with international publishing houses, including IOP Publishing, Elsevier, and Springer Nature. His research interests include biotechnology, biochemistry, epidemiology, nanotechnology, nanobiotechnology, biosensors, and materials science.

Dr. Ravindra Pratap Singh

Dr. Singh received his BSc from Allahabad University, India, and his MSc and PhD in biochemistry from Lucknow University, India. He is currently working as an assistant professor in the Department of Biotechnology, Indira Gandhi National Tribal University, Amarkantak, Madhya Pradesh, India. He has previously worked as a scientist at various esteemed laboratories globally, namely, Sogang University, IGR Paris. His work and research interests include biochemistry, biosensors, nanobiotechnology, electrochemistry, material sciences, and biosensor applications in biomedical, environmental, agricultural, and forensics. He has to his credit several reputed national and international honors/awards. Dr. Singh has authored over 70 articles in international peer-reviewed journals, more than 70 book chapters of international repute, and has edited 14 books. He serves as a reviewer for many reputed international journals and is also a member of many international societies. He is currently also involved in editing various books, which will be published by internationally reputed publication houses, namely, IOP Publishing, CRC Press, Elsevier, and Springer Nature. Moreover, he is a book series editor of "Emerging advances in Bionanotechnology," CRC Press, Taylor and Francis Group. He is also actively involved in guest-editing special issues (SIs) for reputed international journals, and one of the latest edited SI by Dr. Singh is "Smart and Intelligent Nanobiosensors: Multidimensional Applications" for *Materials Letters*, Elsevier, and "Smart and Intelligent Optical Materials for Sensing Applications" for *Luminescence*, Wiley.

Professor Charles Oluwaseun Adetunji

Professor Adetunji is presently a faculty member at the Microbiology Department, Faculty of Sciences, Edo State University Uzairue (EDSU), Edo State, Nigeria, where he utilized the application of biological techniques and microbial bioprocesses for the actualization of sustainable development goals and agrarian revolution, through quality teaching, research, and community development. He is currently the Ag Dean for the Faculty of Science and the Head of the Department of Microbiology at EDSU. He is a visiting professor and the executive director of the Center of Biotechnology, Precious Cornerstone University, Ibadan. He has won several scientific awards and grants from renowned academic bodies. He has published many scientific journal articles and conference proceedings in refereed national and international journals with over 370 manuscripts. He was ranked among the top 500 prolific authors in Nigeria from 2019 till date by SciVal/SCOPUS. His research interests include microbiology, biotechnology, postharvest management, and nanotechnology. He was recently appointed as the President and Chairman Governing Council of the Nigerian Bioinformatics and Genomics Network Society. He is presently the General/Executive Secretary of the Nigerian Young Academy. He is presently a series editor with Taylor and Francis in the United States, editing several textbooks on agricultural biotechnology, nanotechnology, pharmafoods, and environmental sciences. He is an editorial board member of many international journals and serves as a reviewer for many double-blind peer-reviewed journals of renowned publishing houses such as Elsevier, Springer, Francis and Taylor, Wiley, PLOS One, Nature, American Chemistry Society, and Bentham Science Publishers.

List of Contributors

Saman Abrishamkar
Department of Chemistry
University of Isfahan
Isfahan
Iran

Charles Oluwaseun Adetunji
Applied Microbiology, Biotechnology and Nanotechnology Laboratory
Department of Microbiology
Edo University Iyamho
Auchi
Edo State
Nigeria

Alexander Ikechukwu Ajai
Department of Chemistry
Federal University of Technology Minna
Minna
Nigeria

Takashiro Akitsu
Department of Chemistry
Faculty of Science Division II
Tokyo University of Science
Tokyo
Japan

Ghazal Amini
Department of Chemistry
University of Isfahan
Isfahan
Iran

Tinku Basu
Amity Centre for Nanomedicine
Amity University Uttar Pradesh
Noida
India

Esther Bernard
Department of Chemical Engineering
Federal University of Technology Minna
Minna
Nigeria

Chansi
Amity Centre for Nanomedicine
Amity University Uttar Pradesh
Noida
India

Neakanshika Chadha
Department of Applied Sciences
KIET Group of Institutions
Ghaziabad
Delhi
India

Adil Denizli
Department of Chemistry
Hacettepe University
Ankara
Turkey

Manish Kumar Dixit
Department of Metallurgical Engineering and Materials Sciences
Indian Institute of Technology Indore
Indore
India

Mrigendra Dubey
Department of Metallurgical Engineering and Materials Sciences
Indian Institute of Technology Indore
Indore
India

Gregory Edema
Department of Physical Science Laboratory Technology
Federal Polytechnic Auchi
Edo State
Nigeria

Amir Ershad-Langroudi
Color and Surface Coatings Group
Polymer Processing Department
Iran Polymer and Petrochemical Institute (IPPI)
Tehran
Iran

Evbuomwan L.
Department of Microbiology
Edo State University Uzairue
Nigeria

Yves Grohens
University Bretagne Compositic
Sud, Lorient
France

and

Centre de Recherche Christian Huygens
Rue de Saint-Maudé
Lorient
France

Karan Hadwani
Amity Centre for Nanomedicine
Amity University Uttar Pradesh
Noida
India

Abel Inobeme
Department of Chemistry
Edo State University Uzairue
Iyamho
Nigeria

John Olusanya Jacob
Department of Chemistry
Federal University of Technology Minna
Minna
Nigeria

Raghavachari Kavitha
Department of Chemistry
Post-Graduation Studies
Vijaya College
Affiliated to Bengaluru City University
Bengaluru
Karnataka
India

Rout George Kerry
PG Department of Biotechnology
Utkal University
Bhubaneswar
Odisha
India

Seçkin Kılıç
Department of Chemistry
Hacettepe University
Ankara
Turkey

Sampada Koirala
Biomedical Engineering Program
Department of Electrical and Computer Engineering
North Dakota State University
Fargo
ND
USA

Arun Kumar
Department of Chemistry
Doon University
Dehradun
Uttarakhand
India

Devendra Kumar
Department of Applied Chemistry
Delhi Technological University
New Delhi
India

Shivashankar Girish Kumar
Department of Chemistry
RV College of Engineering
Affiliated to Visvesvaraya Technological University
Bengaluru
Karnataka
India

Praveen Kumar
Department of Chemistry
Doon University
Dehradun
Uttarakhand
India

Cansu İlke Kuru
Department of Biotechnology
Graduate School of Natural and Applied Sciences
Ege University
İzmir
Turkey

and

Buca Municipality Science and Art Center
Izmir
Turkey

Kadavil Subhash Lakshmy
School of Energy Materials
Mahatma Gandhi University
Kottayam
Kerala
India

Sneha Lavate
Catalysis & Hydrogen Research Lab
Department of Petroleum Engineering
School of Energy Technology
Pandit Deendayal Energy University
Gandhinagar
Gujarat
India

Sankhula Lokesh
Catalysis & Hydrogen Research Lab
Department of Petroleum Engineering
School of Energy Technology
Pandit Deendayal Energy University
Gandhinagar
Gujarat
India

Abdullahi Mann
Department of Chemistry
Federal University of Technology Minna
Minna
Nigeria

John Tsado Mathew
Department of Chemistry
Ibrahim Badamasi Babangida University
Lapai
Nigeria

Sachin Mishra
Department of Electrical Electronic Engineering and Automation
Rovira i Virgili University
Tarragona
Spain

Sujeet K. Mishra
CCRAS-Central Ayurveda Research Institute
Jhansi
UP
India

Abbas Mohammadi
Department of Chemistry
University of Isfahan
Isfahan
Iran

Anandu M. Nair
School of Energy Materials
Mahatma Gandhi University
Kottayam
Kerala
India

Arunadevi Natarajan
Department of Chemistry
PSGR Krishnammal College for Women
Coimbatore
Tamil Nadu
India

Bangalore Nanjundappa Nagalaxmi
Department of Chemistry
Maharani Lakshmi Ammanni College for Women
Affiliated to Bengaluru City University
Bengaluru
Karnataka
India

Daisuke Nakane
Department of Chemistry
Faculty of Science Division II
Tokyo University of Science
Tokyo
Japan

Efosa John Osarenren
Department of Biochemistry
Federal Polytechnic Auchi
Auchi
Nigeria

Stanley Osita Okonkwo
Department of Chemistry
Federal University of Technology Minna
Minna
Nigeria

Shyam S. Pandey
Graduate School of Life Science and Systems Engineering
Kyushu Institute of Technology
Wakamatsu
Kitakyushu
Japan

Hamidreza Parsimehr
Department of Chemistry
University of New Brunswick
Fredericton
Canada

Ananthu Prasad
International and Inter University Centre for Nanoscience and Nanotechnology (IIUCNN)
Mahatma Gandhi University
Kottayam
Kerala
India

Jong Soo Rhyee
Department of Applied Physics and Institute of Natural Sciences
Kyung Hee University
Yong-In, Gyong-gi
Republic of Korea

Pooja Rawat
Department of Applied Physics and Institute of Natural Sciences
Kyung Hee University
Yong-In, Gyong-gi
Republic of Korea

and

Department of Chemistry
Biochemistry and Forensic Science
Amity School of Applied Sciences
Amity
University Haryana, Gurugram
Haryana, India

Ramakshi Rana
Department of Chemistry
Doon University
Dehradun
Uttarakhand
India

Aman Sahu
Department of Chemistry
LRPG College
CCS University
Ghaziabad
Uttar Pradesh
India

Yoshiyuki Sato
Department of Chemistry
Faculty of Science Division II
Tokyo University of Science
Tokyo
Japan

Mahek Sadiq
Biomedical Engineering Program
Department of Electrical and Computer Engineering
North Dakota State University
Fargo
ND
USA

Elijah Yanda Shaba
Department of Chemistry
Federal University of Technology Minna
Minna
Nigeria

Samaneh Shahsavarifar
Department of Metrology and Optoelectronics
Faculty of Electronics, Telecommunications, and Informatics
Gdansk University of Technology
Gdansk
Poland

Jay Singh
Department of Chemistry
Institute of Science
Banaras Hindu University
Varanasi
Uttar Pradesh
India

Kshitij RB Singh
Graduate School of Life Science and Systems Engineering
Kyushu Institute of Technology
Wakamatsu
Kitakyushu
Japan

and

Department of Chemistry, Institute of Science
Banaras Hindu University
Varanasi
Uttar Pradesh
India

Ravindra Pratap Singh
Department of Biotechnology
Faculty of Science
Indira Gandhi National Tribal University
Amarkantak
Madhya Pradesh
India

Rohit Srivastava
Catalysis & Hydrogen Research Lab
Department of Petroleum Engineering
School of Energy Technology
Pandit Deendayal Energy University
Gandhinagar
Gujarat
India

Channe Gowda Sushma
Department of Chemistry
Maharani Lakshmi Ammanni College for Women
Affiliated to Bengaluru City University
Bengaluru
Karnataka
India

Tedunjaiye O.H.
Edo State University Uzairue
Nigeria

Sushma Thapa
Graduate School of Life Science and Systems Engineering
Kyushu Institute of Technology
Wakamatsu
Kitakyushu
Japan

and

Department of Chemistry, Institute of Science
Banaras Hindu University
Varanasi
Uttar Pradesh
India

Sabu Thomas
International and Inter University Centre for Nanoscience and Nanotechnology (IIUCNN)
Mahatma Gandhi University
Kottayam
Kerala
India

and

School of Energy Materials
Mahatma Gandhi University
Kottayam
Kerala
India

and

Department of Chemical Sciences
University of Johannesburg
Doornfontein
Johannesburg
South Africa

Sukirti Tiwari
Department of Nanotechnology
Amity University
Noida
Uttar Pradesh
India

Aykut Arif Topçu
Medical Laboratory Program
Vocational School of Health Service
Aksaray University
Aksaray
Turkey

Fulden Ulucan-Karnak
Department of Medical Biochemistry
Institute of Health Sciences
Ege University
Bornova
Izmir
Turkey

Danling Wang
Biomedical Engineering Program
Department of Electrical and Computer Engineering
North Dakota State University
Fargo
ND
USA

Suji Mary Zachariah
International and Inter University Centre for Nanoscience and Nanotechnology (IIUCNN)
Mahatma Gandhi University
Kottayam
Kerala
India

and

University Bretagne Compositic
Sud, Lorient
France

and

University of South Brittany (UBS)
Centre de Recherche Christian Huygens
Rue de Saint-Maudé
Lorient
France

Preface

Two-dimensional (2D) nanomaterials have attracted great attention since 2004. Owing to reduced dimension and size, 2D materials have exhibited many intriguing properties that are not found in their bulk counterparts, holding tremendous promise for a host of applications ranging from electronic, biomedical, and optoelectronic devices to electrochemical catalysis. Among them, in 2011, first reported a newly discovered large family of 2D early transition few-atoms-thick layers of transition metal carbides, nitrides, or carbonitrides nanosheets of Ti_3C_2 named "MXenes" is a rapidly rising star in current growing research. The efficient and unique electronic, optical, chemical, and mechanical properties of MXenes coupled with the ease of processing have helped to exhibit their marvelous pledge to revolutionize various science and technological developments. MXenes are exfoliated from their ternary carbide and nitride ceramics and have a common formula in composition: $M_{n+1}X_nT_x$, ($n = 1, 2, 3$, or 4), where M is a transition metal, X presents carbon/nitrogen (C/N), T means the surface group, and x represents the equivalent amount. The strong hydrophilicity, exceptional conductivity, high elastic mechanical strength, large surface-to-volume ratio, chemical stability, and improved electrochemical properties make MXenes open broad prospects for their applications in material science and technology areas, including energy storage, optoelectronics, spintronics, environmental, biomedical, and electronic industries include electro-catalysis, photocatalysis, membrane separation, supercapacitors, hybrid-ion capacitors, batteries, point-of-care devices, hydrogen storage, nanoelectronics, and sensors/biosensors. However, the aggregation, difficulty obtaining a single layer, restacking, and oxidation of MXene nanosheets significantly hinder their performance; and to overcome these issues, an effective and easy strategy is to combine MXene (due to good flexibility) with other materials, including polymers, metal oxides, metal nanoparticles, and carbon, to form MXene hybrid nanocomposites. MXene hybrid nanocomposites provide an opportunity to integrate the outstanding properties of different materials in a complementary way. Therefore, MXenes and MXene-based composites have attracted considerable research interest, holding great promise for numerous applications. Therefore, due to the increased demands of these interesting and fascinating 2D advanced futuristic nanomaterials, there have been increasing concerns regarding the development of high-performance energy storage devices, efficient electrochemical behavior, optoelectronic and flexible spintronics, catalysis industries, and effective management strategies to be urgently nodded. Therefore, the application of nanostructured MXenes and MXene-based composite materials has been discovered as effective techniques that could be sustainable, efficient, effective, and innovative to combat all the challenges mentioned earlier.

This book aims to spotlight the highly promising topic of MXenes, focusing on their fundamental, detailed preparation methods, physicochemical and biological characteristics, functionalization, processing techniques, and multi-domain applications. It presents the current advances in MXenes and their nanocomposites, including electrochemical energy storage, biomedical fields, catalysis and electro/photocatalysis, and environmental applications. In addition, it also provides an overview of the current state-of-the-art progress in the field of gas sensors and electrochemical biosensors for the detection of various biomolecules, pharmaceutical drugs, and environmental pollutants. Finally, this book also highlighted the limitations, advantages, future perspectives, and challenges of MXenes and MXene-based materials.

Jay Singh
Kshitij RB Singh
Ravindra Pratap Singh
Charles Oluwaseun Adetunji

Acknowledgment

It gives us immense pleasure to acknowledge Professor Shri Prakash Mani Tripathi, Honourable Vice-Chancellor of Indira Gandhi National Tribal University, Amarkantak, India, Professor Sudhir K. Jain, Honourable Vice-Chancellor, Banaras Hindu University, Varanasi, Uttar Pradesh, India, Institutes of Eminence (IoE), Mistry of Education, India, Professor Emmanuel O. Aluyor, Vice-Chancellor, Edo State University, Uzairue, Nigeria, and Professor Julius Kola Oloke, Precious Cornerstone University, Ibadan, Nigeria for providing constant assistance in all the possible ways. It is also our great pleasure to acknowledge and express our enormous debt to all the contributors who have provided their quality material to prepare this book. We are grateful to our beloved family members (Babita Singh, Juliana Bunmi Adetunji, Ranjana Verma, Aparajita RB Singh, Emmanuel Ayomiposi Adetunji, Taiwo Jesugbemi Adetunji, Kehinde Jesugbemi Adetunji, and Shanvi Singh), who joyfully supported and stood with us in many hours of our absence to finish this book project. Thanks to Kubra Ameen, Manoj Kumar, Priyadarshini Natarajan, Stefanie Volk, Sakeena Quraishi, and the entire publishing team for their patience and extra care in publishing this book.

Jay Singh
Kshitij RB Singh
Ravindra Pratap Singh
Charles Oluwaseun Adetunji

1

Introduction to MXenes a Next-generation 2D Material

Kshitij RB Singh[1,2], Sushma Thapa[1,2], Jay Singh[2], Shyam S. Pandey[1], and Ravindra Pratap Singh[3]

[1]*Graduate School of Life Science and Systems Engineering, Kyushu Institute of Technology, Wakamatsu, Kitakyushu, Japan*
[2]*Department of Chemistry, Institute of Science, Banaras Hindu University, Varanasi, Uttar Pradesh, India*
[3]*Department of Biotechnology, Faculty of Science, Indira Gandhi National Tribal University, Amarkantak, Madhya Pradesh, India*

1.1 Introduction

Two-dimensional (2D) materials that have sheet-like morphology with lateral magnitude considerably more prominent than the thickness have attracted significant attention due to their unique electronic structure, large surface areas, and various physicochemical properties, which have been widely used and have potential in several fields, for instance, smart wearable sensors, energy storage, agriculture, environment, and biomedical applications [1, 2]. Therefore, the most studied 2D material is graphene because of its successful exfoliation in 2004, which exhibits potential application in nano-composites, electrochemical energy storage, and transparent electrodes owing to its extraordinary mechanical, optical, and electronic properties despite their various properties, it is limited in certain aspects, for example, field-effect transistors [3–5]. Thus, investigation emphasis has deviated to various 2D materials, including black phosphorus (BP), layered double hydroxides (LDHs), transition metal dichalcogenides (TMDs), and 2D hydroxides and oxides with supreme physicochemical properties [2, 6]. Recently, a new class of 2D materials composed of transition metal carbides, nitrides, and carbonitrides, called MXenes, first reported in 2011 by Yury Gogotsi with their coworkers at Drexel University, has developed and brought significant attention to the research community by virtue of their excellent properties (such as surface hydrophilic, magnetic, and semiconductor properties) and unique electronic and structural characteristics compared to the bulk [1].

The general formula of MXene is written as $M_{n+1}X_nT_z$ where "M" is a D-block element, for example, scandium (Sc), titanium (Ti), zirconium (Zr), hafnium (Hf), vanadium (V), and niobium (Nb), X in the formula represents nitrogen (N) or carbon (C), and T_x is a surface functional group such as O, OH, Cl, or F, and n is 1, 2, or 3. To accomplish the exclusive characteristic, MXene is mainly etched from its parent MAX phase as precursors with the chemical formula $M_{n+1}AX_n$, where A represents the main group elements such

MXenes: Next-Generation 2D Materials: Fundamentals and Applications, First Edition.
Edited by Jay Singh, Kshitij RB Singh, Ravindra Pratap Singh, and Charles Oluwaseun Adetunji.

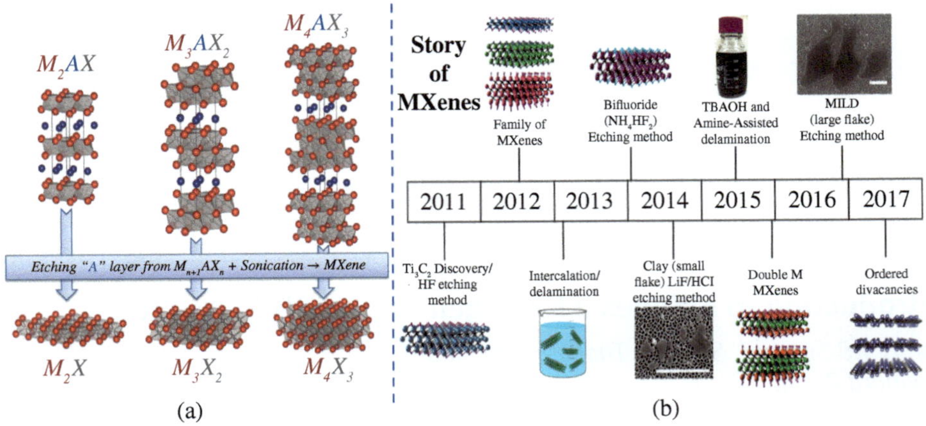

Figure 1.1 (a) Selective etching of MAX phase. Source: Reproduced with permission from [7]/John Wiley & Sons; (b) Timeframe of MXenes fabrication. Source: Reproduced with permission from [18]/ American Chemical Society.

as Al, Si, Ga, etc. (Figure 1.1a) [9, 10]. Thus, the fabricated 2D MXenes are estimated by powder X-ray diffraction (XRD), scanning electron microscopy (SEM), X-ray absorption spectra (XAS), Fourier transform infrared (FTIR), transmission electron microscopy (TEM), energy-dispersive spectrometer (EDS), Raman spectroscopy, and nuclear magnetic resonance (NMR). Also, these techniques evaluate surface termination and intrinsic defects; moreover, elemental stoichiometric ratio and chemical composition are assessed by X-ray photoelectron spectroscopy (XPS) [11, 12]. Multitudinous fabrication approaches have been established to selectively etch the MAX phase, where the M–A bonds in the MAX phase are metallic bonds with comparatively feeble binding strength, whereas the M–X bond is ionic and covalent bonds with high binding strength [12]. Furthermore, in the beginning, the selective etching of the MAX phase was done by fluorine-containing compounds, such as LiF+HCl, bifluoride salts, fluorine-containing molten salt (Figure 1.1b), and HF, which defined the three main surface terminations of MXenes, i.e. O, OH, and F. Later, fluorine-free etching methods (e.g. concentrated alkaline hydrothermal etching and electrochemical etching) were reported to develop fluorine-free MXenes. In addition, during the past two years, for fabricating accordion-like MXenes, a nonaqueous molten salt etching method was used, where Lewis acidic salts were used as the etchants [13–15].

The MAX phases have a hexagonal structure, where the X atoms are located in the octahedral sites formed by M elements, and M and A layers are intercalated. Through computational simulation studies, it has been discovered that 2D MXenes could have six possible morphologies: solid solutions, vacancies randomly distributed, vacancies ordered, mono-M elements, ordered in-plane double-M, and ordered out-of-plane. Similar to the MAX phases, 2D MXenes also have hexagonal morphology, where M atoms are organized in a hexagonal structure and the octahedral sites are filled with X atoms [1, 16, 17]. Currently, above 100 stoichiometric arrangements of ultrathin MXenes have been anticipated. In contrast, more than 30 different configurations of MXenes have been efficiently fabricated through experiments that expand the family of 2D MXenes. Most of the synthesized MXenes are hydrophilic, have high elastic moduli, and are metallic, which exhibit great potential in various applications, for instance, catalysis, agriculture, biomedicine, and energy storage [14, 18]. Moreover, the three primary surface terminations of 2D ultrathin MXenes (OH, O, and F) have a substantial effect on their physicochemical properties. 2D MXenes with surface functionalization show semiconducting behavior, while without termination exhibit metal conductivity; thus, this evidence affirms that surface termination has a direct impact on the electronic state of MXenes [19]. Therefore, most 2D MXenes show metallic behavior, exhibiting electronic conductivity higher than all other 2D materials, which exhibit excellent promise in several applications due to their great potential. Hence, this book chapter briefly covers all areas related to MXenes, such as the synthesis methods, properties, characterization, and applications of MXenes. Accordingly, it focuses on the MXenes challenges, limitations, and prospects.

1.2 Properties

MXene is a 2D material that has acknowledged considerable emphasis compared to other materials (such as graphene) because of the multitudinous interesting chemical, physical, and biological properties owing to excellent hydrophilicity, biocompatibility, high surface area, tunable electronic band gap, and embodied greater elements (like C, Ti, and N). MXenes have fascinating physicochemical, electrical, optical, magnetic, mechanical, and thermal properties. The properties of MXene could be altered for suitable applications by changing MXene's composition. The electronic properties of ultrathin MXene have been comprehensively reviewed at the abstract level, and many works have formerly studied the wide variety of electronic properties. The electronic properties of MXene have an essential role in the effectiveness of supercapacitor cells, which stimulates the research into the electronic properties of MXenes. With respect to other 2D materials, MXene's electronic properties are the most exceptional. Generally, MXene's electronic and structural properties were estimated through the generalized gradient approximation (GGA) framework [20–22].

The electronic properties of 2D MXene rely on the character of X, M, and surface terminations. With the alteration in the elemental composition of M layers, the bandgap of MXene changed. Also, altering the surface termination bandgap of MXene could be modified, which is illustrated by density functional theory (DFT) calculations, and the electronic properties of pristine MXenes vary from metallic to semiconducting upon surface functionalization. Due to the Fermi level shift, most terminated MXenes maintain their metallic character (Ti_2CO_2, Hf_2CO_2, Cr_2CF_2, etc.) and are predicted to be semiconductors. For certain types of activations (for example, the production of reactive oxygen species and bio-catalysis), the semiconducting property of MXenes is essential [1, 23, 24]. The F and OH terminated are semiconductors with bandgap values of 0.1 and 0.05 eV; moreover, by changing their bandgap, these materials are allowed to be utilized in field-effect transistors to semiconductors [25]. Anasori et al. substituted two layers of titanium (Ti) with molybdenum (Mo) in the outer transition metal M layers of M_3C_2 and M_4C_3 MXenes, wherein the Mo atoms impart semiconducting property to the MXene with a fine bandgap of 0.05 eV. Therefore, the outer M layers seriously impact the electronic properties of MXenes [26]. The electronic properties of MXenes are also affected by preparation methods of MXenes because of the various extents of defects and surface termination. Apart from the semiconducting and conductivity properties, the metallic properties of 2D MXenes are also relevant for bio-applicable devices.

The optical properties of MXene could be utilized to demonstrate various applications such as surface Raman scattering (SERS), plasmonic broadband absorbers, photonic diodes, and effective light-to-heat conversion, which are also highly convenient for biomedical applications [27, 28]. The optical properties of materials are closely associated with their electronic and structural properties, for example, band-to-band electronic transition, which improves optical adsorption; moreover, the electronic transition could affect the imaginary dielectric constant. The termination of MXene enormously impacts the electronic and structural properties, so it also has a significant role in optical properties [28]. Lashgari et al. have calculated band structures and density of states (DOS) using DFT formulations, which are used to identify the optical properties of the MXene sheet [29].

Current research has revealed that surface functionalization can largely influence both electronic and structural properties, which also affect optical properties. The hydroxylated/fluorinated terminations could only accept one electron, and two terminations have similar performance in the visible range [30]. Berdiyorov et al. used DFT calculations to examine the consequence of the functionalization on the optical properties of MXene ($Ti_3C_2T_2$, where T signifies a surface functional group). Furthermore, hydroxylated/fluorinated samples manifest lower absorption as against MXene, which makes them more appropriate for transparent electrode applications, whereas oxidized samples exhibit high absorption. The conclusion illustrates that surface functionalization considerably affects optical properties [31]. MXenes show great potential for flexible, transparent electrodes because of their optical transparency in the visible region. Ultrathin $Ti_3C_2T_x$ MXene absorbs light in the UV–vis region from 300 to 500 nm, and the 5 nm-thickness film showed transmittance up to 91.2%, which is essential for PTT applications. By altering ion intercalation, the transmittance values could be optimized. For example, tetramethylammonium hydroxide (NMe_4OH) increased the transmittance of the MXene, while urea, DMSO, and hydrazine reduced the transmittance [32]. However, the investigation of optical properties such as plasmonic, luminescence efficiency, and emission color requires to be exemplified for further augmentations of MXenes applications. Just a few hypothetical reviews have anticipated the magnetic properties of MXenes under the termination-free condition, and until now, the magnetic properties of MXene have not been much examined because of the obstruction in the development of termination-free 2D MXenes [33]. Almost all MAX phases have been established as magnetic containing Mn and Cr, such as Cr_2GeC, Cr_2GaC, Mn_2AlC, Cr_2AlC, Cr_2GeC, C_2GaN, Cr_2AlN, Cr_2TiAlC_2, and Mn_2GaC. Unfortunately, to date, none of the aforementioned MAX phases have been etched into MXene; only Cr_2TiAlC_2 has been exfoliated into Cr_2TiC_2. Furthermore, the ground states of the majority of the functionalized and pristine MXenes are nonmagnetic, which is indicated by the spin-polarized DFT calculations because the covalent bonding between M and X elements as well as the chemically attached group is strong [34–37].

Nevertheless, various MXenes are intrinsically magnetic, for instance, Ti_2N, Cr_2N, Ti_2C, Cr_2C, and Mn_2C, whereas Cr_2N and Mn_2C are antiferromagnetic, and Ti_2N and Ti_2C exhibit approximately half-metallic ferromagnetism with a magnetic moment of 1.00 and 1.91 μB per formula unit. Additionally, at room temperature with F and OH groups attached, the Cr_2C and Cr_2N remain ferromagnetic, and it was also found that in Cr-based MXenes, the energy difference between magnetic and nonmagnetic is so large that these systems withhold their magnetism at room temperature [20, 38, 39]. Khazaei et al. obtained the functionalized ferromagnetic ground states in pristine Cr_2C and Cr_2N MXenes [20]. Subsequently, surface termination groups (for example, OH, F, and O) could be used to obtain desired magnetic properties. He et al. obtained that, after functionalization with OH, F, and Cl, the Mn_2C MXene retains the ferromagnetism ground state [40]. Similarly, several other properties have been explored that augment the MXenes application.

1.3 Synthesis and Functionalization of MXenes

In the past seven years, various fabricating methods have been broadly established for manufacturing novel and additional MXenes, either through bottom–up or top–down fabrication methods. More than 30 types of MXenes have been efficaciously synthesized, and $Ti_3C_2T_x$ was first synthesized in 2011. It has been predicted that additional materials will also subsist and still carry out an investigation, highlighting various opportunities as it seeks high-performance materials. The synthesis parameters significantly impact the uniformity, morphologies, properties, and surface terminal groups of MXenes [41, 42]. MXenes, a new class of 2D (Figure 1.2) transition metal carbides and nitrides, have been researched and reviewed in detail due to their physicochemical properties and various applications [44]. There are primarily two approaches to synthesizing MXenes, including selective etching and exfoliation, which are top–down approaches, and chemical vapor deposition (CVD) and physical vapor deposition (PVD), which are bottom–up

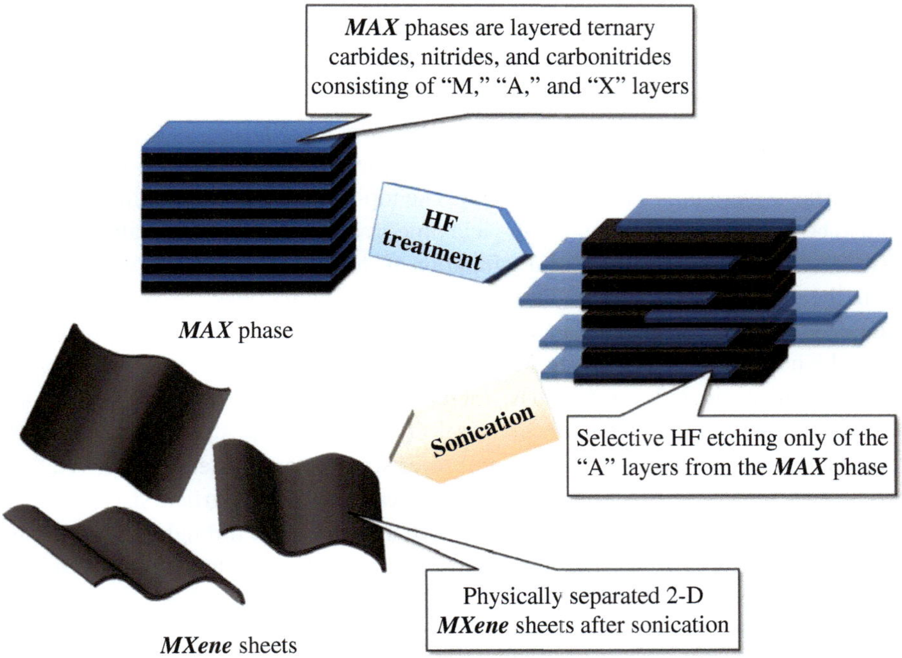

Figure 1.2 Schematic representation of synthesis of 2D MXene. Source: Reproduced with permission from [43]/American Chemical Society.

approaches. In fabricating MXenes, the top–down method is frequently used compared to the bottom–up process. Because of the complicated component configuration, the interaction procedure among the elements is yet unclear [42]. After the first Ti_3C_2 MXene fabrication by utilizing hydrofluoric (HF) acid selective etching of Ti_3AlC2, the MXenes are still in the initial phase. In general, the MXenes are acquired by selectively etching the specific atomic layer (Al layers) from the 3D group of the ternary carbides and nitrides MAX phases and exfoliating them in top–down methods [45, 46].

The M–X and M–A covalent bonds are the significant characteristics of MAX phases. Since the chemical activity of the M–A bond is weaker than that of the M–X bond and could be readily broken. The removal of weakly bonded M–A led to the essential synthesis of nitrides and carbides, which is a necessary step for MXene fabrication. Therefore, the first MXene, multilayered $Ti_3C_2T_{x,}$ was developed by Yury Gogotsi et al. in 2011 by immersing Ti_3AlC_2 in HF [1].

$$Ti_3AlC_2 + 3HF \rightarrow AlF_3 + 3/2H_2 + Ti_3C_2 \tag{1.1}$$

$$Ti_3C_2 + 2H_2O \rightarrow Ti_3C_2(OH)_2 + H_2 \tag{1.2}$$

$$Ti_3C_2 + 2HF \rightarrow Ti_3C_2F_2 + H_2 \tag{1.3}$$

Equation (1.1) describes the step used to extract the Al element from Ti_3AlC_2, yielding Ti_3C2, whereas Eqs. (1.2) and (1.3) illustrate the functionalization steps of the Ti_3C_2 with OH and F elements, respectively, forming $Ti_3C_2(OH)_2$ and $Ti_3C_2F_2$. Also, a series of 2D MXenes were widely investigated experimentally through selective etching (HF) of aluminum layers such as Ti_2C, Mo_2C, Nb_2C, V_2C, Zr_3C_2, and Ta_4C_3. Approximately all sorts of MXenes were achieved with various morphologies of single-stacked nanosheets [47]. However, HF is pernicious to human beings and the atmosphere, and the performance of MXenes will be low because of this type of approach. So, Halim et al. found a substitute for HF to etch the atomic layer from the MAX phase, which is NH_4HF_2 [48].

The top–down fabrication approaches are usually premised on breaking comparatively bulky material to get nano-sized particles [49]. The top–down synthesis method has been affirmed to be the most effortless and high-yield process in fabricating nanoscale and attenuating MXenes, including the direct acid exfoliation of layered solids. Therefore, the conversion from the parent MAX phases to ultrathin 2D MXene follows selective etching employing suitable etchants, including HF etching, HF-forming, alkali etching, and molten-salts etching, which leads to the multi-stacked MXenes, and delamination and intercalation routes to obtain a single-layer MXenes, including organic intercalant and metal-ion intercalant [50]. HF etching is effective in the selective etching of MAX phases and reports as-prepared MXenes with accordion-like multilayer structures. Furthermore, the etching temperature, etching time, MAX phase quality, intercalation agent, particle size, and etching concentration have a substantial impact on MXene fabrication throughout the etching process [14]. Alhabeb et al. fabricated the most reviewed Ti_3AlC_2 by using HF etchant with different concentrations and etching times at room temperature [8]. Besides, a higher concentration of HF etchant and longer etching time are needed for a higher n value in the MAX phase. Different HF concentrations result in different surface terminations. A greater atomic number (M) requires a strong etchant due to the strength of the M–A chemical bond and an extended period of time are essential for the complete exfoliation when different etchants are employed, such as NH_4HF2 instead of HF. Also, fluoride salts (LiF, NaF, and KF) have been utilized together with HCl as an etchant to synthesize MXenes [48]. After selective etching, filtration or centrifugation is performed to separate solid particles. Then MXenes typically undergo a delamination process to separate the MXene sheets. The delamination procedure could be achieved by using an intercalating agent and ultrasonication. The gap between two sequential MXene sheets is usually enhanced by intercalating agents, which cause them to separate effortlessly. The absence of intercalating agents produces smaller-sized MXenes. Some common examples of intercalants are dimethyl sulfoxide (DMSO), isopropylamine, and tetrapropylammonium hydroxide (TPAOH) or tetrabutylammonium hydroxide (TBAOH) [51]. However, an additional delamination step is unnecessary in certain instances because the etching and exfoliation synchronize [52]. The fundamental parameters to corroborate the efficacious synthesis of MXene from the top–down method are the etching time, optimization of etchant, and type of intercalant. Whereas specific, effective bottom–up fabrication has been submitted, it is the alternative method to produce high-quality 2D MXenes through atomic-scale control. Nevertheless, the bottom–up approach has some merit in contrast to the top–down approach, for example, composition, size, morphology, and structure, which are essential to fabricating MXenes.

Bottom–up fabrication commonly commences from small-scale molecules, followed by crystal growth [53]. Barsoum et al. carried out the first bottom–up synthesis of MAX films, from which MXene films were produced, which utilized c-axis-oriented sapphire as the substrate, was cleaned using acetone, dried by nitrogen gas, and finally preheated under high vacuum at 780 °C [48]. The significant fabrication methods premised on the bottom–up approach are CVD and PVD to establish the tiered materials with greater yield. CVD engenders wide-surface and excellent materials that are efficient tools in an application, for example, integrated electronic devices. Ren et al. primarily used the CVD technique in 2015 to achieve excellent-quality ultrafine α-Mo_2C crystals through a Cu/Mo foil as the substrate, and as a carbon source, methane gas is used at a temperature above 1085 °C [54]. In addition to the CVD technique, plasma-enhanced pulsed laser deposition (PELPD) and template method have also been operated for MXene synthesis. Notwithstanding the importance of the bottom–up method for fabricating the ultrathin MXene nanosheets, there are still some restrictions, for example, complicated process, large-scale fabrication, thin-film transmission, and synthesis of monolayer. But through additional enhancements such as purification Cu etching, diminishing waste, and dropping temperatures, MXenes could be fabricated by the bottom–up method at a low cost.

1.4 Characterization of MXenes

Many advanced techniques have been employed, such as XRD, SEM, XAS, FTIR, XPS, TEM, EDS, and NMR, for the characterization of MXene flakes because of the wide variability of MXene composition and morphology, whereas every advanced characterization technique has its own advantages, drawbacks, and characteristics when applied to MXenes [8, 55]. XRD is one of the simplest and first ways to corroborate that MXene is fabricated. The typical layer morphology of the MXenes is observed through SEM after the HF etching [56]. When EDS coupled with SEM technique provides a high-quality information about the M element, although this technique could not be credible for studying the surface terminations and X ratios [57, 58]. The qualitative assessment of the functional groups is accomplished by the FTIR [59]. XAS is another effective technique for analyzing MXene structure; moreover, this technique has been employed to study the fundamental electronic states of materials and surface reactions on MXene sheets (catalysis and adsorption). Surface analysis of MXene XPS has gained immense acclaim due to its surface sensitivity, capability to obtain details about the chemical composition, and low penetration depth. XPS is also employed to evaluate the thermal stability and the composition of the material [60, 61]. Raman spectroscopy studied the vibrational properties of the investigated samples [62]. NMR techniques have been employed to acquire information regarding the effects of the various synthesis paths and surface functionalization of MXenes [11, 63]. Likewise, other techniques have been employed to confirm the successful synthesis of MXenes and determine their structure, properties, and composition.

1.5 Application of MXenes

There have been many proposed applications for 2D MXenes, and energy storage was the foremost application studied for MXenes. The adaptability of MXene has fascinated several applications due to its excellent physicochemical properties, unique 2D layered structure, good electrical conductivity, surface chemistry, and high Young modulus. Some of these MXenes exhibited comparable or greater efficiency than any other material currently being used. Therefore, Mxenes can be used in biomedical, agricultural, environmental, and energy storage applications.

1.5.1 Biomedical

2D MXenes have been intensively applied in biomedical applications on the basis of their remarkable properties, such as magnetic, optical, and electronic. Due to their multiple configurations, MXenes are generally biocompatible and innocuous. MXenes have various functional groups on their surface, which make them hydrophilic, and these surface functional groups are chemically reactive. Additionally, by several approaches, the surface of MXenes could be smoothly altered and furnish additional opportunities for further biomedical applications [53, 64]. Up until now, their specified applications in the biomedical field have been comprising biosensors, antibacterial activity, and drug delivery. To diminish the effects on human health caused by bacterial evolution, several 2D materials have been considered, and among them, MXenes have shown antibacterial activity higher than GO because of their higher conductivity. The surface of MXene enhances the bacteria-cell membrane interaction due to its anion nature and hydrophilicity [24]. However, to date, there are various research reports on the antibacterial activities of ultrathin MXenes. Rasool et al. in 2017 reported the outstanding antibacterial activity of polyvinylidene fluoride (PVDF) supported by titanium carbide ($Ti_3C_2T_x$) MXenes up against both, *B. subtilis* (Gram-positive bacteria) and *E. coli* (Gram-negative bacteria). There exists a decrease in feasibility and suppression of the growth of about 73% *E. coli* and 67% *B. subtilis*. The coated MXene with PVDF has enhanced hydrophilicity, mitigated the presence of huge pores in the membrane, and has a high aspect ratio. Environmental conditions and the morphology of bacterial cell walls affect antibacterial proficiency. By AFM, flow cytometry, and SEM, the interactions between $Ti_3C_2T_x$ MXenes and antibacterials could be investigated [65]. Similarly, Jastrzebska et al. synthesized Ti_3C_2 and Ti_2C from their parent MAX phases (Ti_3AlC_2 and Ti_2AlC) in accordance with the same condition of the classical acidic aluminum extraction method. The biological activity of Ti_3C_2 and Ti_2C MXenes is compared to that of *E. coli* bacteria. Furthermore, the Ti_3C_2 MXene exhibited bactericidal properties while Ti_2C did not affect the feasibility of bacteria, and further investigation presented greater antibacterial activity from downsized nanosheets [66, 67]. The semiconducting property, ease of surface functionalization, and high surface area to volume of MXenes can be implemented and applicable for the killing of bacterial cells [68]. Currently, Li et al. reported that environmentally friendly photoelectrical materials overwhelm the expanding antibiotic resistance and designed an interfacial Schottky junction of $Bi_2S_3/Ti_3C_2T_x$ due to the variation in the contact potential among Bi_2S_3 and $Ti_3C_2T_x$. $Bi_2S_3/Ti_3C_2T_x$ nanocomposite comprehensively enhanced the number of reactive oxygen species (ROS) under 808 nm near-infrared radiation because of the photocatalytic activity and killed 99.92% of *E. coli* and 99.86% of Staphylococcus aureus (*S. aureus*) in 10 minutes [69]. Theoretically, MXenes are supposed to be firm for a sufficiently long time to generate antibacterial properties but also remain biocompatible. Consequently, to increase MXene's application as an antimicrobial agent, further research should be implemented on improving the stability of MXene in its biological conditions.

Drug delivery is identified as one of the relevant approaches to achieving the target drug and alleviating its detrimental impact on patients [70]. MXene-based materials provide numerous opportunities and have enormous potential as a drug delivery system due to their physicochemical properties. Han et al. in 2018 for the first time synthesized chemotherapeutic drugs and photothermal therapy (PTT) from 2D Ti_3C_2 MXenes immobilized by soybean phospholipid (SP), which has been consistently estimated both in vivo and in vitro [71]. Xing et al. fabricated 2D ultrathin Ti_3C_2 MXene nanosheets and cellulose based on hydrogels nanoplatforms for dual-modality chemotherapy against cancer. The synthesized MXene/DOX@cellulose has excellent photothermal properties, a high loading potential of DOX, and biocompatibility. Furthermore, it also exhibited dynamic near-infrared light (NIR)-stimulation and has much greater efficiency for the treatment of tumors than chemotherapy or PTT [72]. Apart from antibacterial and drug delivery, MXenes also have a greater capacity for diagnostic bioimaging. Diagnostic imaging allows the supposition of the tumor site and the monitoring of the PTT. Fluorescence microscopy is one of the most frequently used bioimaging techniques in nanomedicine research due to its user-friendliness, outstanding imaging sensitivity, and the economical apparatus needed. Regardless of the low penetration depth, it is very appropriate to use this method for bioimaging [73]. Liu et al. synthesized in 2017 2D ultrathin Ti_3C_2 MXenes loaded with the fluorescent drug doxorubicin (DOX) and hyaluronic acid (HA) for cancer therapy with a drug loading capacity as high as 84.2%. The as-obtained ultrathin MXene nanosheet manifests efficacious singlet oxygen generation upon 808 nm laser irradiation, excellent photothermal conversion efficiency, and an outstanding mass extinction coefficient [74].

Quantum dots (QDs) have exhibited outstanding properties for fluorescent imaging compared to traditional fluorescein due to their high photostability, tunable wavelength, and required quantum yield. Through the hydrothermal method, MXene nanosheets could be broken into ultrasmall QDs with size defect-induced quantum confinement. Therefore, these MXene QDs with suitable surface modifications (Nb_2C, V_2C, and Ti_3C_2) are proficient in cellular fluorescent imaging [75, 76]. Yang et al. fabricated 2D ultrafine Nb_2C MXene with pristine crystallographic structure by exfoliation assisted by layer cutting through a high-intensity ultrasonication treatment in TPAOH. Because of the high biocompatibility, excellent photobleaching, and unique biodegradation of Nb_2C QDs, which exhibited this, it is a potential application of fluorescence imaging [77]. Similarly, there are other bioimaging modalities of MXenes comprised of photoacoustic (PA), computed tomography (CT), and magnetic resonance (MR) imaging, which are the most useful diagnostic imaging methods in biomedicine due to their low toxicity, nonionizing, spatial resolution, deep tissue penetration, and physicochemical properties [51, 78].

1.5.2 Agricultural

With the advancement of living standards and science and technology, a significant number of researchers have apprehended considerable attention and interest in agricultural activities. In an agricultural environment, rain is a common naturally occurring phenomenon; on the other hand, sometimes endless rain brings adverse impacts on wheat heading, pollination of walnut chestnuts, corn sowing, etc. [79]. Agrotextiles could be extensively used in farming, horticulture, and other agricultural activities to solve this problem due to their several properties, which comprise softness, permeability, stretchability, and breathability. Furthermore, it also has huge potential in sunshade fabric, shields for grass, de-icing, and mulching [80]. Therefore, to ameliorate the yield and quality of agricultural products, Triboelectric nanogenerator (TENG) yarn was knitted into a textile as an agrotextile due to its merits of being stretchable, low-cost, and mechanically strong; moreover, it shields farm animals and crops from environmental circumstances such as continuous rain, sunshade, and anti-icing [81, 82]. Hence, Ping et al. in 2020 developed triboelectric yarn with AgNPs/MXene/fiber consecutively coated with PDMS. The developed triboelectric yarn has a ~400 μm diameter and could be utilized for sensing natural conditions without any external power source owing to its outstanding robustness, stretchability, and good mechanical properties. Besides, the structure and approximate size of MXene could be observed by SEM, TEM, and AFM images [83]. Ammonia (NH_3) is a molecule that is obtaining enormous interest in numerous fields, including agri-food industry, hydrogen production, automotive, agriculture, etc. Despite its utilization, NH_3 is extremely toxic to humankind and environment; it can cause lung edema, nervous disorders, and renal dysfunction. Supremely ammonia is produced from agriculture due to livestock production and fertilizers on a large scale. Hence, to overcome these challenges, an NH_3 sensor is developed, and MXene-polymer nanocomposites (MXP-NCs) has excellent potential for NH_3 sensing in industrial, biomedical, and agriculture due to their excellent conductivity, tunable morphology, great conductivity, and mechanical flexibility [84]. MXene biosensors have also been successfully applied in agriculture for pesticide detection and improving properties. Pesticides including insecticides, fungicides, and herbicides are a threat to mankind and the environment; however, the use of pesticides could effectively protect the crop from weeds, disease, and pests [85, 86]. Ping et al. fabricated multi-dimensional nanocomposites (MXene/Au-Pd) based on electrochemical biosensors for the identification of organophosphorus pesticides (OPs) owing to large surface area and higher conductivity. Through self-reduction on the surface of 2D MXene, bimetallic nanoparticle Au–Pd was synthesized. The fabricated biosensor exhibits a 1.75 ng l^{-1} limit of detection [87]. Likewise, Shen et al. fabricated AV@Ti_3C_2 for pest control, where it was obtained by adsorbing avermectin (AV) on ultrathin Ti_3C_2 MXene nanosheets with a loading capacity of 81.44%. The fabricated AV@Ti_3C_2 reduces the spraying time and dosage of the pesticide, and it also exhibited biosafety, which does not affect the growth and germination of the seed [88].

1.5.3 Environmental

The environment has been severely affected, and there is serious water pollution, which has elevated severe ecological environment concerns due to the ongoing growth of the economy and global industrialization [89]. Additionally, the ineffective process, inappropriate usage, and unintentional discharge of industrial manufacturing of products include the use of heavy metals, pesticides, and a mixture of azo dyes into the ecosystem, which disturbs the equilibrium through their lethal properties affecting the health of the living organism and the environment. Also, chemicals like antibiotic residue, waste gas, and radioactive elements correspondingly pose a threat to human health because the majority of them are toxic [90, 91]. Generally, the pollutants in the environment could be categorized into organic and inorganic compounds. Therefore, to eliminate the toxic impurities and overpower the serious environmental issues, a wide variety of methods have been developed for the removal of environmental pollutants that are environmentally friendly and chemically practical, mainly including adsorption, filtration, electrochemical approaches, chemical oxidation, ion exchange, etc. Due to its high availability, low operating requirements, and low price adsorption is considered to be the most advantageous and economical method in wastewater treatment [92, 93].

For efficient environmental remediation for wastewater treatment, nanomaterials have become an environmentally friendly and economically feasible option because nanomaterial-based devices could permit to achieve a new level of selectivity, efficiency, mechanical stability, and sensitivity [94, 95]. Hence, MXenes and MXenes-based composites have an excellent potential for environmental remediation because of their high surface area, activated metallic hydroxide sites, and high hydrophilicity. As aforementioned, adsorption is frequently implemented in water purification due to its low cost, accessibility of several adsorbents, and ease of operation. One of the most ominous organic pollutants in water are dyes, which are fundamental components in paper, printing, and textile industries. Despite their utilization, they engender considerable damage to the ecological environment and humankind. So, it is necessary and urgent task for all researchers to remove these dyes from wastewater. MXenes have already been broadly employed as adsorbents, which presents a convenient option for removing dyes from wastewater because of the layered structure and high specific area [96–98]. Mashtalir et al. showed that multilayered 2D ultrathin $Ti_3C_2T_x$ MXene is an efficient adsorbent for the cationic dye methylene blue (MB) and anionic dye acid blue 80 (AB80), and by Langmuir and Freundlich isotherms, the adsorption of MB on $Ti_3C_2T_x$ was fitted, with fitting parameters K_F and ν equal to 48.15 and 19.96. The conclusion showed that MB could irreversibly be bonded to

$Ti_3C_2T_x$ MXene with an adsorption capacity of ~ 39 mg g^{-1}, and the concentration of AB80 has not changed [99]. Unlike organic dyes, heavy metal ions are also the utmost threat to the environment and mankind by virtue of their firm nature and nondegradable. 2D MXenes and their composite have attracted immense consideration as desirable adsorbents for the removal of heavy metal ions widely exist in industrial wastewater [100]. Peng et al. in 2014 fabricated hydroxyl-terminated $Ti_3C_2T_x$ ($Ti_3C_2(OH/ONa)_xF_{2-x}$) material by a chemical exfoliation and subsequent alkalization intercalation method exhibiting extraordinary adsorption performance toward lead ion (Pb^{2+}). More hydroxyl groups on the surface of $Ti_3C_2T_x$ MXene form due to the alkalization, which will contribute to the adsorption of lead. The adsorption method discovered that alk-MXene possesses a remarkable lead adsorption capacity of 140 mg g^{-1} [101]. In addition to organic dyes and heavy metals, MXenes and their composites can also treat radioactive ions and gaseous pollutants.

1.5.4 Miscellaneous Field

Like in biomedical, agriculture, and environment, ultrathin MXene nanosheets have also received great attention in energy storage and conversion. The coming generation depends on our competencies to reserve, generate, and use fuels for energy production. Rapid depletion of fossil fuels and global warming are due to augment energy consumption, which demands for highly efficient energy storage and conversion devices comprised fuel cells, alkali metal batteries, supercapacitors, etc. [102, 103]. Therefore, MXene and MXene-based materials with low formula weights, such as Ti_2C, Nb_2C, V_2C, and Sc_2C, have great potential for energy storage and conversion applications owing to their metallic conductivity, excellent hydrophilic surface, large surface area, and mechanical stability [104]. Among all MXenes, Sc_2C has the largest surface area and possesses a storage capacity of 9.0 wt%. Also, the hydrogens are bound to Sc_2C by physical and chemical interaction with binding energies of 4.703 and 0.087, respectively [105]. MXenes have exhibited an enormous promise as electrode for lithium, sodium, potassium, and aluminum ion batteries; however, so far, most of the research has been primarily concerned on Li-ion (LIBs) and Na-ion batteries (SIBs). LIBs consist of a cathode as electron acceptor and an anode as the electron donor, a separator to isolate electrodes, and an electrolyte to connect the two electrodes [106].

Moreover, the LIBs exemplify the significant power source in electric vehicles, hybrid electric vehicles, and small portable electronics, which needed lightweight and high capacity to gratify the increasing energy consumption. Because of their substantial capacities, LIBs are among the most efficacious secondary batteries, which is beneficial for compact volume use [107, 108]. A high capacitance of up to 350 F cm^{-3} has been reported, as a result of intercalation between 2D $Ti_3C_2T_x$ MXene and several ions (Li^+, Na^+, Mg^{2+}, and Al^{3+}) and it has shown that Ti_3C_2 exhibits an excellent Li-ion capacity of 410 mA h g^{-1} at a one cycling rate. Naguib et al. show the possibility of using MXenes as anode materials for LIBs by fabricating 2D ultrathin Ti_2C from Ti_2AlC (MAX phase) at room temperature in aqueous solutions, which showed a reversible capacity about 5 times higher than the Ti_2AlC because of its weaker interlaminar force, weaker bonds between MX layers in Ti_2C, and open structure. The as-produced material exhibited a stable capacity of 225 mAh g^{-1} [109].

Hence, several abstract studies have shown that MXenes are promising for increasing battery performance [110]. On the other hand, supercapacitors also attracted great interest and fall in the category of energy storage systems that store electricity owing to long cycle life, fast charge/discharge capabilities, and high power densities; moreover, they are used in backup power stations and in vehicles [63]. Supercapacitor is classified into three types based on variations in the actual charge storage: pseudocapacitors, asymmetrical supercapacitors, and double-layer capacitors (EDLCs). The three main components of a supercapacitor are electrolyte, anode, and cathode [111]. MXene and MXene-based materials exhibit superior performance in supercapacitors, which are considered alternatives to energy storage devices by virtue of their excellent intercalation effects and superior electrical conductivity. Lukatskaya et al. reported that 2D Ti_3C_2 MXene electrode synthesized from MAX phase in concentrated HCl have volumetric capacitance of over 300 F cm^{-3}; however, by using a solution of HCl–LiF method shows great volumetric capacitance of 900 F cm^{-3} at 2 mVs^{-1} rate with excellent rate performance and cyclability, and this fabrication method avoids the handling of hazardous HCl [112].

Surface chemistry also affects the volumetric capacitance of supercapacitor, and the substitution of F atoms for O terminations could increase MXene performance. In addition, heat treatments, chemical modification, and different MXene synthesis routes also enhance their volumetric capacitance [113–115]. Titanium, vanadium, and molybdenum-based compounds were only synthesized, despite several synthesized MXenes. Hence, to enlarge and establish new supercapacitor electrodes, different compounds have to be considered and assessed.

1.6 Current Scenario, Risk Assessment, and Challenges

Over the past decade, researchers have fabricated and elaborated numerous 2D MXene and MXene-based materials that have significant potential in various applications, for instance, environment, biomedicine (drug delivery, antibacterial, etc.), agriculture, and energy storage because of their unique physicochemical properties; moreover, among all other 2D materials, ultrathin MXene is more dominant. Despite showing great potential in several applications there are still certain challenges affiliated with MXenes, which must be overwhelmed to advance further developments.

So far, most ultrathin MXenes samples have been fabricated through top–down methods from MAX phase that lacks control of reaction process and resultant products. However, only a certain considerable effort has been accomplished in fabricating monolayer MXenes through bottom–up approach with few defects; consequently, the emphasis of researchers in this direction must be devoted to the enormous advancement in the fabrications of excellent-quality MXenes with the appropriate properties and characteristics. Theoretically, a large amount of 2D MXenes have been predicted to have excellent properties, but only a few of them have been experimentally predicted [116]. Studies have reported that MXenes generally have multilayer and accordion-like structures instead of 2D nanosheets, which hinders their properties in various applications [117, 118].

Hitherto majority of 2D MXenes have been synthesized by selectively etching from MAX phase with the etchant HF and fluoric salts, which are extremely toxic, and the severe reaction circumstances inhibit the extensive synthesis of MXenes. Thus, because of HF toxicity other etchants that are fluoride-free and environmentally friendly, for instance, LiF and HCl, electrochemical etching, and molten salt etching method, have been followed for etching A element from MAX phase. Additionally, MXenes with Cl or F terminations without using toxic reagents could be successfully fabricated by molten salt etching method; however, other MXenes with different surface terminations and M should be explored and considered using this method [47, 119]. The structural and electronic properties of 2D MXenes could be tuned by surface termination and by other active components. Notwithstanding the surplus of F and O terminal groups on the surface obstructing the space formation due to the layer structure collapse, which may affect their magnetic and electric properties. Moreover, the affiliation between catalytic performance and surface termination is lacking, so it is necessary to distinguish the relation between MXene with surface termination and catalytic performance due to its increasing application in electrolysis [120, 121]. Likewise, the oxidative and thermal stability of 2D MXenes in aqueous medium is also a decisive challenge. When 2D MXenes are fabricated by utilizing HF-etchant, the produced MXene is less stable and effortlessly oxidized at room temperature. Therefore, currently, such deterioration of MXenes in colloidal solution is being stored at subzero temperature for long-term stability; however, the solution is unfeasible thus improved approaches need to be prospected for prohibiting them from oxidation and for storing MXene solution [122].

Various characterization techniques have been applied, such as XRD, energy-dispersive X-ray spectroscopy (EDS), and XPS, despite that the specific control of surface moieties has been unclear [123]. So far research has been mainly concerned with the optical and electrical properties, with just a few researches on their magnetic properties. Compared to the first two properties, magnetic properties have less application, especially in biomedical applications besides MRI imaging [124]. A deeper comprehension of the mechanism is beneficial to enhance their properties. Mxene has shown tremendous potential for environmental application, but it is only limited to lab-scale experiments or theoretical studies; thus, the potential of environmental applications must be explored to estimate their feasibility under actual conditions. Moreover, more applications of MXenes must be considered in biomedical fields [125, 126]. Therefore, further efforts should be devoted to developing innocuous, great-production, ever-lasting evaluations of the biocompatibility and environmentally friendly synthetic approaches that can effectively control the morphology, surface termination, and stability of MXenes, particularly for those computationally predicted.

1.7 Conclusion and Prospects

MXene, a new 2D material, has been explored through the selective chemical etching of A in $M_{n+1}AX_n$ phases, has gained considerable acclaim and captivated great research interest within a short time span after its discovery due to its peculiar catalytical properties and exclusive layered morphology. MXenes with their various properties, for instance, electronic, mechanical, thermal, magnetic, and optical properties, have shown enormous potential in multifaceted applications for coming-generation energy storage and conversion systems. Also, MXenes have a unique potential for many biomedical applications (such as biosensing, drug delivery, bioimaging, and antibacterial), environmental remediation, and agriculture applications. The MXenes potential applications and properties largely depend on the manufacturing methodologies, the structure, the morphologies, composition, and the type of surface functional groups. Almost all the MXene samples that have been synthesized so far are through top–down approaches, and only a very few attempts have been made to grow MXene through bottom–up methods. Therefore, more efforts should be made to explore more synthetic methods, expand the variety of MXene family, manufacture tunable MXene with properties, and realize large-scale synthesis. The structure, properties, synthetic method, and wide range of applications of MXenes and MXene-based composites in the biomedical field, agriculture, energy storage, and environmental remediations are summarized in this book chapter. Although MXenes have made some achievements in biological imaging, PTT, antibacterial therapy, and other fields, there are still many challenges for MXenes to be solved.

References

1 Naguib, M., Kurtoglu, M., Presser, V. et al. (2011). Two-dimensional nanocrystals produced by exfoliation of Ti_3AlC_2. *Adv. Mater.* 23 (37): 4248–4253.

2 Novoselov, K.S., Mishchenko, A., Carvalho, A., and Castro Neto, A.H. (2016). 2D materials and van der Waals heterostructures. *Science (1979)* 353 (6298).

3 Nicolosi, V., Chhowalla, M., Kanatzidis, M.G. et al. (2013). Liquid exfoliation of layered materials. *Science (1979)* 340 (6139).

4 Zhu, Y., Murali, S., Cai, W. et al. (2010). Graphene and graphene oxide: synthesis, properties, and applications. *Adv. Mater.* 22 (35): 3906–3924.

5 Castro Neto, A.H., Guinea, F., Peres, N.M.R. et al. (2009). The electronic properties of graphene. *Rev. Mod. Phys.* 81 (1): 109–162.

6 Novoselov, K.S., Jiang, D., Schedin, F. et al. (2005). Two-dimensional atomic crystals. *Proc. Nat. Acad. Sci.* 102 (30): 10451–10453.

7 Naguib, M., Mochalin, V.N., Barsoum, M.W., and Gogotsi, Y. (2014). 25th Anniversary article: MXenes: a new family of two-dimensional materials. *Adv. Mater.* 26 (7): 992–1005.

8 Alhabeb, M., Maleski, K., Anasori, B. et al. (2017). Guidelines for synthesis and processing of two-dimensional titanium carbide ($Ti_3C_2T_x$ MXene). *Chem. Mater.* 29 (18): 7633–7644.

9 Barsoum, M.W. (2000). The $M_{N+1}AX_N$ phases: a new class of solids. *Prog. Solid State Chem.* 28 (1–4): 201–281.

10 Eklund, P., Beckers, M., Jansson, U. et al. (2010). The $M_{n+1}AX_n$ phases: materials science and thin-film processing. *Thin Solid Films* 518 (8): 1851–1878.

11 Hantanasirisakul, K. and Gogotsi, Y. (2018). Electronic and optical properties of 2D transition metal carbides and nitrides (MXenes). *Adv. Mater.* 30 (52): 1804779.

12 Sang, X., Xie, Y., Lin, M.-W. et al. (2016). Atomic defects in monolayer titanium carbide ($Ti_3C_2T_x$) MXene. *ACS Nano* 10 (10): 9193–9200.

13 Ghidiu, M., Naguib, M., Shi, C. et al. (2014). Synthesis and characterization of two-dimensional Nb_4C_3 (MXene). *Chem. Commun.* 50 (67): 9517–9520.

14 Anasori, B., Lukatskaya, M.R., and Gogotsi, Y. (2017). 2D metal carbides and nitrides (MXenes) for energy storage. *Nat. Rev. Mater.* 2 (2): 16098.

15 Urbankowski, P., Anasori, B., Makaryan, T. et al. (2016). Synthesis of two-dimensional titanium nitride Ti_4N_3 (MXene). *Nanoscale* 8 (22): 11385–11391.

16 Paul, J.T., Singh, A.K., Dong, Z. et al. (2017). Computational methods for 2D materials: discovery, property characterization, and application design. *J. Phys.: Condens. Matter* 29 (47): 473001.

17 Shein, I.R. and Ivanovskii, A.L. (2012). Planar nano-block structures $Ti_{n+1}Al_{0.5}C_n$ and $Ti_{n+1}C_n$ (n=1, and 2) from MAX phases: structural, electronic properties and relative stability from first principles calculations. *Superlattices Microstruct.* 52 (2): 147–157.

18 Choi, G., Shahzad, F., Bahk, Y.-M. et al. (2018). Enhanced terahertz shielding of MXenes with Nano-metamaterials. *Adv. Opt. Mater.* 6 (5): 1701076.

19 Khazaei, M., Arai, M., Sasaki, T. et al. (2014). Two-dimensional molybdenum carbides: potential thermoelectric materials of the MXene family. *Phys. Chem. Chem. Phys.* 16 (17): 7841–7849.

20 Khazaei, M., Arai, M., Sasaki, T. et al. (2013). Novel electronic and magnetic properties of two-dimensional transition metal carbides and nitrides. *Adv. Funct. Mater.* 23 (17): 2185–2192.

21 Anasori, B., Xie, Y., Beidaghi, M. et al. (2015). Two-dimensional, ordered, double transition metals carbides (MXenes). *ACS Nano* 9 (10): 9507–9516.

22 Gao, G., Ding, G., Li, J. et al. (2016). Monolayer MXenes: promising half-metals and spin gapless semiconductors. *Nanoscale* 8 (16): 8986–8994.

23 Si, C., Zhou, J., and Sun, Z. (2015). Half-metallic ferromagnetism and surface functionalization-induced metal–insulator transition in graphene-like two-dimensional Cr_2C crystals. *ACS Appl. Mater. Interfaces* 7 (31): 17510–17515.

24 Rasool, K., Helal, M., Ali, A. et al. (2016). Antibacterial activity of $Ti_3C_2T_x$ MXene. *ACS Nano* 10 (3): 3674–3684.

25 Xu, J., Shim, J., Park, J.-H., and Lee, S. (2016). MXene electrode for the integration of WSe_2 and MoS_2 field effect transistors. *Adv. Funct. Mater.* 26 (29): 5328–5334.

26 Anasori, B., Shi, C., Moon, E.J. et al. (2016). Control of electronic properties of 2D carbides (MXenes) by manipulating their transition metal layers. *Nanoscale Horiz.* 1 (3): 227–234.

27 Satheeshkumar, E., Makaryan, T., Melikyan, A. et al. (2016). One-step solution processing of ag, au and Pd@MXene hybrids for SERS. *Sci. Rep.* 6 (1): 32049.

28 Chaudhuri, K., Alhabeb, M., Wang, Z. et al. (2018). Highly broadband absorber using plasmonic titanium carbide (MXene). *ACS Photonics* 5 (3): 1115–1122.

29 Lashgari, H., Abolhassani, M.R., Boochani, A. et al. (2014). Electronic and optical properties of 2D graphene-like compounds titanium carbides and nitrides: DFT calculations. *Solid State Commun.* 195: 61–69.

30 Zhang, X., Ma, Z., Zhao, X. et al. (2015). Computational studies on structural and electronic properties of functionalized MXene monolayers and nanotubes. *J. Mater. Chem., A Mater.* 3 (9): 4960–4966.

31 Berdiyorov, G.R. (2016). Optical properties of functionalized $Ti_3C_2T_2$ (T = F, O, OH) MXene: first-principles calculations. *AIP Adv.* 6 (5): 055105.

32 Hantanasirisakul, K., Zhao, M., Urbankowski, P. et al. (2016). Fabrication of $Ti_3C_2T_x$ MXene transparent thin films with tunable optoelectronic properties. *Adv. Electron. Mater.* 2 (6): 1600050.

33 Chandrasekaran, A., Mishra, A., and Singh, A.K. (2017). Ferroelectricity, antiferroelectricity, and ultrathin 2D Electron/hole gas in multifunctional monolayer MXene. *Nano Lett.* 17 (5): 3290–3296.

34 Ingason, A.S., Mockute, A., Dahlqvist, M. et al. (2013). Magnetic self-organized atomic laminate from first principles and thin film synthesis. *Phys. Rev. Lett.* 110 (19): 195502.

35 Sun, Z., Ahuja, R., Li, S., and Schneider, J.M. (2003). Structure and bulk modulus of M_2AlC (M=Ti, V, and Cr). *Appl. Phys. Lett.* 83 (5): 899–901.

36 Liu, Z., Wu, E., Wang, J. et al. (2014). Crystal structure and formation mechanism of $(Cr_{2/3}Ti_{1/3})_3AlC_2$ MAX phase. *Acta Mater.* 73: 186–193.

37 Zhao, S., Kang, W., and Xue, J. (2014). Manipulation of electronic and magnetic properties of M_2C (M = Hf, Nb, Sc, Ta, Ti, V, Zr) monolayer by applying mechanical strains. *Appl. Phys. Lett.* 104 (13): 133106.

38 Hwang, S.K., Kang, S.-M., Rethinasabapathy, M. et al. (2020). MXene: An emerging two-dimensional layered material for removal of radioactive pollutants. *Chem. Eng. J.* 397: 125428.

39 Dillon, A.D., Ghidiu, M.J., Krick, A.L. et al. (2016). Highly conductive optical quality solution-processed films of 2D titanium carbide. *Adv. Funct. Mater.* 26 (23): 4162–4168.

40 Hu, L., Wu, X., and Yang, J. (2016). Mn_2C monolayer: a 2D antiferromagnetic metal with high Néel temperature and large spin–orbit coupling. *Nanoscale* 8 (26): 12939–12945.

41 Huang, K., Li, Z., Lin, J. et al. (2018). Two-dimensional transition metal carbides and nitrides (MXenes) for biomedical applications. *Chem. Soc. Rev.* 47 (14): 5109–5124.

42 Fu, B., Sun, J., Wang, C. et al. (2021). MXenes: synthesis, optical properties, and applications in ultrafast photonics. *Small* 17 (11): 2006054.

43 Naguib, M., Mashtalir, O., Carle, J. et al. (2012). Two-dimensional transition metal carbides. *ACS Nano* 6 (2): 1322–1331.

44 Huang, J., Li, Z., Mao, Y., and Li, Z. (2021). Progress and biomedical applications of MXenes. *Nano Select* 2 (8): 1480–1508.

45 Seredych, M., Shuck, C.E., Pinto, D. et al. (2019). High-temperature behavior and surface chemistry of carbide MXenes studied by thermal analysis. *Chem. Mater.* 31 (9): 3324–3332.

46 Naguib, M. and Gogotsi, Y. (2015). Synthesis of two-dimensional materials by selective extraction. *Acc. Chem. Res.* 48 (1): 128–135.

47 Mashtalir, O., Naguib, M., Mochalin, V.N. et al. (2013). Intercalation and delamination of layered carbides and carbonitrides. *Nat. Commun.* 4 (1): 1716.

48 Halim, J., Lukatskaya, M.R., Cook, K.M. et al. (2014). Transparent conductive two-dimensional titanium carbide epitaxial thin films. *Chem. Mater.* 26 (7): 2374–2381.

49 Lin, H., Chen, Y., and Shi, J. (2018). Insights into 2D MXenes for versatile biomedical applications: current advances and challenges ahead. *Adv. Sci.* 5 (10): 1800518.

50 Wei, Y., Zhang, P., Soomro, R.A. et al. (2021). Advances in the synthesis of 2D MXenes. *Adv. Mater.* 33 (39): 2103148.

51 Soleymaniha, M., Shahbazi, M.-A., Rafieerad, A.R. et al. (2019). Promoting role of MXene nanosheets in biomedical sciences: therapeutic and biosensing innovations. *Adv. Healthcare Mater.* 8 (1): 1801137.

52 Zamhuri, A., Lim, G.P., Ma, N.L. et al. (2021). MXene in the lens of biomedical engineering: synthesis, applications and future outlook. *Biomed. Eng. Online* 20 (1): 33.

53 Huang, K., Li, Z., Lin, J. et al. (2018). Correction: two-dimensional transition metal carbides and nitrides (MXenes) for biomedical applications. *Chem. Soc. Rev.* 47 (17): 6889–6889.

54 Xu, C., Wang, L., Liu, Z. et al. (2015). Large-area high-quality 2D ultrathin Mo_2C superconducting crystals. *Nat. Mater.* 14 (11): 1135–1141.

55 Backes, C., Higgins, T.M., Kelly, A. et al. (2017). Guidelines for exfoliation, characterization and processing of layered materials produced by liquid exfoliation. *Chem. Mater.* 29 (1): 243–255.

56 Karlsson, L.H., Birch, J., Halim, J. et al. (2015). Atomically resolved structural and chemical investigation of single MXene sheets. *Nano Lett.* 15 (8): 4955–4960.

57 Pinto, D., Anasori, B., Avireddy, H. et al. (2020). Synthesis and electrochemical properties of 2D molybdenum vanadium carbides – solid solution MXenes. *J. Mater. Chem., A Mater.* 8 (18): 8957–8968.

58 Shekhirev, M., Shuck, C.E., Sarycheva, A., and Gogotsi, Y. (2021). Characterization of MXenes at every step, from their precursors to single flakes and assembled films. *Prog. Mater Sci.* 120: 100757.

59 Scheibe, B., Tadyszak, K., Jarek, M. et al. (2019). Study on the magnetic properties of differently functionalized multilayered $Ti_3C_2T_x$ MXenes and Ti-Al-C carbides. *Appl. Surf. Sci.* 479: 216–224.

60 Halim, J., Cook, K.M., Naguib, M. et al. (2016). X-ray photoelectron spectroscopy of select multi-layered transition metal carbides (MXenes). *Appl. Surf. Sci.* 362: 406–417.

61 Myhra, S., Crossley, J.A.A., and Barsoum, M.W. (2001). Crystal-chemistry of the Ti_3AlC_2 and Ti_4AlN_3 layered carbide/nitride phases—characterization by XPS. *J. Phys. Chem. Solids* 62 (4): 811–817.

62 Gouadec, G. and Colomban, P. (2007). Raman spectroscopy of nanomaterials: how spectra relate to disorder, particle size and mechanical properties. *Prog. Cryst. Growth Charact. Mater.* 53 (1): 1–56.

63 Pang, J., Mendes, R.G., Bachmatiuk, A. et al. (2019). Applications of 2D MXenes in energy conversion and storage systems. *Chem. Soc. Rev.* 48 (1): 72–133.

64 Chen, C., Boota, M., Xie, X. et al. (2017). Charge transfer induced polymerization of EDOT confined between 2D titanium carbide layers. *J. Mater. Chem., A Mater.* 5 (11): 5260–5265.

65 Rasool, K., Mahmoud, K.A., Johnson, D.J. et al. (2017). Efficient antibacterial membrane based on two-dimensional $Ti_3C_2T_x$ (MXene) nanosheets. *Sci. Rep.* 7 (1): 1598.

66 Jastrzębska, A.M., Karwowska, E., Wojciechowski, T. et al. (2019). The atomic structure of Ti_2C and Ti_3C_2 MXenes is responsible for their antibacterial activity toward *E. coli* bacteria. *J. Mater. Eng. Perform.* 28 (3): 1272–1277.

67 Arabi Shamsabadi, A., Sharifian Gh, M., Anasori, B., and Soroush, M. (2018). Antimicrobial mode-of-action of colloidal $Ti_3C_2T_x$ MXene nanosheets. *ACS Sustainable Chem. Eng.* 6 (12): 16586–16596.

68 Mei, L., Zhu, S., Yin, W. et al. (2020). Two-dimensional nanomaterials beyond graphene for antibacterial applications: current progress and future perspectives. *Theranostics* 10 (2): 757–781.

69 Li, J., Li, Z., Liu, X. et al. (2021). Interfacial engineering of $Bi_2S_3/Ti_3C_2T_x$ MXene based on work function for rapid photo-excited bacteria-killing. *Nat. Commun.* 12 (1): 1224.

70 An, H., Mamuti, M., Wang, X. et al. (2021). Rationally designed modular drug delivery platform based on intracellular peptide self-assembly. *Exploration* 1 (2): 20210153.

71 Han, X., Huang, J., Lin, H. et al. (2018). 2D ultrathin MXene-based drug-delivery nanoplatform for synergistic photothermal ablation and chemotherapy of cancer. *Adv. Healthcare Mater.* 7 (9): 1701394.

72 Xing, C., Chen, S., Liang, X. et al. (2018). Two-dimensional MXene (Ti_3C_2)-integrated cellulose hydrogels: toward smart three-dimensional network nanoplatforms exhibiting light-induced swelling and bimodal photothermal/chemotherapy anticancer activity. *ACS Appl. Mater. Interfaces* 10 (33): 27631–27643.

73 Deng, X. and Gu, M. (2003). Penetration depth of single-, two-, and three-photon fluorescence microscopic imaging through human cortex structures: Monte Carlo simulation. *Appl. Opt.* 42 (16): 3321.

74 Liu, G., Zou, J., Tang, Q. et al. (2017). Surface modified Ti_3C_2 MXene nanosheets for tumor targeting photothermal/photodynamic/chemo synergistic therapy. *ACS Appl. Mater. Interfaces* 9 (46): 40077–40086.

75 Xue, Q., Zhang, H., Zhu, M. et al. (2017). Photoluminescent Ti_3C_2 MXene quantum dots for multicolor cellular imaging. *Adv. Mater.* 29 (15): 1604847.

76 Huang, D., Xie, Y., Lu, D. et al. (2019). Demonstration of a white laser with V_2C MXene-based quantum dots. *Adv. Mater.* 1901117.

77 Yang, G., Zhao, J., Yi, S. et al. (2020). Biodegradable and photostable Nb_2C MXene quantum dots as promising nanofluorophores for metal ions sensing and fluorescence imaging. *Sens. Actuators, B* 309: 127735.

78 Caravan, P., Ellison, J.J., McMurry, T.J., and Lauffer, R.B. (1999). Gadolinium(III) chelates as MRI contrast agents: structure, dynamics, and applications. *Chem. Rev.* 99 (9): 2293–2352.

79 Tian, T., Qiao, G., Deng, B. et al. (2019). The effects of rain shelter coverings on the vegetative growth and fruit characteristics of Chinese cherry (Prunus pseudocerasus Lindl.). *Sci. Hortic.* 254: 228–235.

80 Dierickx, W. and Van Den Berghe, P. (2004). Natural weathering of textiles used in agricultural applications. *Geotext. Geomembr.* 22 (4): 255–272.

81 Lai, Y.-C., Hsiao, Y.-C., Wu, H.-M., and Wang, Z.L. (2019). Waterproof fabric-based multifunctional triboelectric nanogenerator for universally harvesting energy from raindrops, wind, and human motions and as self-powered sensors. *Adv. Sci.* 6 (5): 1801883.

82 Yoo, D., Park, S.-C., Lee, S. et al. (2019). Biomimetic anti-reflective triboelectric nanogenerator for concurrent harvesting of solar and raindrop energies. *Nano Energy* 57: 424–431.

83 Jiang, C., Li, X., Ying, Y., and Ping, J. (2020). A multifunctional TENG yarn integrated into agrotextile for building intelligent agriculture. *Nano Energy* 74: 104863.

84 Chaudhary, V., Gautam, A., Mishra, Y.K., and Kaushik, A. (2021). Emerging MXene–polymer hybrid nanocomposites for high-performance ammonia sensing and monitoring. *Nanomaterials* 11 (10): 2496.

85 Hao, L., Gong, L., Chen, L. et al. (2020). Composite pesticide nanocarriers involving functionalized boron nitride nanoplatelets for pH-responsive release and enhanced UV stability. *Chem. Eng. J.* 396: 125233.

86 Carvalho, F.P. (2017). Pesticides, environment, and food safety. *Food Energy Secur.* 6 (2): 48–60.

87 Zhao, F., Yao, Y., Jiang, C. et al. (2020). Self-reduction bimetallic nanoparticles on ultrathin MXene nanosheets as functional platform for pesticide sensing. *J. Hazard Mater.* 384: 121358.

88 Song, S., Jiang, X., Shen, H. et al. (2021). MXene (Ti_3C_2) based pesticide delivery system for sustained release and enhanced pest control. *ACS Appl. Bio Mater.* 4 (9): 6912–6923.

89 Jasper, J.T., Yang, Y., and Hoffmann, M.R. (2017). Toxic byproduct formation during electrochemical treatment of latrine wastewater. *Environ. Sci. Technol.* 51 (12): 7111–7119.

90 Robinson, T., McMullan, G., Marchant, R., and Nigam, P. (2001). Remediation of dyes in textile effluent: a critical review on current treatment technologies with a proposed alternative. *Bioresour. Technol.* 77 (3): 247–255.

91 Burgess, J.E., Parsons, S.A., and Stuetz, R.M. (2001). Developments in odour control and waste gas treatment biotechnology: a review. *Biotechnol. Adv.* 19 (1): 35–63.

92 Crini, G. (2005). Recent developments in polysaccharide-based materials used as adsorbents in wastewater treatment. *Prog. Polym. Sci.* 30 (1): 38–70.

93 QU, J. (2008). Research progress of novel adsorption processes in water purification: a review. *J. Environ. Sci.* 20 (1): 1–13.

94 Das, R., Vecitis, C.D., Schulze, A. et al. (2017). Recent advances in nanomaterials for water protection and monitoring. *Chem. Soc. Rev.* 46 (22): 6946–7020.

95 Yu, L., Ruan, S., Xu, X. et al. (2017). One-dimensional nanomaterial-assembled macroscopic membranes for water treatment. *Nano Today* 17: 79–95.

96 Sun, S., Jiao, T., Xing, R. et al. (2018). Preparation of MoS_2 -based polydopamine-modified core–shell nanocomposites with elevated adsorption performances. *RSC Adv.* 8 (38): 21644–21650.

97 Wei, Z., Peigen, Z., Wubian, T. et al. (2018). Alkali treated $Ti_3C_2T_x$ MXenes and their dye adsorption performance. *Mater. Chem. Phys.* 206: 270–276.

98 Tunesi, M.M., Soomro, R.A., Han, X. et al. (2021). Application of MXenes in environmental remediation technologies. *Nano Convergence* 8 (1): 5.

99 Mashtalir, O., Cook, K.M., Mochalin, V.N. et al. (2014). Dye adsorption and decomposition on two-dimensional titanium carbide in aqueous media. *J. Mater. Chem. A* 2 (35): 14334–14338.

100 Carolin, C.F., Kumar, P.S., Saravanan, A. et al. (2017). Efficient techniques for the removal of toxic heavy metals from aquatic environment: a review. *J. Environ. Chem. Eng.* 5 (3): 2782–2799.

101 Peng, Q., Guo, J., Zhang, Q. et al. (2014). Unique lead adsorption behavior of activated hydroxyl group in two-dimensional titanium carbide. *JACS* 136 (11): 4113–4116.

102 Reddy, A.L.M., Gowda, S.R., Shaijumon, M.M., and Ajayan, P.M. (2012). Hybrid nanostructures for energy storage applications. *Adv. Mater.* 24 (37): 5045–5064.

103 Chu, S. and Majumdar, A. (2012). Opportunities and challenges for a sustainable energy future. *Nature* 488 (7411): 294–303.

104 Eames, C. and Islam, M.S. (2014). Ion intercalation into two-dimensional transition-metal carbides: global screening for new high-capacity battery materials. *J. Am. Chem. Soc.* 136 (46): 16270–16276.

105 Hu, Q., Sun, D., Wu, Q. et al. (2013). MXene: a new family of promising hydrogen storage medium. *J. Phys. Chem. A* 117 (51): 14253–14260.

106 Wang, Z., Tammela, P., Strømme, M., and Nyholm, L. (2015). Nanocellulose coupled flexible polypyrrole@graphene oxide composite paper electrodes with high volumetric capacitance. *Nanoscale* 7 (8): 3418–3423.

107 Long, W., Fang, B., Ignaszak, A. et al. (2017). Biomass-derived nanostructured carbons and their composites as anode materials for lithium ion batteries. *Chem. Soc. Rev.* 46 (23): 7176–7190.

108 Deng, Y., Fang, C., and Chen, G. (2016). The developments of SnO_2/graphene nanocomposites as anode materials for high performance lithium ion batteries: a review. *J. Power Sources* 304: 81–101.

109 Naguib, M., Come, J., Dyatkin, B. et al. (2012). MXene: a promising transition metal carbide anode for lithium-ion batteries. *Electrochem. Commun.* 16 (1): 61–64.

110 Lukatskaya, M.R., Mashtalir, O., Ren, C.E. et al. (2013). Cation intercalation and high volumetric capacitance of two-dimensional titanium carbide. *Science (1979)* 341 (6153): 1502–1505.

111 Quinlan, R.A., Cai, M., Outlaw, R.A. et al. (2013). Investigation of defects generated in vertically oriented graphene. *Carbon N Y* 64: 92–100.

112 Ghidiu, M., Lukatskaya, M.R., Zhao, M.-Q. et al. (2014). Conductive two-dimensional titanium carbide 'clay' with high volumetric capacitance. *Nature* 516 (7529): 78–81.

113 Okubo, M., Sugahara, A., Kajiyama, S., and Yamada, A. (2018). MXene as a charge storage host. *Acc. Chem. Res.* 51 (3): 591–599.

114 Mashtalir, O., Lukatskaya, M.R., Kolesnikov, A.I. et al. (2016). The effect of hydrazine intercalation on the structure and capacitance of 2D titanium carbide (MXene). *Nanoscale* 8 (17): 9128–9133.

115 Xie, Y., Dall'Agnese, Y., Naguib, M. et al. (2014). Prediction and characterization of MXene nanosheet anodes for non-lithium-ion batteries. *ACS Nano* 8 (9): 9606–9615.

116 Benchakar, M., Loupias, L., Garnero, C. et al. (2020). One MAX phase, different MXenes: a guideline to understand the crucial role of etching conditions on $Ti_3C_2T_x$ surface chemistry. *Appl. Surf. Sci.* 530: 147209.

117 Rasool, K., Pandey, R.P., Rasheed, P.A. et al. (2019). Water treatment and environmental remediation applications of two-dimensional metal carbides (MXenes). *Mater. Today* 30: 80–102.

118 Verger, L., Xu, C., Natu, V. et al. (2019). Overview of the synthesis of MXenes and other ultrathin 2D transition metal carbides and nitrides. *Curr. Opin. Solid State Mater. Sci.* 23 (3): 149–163.

119 Pang, S., Io, W., Wong, L. et al. (2020). Efficient energy conversion and storage based on robust fluoride-free self-assembled 1D niobium carbide in 3D nanowire network. *Adv. Sci.* 7 (10): 1903680.

120 Seh, Z.W., Fredrickson, K.D., Anasori, B. et al. (2016). Two-dimensional molybdenum carbide (MXene) as an efficient electrocatalyst for hydrogen evolution. *ACS Energy Lett.* 1 (3): 589–594.

121 Ling, C., Shi, L., Ouyang, Y., and Wang, J. (2016). Searching for highly active catalysts for hydrogen evolution reaction based on O-terminated MXenes through a simple descriptor. *Chem. Mater.* 28 (24): 9026–9032.

122 Zhang, C.J., Pinilla, S., McEvoy, N. et al. (2017). Oxidation stability of colloidal two-dimensional titanium carbides (MXenes). *Chem. Mater.* 29 (11): 4848–4856.

123 Zhang, C.(J.). and Nicolosi, V. (2019). Graphene and MXene-based transparent conductive electrodes and supercapacitors. *Energy Storage Mater.* 16: 102–125.

124 Jian-Hui, Y., Shao-Zheng, Z., Jia-Lin, J., and Shi-Hao, W. (2015). Adsorption activities of O, OH, F and Au on two-dimensional Ti_2C and Ti_3C_2 surfaces. *Acta Phys. Chim. Sin.* 31 (2): 369–376.

125 Zhang, X., Xue, M., Yang, X. et al. (2015). Preparation and tribological properties of $Ti_3C_2(OH)_2$ nanosheets as additives in base oil. *RSC Adv.* 5 (4): 2762–2767.

126 Ronchi, R.M., Arantes, J.T., and Santos, S.F. (2019). Synthesis, structure, properties and applications of MXenes: current status and perspectives. *Ceram. Int.* 45 (15): 18167–18188.

2

Structure, Composition, and Functionalization of MXenes

Praveen Kumar[1], *Ramakshi Rana*[1], *Arun Kumar*[1], *Pooja Rawat*[2,3], *and Jong Soo Rhyee*[2]

[1]*Department of Chemistry, Doon University, Dehradun, Uttarakhand, India*
[2]*Department of Applied Physics and Institute of Natural Sciences, Kyung Hee University, Yong-In, Gyong-gi, Republic of Korea*
[3]*Department of Chemistry, Biochemistry and Forensic Science, Amity School of Applied Sciences, Amity University Haryana, Gurugram, Haryana, India*

2.1 Introduction

MXenes are considered to be a member of huge as well as rapidly expanding category of two-dimensional (2D) materials. MXene with transition elements like nitride/carbonitride/sulfide/carbide is a recent family member of 2D materials that are feasibly shown with a typical formulation as $M_{n+1}X_nT_x$ (in which M represents metal atoms, like Nb, Ti, V, and Mo; T_x exemplifies surface terminations functional group, i.e., –O, –OH, as well as –F and X_n refers to nitrogen /carbon, where $n = 1$, 2, or 3) [1–6]. In 2D MXene structure, $n + 1$ layer, where $n = 1$–3 of initial transition element (M, red representation in Figure 2.1), is inserted within n layers of nitrogen and carbon (represented by X elements in gray in Figure 2.1), having typical formulation $M_{n+1}X_nT_x$. Here, T_x exemplifies the surface terminal group, namely, O, F, Cl, or OH, that is adhered to the exterior M surface [3]. The different forms of MXenes displaying their atomic structures are a demonstration about chemistry rudimentary tenet, representing that in what way there is utilization of elements as key components to create a numerous diversity of nanomaterials. The most common examples of MXene composition include Mo_2CT_x, $Nb_4C_3T_x$, and $Ti_3C_2T_x$, where the X layers are enclosed by the M layers, like the system of $[MX]_nM$ alignment. MXenes with composition $M_{n+1}TX_n$ are obtained via scrupulous engraving of "T" elements from their ternary layered nitride or carbide (MAX phases) precursors [6]. During the etching reaction of T elements using an acidic solution of hydrogen fluoride, resulting in M cover being abolished alongside end groups (–O, –OH, or –F), which have a strong affinity toward water. Because of distinctive characteristics such as structure modification, metallic conductivity (10,000 Scm^{-1}), hydrophilic in nature, and large surface area, MXene has been widely studied and used in several applications like environmental as well as water purification [6, 7], energy storage [8, 9], sensors [10, 11], electromagnetic interference shading [12, 13], catalysis [14, 15] then [16–18], etc.

MXenes: Next-Generation 2D Materials: Fundamentals and Applications, First Edition.
Edited by Jay Singh, Kshitij RB Singh, Ravindra Pratap Singh, and Charles Oluwaseun Adetunji.

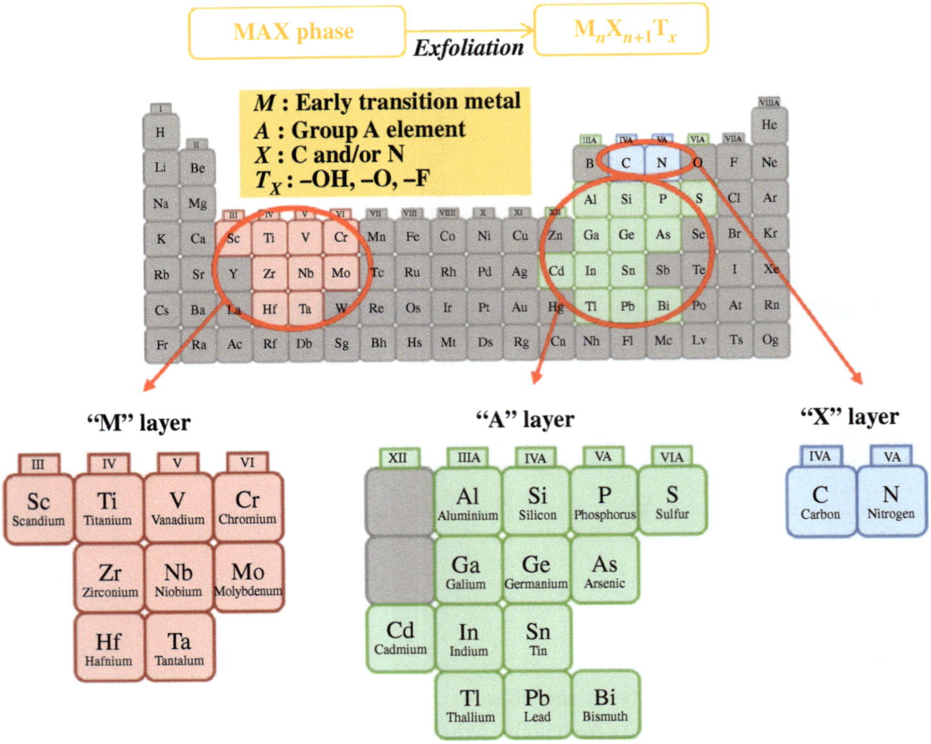

Figure 2.1 MAX phase and MXene compositions are shown by using periodic table. Schematic representation of periodic table with different color elements showing that element used for the different compositions of MXenes, MAX phases, and its intercalated ions. M: early transition metal, A: Group A element, X: C and/or N, T_x: surface functional group. Source: Zamhuri et al. [3]/Springer Nature/CCBY 4.0.

In literature, more than hundreds of stoichiometric MXene frameworks are reported, as well as a very large count pertaining to solid solution attributes, a combination of distinct properties, and also gives the opportunity to tune them via the different ratios of M or X elements [19]. Since MXene is the inherited part of the structural arrangement and framework of the radix MAX phase, therefore, the research in the MAX phase governs progress in MXenes to a large extent. Other than the haphazard solid solutions onto M position [20], hitherto, a considerable heed is being paid toward MAX structures along with ordered M elements alloying taking place. The pioneering discovery related to the initial chemically ordered MAX phase alloy $(Cr_2/_3Ti_{1/3})_3AlC_2$ was found I 2014 [21]. Basically, this assembly is suppressed by M_3AX_2 when duplet of M elements is taken as ratio of 2:1 and forms a phase out-of-plane chemical ordering via only one type of M element arranged in alternating layers.

Near 2017, ordering betwixt M elements as 2:1 scale was found within the M_2AX phase, $(Mo_{2/3}Sc_{1/3})_2AlC$ [22], although at that period, the in-plane chemical organization was suggesting the annotation i-MAX [23]. Several notable attributes among such phases are the small M element via the M-layer to the A-layer, as well as rearrangement of the A, creating a lattice similar to Kagome. Further, a comprehensive elucidation regarding i-MXene's assembly amid conceptual modeling is accessible in the earlier reported literature [24]. For enhancement of MXenes properties, one of the important factors is surface terminations, which are autonomous of existing as conventional style or vacancy or chemically ordered type. The function of the i-MAX phases furthermore promoted the notion regarding "targeted etching," which describes the process of chemical synthesis for determining the space of MXene; otherwise, MXene is chemically organized out of alike radix substance, indicated by $(Mo_{2/3}Y_{1/3})_2AlC$ [25]. The differential strength of the MAX phases that serve as MXenes harbinger is apparent, while contrasting strength of the Sc-MX as well as Y–MX bonds to $(Mo_{2/3}Sc_{1/3})_2AlC$ as well as $(Mo_{2/3}Y_{1/3})_2AlC$, sequentially. In order to find new mixed group B MXenes, Anasori et al. studied comparative stability for seventeen $M'_2M''C_2$ as well as fifteen $M'_2M''_2C_3$ MXenes out of assumed o-MAX phases having diverse chemical configurations within a shape of completely oriented as well as incompletely oriented arrangement [26]. In continuation, Tan et al. examined the structural stability relationships of the eight MXene hybrid systems through modeling the order level against temperature using advanced calculations [27].

From the past few ages, numerous surface functionalization approaches, namely polymerization that began at the surface as well as individual heteroatom doping, have dexterously evolved toward broadening the potentiality appertaining to MXenes in individual domains. Typically, surface properties appertaining to MXenes maybe permuted and enhanced based on kind and proportion of the T grouping throughout the course of production [28]. Lately, immaculate MXenes have been modified on their surfaces with covalent and non-covalent alterations to create functionalized MXenes. Non-covalent surface alteration is acquired by an associated effect of van der Waals forces, hydrogen bonding, as well as electrostatic attraction [7, 29–31]. There are three categories for covalently surface functionalization techniques [32–34]: (i) the individual incorporates tiny compounds like organic amines, alkali metal hydroxides, acid halides, as well as acid anhydrides [35–38], (ii) the individual incorporates macromolecules that start polymerization on the surface [39, 40], and (iii) the individual incorporates solitary heteroatoms [41].

Three methods exist to refine peculiarities appertaining to MXenes: modifying the composition, surface cessation, as well as framework/morphology. Because of their potential for structural adjustment, alloying, as well as the availability of surface cessation, MXenes are a diverse and evolving category of 2D materials that have been investigated in this review. MXenes belongs to a remarkable emergent category in 2D material due to broad range of variables available for property adjustment.

2.2 MXenes Composition

Multilayer (ML) dust or desquamated solitary flakes are the two main shapes that MXenes commonly take. Currently, more than 150 MXene compositions exist [42–46]. MAX phase (M_2AX), i.e., layered ternary metal carbide and nitride, is precursor used for MXenes, and A is an element such as Al or Si. The first MXene was produced in 2011 ($Ti_3C_2T_x$) at Drexel University by immersing or etching MAX phase in hydrofluoric acid (HF) by removing one of the atomic layers and replacing it with surface termination group at room temperature and etching with $ZnCl_2$. Fully chlorinated MXenes are seized, and by electrochemical and hydrothermal methods, fluorine-free MXenes are attained [43]. Ever since, >30 MXene compositions appeared to be published (illustrated by blue in Figure 2.2), as well as

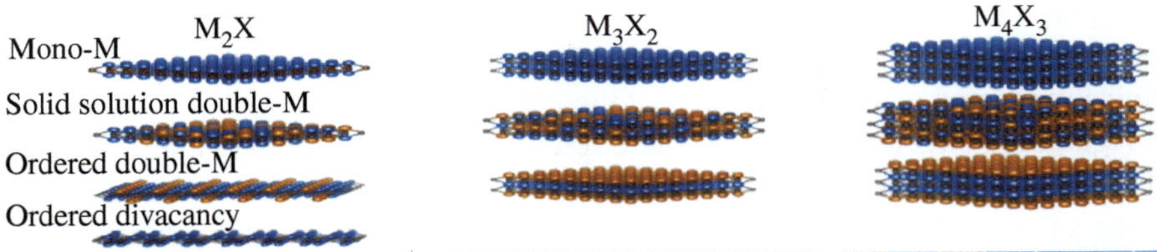

Sc_2C	Ti_2C	Ti_2N	Zr_2C
Zr_2N	Hf_2N	Hf_2N	V_2C
V_2N	Nb_2C	Ta_2N	Cr_2C
Cr_2N	Mo_2C	$Mo_{1.33}C$	$Cr_{1.33}C$
$(Ti,V)_2C$	$(Ti,Nb)_2C$	W_2C	$W_{1.33}C$
Mo_2N	$Nb_{1.33}C$	$Mo_{1.33}Y_{0.67}C$	

Ti_3C_2	Ti_3N_2	Ti_3CN	Zr_3C_2
$(Ti,V)_3C_2$	Ti_2NbC_2	Ti_2TaC_2	Ti_2MnC_2
Hf_3C_2	Hf_2VC_2	Hf_2MnC_2	V_2TiC_2
Cr_2TiC_2	Cr_2VC_2	Cr_2NbC_2	Cr_2TaC_2
Mo_2ScC_2	Mo_2TiC_2	Mo_2ZrC_2	Mo_2HfC_2
Mo_2ZrC_2	Mo_2NbC_2	Mo_2TaC_2	W_2TiC_2
W_2ZrC_2	W_2HfC_2		

Ti_4N_3	V_4C_3	Nb_4C_3	Ta_4C_3
$(Ti,Nb)_4C_3$	$(Nb,Zr)_4C_3$	$Ti_2Nb_2C_3$	$Ti_2Ta_2C_3$
$V_2Ti_2C_3$	$V_2Nb_2C_3$	$V_2Ta_2C_3$	$Nb_2Ta_2C_3$
$Cr_2Ti_2C_3$	$Cr_2V_2C_3$	$Cr_2Nb_2C_3$	$Cr_2Ta_2C_3$
$Mo_2Ti_2C_3$	$Mo_2Zr_2C_3$	$Mo_2Hf_2C_3$	$Mo_2V_2C_3$
$Mo_2Nb_2C_3$	$Mo_2Ta_2C_3$	$W_2Ti_2C_3$	$W_2Zr_2C_3$
$W_2Hf_2C_3$			

Figure 2.2 Chemical versatility in MXene compositions reported so far. The tabular data comprises compositions of MXenes that have been conceptually as well as empirically investigated (represented in grey and blue, sequentially). Surface cessation is not depicted. Source: Reproduced with permission from [46]/American Chemical Society.

hundreds more seemed to be investigated using computational techniques (illustrated by gray in Figure 2.2) [46]. While two transition metals are being homogenized within an MXene assembly, a distinct characteristic regarding MXenes is brought to the forefront. Along with the development of anticipated solid solutions, including $(Ti,Nb)CT_x$ (illustrated by green in Figure 2.2), only a sole 2D MXene flake of transition metals is capable of forming organized formations, indeed, by designing atomic sandwiches appertaining to transition metal planes (for $n \geq 2$) like $Mo_2TiC_2T_x$, or organized structures that are in plane, such as $(Mo_{2/3}Y_{1/3})_2CT_x$ [44]. Many new compositions of ordered MXene were generated within MXene subclass (Illustrated by red in Figure 2.2). MXenes are produced with certain compositions of M_2X, M_3X_2, and M_4X_3. As recently synthesized carbides, MXenes are composed of two transition metals. These two (double) transition metal MXenes are composed of $M'_2M''C_2$ or $M'_2M''_2C_3$, where M'(outer layer metal) as well as M"(inner layer metal) are different transition metals. For $M'_2M''_2C_3$ compositions, two completely organized frameworks of MXenes with formulations of $M'_2M''_2C_3$ as well as $M''_2M'_2C_3$ [45, 46]. There are four typical structures of MXenes, i.e., M_2XT_x, $M_3X_2T_x$, $M_4X_3T_x$, and $M_5X_4T_x$. Here M is synthesized MXenes, i.e. Sc, Ti, V, Cr, Y, Zr, Nb, Mo, Hf, Ta, W; whereas X signifies element carbon as well as nitrogen; T typifies surface cessation elements, i.e., —O, —OH, —S, —F, —Cl, —Se, —Te, —I, —Br, —H, $—NH_2$. According to a study regarding MXene composition, the arrangement of surface termination greatly affects its physical as well as chemical characteristics [46].

A limitless number of solid solutions and 100 stoichiometric MXene compositions create properties of unique combinations [47]. In the 20th century, group 4 to 6 transition metal carbides as well as nitrides have been investigated as high-temperature, chemically stable, and resistant wearable materials and started disclosing their energy storage utilization [48]. As we talk about adherence among MXenes and related 2D materials, it depends upon number of substance monolayers within the stacking, taking into consideration 40 interfacing pairings for 2D materials, the least adherence energy is for interfaces involving $MoSe_2$, i.e. $0.27\,J\,m^{-2}$, and the highest for $Ti_3C_2T_x$ value similar to $1.23\,J\,m^{-2}$ [49].

2.2.1 Group IV Elemental Analog

As we are going to explore MXenes, group IV elements play a very significant role. Elements such as Ti, Zr, and Hf (Group IV) transition metals form carbide and nitride [48, 50]. The MAX phases are formulated as $M_{n+1}AX_n$ ($n = 1, 2$, or 3). Here M typifies a transition metal, and A suggests A-group element mainly from 13 to 16 of the periodic table, such as Al, Ga, In, Si, Ge, Sn, P, As, S, Cd, Au, Ir, Tl, Pb, along with X is carbon/nitrogen [51]. 2D transition metal carbide ($Ti_3C_2T_x$) in 2011 led to exploration of the first MXenes, it reveals that A layer remains particularly engraved as of MAX phase (Ti_3AlC_2), while surface termination is generally mixture of —O, —F, —OH. Its relative stability is $Ti_3C_2O_2 > Ti_3C_2F_2 > Ti_3C_2(OH)_2$. There is also a series of halogenated $Ti_3C_2T_x$. Here T (surface termination group) is Br, I, or a combination of Br and I [46, 52]. As from above, we understand that M–X bond has more stability than M–A bond, which eases engraving (etching) layer A from MAX phase leading to the intercalation of appropriate etching agent into to generation of single M–X layer [53, 54]. Lately, HF–HCl (mixed acid) came to be utilized for engraving, subsequently incorporating the salt lithium chloride (LiCl) for facile interpolation and separation into constituent layers [48]. Till 2016, family of MXenes was narrower to carbides and carbonitrides but after that the first MXene nitride was outlined, i.e., Ti_4N_3. Another group IV element forms 2D compound MXenes by engraving MAX phase such as $Zr_3Al_3C_5$ are etched to obtain 2D $Zr_3C_2T_z$ and $Hf_3(Al,Si)_4C_6$ etched to get $Hf_3C_2T_z$ crumb. Recently, $Ti_3C_2T_z$ was prevailed from Ti_3SiC_2 via engraving of Si into combination of HF along with H_2O_2 used as an oxidant. Different etching methods to obtain MXenes, like acid with fluorine, molten salts, hydrothermal, and electrochemical, Lewis's acid only functions on carbides with only anomaly nitrides, $Ti_4N_3T_z$, which is synthesized through molten salt method, while Ti_2NTz is only nitride synthesized by wet chemical etching method [55]. Double transition metal MXenes were also synthesized, namely, Mo_2TiC_2, $Mo_2Ti_2C_3$, and Cr_2TiC_2 [45, 56]. A cost-effective MXene is explored with rutile, Al, as well as C. The indicated reaction progresses by reacting Al and C with rutile to produce Ti_3AlC_2 or Ti_2AlC and alumina (By product). In context of electrical properties of MAX phase metallic conductors, resistivity at room temperature is of the order of several tens of μΩ-cm. It has low Seebeck coefficients (S), but exceptionally, Ti_3SiC_2 reveals negligible S over an extensive temperature range of 300 to 850 K. As we see the extraordinary behavior of MXenes to actively capture CO_2, mostly group IV (Ti_2X besides Zr_2X) reveals elevated adsorption energies as well as charge shift via M_2X to CO_2 [57, 58]. MXene Ti_2C shows a hydrogen adsorption capacity of 8.6 wt%, which is much higher than metal-based complex hydride's gravimetric capacity, i.e. similar to 5.5 wt% [53].

2.2.2 Group V Elemental Analog

Group V elements include V, Nb, and Ta in MXene formation. MAX phases (V_2AlC and Cr_2AlC) have a much higher degree of anisotropy. Generally, group V MXenes synthesize by etching Al layers from different MAX phase V_2AlC, Nb_2AlC, Nb_4AlC_3, Ta_4AlC_3, $Ti_{0.5}Nb_{0.5})_2AlC$, $(V_{0.5}Cr_{0.5})_3AlC_2$, $(Ti_{0.5}V_{0.5})_2AlC$, $(Ti_{0.5}V_{0.5})_3AlC_2$, and $(Nb_{0.5}V_{0.5})_4AlC_3)$ by using aqueous HF at room temperature [59]. A fully ordered MXene from group V element at 0 K is following: Mo_2VC_2, V_2TaC_2, Mo_2TaC_2, Cr_2VC_2, Mo_2NbC_2, Ti_2NbC_2,

Cr_2TaC_2, Cr_2NbC_2, Ti_2TaC_2, as well as V_2TiC_2. But some MXenes, such as Nb_2VC_2, Ta_2VC_2, Nb_2TiC_2, and Ta_2TiC_2, are extra sturdy in their partly organized state as compared to their fully ordered state [45, 60]. V_2NT_z is nitride MXene, starts with carbide (V_2CT_z), and then at 550 °C, NH_3 gas is used to drive the ammonization process. Besides, MXenes have very promising applications like electromagnetic interference (EMI) screening, water purification, biosensors, and transparent conductive electrodes [61]. Nb_2C MXenes affected by the surface termination group show surface-group-dependent superconductivity [62].

2.2.3 Group VI Elemental Analog

Group VI elements such as Cr and Mo have lots of functions in the construction of MXene as a significant material. For example, nitride MXene as Mo_2NT_z is obtained from Mo_2CT_z by ammonization treatment with NH_3 gas at a certain temperature. MAX phase such as Mo_2GaC is resistant to etching, while Mo_2Ga_2C gives Mo_2CT_z (MXenes). MAX phase like Cr_2AlC is not stable it dissolves away when undergoing etching process, but ordered MAX phase such as Cr_2TiAlC_2 (M_3AX_2 structure) has been successfully etched into $Cr_2TiC_2T_z$ with a Ti layer intercalated betwixt two Cr carbide layers [55, 63].

The expeditious desire to generate pristine MXenes had apprehended the MAX community's efforts to come across with pristine MAX phase compositions. Numerous isostructural MAX phases with solid solutions have been outlined in this area. [64], with even more latest research on chemically ordered quaternaries, in which the arrangement of the two M components yields an out-of-plane ordered (o-MAX) [21] or in-plane ordered (i-MAX) arrangement [65, 66]. Fascinatingly, the present ordering legitimizes the incorporation of conventional MAX phase elements, namely, Sc, Y, and W, as well as scarce-earth elements [67–70]. As majority of lately ascertained MAX phases being Al-focused, the clan appertaining to 2D MXenes is looking forward to substantially grow in foreseeable time. Currently, some o-MAX phases are being desquamated to create double transition metal MXenes that are organized out-of-plane, like $Cr_2TiC_2T_z$ [26], $Mo_2TiC_2T_z$ [26], $Mo_2ScC_2T_z$ [71], as well as $Mo_2Ti_2C_3T_z$ [26]. Most intriguingly, i-MAX phases maybe desquamated to create 2 distinct forms of MXenes, depending on the etching circumstances as exemplified in Table 2.1.

Table 2.1 Various etchants used in various synthetic methods and their reaction conditions.

Acid alongside fluorine	**HF**	**Room temperature (RT) near 55**
	H_2O_2 + HF	40
	HCl + LiF	35–55
	HCl + (Na, K, or NH_4F)	30–60
	NH_4HF_2	RT
Molten salts	LiF + NaF + KF	550
Hydrothermal	NaOH	270
	$NaBF_4$, HCI	180
Electrochemical	NH_4Cl/TMAOH	RT
	HCl	RT
Lewis's acid	$ZnCl_2$	550

Furthermore, Halim et al. also observed that the solid solution of $(Nb_{2/3}Sc_{1/3})_2AlC$ desquamated into $Nb_{1.33}CTz$ accompanied by disorganized voids [72]. In recent years, chemically organized alloy MXenes, a novel subtype belonging to the MXene group, has evolved. Both out-of-plane and in-plane displays of the chemical sequence are possible, using alternate sheets of single elements in sandwiched structures, otherwise two M-elements by a 2:1 ratio arranged along a chemically organized positioning. Due to such results, radix materials are now often denoted by the letters o-MAX as well as i-MAX, with comparative designation used for correspondingly generated MXenes. Prominently, i-MAX phases too, made it possible to create MXenes with orderly voids. Overall, these fresh foliates significantly increased the number of metals that may be added to pristine MAX phases as well as MXenes, including Sc, Y, as well as W. Such organized phases would probably soon comprise a sizable portion of such recognized MAX phases in addition to MXenes as quantity of distinct phases continues to grow unremittingly. They would also certainly have quite an effect on the evolution of 2D materials in particular.

2.3 Structural Analysis Regarding MXenes

MXenes refer to neoteric category appertaining to 2D transition metal carbides as well as nitrides having typical formulations as- $M_{n+1}AX_{n,}$ ($n = 1, 2$, and 3). Whereas "M" constitutes of initial transition metals (M=Sc, V, Ti, Cr, Zr, Nb, Hf, Mo, Ta), "A" constitutes of group 13 as well as 14 elements, along with "X," which represents C/N [1-4]. If we look up for perfect elucidation of MXenes, then we represent them as $M_{n+1}X_nT_x$, where about "T_x" suggests surface cessations (like –F, –OH =O, –Cl) [1, 70]. The width of a single layer of MXene would be around 1 nm and differ with the magnitude of "n" as well as the type of terminal grouping [73]. MXenes have been generated through chemical desquamation appertaining to the radix MAX phases [74]. From typical elucidation of MXenes, it is prominent that modifications can be introduced in it by 3 possible ways- (i) by selection and choosing "M" and "X" elements, so as to dodge the composition, (ii) by width of sheet, which is affected by value of "n" and, (iii) terminal grouping, i.e., "T_x" [74]. Since, MXenes are consistent in structure with their parent MAX phase, so it becomes predominant to analyze the arrangement of MAX phases [73]. MAX phases are used as MXene's harbinger, with a definite ratio as $M_{n+1}AX_n$, having a hexadic assembly, followed by insinuation of "M" as well as "A" layers [75]. The "X" atoms tend to take over the octahedral lots created by "M" atoms [73, 76]. The "n" value can range from 1–3 while MAX phases can be categorized as 211, 312, and 413 having M_2AX, $M_3AX_{2,}$ and M_4AX_3 assembly and the MXenes originated from them having M_2X, $M_3X_{2,}$ M_4X_3 sequentially with M_2X containing 3, M_3X_2 containing 5, and M_4X_3 having 7 atomic layers depicted in Figure 2.3 [70, 77].

The 211,312, and 413 MAX phases correspond to 2, 3, and 4 Ti layers betwixt Al layers and for number of "C" layers being [number of Ti layer –1], where "M" represents initial transition metals, like Ti, "A" comprising of 13 as well as 14 group elements, like Al/Si along with "X," which refers to C/N [77]. Indeed, >70 MAX phases (like, Ti_2AlC, Ti_3AlC_2) have been perceived [75, 78]. As the generation of MXenes involves shifting "A" layers as of MAX phases, this indicates that MXenes do not have a plain-sailing 3D harbinger, unlike graphene as well as phosphorene, which uses graphite and black phosphorus as their direct precursor sequentially [77]. While focusing on MXene, the 1st discovered MXene, i.e., Ti_3C_2

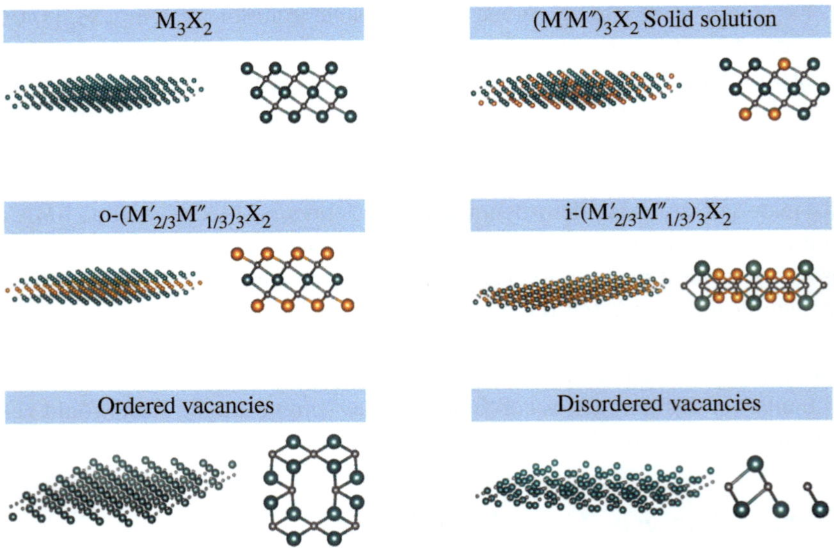

Figure 2.3 Illustrative representation of different MXene structures. Source: Reproduced with permission from [74]/Elsevier.

was derived from their parent MAX phase Ti_3AlC_2, followed by drenching its MAX dust in HF acid solution [79]. The M- X as well as M–A bond are present during development and shaping of MXenes from MAX phases and the contrast in strength of both bonds show considerable disparity of MAX phases [75]. There exists an intense metallic bond betwixt the M and A of MXene sheet [80]. The pre-eminent step in MXene formation is the elimination of feebly bonded A atom layers, so as to form carbides and/or nitrides [81, 82]. $Ti_3C_2T_x$, which is generally the most analyzed one, is prepared by effective etching of parent MAX phases, i.e., Ti_3AlC_2 MAX phase in HF solution [83]. This step helps to banish the M–A bond and formation of multi-layer assembly with various functional groups (like, –O, –F, –OH) [83–85]. Thereafter, the step is to intumesce the space betwixt the $Ti_3C_2T_x$ MXene, which is achieved by various interpolating compounds, like DMSO, and therefore register the exfoliation of sundered 2D sheets as a result of outer shear force [86, 87]. The MXene nanosheets formed are comparable to graphene and due to the presence of often restrained functional groupings (–O,–OH,–F) on the exterior side, origination concerning semiconducting functionalized MXene occurs [77]. Intriguingly, HF is the preferred choice as an etchant, as MAX phases become immobile in the existence of quotidian acids (HCl, H_2SO_4, HNO_3). Quotidian bases and salts (NaCl, Na_2SO_4) [77]. There is sundry criterion that may affect the etching specifications, like the standard of the MAX phase being used, it is constitution as well as particle's measurement (i.e. size), interpolating agents, the etching mixture as well as temperature and time [73, 75, 88]. This can be understood by an example- if we consider "M," i.e., the transition metal used, of greater atomic no., then it will be in need of additional time for exfoliation as well as a powerful etchant [8]. So, the benefit of HF being used as an etchant, is the prominent incomplexity in its route as well as its versatileness, as it is copiously used in MXene synthesis [73]. However, the stumbling block in HF etching technique is that it engages perilous chemical mixtures, causing obstruction in its scale-up processes [75]. Forbye, the MXenes produced by HF-etching methodology, are manifold sheet that are stockpiled by feeble Van-der walls bond [79]. The in-situ induced HF was produced by reacting Ti_3AlC_2 with HCl accompanied by use of a fluoridic salt [75], as found out by Ghidiu and co-workers [89]. If we make a point of the benefits of this method, then it is found that this route uses harsh reactants to a smaller extent as compared to HF, secondly, the time for sonication is less, and promising exfoliation yield, followed by lesser quantity of voids as well as formation of pliable MXenes [86]. This route showed its accomplishment as it led to synthesis of Ti_3CNT_x [90], Mo_2CT_x [91], $Mo_{1.33}CT_x$ [1]. Thus, theoretical perspective is very important while exploring stable as well as up to date MXenes [73].

2.3.1 Theoretical Studies

MXenes, so obtained, are consistent in structure with their parent MAX phases and organized in a way such as the "M" atoms being organized in a hexadic manner as well as "X" atoms occupying octahedral lots [73, 76]. But this atomic pattern can be modified with respect to "n" value, like M_2X showing AB packing (hexagonal), while M_3X_2 and M_4X_3 show ABC packing (fcc) [75]. The high "n" value suggests the instability in MXenes, like the hexadic assembly of Mo as well as C atom [75], suggests the stability in Mo_2C [44], and instability in Mo_3C_2 as well as Mo_4C_3 [92]. A single layer of MXene possesses a hexadic latticework with a sextuple symmetry and lozenge cell from upper view, and as observed from a side view, it shows the arrangement of "X" layer placed betwixt two "M" layers [77]. In o-MXenes, with typical formulation- $M'_2M''X_2$, $M'_2M''_2X_3$, hexadic assembly is observed with outside and inside layers of M' and M'', respectively [73]. The two unalike transition metals in each layer cause an out-of-plane reposition regarding greater M'' atoms out of their fixed locations, which eventually causes the structure to switch to monoclinic symmetry from a hexadic symmetry [93]. Functionalized MXenes are thermodynamically stable, and the stability is upgraded with MXenes accomplishing complete termination [94]. Immaculated MXene is metallic alike its MAX phases, with rich electron density at the Fermi level [75]. The outermost transition metal sheets present in ordered double-M MXenes show their escalated impact on electronic properties in juxtaposition to internal sheet of metal [92, 95]. As Khazei et al. propounded [94], with comparative site of terminal grouping, various functionalization paradigm are suggested: the 1st model takes similar kind of two functional groups being placed on vertex of two transition metal atoms, the subsequent model, the two functional groups are being allocated at tip of the vacant lot, the ternary model contemplates that on top of vacant site "a," one functional group is placed on the tip and likewise on the vacant site "b," other functional group is placed on tip, and in the last paradigm, it suggests the placement of two functional groups on tip of only vacant lot "b" [73] . As a common fashion observed in functionalized M_2X complex, the 1st model lacks stability in contrast to the rest three paradigm [73]. The type of group attached to the terminal site and the ionic state of the transition metal atom cause either of the three models to attain almost all stability [94]. The heterogeneously – terminated V_2CT_2 was created and studied with the help of the aforementioned four models, and it was established that the tantamount of 2nd model is the most preferred configuration [96]. However, not a great deal of records are there for such terminations but the stability of Ti_2CT_z, $Ti_3C_2T_z$ and $Nb_4C_3T_z$ assemblies carried with heterogenous terminations, was looked into by [73], Hu et al. [76]. Highlighting the electronic properties of MXenes, they require to be inspected individually with the introduction of functional groups [75]. For majority of MXenes, p bands present in "X" appear underneath d bands, followed by little bandgap betwixt them [77]. The termination causes a heave in d band beyond the Fermi level, generating a band space and curtailment of the density of states (DOS) [97]. This came about due to hybridization betwixt functional groups of p and d orbitals in the "M" metal [98]. Substantially, M_2XT_z compounds are semiconductors, as connoted by the DFT studies [75] as well as enhancements in their bandgap with reference to atomic number of "M" metal [99]. MXene bandgap rivals to solar spectrum, hence rendering scope for bandgap arrangement in heterostructures for optoelectronics [77]. The Ti_2CT_x surface group adjustment causes change in its bandgap, but it causes no cutback in the carrier portability, which is dissimilar from graphene [100]. Based on spin-orbital coupling, MXenes are classified as either topologically trivial or nontrivial, and based on the electrical conductivity, MXenes are grouped into metallic, semi-metallic, as well as semiconducting [77].

Topologically trivial metallic as well as semiconducting MXenes-bandgap for semiconducting MXenes, as shown by studies executing the generalized gradient approximation (GGA) was 0.24 eV, 0.88 eV, 1.0 eV, along with 1.8 eV, 1.03 eV, and 0.45 eV sequentially [77]. The majority of the semiconductors possess roundabout bandgap, omitting $Sc_2C(OH)_2$ that possesses an exact band space [77]. –F and –OH groups, both just take an individual electron and thus show kindred consequences on MXene's electronic structure, whereas oxygen takes two electrons and thus acts disparately [97]. Consequently, electronic structures concerning MXenes rely on the type of "M" besides "X" atoms being used along with the sort of terminal group taken [101]. Remarkably, $TiCr_2C_2$ was the first and uncommonly basic MXene that was anticipated as semiconductor [102].

Topologically nontrivial metallic along with semiconducting MXenes-Numerous MXenes be made up of 4d as well as 5d hefty transition metals, consequently, their electronic form being remarkably affected by respective spin-orbital coupling [77]. Lately, the affiliation of bandgap to modulation strain as well as material doping has been indicated [75]. For example, niobium atoms doped with Sc_2CO_2 cut back the band space from 2.0 to 1.8 eV, which could additionally remain dwindled toward 1.4 eV by 3% uniaxial strain [103]. Besides, few MXenes being topological insulators, as manifested by $Sc_2C(OH)_2$ compound, which changes from trivial to a topological insulator extending toward metal being prompted with the effect of electric field amplitude [75, 104].

The C/N proportion may be utilized to alter the bandgaps in MXenes and the development of new compounds, as manifested in $Sc_3(CN)F_2$ carbonitride [105]. Howbeit, the metallic character of compounds (like Ti_3C_2, Zr_3C_2, Ti_4C_3, Nb_4C_3, and Ta_4C_3) with greater extent of "n" value, persists nonetheless of the fact that the functional groups exist; for example, the electron density of $Ti_3C_2T_x$ is 10^{22} cm^{-3} [75]. Consequently, the metallic nature was shown by experimentally formulated $Ti_3C_2T_x$ and encouraged the remarkable rise of electrical conductivities in polymers [75] along with ceramic matrixes [106]. Collated in the direction of carbide-MXenes, final higher DOS appertaining to MXenes nitrides as well as carbonitride, which arises from strong Ti–N bond (as contrasted to Ti-C) with additional nitrogen electron, which upshots in elevated electrical conductivity [75]. Consequently, they can contemplate for numerous propitious utilizations, like in nanoelectronics devices as well as sensors [24, 107]. Howbeit, experimental studies concerning nitride MXenes are very scanty as collated to carbide part [75].

2.3.2 Computational Studies

Computational archetype is widely employed to appreciate the construction and the feasible stable compounds of MXenes, which helps in the experimental build out [75]. These studies specifies that the 6 plausible structures of MXenes are likely – (i) single- M elements, (like Ti_2C, Nb_4C_3), (ii) solid solution (like $(Ti, V)_3C_2$), (iii) orderly out-of-plane double-M elements, that have a solitary transition metal inhabiting the outer coat (generally, Cr as well as Mo), although the medial units are infused by other metal(like, Nb as well as Ta), for instance, Mo_2TiC_2 as well as $Mo_2Ti_2C_3$ [92], (iv) orderly in-plane double-M elements, that possess unlike "M" elements that marshal within base plane (like, $(Mo_{2/3}Y_{1/3})_2$ AlC) [70], (v) void ordered (like $Mo_{1.33}CT_x$) [84] as well as $(W_{1.33}CT_x)$ [108], (vi) void desultory distributed (like, $Nb_{1.33}CT_x$) [75]. Computational estimation had pointed out the classic 6-coordination of transition metals due to 6 bonds existing betwixt metal "M" and "X" atoms, leading to M_2XT_2 compound formation [97]. Furthermore, it was established that the configuration with the highest stability normally takes place with "T," i.e. the termination, is placed at a place other than "M" and "X" atoms [75]. Howbeit, in somewhat compounds, it was observed that "T" atoms are overhead of "X" atoms, so as to augment their electronic interaction [109, 110]. Experimentally, the desultory dispersal of terminal groups in "$Ti_3C_2T_x$" and "V_2CT_x" was pointed out as predicted by DFT [75, 111]. Yet, further experimental research concentrated on surface characterization as well as control is needed to consummate confirmation of counterfeit or simulation forecast [112, 113]. Test studies determined MXenes framework by high-resolution transmission electron microscopy, atomic-resolution scanning transmission electron microscopy (STEM) along with electron energy-loss spectroscopy (EELS) [75, 77]. The aforesaid techniques render details concerning the atomic structure, natural defects, as well as surface terminations [95, 97, 114]. Moreover, X-ray photoelectron (XPS) is often exercised toward recognizing MXenes constitution as well as elemental stoichiometric proportion [95, 97]. Ultimately, few studies employed nuclear magnetic resonance (NMR) spectroscopy so as to acquire data concerning the surface functionalization in MXenes as well as the effect of various synthesis pathway (like HF, LiF–HCl) [75, 95, 97]. The Density function theory was carried out by Vienna ab initio simulation package (VASP) accompanied by projector-augmented plane wave pseudopotentials [115]. In the simulated research, Zhang et al. employed the MXene, which has 1 carbon film packed betwixt 2 metal films (metal = Zr, Ti, as well as Hf) [116]. The electron exchange-correlation functional remains studied by GGA in a fashion presented through Perdew-Burke-Ernzerhof (PBE) [116]. The cutoff energy being 520 eV with atomic position optimizations accomplished by $12 \times 12 \times 1$ F-centered K-mesh, which is carried on upto the point when remaining force decreases below 0.02 eV $Å^{-1}$ [117].

The chosen MXene substances were subsequently functionalized via the surface groups of –F, –O, as well as –OH in attempt to widen the spaces. Furthermore, to demonstrate the three various morphologies of surface functionalized MXenes, illustrated in Figure 2.4, we used Ti_2CO_2 as an exemplar(c) [96]. Figure 2.4 illustrates how the calculated band configurations indicate that practically all of the M_2CT_2 materials (M = Ti, V), exhibit

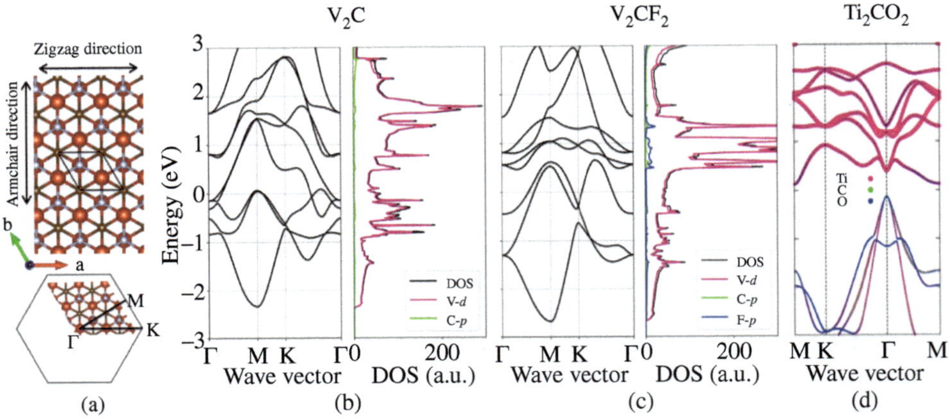

Figure 2.4 (a) MXene atomic structure, unit cell, and Brillouin zone, including Γ, M, and K points, with Γ-M and Γ-K, respectively, along the armchair and zigzag direction. DOS and electronic band structure of (b) pristine V_2C, and (c) functionalized-V_2CF_2 computed within GGA, and (d) projected band structure of Ti_2CO_2. The Fermi level is set as the reference of zero energy. Source: (a)–(c) Reproduced with permission from [96]/American Physical Society.

metallic properties. These M_2CT_2 substances under investigation should exhibit a diversity of structural shapes and chemical compositions, which should be used to identify their morphologies.

For computing electronic structure of PBE sizeable K-mesh of $42 \times 42 \times 1$ was employed, whereas K-mesh of $12 \times 12 \times 1$ used for HSE06 because of exorbitant computational cost [117]. Experimental computation carried with $30 \times 30 \times 1$ for PBE indicated that bandgap for Sc_2CF_2 as well as Sc_2CCl_2 scarcely variation (<4 meV), proposing authenticity of K-grid being employed in $HSE0_6$ [117]. Meant for single-layer MXene, a vacuum space of 20 Å was adopted so as to avoid and banish interactions betwixt adjacent layers. As researches have denoted that deterioration of MXenes may occur when subjected to moisture or high temperatures [118, 119]. For example, an expounded Nb_2CT_x MXene study indicated positioning of Nb ad-atoms upon hcp surface sites, which bonds with circumambient oxygen, while exposure, which causes destabilization, which ultimately deteriorates the complex's structure [120]. Moreover, when $Ti_3C_2T_x$ FET was subjected to air for 70 hours, then almost 20% reduction in its inceptive conductivity was manifested [118]. Following their synthesis, it is commended to hoard the colloids in an oxygen-deprived ambience accompanied by less exposure to illumination, so as to complement their stability [73, 119]. It is done in a strong polar solvent, (like DMSO) [121] or by using solution filtration methods (i.e. MXene films) [119]. Furthermore, the most effective etching methods, which lessens the adsorbed atom on MXene surface might enhance their stability too [120]. In the initial test study concerning the thermal stability of MXene, [122] the observation made was that at 3 steps of $Ti_3C_2T_x$, mass drop around argon surrounding- (i) 0.38% after circumambient temperature toward 200 °C, (ii) 4.48% betwixt 200° as well as 800 °C as well as, (iii) 2.47% beginning with 800° till 1000 °C [75]. The earliest step shows loss of adsorbed water as well as HF, whereas the subsequent step is consequence of eviction of –OH groups, and the last step indicates the liberation of –F atoms [75]. Same conduct was noted by some studies [75] and authenticated by thermogravimetric accompanied by mass spectrometry (TG-MS) analysis [123]. Consequently, such researches suggested that thermal analysis can be an efficacious strategy to eliminate –OH or –F cessations, as everyone being eliminated in a particular temperature limit, but the end substance is not an immaculate MXene, but oxygen-terminated MXene [8]. Lately, subjection of $Ti_3C_2T_x$ at 750 °C under vacuum, eliminates preponderance toward o-terminations, insinuating utilization concerning this strategy for deriving immaculated MXenes [124]. Nevertheless, pragmatic applications demand comprehension about thermal properties in air [125] along with oxygen aerosphere [122, 126]. In oxygen aerosphere, notwithstanding rutile crystal forming at 500 °C, MXenes phase change occurs at extortionate temperatures (for example, $Ti_3C_2T_x$ change to TiC occurs at temperature more than 1000 °C for at minimum 2 hours) [122, 126]. Lately, Mo-based exhibits thermal stability up till 530 °C (H_2/N_2 atmosphere) [127], together with $Zr_3C_2T_x$ exhibiting preferable thermal stability compared to $Ti_3C_2T_x$, being auspicious for high-temperature applications [128].

During the attachment process from MAX to MXene, the newly developed and most active M-elements are actively depleted by cutting-edge models from etchant and outside the MAX parent category. Appropriately, MXene formulation- $M_{n+1}X_nT_x$ is underlined to emphasize importance about a higher surface termination T_x phase. It is clear that etchant combines the first and easiest choices in the planned cut, and there is a specific function in this field [93, 129–132]. Additionally, K+ ion intercalation is one post-etching procedure that may be used to compensate for extinct species [93]. Although it shows promise of adjusting the terminating species as well as the MXene structures and properties, the obtainable cessating species persist to be inadequate toward these found in the etchant. We anticipate that as MXenes make a remarkable impression upon upcoming 2D material studies, it will lead to extraordinary properties along with applicability of such substances over the upcoming future. With MXenes, 2D material science facilitates from solid structures to atoms with an atomic-level design of a 2D sheet.

2.4 Structure Functionalization of MXene

Chemically modifying MXenes as a successful method toward improving electrical compatibility alongside other materials provides significant inducement for its practical commercial application. Numerous analyses have revealed the predilection toward stopping at the bare MXene surface in first-principles calculations. As a result of the creation of sturdy bonds betwixt the surface regarding M atoms along with the terminating species, MXenes gained notable negative formation energy upon surface cessation [133, 134]. The nonmetal adsorption or replacement of MXenes by nonmetals is the subject of numerous theoretical investigations nowadays [135]. The structural and electrical characteristics of Ti_2C and Ti_3C_2 can be changed by applying density functional theory calculations to surface adsorbed groups, such as O, F, OH [110, 136–138], and OCH_3 [37]. Balci et al. carefully examined the C atom substitution in $Sc_2C_2F_2$ MXene through Si, B, Sn, S, F, Ge, and N, along with B+N. They found how S, F, and N with B + N swap are energy-promising and may modify the band lacuna concerning $Sc_2C_2F_2$ [139]. According to Schwingenschlögl et al., replacing the surface groups (F, OH) of Ti_2C besides V_2C with P and Si atoms is energy-favorable as well as increasing Li/Na/K capacity [140]. The charge density contrast caused via adsorbed single metal adatoms upon Ti_3C_2 for the purpose to better comprehend the bonding process betwixt such metal atoms along with Ti_3C_2, as well as the inherent principles for adsorption energies. MXenes possessing structure in general, the distribution of functional groups in $M_{n+1}X_nT_x$ is unaffected by the kind of metal, the presence of carbon/nitrogen species, and the number of atomic layers. Properties of MXenes, like as work function, hydrophilic conduct, along with catalytic activity relate to surface functionalization [137]. The density functional theory tight-binding approach is used to study the MXenes like Ti_2C or Ti_3C_2 that are operationalized using methoxy groups. Because surface functional groups are present, MXenes' properties undergo drastic modifications. As an example, the functionalized derivatives of pure Ti_2C and Ti_3C_2 act as magnetic metals that can either be nonmagnetic tapered-bandgap semiconductors or metals contingent upon the kind of surface cessation [141]. Due to the intercalation of C atoms between 2 layers of transition metal (Ti, Zr, Hf) atoms functionalized with oxygen in several MXenes, such as Ti_2CO_2, Zr_2CO_2, and Hf_2CO_2 [142]. In this particular MXene $Ti_3C_2T_x$, surface functionalization with carboxyl or glycine groups and silane coupling agents results in gaining stability and charge percolation of $Ti_3C_2T_x$, and that concedes application of thin films like antibacterial coverings, fuel cells, surface-initiated polymerization, etc. [143]. As we talk about Ti_3C_2 sheets (damage tolerance), which lead to ion implantation, they create an abundance of extended defects of both Ti and C that promote oxygen functionalization. For unforced functionalization of defective titanium and carbon sites with oxygen disorganized nature of $Ti_3C_2T_x$ effect by ion implantation is extremely beneficial, hence increasing the amount of oxygen in MXene peeling that alters their electronic structure that adjusts bandgap in different MXene chemistry [144]. In contrast to fluorine and hydroxyl-functionalized MXene, oxygen-functionalized MXene assemblies have fewer lattice parameters and display unequalled mechanical stability [98, 145]. According to a renowned scientist, O-surface functionalized MXene Ti_2C (Ti_2CO_2) enhances photocatalytic performance for CO_2 reduction to formic acid. Compounds like Sc_2C function as monolayers along with their functionalized derivatives. When contrasted with effects regarding functionalization on electrical, thermal, and thermoelectric properties, it was clear that the latter had a significant impact on the former [145]. As we functionalize MXenes such as Sc_2C, which have ability to gain semiconducting states toward diverse functionalizations together with O, F, and OH Atomic arrangement: Lateral and upper views of monolayer Sc_2C functionalized by (a) O, (b) F, and (c) OH were depicted as in Figure 2.5.

Evaluating –O, –OH, and –F functional interactions of Sc_2C's bilateral, one observes that $T = O$ has a considerably distinctive configuration than $T = F/OH$ has a teeny C-T separation. As we study stability of the functionalizations. $T = O$ shows lowered frequency than $T = F/OH$ [146].

2.4.1 Different Group Used for Structural Functionalization

2.4.1.1 Oxygen-Functionalized MXene

Selectivity of MXenes like M_2C is improved by oxygen functionalization. Ti, V, Nb, and Mo are referred to as M in this instance. We are aware from a number of studies that non-functionalized surfaces with high reactivity are more suited for catalytic applications. In this instance, reactivity falls but selectivity rises. Mo_2CO_2 and V_2CO_2, for example,

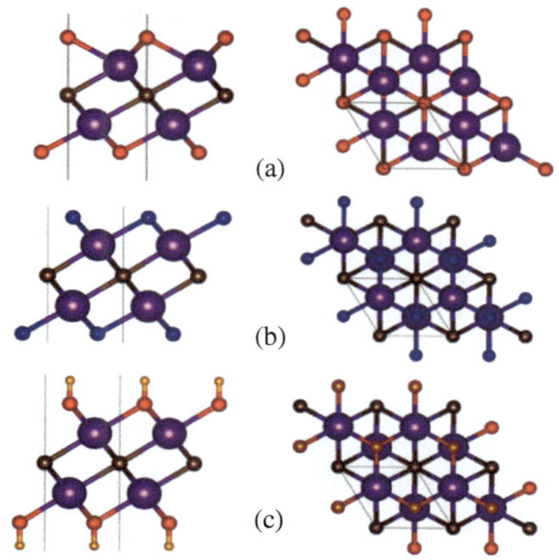

Figure 2.5 Illustrative representation regarding functionalization of MXene. Where Sc represents pink, O represents red, C represents brown, H represents yellow, and F represents blue Source: Reproduced with permission from [146]/American Physical Society.

have high selectivity against NO molecules, whereas Nb_2CO_2 in addition to Ti_2CO_2 have notable selectivity regarding NH_3. Since few pristine MXenes display both metallic and semiconductor features during functionalization, their electrical properties are regulated. O-functional groups in MXene, such as those found in Ti_2C, Hf_2C, and Zr_2C, produce semiconducting properties, while –F/–OH groups hold their metallic properties as they do within non-functionalized MXenes. Plane wave-based DFT computations are used to examine the gas molecules's adsorption upon 4 MXenes (M_2C, M = V, Ti, Mo, and Nb) as well as its O-functionalized surfaces. By certain investigations, Ti_2CO_2 with oxygen vacancies come up with good catalytic execution to convert CO_2 to HCOOH [147]. 2D MXenes with hexagonal lattices, such as Ti_2CO_2, Zr_2CO_2, and Hf_2CO_2, have one layer of C atoms intercalated between two layers of transition metal atoms functionalized with oxygen, as shown in Figure 2.6, and it is already proven that below structure is the most stable oxygen functionalized structure of MXenes.

2.4.1.2 Sulfur-Functionalized MXenes

Surface-based surface cessation of MXenes regulates ion kinematics as well as electrochemical characteristics. As of yet –O, –F, and –OH groups have terminated MXenes utilized to study potential uses for storage of energy. In recent times, sulfur-functionalized MXene assemblies have come into frame as contender that enhance battery performance because it is having low diffusion barriers .Some of Li-adsorbed sulfur-functionalized transition metal (M)-stationed MXenes M_2CS_2, here M belongs to elements, like Zr, Sc, Ti, V, Hf, Nb, Cr, Ta, W, and Mo. As we are going to explore the lightest S-terminated MXenes (i.e. TiC) by intercalating elements of group I and II (alkali and alkali earth-metal) like Na, Li, K, Mg, and Ca [148]. In recent discovery, S-functionalized nitride MXenes have been explored named V_2NS_2 and Ti_2NS_2 functions like anodes for Li/Na ion batteries have been investigated through first principle DFT calculations. Some more MXenes are also explored as battery application carbide MXenes, such as Ti_2C, Ti_2C_3 and whereas nitride MXenes, like Ti_2N [137].

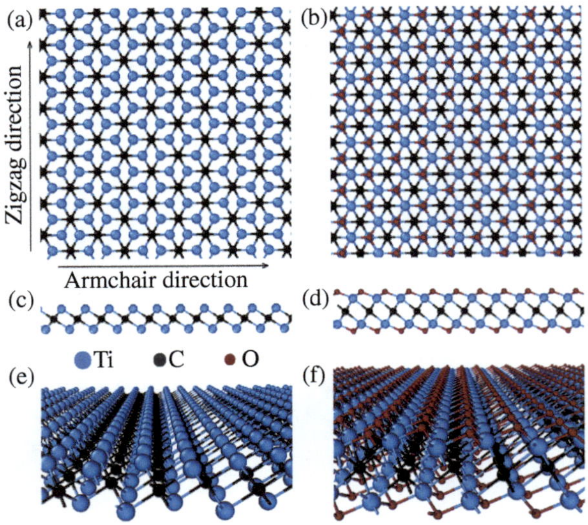

Figure 2.6 Structural representation of monolayers; where (a) bare MXenes and (b) oxygen functionalized MXenes. Source: Reproduced with permission from [147]/American Physical Society.

2.4.1.3 Methoxy Group-Functionalized MXenes

Ti_2C and Ti_3C_2, which have had methyl groups functionalized, are studied using density functional theory tight-binding approach, which primarily focuses upon their structural, electrical, along with comparative stability characteristics. The lattice structure of hydroxylated MXenes like $Ti_2C(OH)_2$ along with $Ti_3C_2(OH)_2$ fully depends upon the kind of OH casing and either increases or lowers in comparison to Ti_2C and Ti_3C_2's lattice structures. Lattice parameters rise when surfaces' OH groups are fully substituted by methoxy groups ($Ti_2C(OCH_3)(OH)$) or partially ($Ti_2C(OCH_3)_2$ and $Ti_3C_2(OCH_3)_2$), as opposed to hydroxylated MXenes with similar types of surface covering [149]. Yang et al. reported the surface functionalization regarding MXene alongside chitosan by means of in-situ formation of polyimidazoles [150]. Imidazoles-MXene hybrid composite as polyimidazoles chain overlaps upon the exterior layer of MXene, i.e. Ti_3C_2@IMIZ, being processed through simplicated technique illustrated in Figure 2.7.

This method involves the growth of imidazols in situ upon the exterior layer of MXenes by simple multicomponent process utilizing chitosan, just like a renewable reactant. The adsorption fulfillment of Ti_3C_2@IMIZ regarding eliminating environmental contaminants was accomplished by using Cr(VI) heavy metal ions as adsorbate. XPS analysis depicts Cr(VI) changing to Cr(III) having low toxicity during adsorption process. According to XPS analysis, during the adsorption process, Cr(VI) converted into low-toxic Cr(III), which subsequently resulted in the eradication of Cr(VI) species that further come up with elimination of Cr(VI) species.

2.4.2 Factor Affecting the Structure Functionalization

2.4.2.1 Electric and Optical Properties

Electrical properties of Ti_3C_2 MXenes getting through heat treatment of the material to alter the surface. As we perform alkaline treatment in diluted NaOH succeeded by calcination of the respective MXene around 400 °C as well as 600 °C, that increases conductivity of

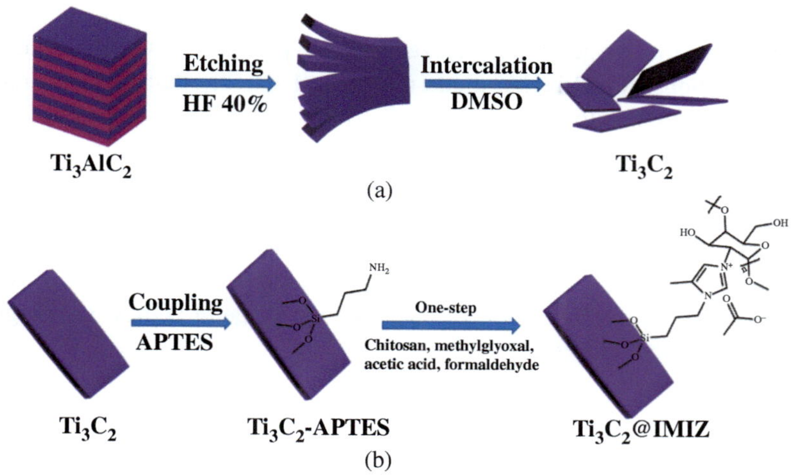

Figure 2.7 Illustrative description of the synthesis of Ti3C2@IMIZ composites through a blend of coupling reactions with Radziszewski reaction. Source: Reproduced with permission from [150]/SAGE Publications / CCBY 4.0.

MXene via 850 S cm^{-1} up to 2410 S cm^{-1} correspondingly, while optical properties, MXenes are having high photo-thermal conversion efficiency with certain surface variations [145].

2.4.2.2 Thermal Conductivity

The thermal conductivity of the lattice is low for –F along with –O-functionalized MXenes, it strongly increases via reducing the lattice thermal conductivity because power factor is high [151].

2.4.2.3 Electrochemical Properties

As we talk about electrochemical characteristics of S-factionalized MXenes. The following expression describes the chemical process that adsorbs Li upon $M_2CS_2Li_{x1}$ to create

$$M_2CS_2Li_{x2} : -(x_2 - x_1)Li^+ + (x_2 - x_1)e^- + M_2CS_2Li_{x1} \leftarrow - - \rightarrow M_2CS_2Li_{x2}$$

Here x_1 and x_2 represent the Li content prior to and post the reaction.

2.4.2.4 Thermoelectric Property

When fluorine is added to Cr-group MXenes, the most stable compounds are B-type Cr_2CF_2 and Mo_2CF_2, but A-type W_2CF_3, Cr_2CF_2, and Mo_2CF_2 are semiconductors having identical bandgaps related to 0.227 along with 0.224 eV. The material's Seebeck coefficients (S) are expressed as V/T, wherein V constitutes the induced voltage running through its material along with T that represents temperature gradient. Seebeck coefficients are temperature-sensitive, with smaller coefficients at higher temperatures. But some MXenes, like Cr_2CF_2 and Mo_2CF_2 show high Seebeck coefficients at 500 K (high temperature) [152].

Zha et al. explored the structural, mechanical, along with electrical features regarding M_2CT_2 (M = Sc, Ti, V, Cr, Zr, Nb, Mo, Hf, Ta, W) MXenes anatomy [152]. They demonstrated how the surface functional species (T = F, O, and OH) and the transition metal atom have noteworthy impression upon the crystal anatomy of these materials. Additionally, they have shown that the oxygen-functionalized MXene assemblies, which acquire tetchy lattice specifications, reveal higher-grade mechanical stability than fluorine and hydroxyl-functionalized ones. Numerous other studies have outlined how functionalization affects the structural, electrical, along with thermal characteristics regarding MXene structures besides the structural ones. It is worth noting that, based on experimental findings, the bare MXenes are arduous to procure, as well as surface-functionalized groups with −OH, −O, and −F species, which are frequently noticed. Until now, functionalization has been illustrated for MXenes Ti_2C, Ti_3C_2, Ti_3CN, Ta_4C_3, Nb_2C, V_2C, $(Ti_{1/2}Nb_{1/2})_2C$, along with $(V_{1/2}Cr_{1/2})_2C$, becoming a concern regarding the range of aptness in energy storage as well as capacitative widget.

2.5 Conclusion and Future Prospects

In this book chapter, we have discussed about the different compositions of MXenes and their existence in various structural forms. We also traced the fabulous obstacles as well as opportunities regarding the functionalization of MXenes. This incorporates perception toward the recognition of as well as the stability of surface terminations and comprehension of the level of surface saturation apropos terminating species. From one angle, it is possible to manage the electronic structures of MXenes and increase their inherent activity by doping them using heteroatoms like N, P, Pt, and Ru, along with others, or else moderating their terminations (–O, –OH, and –F). MXenes obtained through quaternary in-plane as well as out-of-plane chemically organized MAX phases would substantially grow the family by expanding the range of tunable compositions and structures. Despite significant progress made in MXenes and their beneficial uses, it is still difficult to see a manageable surface termination on MXenes. In addition, the primary etchant used today is HF, which is detrimental toward the natural world. In view of the intense bonding found in MAX stages, nevertheless, it is not plausible for exfoliating the stratified structure using quotidian etchants. It is not possible for exfoliating the stratified assembly using customary etchants, however, because of the strong bonding seen in MAX stages. It raised the challenge of how to stabilize the atmosphere with gentle chemical etching as a result. More synchronous synthesis, functional change, and profitable as well as superficial processing strategies should be present in the next generation of MXene families. Analyzing the impact of composition as well as thickness on chemical characteristics can help us better understand the fundamental benefits of MXenes. It appears that by the upcoming ears, plumb along with lateral heterostructures of MXene will be created in combination with other 2D materials.

Acknowledgment

The National Research Foundation of Korea (NRF), which is sponsored by the Ministry of Science and ICT, provided further assistance for this work through its Brain Pool project. (2019H1D3A1A01070741).

References

1 Anasori, B., Lukatskaya, M.R., and Gogotsi, Y. (2017). *Nat. Rev. Mater.* 2: 1–17.
2 Naguib, M., Mashtalir, O., Carle, J. et al. (2012). *ACS Nano* 6: 1322–1331.
3 Zamhuri, A., Lim, G.P., Ma, N.L. et al. (2021). *Biomed. Eng. Online* 20: 33.
4 Wang, H. and Lee, J.M. (2020). *J. Mater. Chem. A* 8: 10604.
5 Xie, X., Chen, C., Zhang, N. et al. (2019). *Nat. Sustainability* 2: 856–862.
6 Ding, L., Li, L., Liu, Y. et al. (2020). *Nat. Sustainability* 3: 296–302.
7 Xia, Y., Mathis, T.S., Zhao, M.Q. et al. (2018). *Nature* 557: 409–412.
8 Lukatskaya, M.R., Mashtalir, O., Ren, C.E. et al. (2013). *Science* 341: 1502–1505.
9 Kim, S.J., Koh, H.J., Ren, C.E. et al. (2018). *ACS Nano* 12: 986–993.
10 Zhang, Y.Z., Lee, K.H., Anjum, D.H. et al. (2018). *Sci. Adv.* 4: eaat0098.
11 Shahzad, F., Alhabeb, M., Hatter, C.B. et al. (2016). *Science* 353: 1137–1140.
12 Liu, J., Zhang, H.B., Sun, R. et al. (2017). *Adv. Mater.* 29: 1702367.
13 Zhang, J., Zhao, Y., Guo, X. et al. (2018). *Nat. Catal.* 1: 985–992.
14 Zhao, D., Chen, Z., Yang, W. et al. (2019). *J. Am. Chem. Soc.* 141: 4086–4093.
15 Li, R., Zhang, L., Shi, L., and Wang, P. (2017). *ACS Nano* 11: 3752–3759.
16 Sarycheva, A., Polemi, A., Liu, Y. et al. (2018). *Sci. Adv.* 4: eaau0920.
17 Xu, B., Zhu, M., Zhang, W. et al. (2016). *Adv. Mater.* 28: 3333–3339.
18 Persson, P.O.Å. and Rosen, J. (2019). *Curr. Opin. Solid State Mater. Sci.* 2: 100774.
19 Naguib, M. et al. (2012). *ACS Nano* 6: 1322.
20 Liu, Z., Wu, E., Wang, J. et al. (2014). *Acta Mater.* 73: 186.
21 Tao, Q., Dahlqvist, M., Lu, J. et al. (2017). *Nat. Commun.* 8: 14949.
22 Dahlqvist, M., Lu, J., Meshkian, R. et al. (2017). *J. Rosen Sci. Adv.* 3: e1700642.
23 Lind, H., Halim, J., Simak, S., and Rosén, J. (2017). *Phys. Rev. Mater.* 1: 044002.
24 Persson, I., el Ghazaly, A., Tao, Q. et al. (2018). *Small* 4: 1703676.
25 Anasori, B., Xie, Y., Beidaghi, M. et al. (2015). *ACS Nano* 9: 9507.
26 Tan, T.L., Jin, H.M., Sullivan, M.B. et al. (2017). *ACS Nano* 11: 4407.
27 Rajan, A.C., Mishra, A., Satsangi, S. et al. (2018). *Chem. Mater.* 30: 4031–4038.
28 Li, Z., Zhuang, Z., Lv, F. et al. (2018). *Adv. Mater.* 30, 1803220.
29 Huang, Z., Wang, S., Kota, S. et al. (2016). *Polymer* 102: 119–126.
30 Shao, W., Tebyetekerwa, M., Marriam, I. et al. (2018). *J. Power Sources* 396: 683–690.
31 Chen, J., Liu, M., Huang, H. et al. (2018). *J. Mol. Liq.* 259: 179–185.
32 Zhang, X., Wang, K., Liu, M. et al. (2015). *Nanoscale* 7: 11486–11508.
33 Liu, M., Zeng, G., Wang, K. et al. (2016). *Nanoscale* 8: 16819–16840.
34 Wang, H., Zhang, J., Wu, Y. et al. (2016). *Appl. Surf. Sci.* 384: 287–293.
35 Zhu, X., Liu, B., Hou, H. et al. (2017). *Electrochim. Acta* 248: 46–57.
36 Tian, Y., Que, W., Luo, Y. et al. (2019). *J. Mater. Chem. A* 7: 5416–5425.
37 Liu, P., Ding, W., Liu, J. et al. (2020). *J. Alloys Compd.* 829: 154634.
38 Boota, M., Anasori, B., Voigt, C. et al. (2016). *Adv. Mater.* 28: 1517–1522.
39 Qin, L., Tao, Q., Liu, X. et al. (2019). *Nano Energy* 60: 734–742.
40 Chen, J., Huang, Q., Huang, H. et al. (2020). *Nanoscale* 12: 3574–3592.
41 Wei, Y., Zhang, P., Soomro, R.A. et al. (2021). *Adv. Mater.* 33: 2103148.
42 Wang, C., Shou, H., Chen, S. et al. (2021). *Adv. Mater.* 33: 2101015.
43 Anasori, B., Lukatskaya, M.R., and Gogotsi, Y. (2017). *Nat. Rev. Mater.* 2: 16098.

44 Anasori, B., Xie, Y., Beidaghi, M. et al. (2015). *ACS Nano* 9 (10): 9507–9516.
45 Björk, J. and Rosen, J. (2021). *Chem. Mater.* 33: 9108–9118.
46 Gogotsi, Y. and Anasori, B. (2019). *ACS Nano* 13 (8): 8491–8494.
47 Anasori, B. and Gogotsi, Y. (2019). *Springer Cham.* 3–12.
48 Li, Y., Huang, S., Wei, C. et al. (2021). *ACS Appl. Mater. Interfaces* 13 (3): 4682–4691.
49 Xie, S.Y., Su, J.H., and Zheng, H. (2019). *Solid State Commun.* 291: 51–53.
50 Barsoum, W.M. and Eklund, P. (2019). *Springer, Cham* 15–35.
51 Li, M., Li, X., Qin, G. et al. (2021). *ACS Nano* 15: 1077–1085.
52 Kumar, P., Singh, S., Hashmi, S.A.R., and Kim, K.H. (2021). *Nano Energy* 85: 105989.
53 Abraham, B.M., Parey, V., and Singh, J.K. (2022). *J. Mater. Chem. C* https://doi.org/10.1039/D1TC06029E.
54 Verger, L., Natu, V., Carey, M., and Barsoum, M.W. (2019). *Trends Chem.* 1 (7): 656–669.
55 Hong, W., Wyatt, B.C., and Anasori, S.K.N.B. (2020). *MRS Bull.* 45: 850–861.
56 Parey, V., Abraham, B.M., Mir, S.H., and Singh, J.K. (2021). *ACS Appl. Mater. Interfaces* 13 (30): 35585–35594.
57 Chen, Y., Liu, C., Guo, S. et al. (2022). *Green Energy Environ.* 7: 394–410.
58 Naguib, M., Barsoum, M.W., and Gogotsi, Y. *Adv. Mater.* 2103393.
59 Biswas, S. and Alegaonkar, P.S. (2022). *Surfaces* 5: 1–34.
60 Kumar, J.A., Prakash, P., Krithiga, T. et al. (2022). *Chemosphere* 286: 131607.
61 Gogotsi, Y. and Huang, Q. (2021). *ACS Nano* 15 (4): 5775–5780.
62 Hantanasirisakul, K., Anasori, B. et al. (2020). *Nanoscale Horiz.* 5: 1557–1565.
63 Mockute, A., Lu, J., Moon, E.J. et al. (2015). *Mater. Res. Lett.* 3: 16–22.
64 Liu, Z. et al. (2014). *Acta Mater.* 73: 186–193.
65 Dahlqvist, M., Lu, J., Meshkian, R. et al. (2017). *Sci. Adv.* 3: 1–9.
66 Tao, Q. et al. (2018). *Phys. Rev. Mater.* 2: 114401.
67 Tao, Q. et al. (2019). *Chem. Mater.* 31: 2476–2485.
68 Petruhins, A., Lu, J., Hultman, L., and Rosen, J. (2019). *Mater. Res. Lett.* 7: 446–452.
69 Petruhins, A., Dahlqvist, M., Lu, J. et al. (2020). *Cryst. Growth Des.* 20: 55–61.
70 Meshkian, R. et al. (2018). *Adv. Mater.* 30: 1706409.
71 Halim, J. et al. (2018). *ACS Appl. Nano Mater.* 1: 2455–2460.
72 Naguib, M., Mochalin, V.N., Barsoum, M.W., and Gogotsi, Y. (2014). *Adv. Mater.* 26: 992–1005.
73 Ronchi, R.M., Arantes, J.T., and Santos, S.F. (2019). *Ceram. Int.* 45: 18167–18188.
74 Champagne, A. and Charlier, J.C. (2020). *J. Phys.: Mater.* 3: 032006.
75 Shein, I.R. and Ivanovskii, A.L. (2012). *Superlattices Microstruct.* 52: 147–157.
76 Pang, J., Mendes, R.G., Bachmatiuk, A. et al. (2019). *Chem. Soc. Rev.* 48: 72–133.
77 Pietzka, M.A. and Schuster, J.C. (1994). *J. Phase Equilib.* 15: 392–400.
78 Coleman, J.N., Lotya, M., O'Neill, A. et al. (2011). *Science* 331: 568–571.
79 Mishra, A., Srivastava, P., Mizuseki, H. et al. (2016). *Phys. Chem. Chem. Phys.* 18: 11073–11080.
80 Enyashin, A.N. and Ivanovskii, A.L. (2013). *J. Solid State Chem.* 2013 (207): 42–48.
81 Chaudhari, N.K., Jin, H., Kim, B. et al. (2017). *J. Mater. Chem. A* 5: 24564–24579.
82 Han, F., Luo, S., Xie, L. et al. (2019). *ACS Appl. Mater. Interfaces* 11: 8443–8452.
83 Tao, Q., Dahlqvist, M., Lu, J. et al. (2017). *Nat. Commun.* 8: 1–7.

84 Zhou, J., Zha, X., Zhou, X. et al. (2017). *ACS Nano* 11: 3841–3850.
85 Ghidiu, M., Lukatskaya, M.R., Zhao, M.Q. et al. (2014). *Nature* 516: 78–81.
86 Mashtalir, O., Naguib, M., Mochalin, V.N. et al. (2013). *Nat. Commun.* 4: 1–7.
87 Zhang, T., Pan, L., Tang, H. et al. (2017). *J. Alloys Compd.* 69: 818–826.
88 Ghidiu, M., Naguib, M., Shi, C. et al. (2014). *Chem. Commun.* 50: 9517–9520.
89 Du, F., Tang, H., Pan, L. et al. (2017). *Electrochim. Acta* 235: 690–699.
90 Halim, J., Kota, S., Lukatskaya, M.R. et al. (2016). *Adv. Funct. Mater.* 26: 3118–3127.
91 Meshkian, R., Näslund, L.A., Halim, J. et al. (2015). *Scr. Mater.* 108: 147–150.
92 Dahlqvist, M., Petruhins, A., Lu, J. et al. (2018). *ACS Nano* 12: 7761–7770.
93 Khazaei, M., Arai, M., Sasaki, T. et al. (2013). *Adv. Funct. Mater.* 23: 2185–2192.
94 Hantanasirisakul, K. and Gogotsi, Y. (2018). *Adv. Mater.* 30: 1804779.
95 Champagne, A., Shi, L., Ouisse, T. et al. (2018). *Phys. Rev. B.* 97: 115439.
96 Hu, T., Hu, M., Gao, B. et al. (2018). *J. Phys. Chem. C.* 122: 18501–18509.
97 Sinha, A., Zhao, H., Huang, Y. et al. (2018). *Trends Anal. Chem.* 105: 424–435.
98 Sarikurt, S., Çakır, D., Keçeli, M., and Sevik, C. (2018). *Nanoscale* 10: 8859–8868.
99 Lai, S., Jeon, J., Jang, S.K. et al. (2015). *Nanoscale* 7: 19390–19396.
100 Tang, X., Guo, X., Wu, W., and Wang, G. (2018). *Adv. Energy Mater.* 8: 1801897.
101 Sun, W., Xie, Y., and Kent, P.R. (2018). *Nanoscale* 10: 11962–11968.
102 Guo, J., Sun, Y., Liu, B. et al. (2017). *J. Alloys Compd.* 712: 752–759.
103 Balcı, E., Akkuş, Ü.Ö., and Berber, S. (2018). *Appl. Phys. Lett.* 113: 083107.
104 Luo, K., Zha, X.H., Zhou, Y. et al. (2018). *RSC Adv.* 8: 22452–22459.
105 Guo, J., Legum, B., Anasori, B. et al. (2018). *Adv. Mater.* 30: 1801846.
106 Zhang, N., Hong, Y., Yazdanparast, S., and Zaeem, M.A. (2018). *2D Mater.* 5: 045004.
107 Zhang, X., Zhang, Z., and Zhou, Z. (2018). *J. Energy Chem.* 27: 73–85.
108 Li, L. (2016). *Comput. Mater. Sci* 124: 8–14.
109 Khazaei, M., Ranjbar, A., Arai, M., and Yunoki, S. (2016). *Phys. Rev. B.* 94: 125152.
110 Srivastava, P., Mishra, A., Mizuseki, H. et al. (2016). *Mater. Interfaces* 8: 24256–24264.
111 Sharma, G., Naguib, M., Feng, D. et al. (2016). *J. Phys. Chem. C.* 2016 (120): 28131–28137.
112 Magnuson, M., Halim, J., and Näslund, L.Å. (2018). *J. Electron. Spectrosc. Relat. Phenom.* 224: 27–32.
113 Karlsson, L.H., Birch, J., Halim, J. et al. (2015). *Nano Lett.* 15: 4955–4960.
114 Ivanov, B.A. and Tartakovskaya, E.V. (1996). *Phys. Rev. Lett.* 77: 386.
115 Zhang, H., Yang, G., Zuo, X. et al. (2016). *J. Mater. Chem. A* 4: 12913–12920.
116 Liu, J.H., Kan, X., Amin, B. et al. (2017). *Phys. Chem. Chem. Phys.* 19: 32253–32261.
117 Lipatov, A., Alhabeb, M., Lukatskaya, M.R. et al. (2016). *Adv. Electron. Mater.* 2: 1600255.
118 Zhang, C.J., Pinilla, S., McEvoy, N. et al. (2017). *Chem. Mater.* 29: 4848–4856.
119 Palisaitis, J., Persson, I., Halim, J. et al. (2018). *Nanoscale* 10: 10850–10855.
120 Maleski, K., Mochalin, V.N., and Gogotsi, Y. (2017). *Chem. Mater.* 29: 1632–1640.
121 Li, Z., Wang, L., Sun, D. et al. (2015). *Mater. Sci. Eng., B* 191: 33–40.
122 Li, J., Yuan, X., Lin, C. et al. (2017). *Adv. Energy Mater.* 7: 1602725.
123 Persson, I., Halim, J., Lind, H. et al. (2019). *Adv. Mater.* 31: 1805472.
124 Kong, F., He, X., Liu, Q. et al. (2018). *Electrochim. Acta* 2018 (265): 140–150.
125 Wang, K., Zhou, Y., Xu, W. et al. (2016). *Ceram. Int.* 42: 8419–8424.

126 Kim, H., Anasori, B., Gogotsi, Y., and Alshareef, H.N. (2017). *Chem. Mater.* 29: 6472–6479.

127 Zhou, J., Zha, X., Chen, F.Y. et al. (2016). *Angew. Chem. Int. Ed.* 55: 5008–5013.

128 Ren, C.E., Zhao, M.Q., Makaryan, T. et al. (2016). *Chem. Electro. Chem.* 3: 689–693.

129 Tang, Q., Zhou, Z., and Shen, P. (2012). *J. Am. Chem. Soc.* 134: 16909–16916.

130 Magne, D., Mauchamp, V., Célerier, S. et al. (2016). *Phys. Chem. Chem. Phys.* 18: 30946–30953.

131 Wang, H.W., Naguib, M., Page, K. et al. (2015). *Chem. Mater.* 28: 349–359.

132 Halim, J., Persson, I., Eklund, P. et al. (2018). *RSC Adv.* 8: 36785–36790.

133 Ashton, M., Mathew, K., Hennig, R.G., and Sinnott, S.B. (2016). *J. Phys. Chem. C* 120: 3550–3556.

134 Meng, Q., Ma, J., Zhang, Y. et al. (2018). *Nanoscale* 10: 3385–3392.

135 Xie, Y., Naguib, M., Mochalin, V.N. et al. (2014). *J. Am. Chem. Soc.* 136: 6385–6394.

136 Hu, T., Wang, J., Zhang, H. et al. (2015). *Phys. Chem. Chem. Phys.* 17: 9997–10003.

137 Enyashin, A.N. and Ivanovskii, A.L. (2013). *J. Phys. Chem. C* 117: 13637–13643.

138 Balcı, E., Akkuş, Ü.Ö., and Berber, S. (2017). *J. Mater. Chem. C* 5: 5956–5961.

139 Zhu, J. and Schwingenschlögl, U. (2017). *2D Mater.* 4: 025073.

140 Ibragimova, R., Erhart, P., Rinke, P., and Komsa, H.P. (2021). *J. Phys. Chem. Lett.* 12 (9): 2377–2384.

141 Khan, S.A., Amin, B., Gan, L.Y., and Ahmad, I. (2017). *Phys. Chem. Chem. Phys.* 19 (22): 14738–14744.

142 Nie, Y., Huang, J., Ma, S. et al. (2020). *Appl. Surf. Sci.* 527: 146915.

143 Pazniak, H., Benchakar, M., Bilyk, T. et al. (2021). *ACS Nano* 15 (3): 4245–4255.

144 Yengejeh, S.I., Kazemi, S.A., Wen, W., and Wang, Y. (2021). *Mech. Mater.* 160: 103957.

145 Kumar, S. and Schwingenschlögl, U. (2016). *Phy. Rev. B* 94 (3): 035405.

146 Yu, H., Xu, K., Zhang, Z. et al. (2021). *J. Mater. Chem. C* 9: 2416–2425.

147 Siriwardane, E.M., Demiroglu, I., Sevik, C. et al. (2020). *J. Phys. Chem. C* 124 (39): 21293–21304.

148 Shukla, V., Jena, N.K., Naqvi, S.R. et al. (2019). Modelling high-performing batteries with Mxenes: the case of S-functionalized two-dimensional nitride Mxene electrode. *Nano Energy* 58: 877–885.

149 Yang, G., Hu, X., Liang, J. et al. (2021). *J. Hazardous Mater.* 419: 126220.

150 Michael, J., Qifeng, Z., and Danling, W. (2019). *Nanomater. Nanotechnol.* 9: https://doi.org/10.1177/1847980418824470.

151 Zou, X., Li, G., Wang, Q. et al. (2018). *Comput. Mater. Sci* 150: 236–243.

152 Zha, X.-H., Luo, K., Li, Q. et al. (2015). *Europhys. Lett.* 111 (2): 26007.

3

Synthesis of MXenes

Manish Kumar Dixit and Mrigendra Dubey

Department of Metallurgical Engineering and Materials Science, Indian Institute of Technology Indore, Indore, India

3.1 Introduction

Since the discovery of 2D transition metal carbides/nitrides and/or carbonitrides in 2011, [1, 2] more than 40 different types [3] of lamellar MXenes have been developed which are promising host for variety of applications such batteries, electromagnetic interference shielding, purification of water, electrocatalysis, photocatalysis, supercapacitors, and biological sensor [4–10] These large number of interdisciplinary properties have been achieved possibly due to their highly variable elemental compositions and nonideal surface terminations [3]. In general, MXenes have been engineered by selective etching of "A" layers (mostly Al or Si) from the layered ternary carbides or nitrides containing parental MAX phase with stoichiometric formula represented as $M_{n+1}AX_n$ ($n = 1$–4) where M is defined as an early transition metals such as Nb, V, Ta, Cr, Zn, Sc. A designated primarily for group 13 and 14 group element, X stands for carbon (C) and/or nitrogen (N), and n varies from 1 to 3 (Figure 3.1) [1, 2]. MXenes primarily comprised of three common structures viz; M_2X, M_3X_2, and M_4X_3 but recently this fast growing family garnered with some more structures e.g. double-M MXenes ($M_2M'C_2$ and $M_2M'_2C_3$). Interestingly, after exfoliation of MAX phase, surface of MXenes were found to be covered with a mixture of hydroxyl functional group (i.e. –OH), oxygen and fluorine in turn provides a hydrophilic surface leading to easy processability of the materials in aqueous or organic solutions.

3.2 Fabrication of MXene

In the earlier days of MXenes, scientists preferred water as a solvent along with fluoride containing etchants for selective etching of Al layer, but, this approach could lead to the release of hazardous gases, exfoliation with water resulting in –OH functionality terminals on the MXene surfaces [11, 12] (Figure 3.2). In order to rectify these drawbacks, some novel methods have been engineered for exfoliation and delamination of MAX phase and

MXenes: Next-Generation 2D Materials: Fundamentals and Applications, First Edition.
Edited by Jay Singh, Kshitij RB Singh, Ravindra Pratap Singh, and Charles Oluwaseun Adetunji.

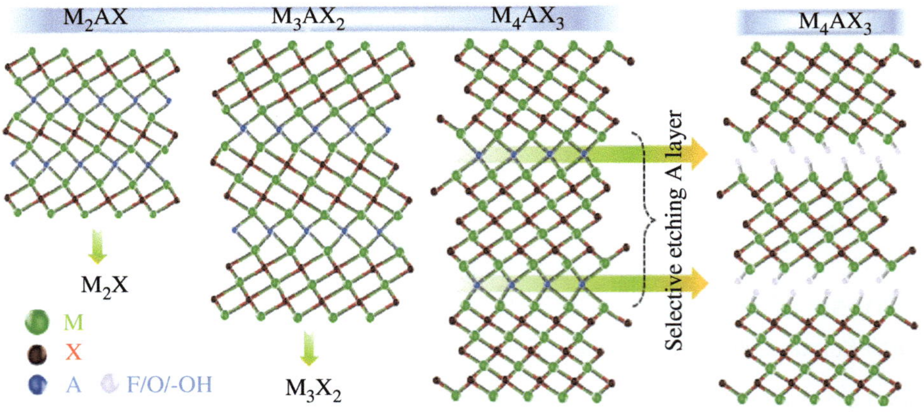

Figure 3.1 Pictorial presentation of original MAX phases and their respective three common MXenes as M_2X, M_3X_2 and M_4X_3.

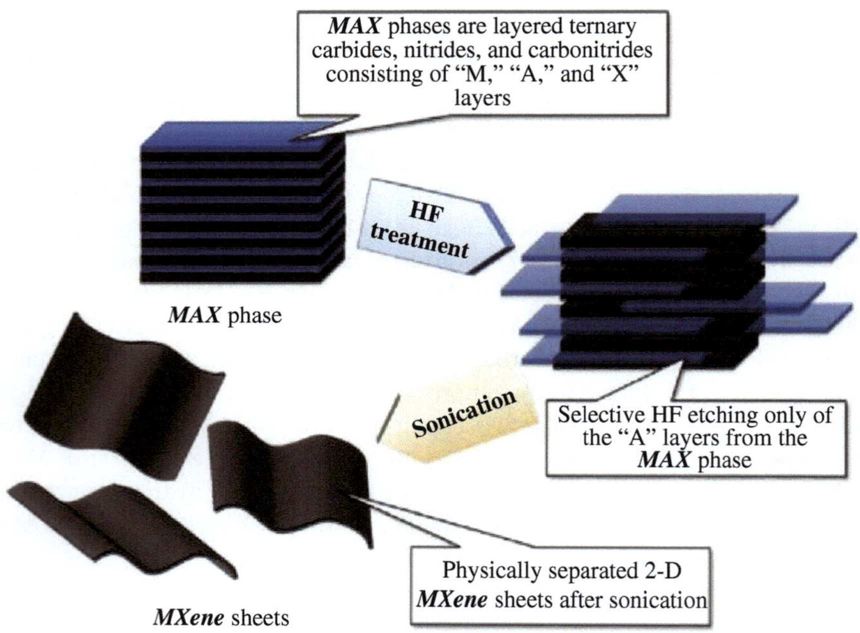

Figure 3.2 A general synthetic scheme for MXenes. Source: Reproduced with permission from Ref. [12]. © 2015, American Chemical Society.

MXene, respectively, such as, fluoride-containing acidic solutions and fluoride generating chemicals [1, 13, 14], hydrothermal synthesis [14, 15], chemical vapor deposition methods [16, 17], and alkali based etchants [18, 19]. As the etchants and synthetic procedure adapted for the synthesis of MXenes are well corroborated with the resulting properties, in turn their applicability, the fluoride-based acid and other fluoride generating derivatives are still the most popular choice in current research.

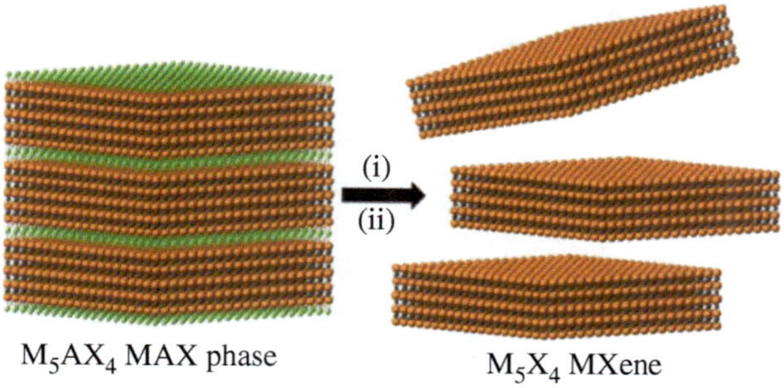

Figure 3.3 (i) Exfoliation and (ii) delamination of MAX phase in to respective MXene. Source: Reproduced with permission from Ref. [21]. © 2015, American Chemical Society.

3.2.1 Fabrication Through Etching Agents

The properties of resulting MXene are strongly path dependent in nature which is directly proportional to its applicability [20]. Generally, the preparation of the MXenes can be performed within hours, but it can vary to few days too. The general synthetic procedures adapted for synthesis of MXenes have been displayed in Figures 3.2 and 3.3. Usually, the time taken in the fabrication process depends on the concentration of fluoride containing etching agent such as, (i) HF or a mixture of lithium fluoride (LiF) and hydrochloric acid (HCl) or NH_4HF_2, and (ii) temperature of etching step. Thus, in this section, we will discuss various etchants along with the reaction conditions used for the synthesis of a variety of MXenes throughout the time scale (Figure 3.4).

3.2.1.1 HF Etchants

In 2011, the first MXene ($Ti_3C_2T_x$) was developed by Naguib et al. by gradual addition of Ti_3AlC_2 [11] powder in 50% concentrated HF with the help of teflon coated bar at room temperature (RT) for two hours which could be represented in a Eq. (3.1) [7] as,

$$Ti_3AlC_2 + 3HF \longrightarrow Ti_3C_2 + 3/2\, H_2 + AlF_3 \quad (3.1)$$

The conversion of Ti_3AlC_2 to $Ti_3C_2T_x$ could be evidenced by the PXRD study where the main peak at $2\theta = {\sim}40°$ shifted to ${\sim}10°$. Further, the presence of surface functional groups was confirmed by matching the simulated PXRD pattern of $Ti_3C_2F_2$ and $Ti_3C_2(OH)_2$, and the experimental PXRD pattern after HF etching, which could be represented by Eqs. (3.2) and (3.3).

$$Ti_3C_2 + 2H_2O \longrightarrow Ti_3C_2(OH)_2 + H_2 \quad (3.2)$$

$$Ti_3C_2 + 2HF \longrightarrow Ti_3C_2F_2 + H_2 \quad (3.3)$$

Further, the centrifuge performed over resulting MXene, would isolate it from impurities, and the obtained solid would be washed with deionized water till the achievement of pH in the range of 4 and 6.

In a recent case study, Gogotsi et al. used HF as an etchant for synthesis of $Mo_4VC_4T_x$ MXenes from its precursor Mo_4VAlC_4 MAX phase (Figure 3.5) [21]. In another report, very famous $Ti_3C_2T_x$ synthesized from Ti_3AlC_2 using, HF and in situ HF as an etchant

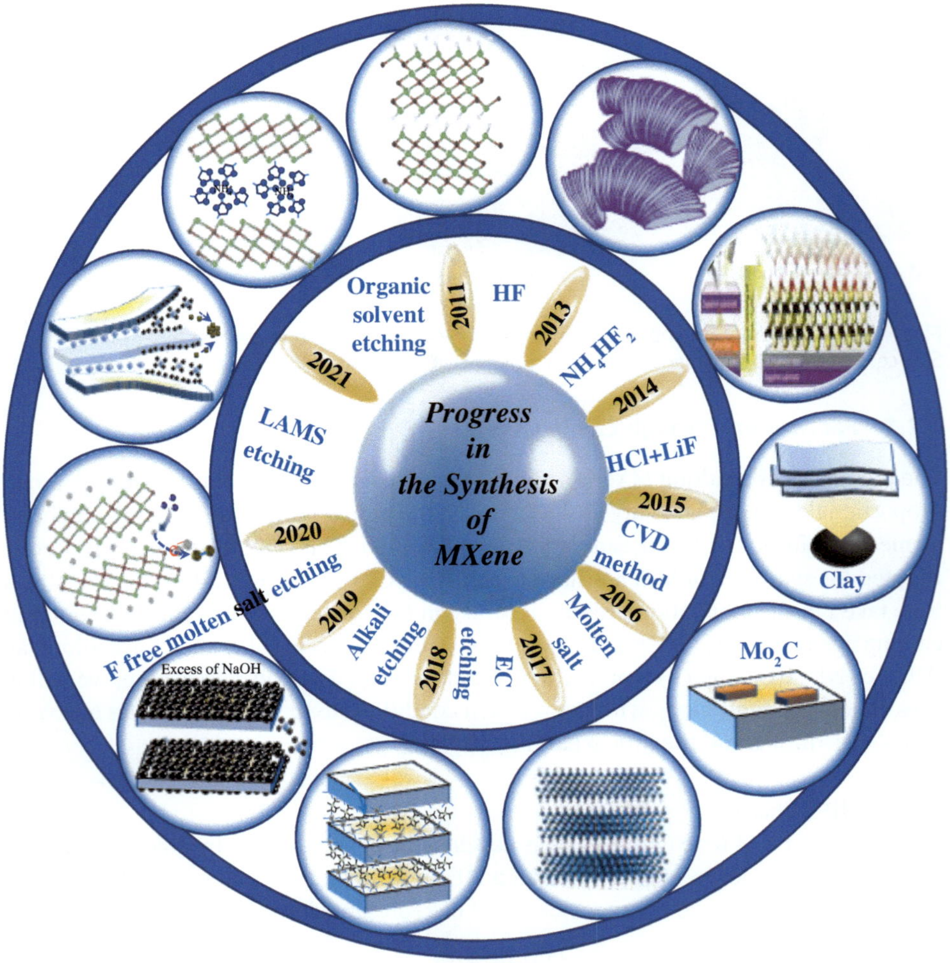

Figure 3.4 A summary picture for the progress in the synthesis of various MXenes through the years 2011–2021. For the first time in 2011, MXene ($Ti_3C_2T_x$) was fabricated through HF etching. DMSO delaminated shaped MXene was synthesized in 2013. In 2014, NH_4HF_2 as bifluoride salt and HCl+LiF based etchants were came into existence to obtain MXenes. Source: Reproduced with permission from Ref. [22]. © 2015, American Chemical Society. CVD process was bloomed in 2015 for producing unconventional MXenes. In the very next year, F-containing molten salt etching method was successfully employed to coin out the first nitride MXenes. Reproduced with permission [23] Copyright 2016, Royal Society of Chemistry. To avoid the use of HF, in 2017, electrochemical etching process was executed to form conventional MXenes. In 2018, MXenes with no F-terminals were achieved through Fluoride free alkali etching. Fluorine-free late transition metal $ZnCl_2$ molten salt etching was utilized to produce MXenes in 2019. Reprinted with permission from Ref [24]. Copyright 2015, American Chemical Society. In 2020, a breakthrough study has been reported via Lewis acid molten salt (LAMS) method. In the year 2021, MXene was synthesized using polar organic solvents.

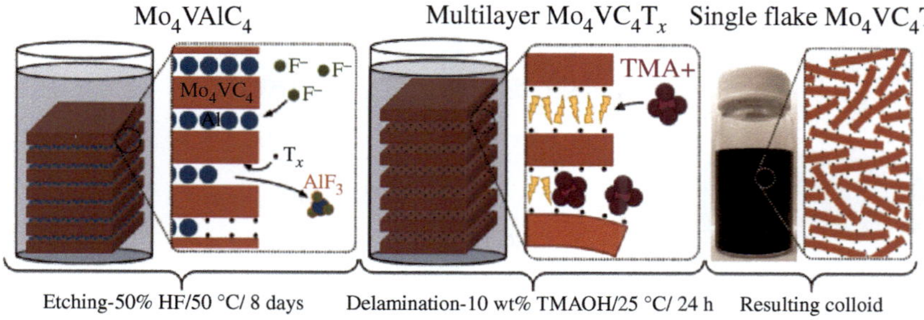

Figure 3.5 MXene Mo_4VC_4 has been fabricated from Mo_4VAlC_4 MAX phase using HF etachants to produce MXene. Delamination was performed by the treatment of MXene with tetramethylammonium (TMA^+) ions. Source: Reproduced with permission from Ref. [21]. © 2015, American Chemical Society.

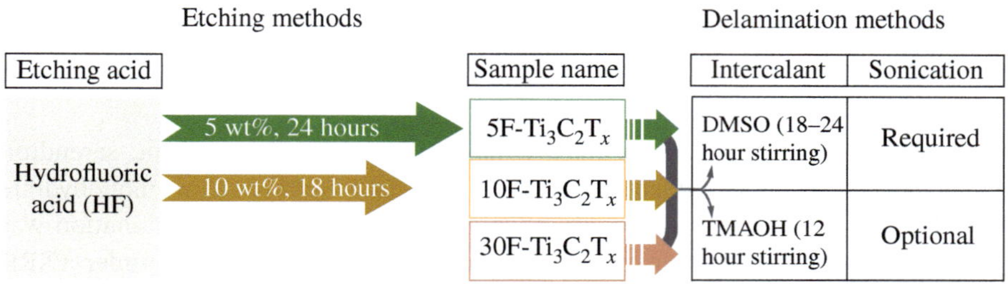

Figure 3.6 General synthetic procedure adapted for synthesis of $Ti_3C_2T_x$ MXene under three different conditions. Source: Reproduced with permission from Ref. [25]. © 2015, American Chemical Society.

(Figure 3.6) [25]. On moving through HF route, the author varies HF concentrations and exposure time, and observed that all of them produced the desired $Ti_3C_2T_x$. In contrast to other etchants, HF produced $Ti_3C_2T_x$ behaved as clay instead of powder. Further, delamination of obtained $Ti_3C_2T_x$ was carried out using dimethyl sulfoxide (DMSO) or tetramethyl-ammonium hydroxide (TMAOH) in presence of ultrasound.

It was observed that the formation and stability of HF etched multilayered MXenes required very delicate balance various non covalent weak interactions which helped it to retain its multi layered nature i.e. accordion like morphologies, but in absence of intercalation species and sonication, yield for single or few layered MXenes were very low. One of the frontline work for the synthesis of accordion shaped HF-etched delaminated $Ti_3C_2T_x$ was achieved under sonication for five minutes in methanol as a solvent, but, were only able to achieve low yield of few layered $Ti_3C_2T_x$ with even lower yield of single-layered $Ti_3C_2T_x$ [1].

In progress to the above report, Mashtalir et al. reported a very first high yield synthetic procedure for the fabrication of accordion like single to few layered delaminated $Ti_3C_2T_x$. Under constant stirring of 18 hours at RT, an organic moiety DMSO was used as intercalant for the multilayered $Ti_3C_2T_x$ with additional ultrasound for six hours in DI water [26]. Achieving successful delamination of multilayered $Ti_3C_2T_x$ through DMSO encouraged the use of other such organic intercalants (e.g. THF, ethanol,

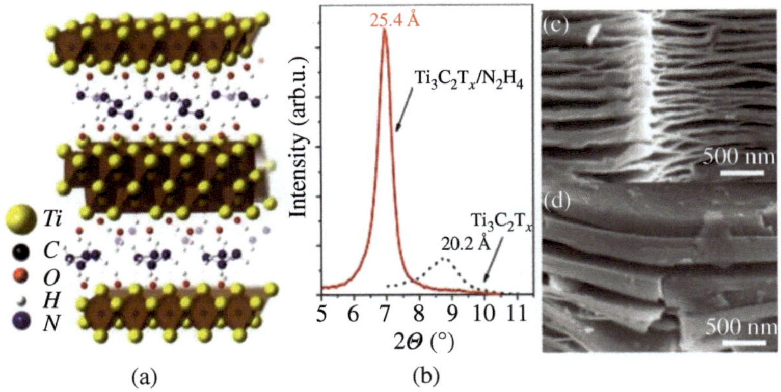

Figure 3.7 (a) Pictorial presentation of Ti_3C_2OH with intercalated N_2H_4. (b) a comparative PXRD patterns of $Ti_3C_2T_x$ (0002) peak for the N_2H_4 untreated and treated sample around c lattice parameter and their respective (c, d) SEM images at 500 nm scale. Source: Reproduced with permission from Ref. [27]. © 2016, Royal Society of Chemistry.

acetone, HCHO, $CHCl_3$) but, unfortunately, none of them worked, except, serendipitously obtained *N,N* dimethylformamide (DMF), solution of hydrazine monohydrate (N_2H_4 H_2O), and urea. Almost complete intercalation of N_2H_4 after delamination was well depicted through shift in major intense peak from 20.2 to 25.4 Å, under PXRD analysis (Figure 3.7) [27]. In order to obtain clear insights of intercalation of N_2H_4, Mashtalir et al. performed inelastic neutron scattering (INS) experiments over it before and after intercalated samples. Through careful analysis of proton stretching frequencies in OH, H_2O, and N_2H_4, it was shown that the OH and H_2O were present in abundance before intercalated sample while N_2H_4 was present as a whole layer in intercalated sample. Despite of positive results obtained using DMSO, its use as an intercalant is very limited because it was only found useful for the delamination of $Ti_3C_2T_x$ along with the difficulty in its removal. In addition to this, large size of DMSO molecule also enforces the larger lamellar thickness than expected or desired. Furthermore, the specific intercalation activity of DMSO toward multilayered $Ti_3C_2T_x$ is still needing serious scrutiny.

Learning from the previous experience and the abovementioned report, Naguib et al. utilized some other organic molecules (tetrabutylammonium hydroxide [TBAOH], choline hydroxide or n-butylamine) with comparatively high molar mass for the delamination of HF-etched multilayered carbonitrides and carbides e.g. Ti_3CNT_x, V_2CT_x and Nb_2CT_x at RT with mechanical shaking or ultrasound (Figure 3.8) [28]. Due to the use hydroxide and amine containing intercalants, lesser F-terminations for resulting MXene has been observed in turn resulted in different surface chemistry.

In continuation with the previous report, for the first time, Urbankowski et al. utilized the TBAOH to delaminate a high temperature HF-etched multi layered MXene $Ti_4N_3T_x$ into a mixture of single- to few-layered $Ti_4N_3T_x$ flakes using ultrasound method (Figure 3.9) [23]. In this method, molten salts of 59%KF, 29% LiF, 12% NaF was utilized at elevated temperature i.e. 550 °C (Figure 3.10).

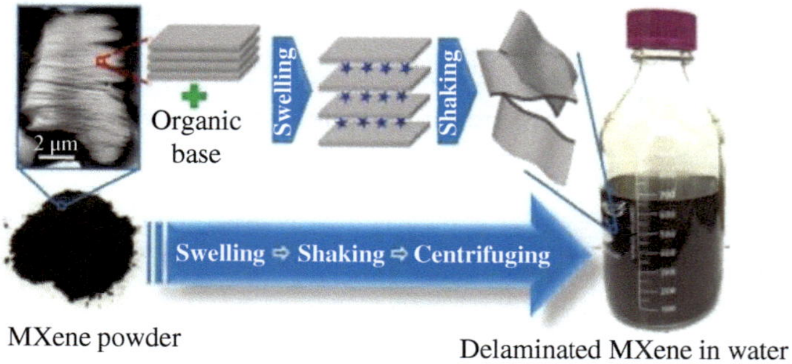

Figure 3.8 Pictorial presentation of delamination process for MXene using an organic base e.g. amine or hydroxide leads to increase in the interlamellar distance significantly. A colloidal solution was obtained through simple mechanical hand shaking or short duration ultrasound (right). A SEM image revealed the characteristic lamellar Ti_3CNT_x MXene (top left corner). Source: Reproduced with permission from Ref. [28]. © 2015 / Royal Society of Chemistry.

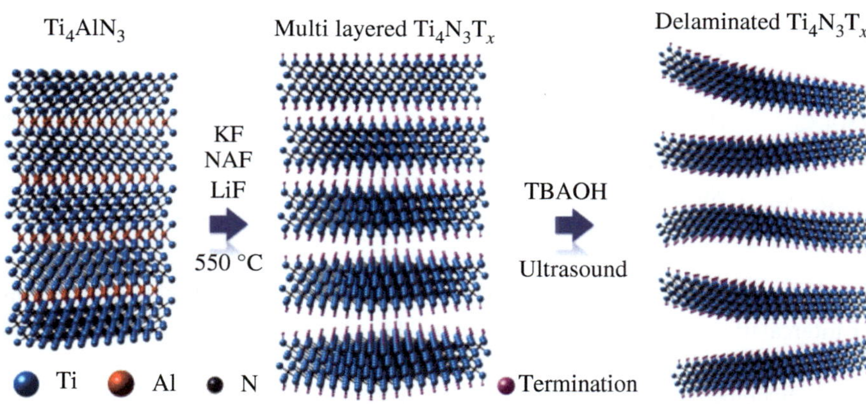

Figure 3.9 Model presentation for the synthesis of multi-layered $Ti_4N_3T_x$ first nitride MXene using LiF/NaF/KF molten salts. Source: Reproduced with permission from Ref. [23]. © 2016, Royal Society of Chemistry.

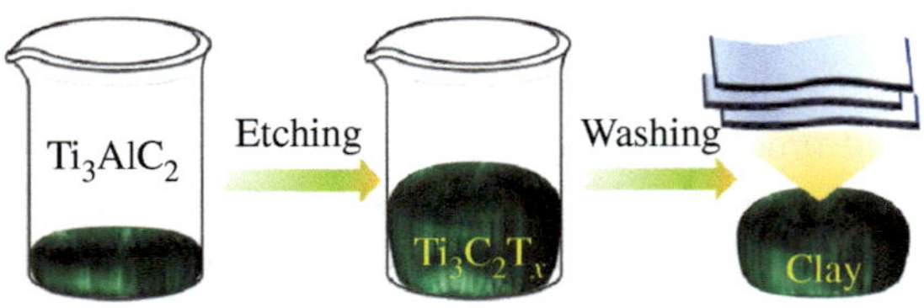

Figure 3.10 Synthetic procedure adapted for clay formation through etching and washing.

The similar sonication strategy has been used by Mashtalir et al. [29, 30] and Wang et al. [31] separately. In first case, for delamination of HF-etched multilayered Nb_2CT_x, the aqueous isopropylamine was used as intercalant instead of TBAOH [29, 30]. The HF-etched multi layered Nb_2CT_x is believed to be negatively charged on the surface which might electrostatically attract the positively charged $R\text{-}NH_3^+$ from aqueous isopropylamine. In

Table 3.1 Crucial information for the synthesis of MXene from respective MAX phases using HF etchant is summarized in the table.

MAX Phase	MXene	% of HF etchant	Time (hr)	Temperature	Ref.
Ti_3AlC_2	$Ti_3C_2T_x$	50	2	RT	[1]
Ti_4AlC_3	$Ti_4C_3T_x$	50	72	RT	[2]
Ti_3AlCN	Ti_3CNT_x	30	18	RT	[2]
Ti_2AlC	$Ti_2C\,T_x$	10	10	RT	[2]
Nb_2AlC	Nb_2CT_x	50	40	55 °C	[32]
V_2AlC	V_2CT_x	50	90	RT	[33]
Nb_4AlC_3	$Nb_4C_3T_x$	48–51	96	Ambient temperature	[33]
Ti_3AlC_2	$Ti_3C_2T_x$	40	168	15 °C	[34]
Nb_2AlC	Nb_2CT_x	50	90	RT	[35]
Mo_2TiAlC_2 or $Mo_2Ti_2AlC_3$	Mo_2TiC_2 or $Mo_2Ti_2C_3T_x$	48–51	48 90	Ambient temperature/55 °C	[36]
Ti_3AlC_2	$Ti_3C_2T_x$	40	10	40 °C	[37]
$Hf_3[Al(Si)]_4C_6$	$Hf_3C_2T_x$	35	60	RT	[38]
Mo_4VAlC_4	Mo_4VC_4	48–51	8 days	50 °C	[21]
$(Nb_{2/3}Sc_{1/3})_2AlC$	$Nb_{1.33}CT_x$	48	30	RT	[39]

the second approach to achieve $Ti_3C_2T_x$ flakes by Wang et al. dilute NaOH has been utilized for intercalation of HF-etched multilayered $Ti_3C_2T_x$. Further, ice cooled diazonium salts of suphanilic acid were used as surface modifier followed by mild sonication [31]. NaOH not only helped for intercalation but also maximize the entry of diazonium salts of suphanilic acid after intercalation. One of the key benefits of this method is the easy removal of highly water soluble unreacted diazonium salts. Similarly, few of the reports using HF as an etchants have been summarized in Table 3.1.

3.2.1.2 In situ HF Etchants

Due to hazardous effect of HF over the surrounding including humans, Ghidiu et al. opted for a much safer strategy and high yield process to produce in situ HF by the reaction of HCl and LiF. Ti_3AlC_2 was further dispersed in the resulting solution (LiF:Ti_3AlC_2, 5:1) mixture for 24 hours at 35 °C to obtain $Ti_3C_2T_x$ MXene after moderate ultrasound (Figure 3.10) [13]. Due to different synthetic approach, fewer defects were observed in $Ti_3C_2T_x$ MXene which was evident from the transmission electron microscopy (TEM) images. Further, the presence of hydrated Li^+ ion provided the extra edge over rheological properties of material as it can be molded in to different shapes. In addition to this, the reported method also reduced the steps involved in delamination of HF-etched MXenes. As the mixture of HCl and LiF solution contains the same hydrogen and fluoride ions, it also follows the same etching mechanism as of the MAX phases with HF. Observing the positive response in case of HCl and LiF, few of the other combinations were tried to synthesize MXenes with acids such as H_2SO_4 and salts e.g. NaF, FeF_3, KF, CsF, CaF_2 [13]. Starting from $Ti_3C_2T_x$ MXene, this single-step methodology has also been applied to produce various multilayered Nb_2CT_x, Ti_2CT_x, $Cr_2TiC_2T_x$, $Mo_2TiC_2T_x$, $Mo_2Ti_2C_3T_x$, Mo_2CT_x, and $(Nb_{0.8}Zr_{0.2})_4C_3T_x$ [2, 32, 33, 36, 40].

After the research of two years, Shahzad et al. used a very similar strategy for the synthesis of $Ti_3C_2T_x$ [41]. In order to enhance the molar ratio of Li^+ used for intercalation, authors altered the molar ration of LiF to Ti_3AlC_2 from 5:1 to 7.5:1. Further, for the etching of all the Aluminium, the molar ration of HCl to LiF was also doubled. Except the absence of ultrasound and 9M HCl instead of 6M in the current procedure, all steps remains the same to the method reported by Ghidiu et al. such as 35 °C for 24 hours.

In the same year, Lipatov et al. took MXenes synthesized from both the molar ratio of LiF to Ti_3AlC_2 (5:1 and 7.5:1) in 6 M HCl solution under similar conditions of 35 °C for 24 hours [42]. In first case of molar ratio of LiF to Ti_3AlC_2 of 5:1, the 200–500 nm diameter of $Ti_3C_2T_x$ flakes were observed with uneven edges and pin holes. In contrast to this, the absence of pin holes with well-defined edges in the exfoliated $Ti_3C_2T_x$ flakes (4–15 μm) obtained from the molar ratio of LiF to Ti_3AlC_2 of 7.5:1 was well examined under high resolution transmission electron microscopy (HRTEM). In addition, the larger $Ti_3C_2T_x$ flakes were found to be more orderly stacked, resulting in enhanced mechanical robustness.

In the quest to find milder and less toxic in situ HF etchants, Halim et al. developed an Ammonium Bifluoride (NH_4HF_2) salt for epitaxial Ti_3AlC_2 thin films (Figure 3.11) [22]. In this study, the efficiency of two different etching methods was compared. Experiment was setup by the deposition of Ti_3AlC_2 thin films on sapphire substrate at 780°C. In the first traditional experiment, 50% concentrated HF was applied on samples of thickness 15, 28, 43, and 60 nm for 10, 15, 60, and 160 minutes, respectively. Second experiment in parallel used 1M of NH_4HF_2, with the similar sample thicknesses and enhanced exposure time as 150, 160, 420, and 660 minutes, respectively. The second method involves simultaneous intercalation of NH_4^+ ions during etching step itself. The chemical reaction of second method can be written as:

$$Ti_3AlC_2 + 3NH_4HF_2 \Rightarrow (NH_4)_3AlF_6 + Ti_3C_2 + 3/2\,H_2 \tag{3.4}$$

$$Ti_3C_2 + aNH_4HF_2 + bH_2O \Rightarrow (NH_3)_c(NH_4)_dTi_3C_2(OH)_xF_y \tag{3.5}$$

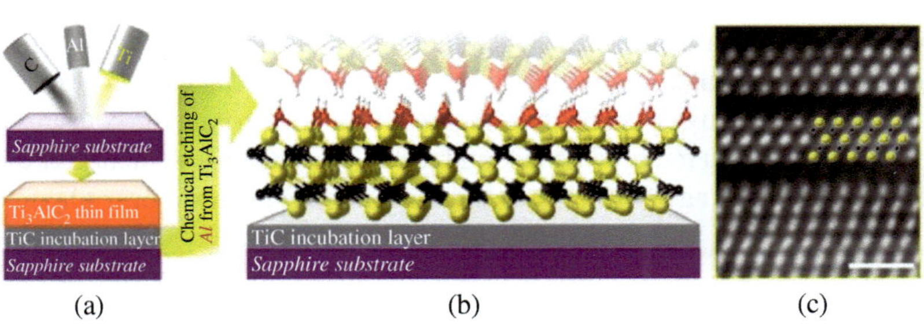

Figure 3.11 (a) Procedure adapted for synthesis of thin films of Ti_3AlC_2 through magnetron sputtering. (b) Pictorial model displayed the selective etching of Al layer from MAX phase (c) STEM micrographs for $Ti_3C_2T_x$ layers at scale bar 1 nm. (Ti atoms with yellow and C atoms in black color). Source: Reproduced with permission from Ref. [22]. © 2014, American Chemical Society.

Since this work, the hazardous HF has found a suitable replacement in the form of NH_4HF_2 and it has preferentially been used depending upon the desired application. Reaction (3.5) simply displayed the intercalation of ammonia and ammonium ions between the $Ti_3C_2T_x$ layers leading to the swelling of the resulting MXene. Further, the NH_4HF_2 method was not only applied to such thin films but also for the Ti_3AlC_2 powder samples [43]. In this work, $Ti_3C_2T_x$ MXene has been prepared by adding freshly prepared Ti_3AlC_2 powder to 1M of NH_4HF_2 solution and kept it for five days which upon several washing of water and centrifugation produced $Ti_3C_2T_x$ MXene [44]. Another classic example for benefits of NH_4HF_2 over HF has been demonstrated by Feng et al. [45]. The work has effectively summarized the optimal condition for synthesis of $Ti_3C_2T_x$ MXene with the immersion of Ti_3AlC_2 in NH_4HF_2. Interestingly, the larger lattice parameter c has been observed for NH_4HF_2 (24.9 Å) than for HF etched Ti_3AlC_2 samples. In addition to this, multilayered Ti_3C_2 was found to be more heat resistant in case of NH_4HF_2 (900 °C) while only 350 °C for HF exfoliated Ti_3C_2. Thus, the important aspect of NH_4HF_2 mediated method is that the MXene with tunable properties could easily be synthesized in just a single step by mere change in precursor fluoride salts [13]. A progressive report over the use of NH_4HF_2 has been provided by Natu et al. where, in contrast to most of the other works, bifluoride salt has been used in absence of DI water to produce F-terminal MXenes [46]. It was also demonstrated that propylene carbonate etched $Ti_3C_2T_z$ samples displayed twice the efficiency toward Na-ion battery than water etched MXene.

3.2.1.3 MXenes Preparation Through Fluoride Free Routes

Over the years, it has been realized that fluoride-based etching methods are the most promising candidate for the synthesis of MXene but these chemicals are highly hazardous to surroundings including human life. In addition to this, electrochemical studies over fluoride based etching methods reveals that the abundance of F-end group in MXene limits its use as an advanced electrode material. Thus, the mentioned points collectively propel the scientists to explore new synthetic methods for fluoride-free MXenes. Few important methods have been introduced in between 2015–2020.

3.2.1.3.1 Chemical Vapor Deposition Process In order to develop an ultrathin ranged α-Mo_2C crystals, Xu et al. successfully designed a new chemical vapor deposition (CVD) process in 2015 (Figure 3.12). At elevated temperature around 1100 °C, a copper foil mounted on Mo foil was taken as the initial substrate and C supply had been obtained from methane [16]. A very similar approach has been adapted by Geng et al. for the 2D regulatory nucleation of Mo_2C using copper foil (Figure 3.13) [47]. Further, the concentration of CH_4, and thickness of Cu foil used were found to be the regulatory elements for the shape and height of the resulting Mo_2C [47]. Despite of enhanced control over physical

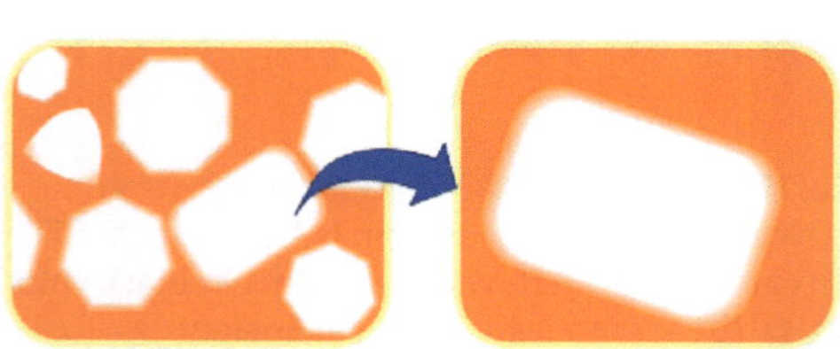

Figure 3.12 Representative depiction of Mo_2C crystal (right) with enlarged view of Mo_2C single crystal.

Figure 3.13 A representative images for the nucleation process of Mo_2C crystals.

parameters (such as, height and width profile) of produced MXene, the use of CVD method has been very limited due to the low yield of synthesized MXene and requirement of high temperature. In addition to this, the deposition of amorphous C produced from CH_4 usually deactivates the catalytic sites. To overcome the mentioned drawbacks, in the same year, Ma et al. synthesized Mo_2C and Mo_2N through "urea glass" method for catalysing hydrogen evolution reaction (HER) [48]. In this method, initially Mo-orthoesters were produced by the 1.45M solution of solid $MoCl_5$. Further, urea was added to the above ethanolic solution and stirred for 1–2 hours to achieve homogenous solution. Added urea was not only used as source for C/N but also stabilize the ethanolic $MoCl_5$ precursor during heat treatment for its conversion to Mo_2C or Mo_2N [49]. Further, it produced negligible amount of amorphous/deactivated C or N [49] which enhanced the catalytic activity and were able to produce significantly more crystalline, pure, and mono-dispersive Mo_2C and Mo_2N nanoparticles [50, 51]. The coordination polymeric gel like precursors were placed in different crucibles for calcination (800 °C for 3 hours) to obtain silvery black end product of α-Mo_2C and γ-Mo_2N which highlighted a new path for the development of a new class of highly active and stable Mo-based electrocatalysts for HER.

As discussed in the above paragraph, in contrast to the chemical etching methods for the synthesis of traditional MXenes, CVD method has been used to synthesize unconventional crystalline, pure, and ultrathin α-Mo_2C, WC, and TaC nanosheets [16]. Additionally this method paved the route for the synthesis of less defective surface and less number of surface terminations which in turn would be a benefit for decoding the intrinsic properties of MXenes. A low yield process, control over number of layer formation, and economically viable synthetic modification are needed to be address urgently for the synthesis of MXene using CVD method.

3.2.1.3.2 Ionic Exchange Method As most of the methods discussed till now are primarily focused toward the conventional MAX phases containing Al contributed as "A" part but, variety of unconventional MXenes could only be synthesized once Al would be replaced with another metal ion in MAX phase. Inspired from this fact, Huang et al. discovered a method to produce unconventional MXene from MAX phases containing A as Si, Zn, and Ga which were etched by Lewis acid molten salts (LAMS) [52]. Experimental results established that it also follows very similar etching mechanism to the one shown by etching via HF. In this method, the role of F^- was very well played by cations present in salts such as H^+, Cl^- ions which coordinate to M atoms of MAX phase as the F^- in HF (Figure 3.14).

3.2.1.3.3 Hydrothermal Method Keeping an idea to reduce the complications involved in synthetic methods, Wang et al. deduced a new one pot synthetic procedure for the fabrication of conventional $Ti_3C_2T_x$. MAX phase was etched using NH_4F via hydrothermal

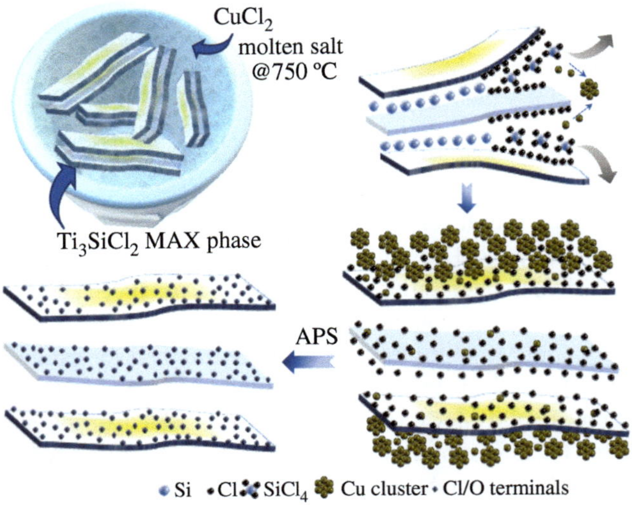

Figure 3.14 (a–d) Stepwise model representation for the synthesis of $Ti_3C_2T_x$ MXene preparation where (APS) represents ammonium persulfate solution.

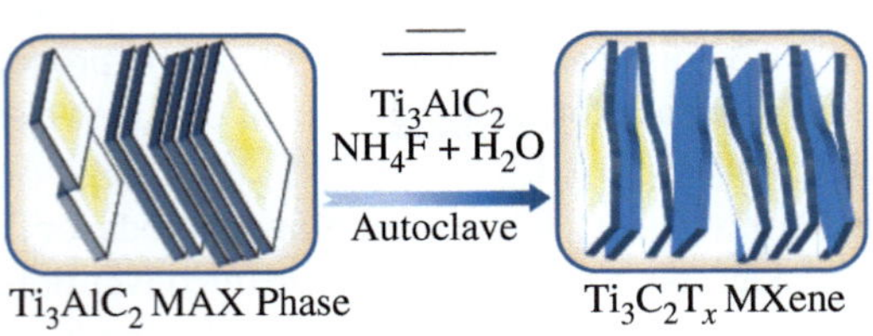

Figure 3.15 A model representation for the synthesis of $Ti_3C_2T_x$ (right) via Hydrothermal process from MAX phase (left).

process (Figure 3.15) [14]. A fine powdered Ti_3AlC_2 was premade under Ar atmosphere. Under vigorous magnetic stirring, 0.5 g of Ti_3AlC_2 fine powder was homogeneously added into a 60 ml aqueous solution of 5 g of NH_4HF. The resulting solution mixture was tightened up into an 80 ml autoclave at 150 °C for 24 hours. After the completion of reaction the mixture was kept at RT for natural cooling. Further, the synthesized solution mixture was centrifuged to obtain a black precipitate which was thoroughly washed with DI water and EtOH, simultaneously. The solid precipitate was then kept in vacuum at 60 °C for 12 hours to collect the dried product. The work has showcased the effect of concentration of NH_4F, reaction temperature, and time over the synthesized nanomaterial. In 2018, extending the dimensions of the above work, Peng et al. developed a new hydrothermal method for the fabrication of h-Ti_3C_2 MXene, using comparatively very low toxic etchants, $NaBF_4$ and HCl [53]. h-Ti_3C_2 was achieved with higher c-lattice parameter when compared to the one prepared by traditional HF etchants which was due to the slow release of etchants during controlled hydrothermal method. This hydrothermal process was introduced not only to produce more efficient Ti_3C_2 but also to restrict the release of toxic HF gas in the atmosphere. The high purity Ti_3AlC_2 powders was obtained from ball-milling method whereas Nb_2AlC via hot-pressing synthetic method. Further, in this work, conventional HF etched t-Ti_3C_2, and hydrothermally obtained h-Ti_3C_2, explored in comparative manner. Further, the delamination was performed using organic intercalant i.e. DMSO in the presence of mild ultrasound. The chemical reactions involved in process are given below:

$$NaBF_4 + HCl \longrightarrow HBF_4 + NaCl \quad (3.6)$$

$$HBF_4 \longrightarrow HF + BF_3 \quad (3.7)$$

$$BF_3 + 3H_2O \longrightarrow 3HF + H_3BO_3 \quad (3.8)$$

$$Ti_3AlC_2 + 3HF \longrightarrow AlF_3 + 3/2\,H_2 + Ti_3C_2 \quad (3.9)$$

$$Ti_3C_2 + 2H_2O \longrightarrow Ti_3C_2(OH)_2 + H_2 \quad (3.10)$$

$$Ti_3C_2 + 2HF \longrightarrow Ti_3C_2F_2 + H_2 \quad (3.11)$$

$$Nb_2AlC + 3HF \longrightarrow AlF_3 + 3/2\,H_2 + Nb_2C \quad (3.12)$$

$$Nb_2C + 2H_2O \longrightarrow Nb_2C(OH)_2 + H_2 \quad (3.13)$$

The above mentioned reactions displayed that the Ti_3C_2 and Nb_2C both underwent the same chemical changes. The precursor $NaBF_4$ and HCl first produced HBF_4 (Eq. 3.6) which was converted into BF_3 and HF (Eqs. 3.7 and 3.8). Thus, the in-situ HF was utilized for slow etching of Ti_3AlC_2 resulting in Ti_3C_2 with AlF_3 as by-product (Eq. 3.9).

3.2.1.3.4 Alkali Metal Ion Mediated MXene Synthesis As clearly seen in the Figure 3.16, Li et al. fabricated an alkali base mediated fluorine free accordion like $Ti_3C_2T_x$ where -OH and -O terminal were achieved due to the presence of NaOH [19]. In contrast to F^- in

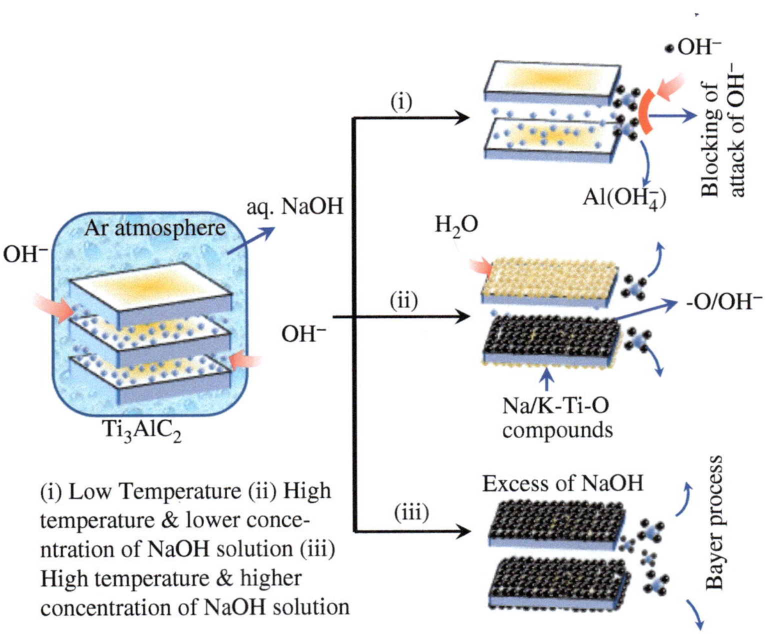

Figure 3.16 The schematic model representation of the reaction between Ti_3AlC_2 and aq. NaOH under different reaction conditions such as temperature and concentration of NaOH.

HF mediated synthesis of MXene, OH^- is utilized for etching process which react to Al of MAX phase and form $Al(OH)_4^-$. Thus, this method introduced a much safer etching technique. In addition to this, it produced the same accordion like MXene but it possessed high energy storage ability than the convention MXene, which is attributed to the absence of –F terminal groups of the surface. In conclusion, through alkali mediated etching methods, the harmful HF and –F terminals over synthesized MXene can both be restricted. But, this method usually required large amount of energy in the form of high temperature along with high concentration of alkali bases which is again hazardous resulting in limited use of this method towards large scale production of MXenes.

3.2.1.3.5 Electrochemical (EC) Process for Synthesis of MXenes In the recent past, Feng et al. discovered a good alternative electrochemical approach for the synthesis of $Ti_3C_2T_x$ to avoid the use of harmful HF etchants for selective removal of Al from Ti_3AlC_2 MAX phase [54]. This method used a binary aqueous electrolyte comprised of 1M NH_4Cl and 0.2M TMAOH at pH>9) for intercalation and subsequent delamination (Figure 3.17). The in situ generated Cl acted as an etching agent whereas intercalation through OH produced single to few layered MXenes with quite high average lateral size than using conventional HF etching process.

Furthermore, Yang and co-workers opted for a new electrochemical approach "intercalation–alloying–expansion–microexplosion mechanism" for the synthesis of fluoride free $Ti_3C_2T_x$ in high yield which displayed the enhanced supercapacitor properties with -OH and -O terminals (Figure 3.18) [55]. Under this process, Lithium foil as cathodic material was used for the intercalation of anodic "A" layer from MAX phase under the applied electric field which would ultimately expand the Al–Al layer. A fast hydrogenation reaction took place upon immersion of the freshly obtained lithiated MAX phase upon immersion into water. This hydrogen evolution leading to a microexplosion helped for the exfoliation of lithiated MAX phase to nanosheets in high yields (Figure 3.16). Due to the

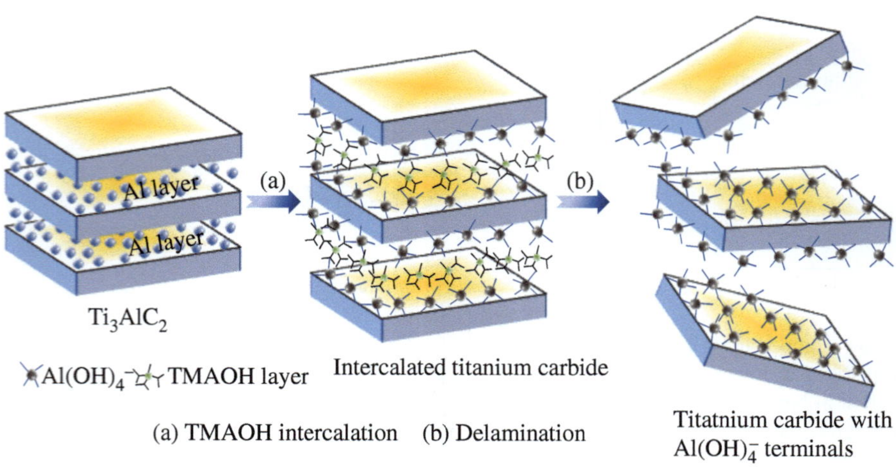

Figure 3.17 A Pictorial presentation of intercalation and delamination process for the synthesis of $Ti_3C_2T_x$ flaks using organic intercalant TMAOH.

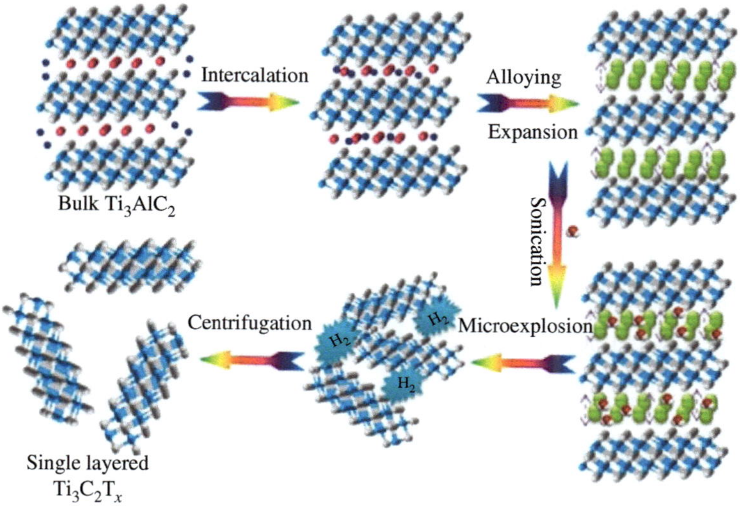

Figure 3.18 Schematic of the lithiation intercalation–expansion–microexplosion process. Source: Reproduced with permission from Ref. [55]. © 2019, American Chemical Society.

presence of water, this method did not produce any toxic materials as end products. This high yield method was found to be an excellent approach except for one limitation that "A" layer should work as an anodic material.

3.2.1.3.6 Fluoride-free Molten Salt Method In 2019, M. Li et al. discovered a variety of unconventional MAX phases such as Ti_3ZnC_2, Ti_2ZnC, Ti_2ZnN, and V_2ZnC, and corresponding possible Cl-terminated MXenes $Ti_3C_2Cl_2$ and Ti_2CCl_2 by a replacement reaction [24]. In this method, Al of A layer is replaced from MAX phase by Zn sourced from molten $ZnCl_2$. Thus, this method opened up a new strategy for synthesis of a large number of unexplored MAX phases e.g. Al could be replaced for Fe to obtain magnetic MAX phases (Figure 3.19). $ZnCl_2$ was not only chosen for comparatively lower melting point but it also displayed strong Lewis acidic characteristics which helped in replacement of Al atoms

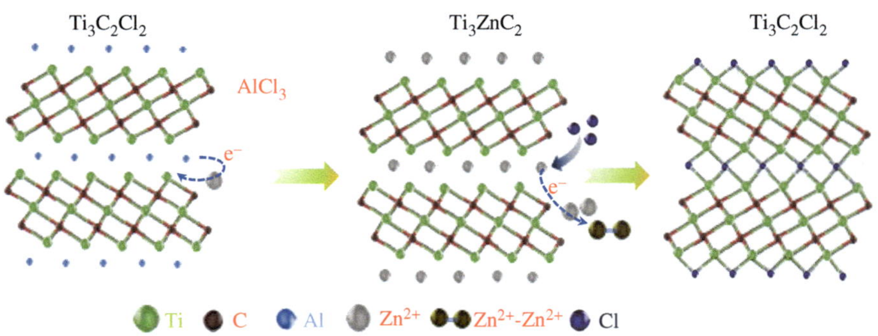

Figure 3.19 Schematic model for synthesis of Ti_3ZnC_2 and respective $Ti_3C_2Cl_2$.

from the “A” layer in the MAX phase under moderate temperature range. Reaction for MAX phase could be presented as

$$Ti_3AlC_2 + 3/2\,ZnCl_2 \rightarrow Ti_3ZnC_2 + ½\,Zn + AlCl_3 \quad (3.14)$$

$$Ti_3AlC_2 + 3/2\,ZnCl_2 \rightarrow Ti_3C_2 + 3/2\,Zn + AlCl_3 \quad (3.15)$$

$$Ti_3C_2 + 3/2\,Zn \rightarrow Ti_3ZnC_2 \quad (3.16)$$

Further, volatile nature of produced Al halides helped in its diffusion, and replacement by Zn^{2+} through infusion which can be represented by following equations,

$$Ti_3ZnC_2 + ZnCl_2 \rightarrow Ti_3C_2Cl_2 + 2\,Zn \quad (3.17)$$

$$Ti_3ZnC_2 + Zn^{2+} \rightarrow Ti_3C_2Cl_2 + {Zn_2}^{2+} \quad (3.18)$$

$$Ti_3C_2 + 2Cl^- \rightarrow Ti_3C_2Cl_2 + 2e^- \quad (3.19)$$

$${Zn_2}^{2+} + 2e^- \rightarrow 2Zn \quad (3.20)$$

Exfoliation of MAX phases Ti_3ZnC_2 and Ti_2ZnC lead to the formation of respective Cl- containing $Ti_3C_2Cl_2$ and Ti_2CCl_2 MXenes in the presence of $ZnCl_2$ molten salt. Unfortunately, Ti_2NCl_2 and V_2CCl_2 could not be synthesized from their respective Zn-MAX phases as Zn-Ti bonds were stronger than Ti-N bonds.

3.2.1.4 Molten Fluoride Salt as Etchants

Until 2015, HF etching along with other etching methods have been well explored for the synthesis of carbide- based MXenes whereas synthesis of nitride- based MXene was very rarely reported by any such etching methods due to the higher affinity of Al atoms in $Ti_{n+1}AlN_n$ than $Ti_{n+1}AlC_n$ MAX phases. In the year of 2016, Urbankowski et al. fabricated a 2D $Ti_4N_3T_2$ via molten salt method which is discussed earlier in the book chapter at page number 10 (Figure 3.9) [23].

3.2.1.5 MXenes Prepared from Unconventional Al-MAX Phases

In 2016, for the first time, Zhou et al. showed the selective etching of Al-C groups from an alternatively stacked ternary $Zr_3Al_3C_5$ MAX phase to obtain a Zr-incorporated 2D carbide ZrC_2T_x [56]. Similar to the conventional HF etching method ZrC_2T_x was obtained from the precursor $Zr_3Al_3C_5$ powder by its immersion in 50% concentrated HF solution and followed the below mentioned etching reactions:

$$Zr_3Al_3C_5 + HF \rightarrow AlF_3 + CH_4 + Zr_3C_2 \quad (3.21)$$

$$Zr_3C_2 + H_2O \rightarrow Zr_3C_2(OH)_2 + H_2 \quad (3.22)$$

$$Zr_3C_2 + HF \rightarrow Zr_3C_2F_2 + H_2 \quad (3.23)$$

In contrast to the $Zr_3C_2F_2$, synthesis of $Hf_3C_2T_x$ from $Hf_3Al_3C_5$ precursor was quite difficult as the bonds between Al-C and Hf-C groups were found to be strong enough to inhibit any intercalation. Further to overcome this problem, the material was first doped with Si to obtain $Hf_3[Al(Si)]_4C_6$ and then treated HF for the synthesis of MXene. Further, the DFT calculation performed over Zr_3C_2T (T = OH, O, F) revealed its metallic characteristics, and $Zr_3C_2O_2$ turned out to be exceedingly high mechanical strength.

In 2015, first attempt has been made by Meshkian et al. to synthesize an MXene from Ga-based non-Al MAX precursor [57]. Initially thin films of Mo_2Ga_2C were obtained through deposition of elemental Mo, Ga, and C using DC magnetron sputtering technique. Mo_2Ga_2C films were further immersed into 50% HF solution for 3 hours at 50 °C temperature leading to the extraction of Ga from unconventional MAX phase. The obtained sample was collected and washed thoroughly in DI water and EtOH. Thus, the report provided first Mo-based MXene which provided a new member in the family of 2D layered materials.

3.3 Conclusion

Various unique properties, wide range of promising applications, and comparatively easy scalable synthesis of MXenes made them a perfect choice for researchers, which are well reflected by the sudden spike in research papers and quality patents related to MXenes. Their composition can easily be modified by replacement of M, X, n, T_x and the intercalants. Due to enormous possibilities of variation by replacement of both M and X phases, MXenes are a rapidly growing material and promising to be one of the largest families of 2D materials.

Even after several theoretical possibilities for the synthesis of MXenes by replacement of C with N or Boron, majority of the MXenes reported till date are carbides. Very few of the MXenes are reported in the form of nitrides and carbonitrides. Computationally predicted MXenes generally faces their biggest hurdle in the form of their stability in the etchants. Therefore, a constant progress toward various suitable synthetic approaches is crucial for achieving new MXenes with different compositions.

Notably, as the rapid development of MXenes is highly anticipated, it is equally important to keep few factors always on the check such as environmental friendliness, safety, efficiency, and easy synthetic procedures, control on terminal groups, and on the resulting physical properties. These factors should be on the main stream of discussion for further development in this field and moving ahead for further practical applications especially for water involved applications such as desalination.

References

1 Naguib, M., Kurtoglu, M., Presser, V. et al. (2011). Two-dimensional nanocrystals produced by exfoliation of Ti_3AlC_2. *Adv. Mater.* 23: 4248–4253.

2 Naguib, M., Mashtalir, O., Carle, J. et al. (2012). Two-dimensional transition metal carbides. *ACS Nano* 6: 1322–1331.

3 Khazaei, M., Ranjbar, A., Arai, M. et al. (2017). Electronic properties and applications of MXenes: a theoretical review. *J. Mater. Chem. C* 5: 2488–2503.

4 Hong Ng, V.M., Huang, H., Zhou, K. et al. (2017). Recent progress in layered transition metal carbides and/or nitrides (MXenes) and their composites: synthesis and applications. *J. Mater. Chem. A* 5: 3039–3068.

5 Lukatskaya, M.R., Kota, S., Lin, Z. et al. (2017). Ultra-high-rate pseudocapacitive energy storage in two-dimensional transition metal carbides. *Nat. Energy* 2: 17105.

6 Ying, G., Kota, S., Dillon, A.D. et al. (2018). Conductive Transparent V_2T_x (MXene) Films. *FlatChem* 8: 25–30.

7 Naguib, M., Mochalin, V.N., Barsoum, M.W., and Gogotsi, Y. (2014). 25th Anniversary article: MXenes: a new family of two-dimensional materials. *Adv. Mater.* 26: 992–1005.

8 Anasori, B., Lukatskaya, M.R., and Gogotsi, Y. (2017). 2D metal carbides and nitrides (MXenes) for energy storage. *Nat. Rev. Mater.* 2: 16098.

9 Eklund, P., Rosen, J., and Persson, P.O.Å. (2017). Layered ternary $M_{n+1}AX_n$ phases and their 2D derivative MXene: an overview from a thin-film perspective. *J. Phys. D: Appl. Phys.* 50: 113001.

10 Sun, S., Liao, C., Hafez, A.M. et al. (2018). Two dimensional MXenes for energy storage. *Chem. Eng. J. (Amsterdam, Neth.)* 338: 27–45.

11 Xie, Y., Naguib, M., Mochalin, V.N. et al. (2014). Role of surface structure on Li-ion energy storage capacity of two-dimensional transition-metal carbides. *JACS* 136 (17): 6385–6394.

12 Naguib, M. and Gogotsi, Y. (2015). Synthesis of two-dimensional materials by selective extraction. *Acc. Chem. Res.* 48: 128–135.

13 Ghidiu, M., Lukatskaya, M.R., Zhao, M.Q. et al. (2014). Conductive two-dimensional titanium carbide 'clay' with high volumetric capacitance. *Nature* 516 (7529): 78–81.

14 Wang, L.B., Zhang, H., Wang, B. et al. (2016). Synthesis and electrochemical performance of $Ti_3C_2T_x$ with hydrothermal process. *Electron. Mater. Lett.* 12 (5): 702–710.

15 Guo, Y.T., Jin, S., Wang, L.B. et al. (2020). Synthesis of two-dimensional carbide Mo_2CT_x MXene by hydrothermal etching with fuorides and its thermal stability. *Ceram. Int.* 46 (11): 19550–19556.

16 Xu, C., Wang, L.B., Liu, Z. et al. (2015). Large-area high-quality 2D ultrathin Mo_2C superconducting crystals. *Nat. Mater.* 14 (11): 1135–1141.

17 Yang, Q., Wang, Y.K., Li, X.L. et al. (2018). Recent progress of MXene-based nanomaterials in flexible energy storage and electronic devices. *Energy Environ. Mater.* 1 (4): 183–195.

18 Xie, X.H., Xue, Y., Li, L. et al. (2014). Surface Al leached Ti_3AlC_2 as a substitute for carbon for use as a catalyst support in a harsh corrosive electrochemical system. *Nanoscale* 6: 11035–11040.

19 Li, T.F., Yao, L.L., Liu, Q.L. et al. (2018). Fluorine-free synthesis of high-purity $Ti_3C_2T_x$ (T=OH, O) *via* alkali treatment. *Angew. Chem. Int. Ed.* 57 (21): 6115–6119.

20 Wang, X.H. and Zhou, Y.C. (2010). Layered machinable and electrically conductive Ti_2AlC and Ti_3AlC_2 ceramics: a review. *J. Mater. Sci. Technol.* 26: 385–416.

21 Deysher, G., Shuck, C.E., Hantanasirisakul, K. et al. (2020). Synthesis of Mo_4VAlC_4 MAX phase and two dimensional Mo_4VC_4 MXene with five atomic layers of transition metals. *ACS Nano* 14: 204–217.

22 Halim, J., Lukatskaya, M.R., Cook, K.M. et al. (2014). Transparent conductive two-dimensional titanium carbide epitaxial thin films. *Chem. Mater.* 26: 2374–2381.

23 Urbankowski, P., Anasori, B., Makaryan, T. et al. (2016). Synthesis of two-dimensional titanium nitride Ti_4N_3 (MXene). *Nanoscale* 8: 11385–11391.

24 Li, M., Lu, J., Luo, K. et al. (2019). Element replacement approach by reaction with Lewis acidic molten salts to synthesize nanolaminated MAX phases and MXenes. *JACS* 141: 4730.

25 Alhabeb, M., Maleski, K., Anasori, B. et al. (2017). Guidelines for synthesis and processing of two-dimensional titanium carbide ($Ti_3C_2T_x$ MXene). *Chem. Mater.* 29: 7633–7644.

26 Mashtalir, O., Naguib, M., Mochalin, V.N. et al. (2013). Intercalation and delamination of layered carbides and carbonitrides. *Nat. Commun.* 4: 1716.

27 Mashtalir, O., Lukatskaya, M.R., Kolesnikov, A.I. et al. (2016). The effect of hydrazine intercalation on the structure and capacitance of 2D titanium carbide (MXene). *Nanoscale* 8: 9128–9133.

28 Naguib, M., Unocic, R.R., Armstrong, B.L., and Nanda, J. (2015). Large-scale delamination of multi-layers transition metal carbides and carbonitrides "MXenes". *Dalton Trans.* 44: 9353–9358.

29 Lukatskaya, M.R., Mashtalir, O., Ren, C.E. et al. (2013). Cation intercalation and high volumetric capacitance of two-dimensional titanium carbide. *Science* 341: 1502–1505.

30 Mashtalir, O., Cook, K.M., Mochalin, V.N. et al. (2014). Dye adsorption and decomposition on two-dimensional titanium carbide in aqueous media. *J. Mater. Chem. A* 2: 14334–14338.

31 Wang, H., Zhang, J., Wu, Y. et al. (2016). Surface modified MXene Ti_3C_2 multilayers by aryl diazonium salts leading to large-scale delamination. *Appl. Surf. Sci.* 384: 287–293.

32 Naguib, M., Halim, J., Lu, J. et al. (2013). New two-dimensional niobium and vanadium carbides as promising materials for Li-ion batteries. *JACS* 135: 15966–15969.

33 Ghidiu, M., Naguib, M., Shi, C. et al. (2014). Synthesis and characterization of two-dimensional Nb_4C_3 (MXene). *Chem. Commun.* 50: 9517–9520.

34 Li, Z., Wang, L., Sun, D. et al. (2015). Synthesis and thermal stability of two-dimensional carbide MXene Ti_3C_2. *Mater. Sci. Eng., B* 191: 33–40.

35 Zhang, C., Beidaghi, M., Naguib, M. et al. (2016). Synthesis and charge storage properties of hierarchical niobium pentoxide/carbon/niobium carbide (MXene) hybrid materials. *Chem. Mater.* 28: 3937–3943.

36 Anasori, B., Xie, Y., Beidaghi, M. et al. (2015). Two-dimensional, ordered, double transition metals carbides (MXenes). *ACS Nano* 9 (10): 9507–9516.

37 Peng, Q., Guo, J., Zhang, Q. et al. (2014). Unique lead adsorption behavior of activated hydroxyl group in two-dimensional titanium carbide. *JACS* 136: 4113–4116.

38 Zhou, J., Zha, X., Zhou, X. et al. (2017). Synthesis and electrochemical properties of two-dimensional hafnium carbide. *ACS Nano* 11: 3841–3850.

39 Halim, J., Palisaitis, J., Lu, J. et al. (2018). Synthesis of two-dimensional $Nb_{1.33}C$ (MXene) with randomly distributed vacancies by etching of the quaternary solid solution $(Nb_{2/3}Sc_{1/3})_2AlC$ MAX phase. *ACS Appl. Nano Mater.* 1: 2455–2460.

40 Yang, J., Naguib, M., Ghidiu, M. et al. (2016). Two-dimensional Nb-based M_4C_3 solid solutions (MXenes). *J. Am. Ceram. Soc.* 99: 660–666.

41 Shahzad, F., Alhabeb, M., Hatter, C.B. et al. (2016). Electromagnetic interference shielding with 2D transition metal carbides (MXenes). *Science* 353: 1137–1140.

42 Lipatov, A., Alhabeb, M., Lukatskaya, M.R. et al. (2016). Effect of synthesis on quality, electronic properties and environmental stability of individual monolayer Ti_3C_2 MXene flakes. *Adv. Electron. Mater.* 2: 1600255.

43 Karlsson, L.H., Birch, J., Halim, J. et al. (2015). Atomically resolved structural and chemical investigation of single MXene sheets. *Nano Lett.* 15: 4955–4960.

44 Halim, J., Kota, S., Lukatskaya, M.R. et al. (2016). Synthesis and characterization of 2D molybdenum carbide (MXene). *Adv. Funct. Mater.* 26: 3118–3127.

45 Feng, A., Yu, Y., Jiang, F. et al. (2017). Fabrication and thermal stability of NH_4HF_2-etched Ti_3C_2 MXene. *Ceram. Int.* 43: 6322–6328.

46 Natu, V., Pai, R., Sokol, M. et al. (2020). 2D $Ti_3C_2T_z$ MXene synthesized by water-free etching of Ti_3AlC_2 in polar organic solvents. *Chem* 6: 616.

47 Geng, D., Zhao, X., Li, L. et al. (2016). Controlled growth of ultrathin Mo_2C superconducting crystals on liquid cu surface. *2D Mater.* 4: 011012.

48 Ma, L., Ting, L.R.L., Molinari, V. et al. (2015). Efficient hydrogen evolution reaction catalysed by molybdenum carbide and molybdenum nitride nanocatalysts synthesized *via* the urea glass route. *J. Mater. Chem. A* 3: 8361–8368.

49 Giordano, C., Erpen, C., Yao, W.T., and Antoniett, M. (2008). Synthesis of Mo and W carbide and nitride nanoparticles *via* a simple "urea glass" route. *Nano Lett.* 8: 4659–4663.

50 Giordano, C. and Antonietti, M. (2011). Synthesis of crystalline metal nitride and metal carbide nanostructures by sol–gel chemistry. *Nano Today* 6: 366–380.

51 Giordano, C., Erpen, C., Yao, W.T. et al. (2009). Metal nitride and metal carbide nanoparticles by a soft urea pathway. *Chem. Mater.* 21: 5136–5144.

52 Li, Y., Shao, H., Lin, Z. et al. (2020). A general Lewis acidic etching route for preparing MXenes with enhanced electrochemical performance in non-aqueous electrolyte. *Nat. Mater.* 19: 894.

53 Peng, C., Wei, P., Chen, X. et al. (2018). A hydrothermal etching route to synthesis of 2D MXene (Ti_3C_2, Nb_2C): enhanced exfoliation and improved adsorption performance. *Ceram. Int.* 44: 18886–18893.

54 Xuan, J., Wang, Z., Chen, Y. et al. (2016). Organic base driven intercalation and delamination for the production of functionalized titanium carbide nanosheets with superior photothermal therapeutic performance. *Angew. Chem.* 128: 14789.

55 Sun, Z., Yuan, M., Lin, L. et al. (2019). Selective Lithiation–expansion–microexplosion synthesis of two-dimensional fluoride-free Mxene. *ACS Mater. Lett.* 1: 628.

56 Zhou, J., Zha, X., Chen, F.Y. et al. (2016). A two-dimensional zirconium carbide by selective etching of Al_3C_3 from nanolaminated $Zr_3Al_3C_5$. *Angew. Chem. Int. Ed.* 55: 5008–5013.

57 Meshkian, R., Tao, Q., Dahlqvist, M. et al. (2017). Theoretical stability and materials synthesis of a chemically ordered MAX phase, Mo_2ScAlC_2, and its two-dimensional derivate Mo_2ScC_2 MXene. *Acta Mater.* 125: 476–480.

4

Physicochemical and Biological Properties of MXenes

Neakanshika Chadha[1], Aman Sahu[2], Kshitij RB Singh[3], and Jay Singh[3]

[1]*Department of Applied Sciences, KIET Group of Institutions, Ghaziabad, Delhi, India*
[2]*Department of Chemistry, LRPG College, CCS University, Ghaziabad, Uttar Pradesh, India*
[3]*Department of Chemistry, Institute of Science, Banaras Hindu University, Varanasi, Uttar Pradesh, India*

4.1 Introduction

MXenes, 2D material, because of its exceptional qualities, have gained attention from various disciplines of science and technology for multidimensional applications [1–3]. They belong to the family of metal carbides and carbo-nitrides. MXenes, with general formula of $M_{n+1}AX_n$, where M belongs to transition metals, A to main-group elements, and X is either C or N atoms or both [4, 5]. They become the study of interest due to their high surface area, excellent mechanical strength, hydrophilicity, high conductivity, and photocatalytic property [6]. The different negatively charge surface functionalities (-halides, hydroxides) present on the edges makes it suitable for energy storage, optoelectronics, electromagnetic interference shielding, biosensors, and biomedical applications [7–9]. Commonly 2D layered MXenes are obtained by wet chemical etching method from multilayered MAX phase using acidic solutions [10]. Naguib and his co-workers discovered MXenes in 2011, where they produced 3D Ti_3C_2 nanosheets using hydrofluoric acid as etchant [11]. Afterwards, numerous studies have been devoted to synthesize MXenes by different routes including molten salt etching, hydrothermal, electrochemical methods [12]. Physiochemical properties and performance of MXenes for numerous applications are fundamentally influenced by their stoichiometric compositions, synthesis route, surface functionalities, flake size, and thickness [13]. Different synthesis routes affect differently on the lateral size of nano-flakes and precise control over it remains a challenge till date. These parameters are investigated by tremendous volume of research for correlating the link between these and their size depending physiochemical properties such as mechanical, electrochemical, and opto-electronic properties of MXenes [14]. Controlling the size of MXene flakes, synthesized through mechanical, and sonication process, poses a challenge, as it disorganize the 2D morphology and the physicochemical properties of flakes [15]. Due to the presence of surface terminations (–F, –OH, =O), the MXenes becomes hydrophilic in nature which enhances its capability to form stable dispersions and MXene/composites [16]. Ultrafast water flux ability allows its utility in filter membranes and antimicrobial systems. In order to find the microbial activity of any compound, examination of its

MXenes: Next-Generation 2D Materials: Fundamentals and Applications, First Edition.
Edited by Jay Singh, Kshitij RB Singh, Ravindra Pratap Singh, and Charles Oluwaseun Adetunji.

cytotoxicity is of prime importance [17, 18]. Explicitly, the toxicity depends on many extrinsic parameters such as physiochemical properties that include solubility, flake size, thickness, chemical linkages, shape, surface modifications, and aggregation [19]. Good flexibility and versatile chemistry combined with 2D layered structure makes it a promising candidate to form composite, which provides an opportunity to integrate state-of-the-art materials. Translation of lab-scale designed MXenes to commercial or large-scale level needs careful investigation of physicochemical and biomedical properties [20, 21]. Also, the interconnection between the properties and possible applications is still not extensively studied. This chapter provides an insight of the synthesis methods, physiochemical and biological properties of MXenes, and current situation in related applications.

4.2 Structure and Synthesis of MXenes

New MXenes are synthesized by compositional modifications in the existing ones. It comprises more than 60 members of the class of layered and hexagonal geometric carbides, nitrides or carbo-nitrides of Ti, Mo, and W. M–A metallic bond is quite strong so the isolation of the $M_{n+1}X_n$ layers from MAX phases (inorganic ternary carbides or nitrides) by mechanical force/shearing becomes challenging and difficult [22, 23]. However, on the treatment of more chemically active M–A bonds than the stronger M–X bonds using the etching technique each layer can be isolated selectively [24]. Figure 4.1 demonstrates the preparation of MXenes by selectively etching each layer from parent MAX phases; various other MXenes can be generated by the same route [25].

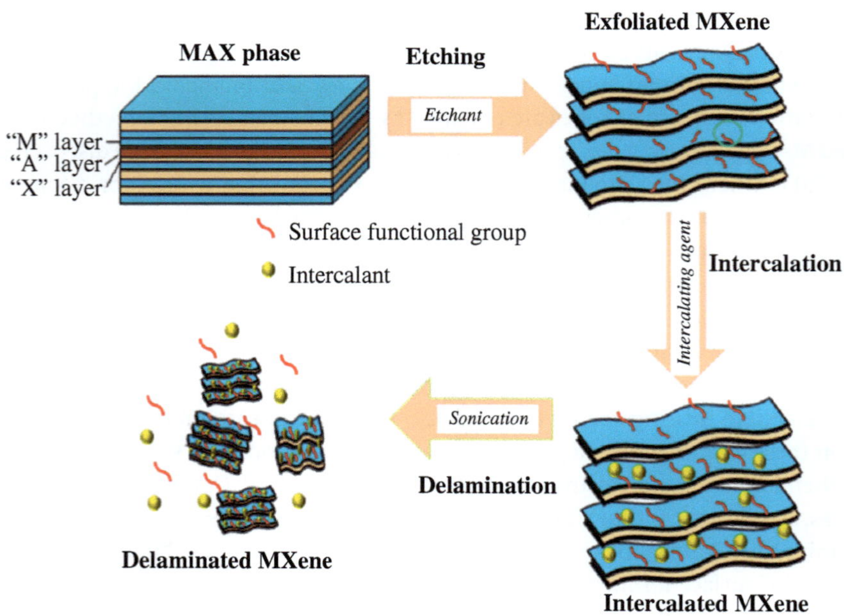

Figure 4.1 MXene synthesis using etching method by delamination of MAX phase (Reproduced with permission from reference [19])/Elsevier).

Multilayered $M_3C_2T_x$ MXenesis was synthesized by the treatment of multilayered ternary carbide (M_3AC_2) with an acidic solution of HF (e.g. HCl-LiF). In this process, the A layer (made of aluminium) was etched out selectively, and hydroxyl groups, oxygen, or fluorine atoms terminate the surface groups [26]. The first MXene, $Ti_3C_2T_x$, was prepared by Ti_3AlC_2 powders. Naguib et al. sank this powder in hydrofluoric acid (HF) of 50% concentration at room temperature (RT) for 2 hours [27]. Afterwards, this synthesis route was employed to synthesize different MXenes and their combinations e.g. V_2CT_x [28], Ti_2CT_x [29], Nb_2CT_x [30], Ti_2NT_x [31], $(V_{0\cdot5}Cr_{0.5})_3C_2T_x$ [32], Ti_3CNT_x [33], $Ta_4C_3T_x$ [34], $V_4C_3T_x$ [35], $Mo_2TiC_2T_x$ [36], and $Cr_2TiC_2T_x$ [37] successfully. Safer mixture such as hydrochloric acid (HCl) and lithium fluoride (LiF) solution were also employed by Ghidiu et al. They soaked Ti_3AlC_2 powders at 35 °C for 24 hours in this solution to obtain $Ti_3C_2T_x$ [38]. In fact same approach has also been used to synthesize variety of MXenes, where other combinations of acids and fluoride containing salts used as etchant. Nitride and carbonitride MXenes are not as explored as carbon-based MXenes, since very limited MAX phases with N are available. Besides this, delamination of nitride layers is also difficult. Ultrathin large-area Mo_2C, WC and TaC films were reported to be synthesized using chemical vapor deposition (CVD) [39, 40]. Xu et al. prepared a 2D ultrathin film (Mo_2C) with CVD method. They utilized methane and molybdenum foil as a carbon source and copper as substrate under a temperature of ~1000 °C for synthesizing this film [39].

Preparation of 2D nitrides of Mo, W, and V was reported by Xiao et al. using salt-templated synthesis method. In this method hexagonal oxides are reduced by the surface termination [41]. Among the few, Kamysbayev et al. synthesized Li^+ intercalated nanosheets, they intercalate $Ti_3C_2T_x$ MXene using the n-BuLi using molten salt method at temperature of 50 °C. Intercalating various species such as cations and organic moities in the MXene sheets, have enriched the versatility of these compounds [42]. Brady et al. prepared sodium intercalated Ti_3C_2 sheets with multilayer $Ti_3C_2T_x$. Solution of Na-biphenyl radical anion complex is reacted with this multilayer $Ti_3C_2T_x$ at RT [43]. Beyond different synthesis routes, theoretical approaches have also caught high interest in developing new, stable MXene structures as depicted in Figure 4.2. In 2016, Khazaei et al. using first- principle calculations, demonstrated that many M_2C and M_2N systems having fluorine, hydroxyl, and oxygen as surface terminal groups could be formed with stability [44]. Anasoriet al. studied 32 ordered-MXenes for their relative stability on the basis of chemical ordering for fully-ordered and partially-ordered configurations [45]. Dahlqvist et al. reported a well systematic procedure for investigating several i-MAX and o-MAX phases for their phase stability and chemical positioning. Intercalation of sheets have increased the potential of already chemically versatile MXenes for optimizing them for their particular applications [46]. Consequently, MXenes with versatile chemical and physical properties, become the promising fundamental element for applications such as energy storage devices, thermoelectric materials, electromagnetic interference shields, conducting materials, structural composites, bio-catalysts, and sensing devices [46–48].

MXenes' potential in various arenas has also been explored through different studies including theoretical studies due to its diverse electronic, magnetic, optical, and electrochemical properties (Figure 4.3). For simulation and production of MXenes' single-flakes with homogeneous terminations, experimental and theoretical both types of studies are required [50]. For this, chemistry and structure are the crucial parameters that determine

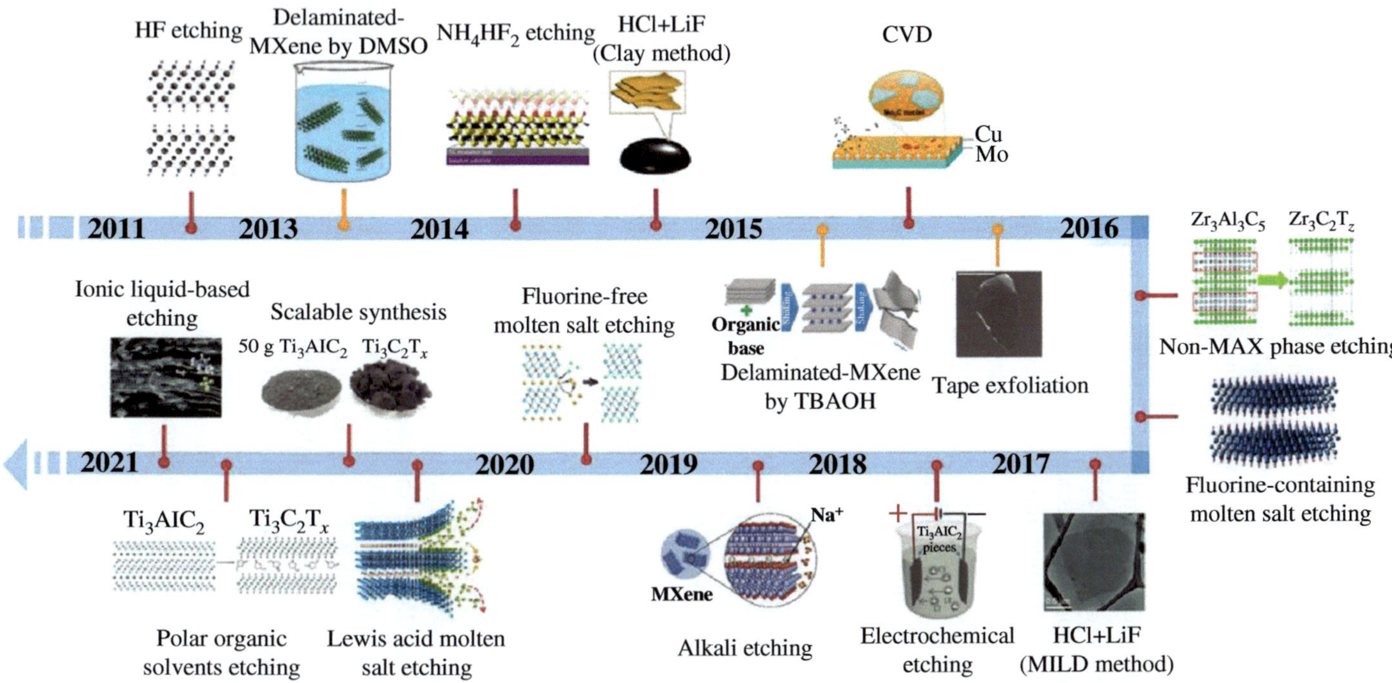

Figure 4.2 Advances in synthesis technique over the year (Reproduced with permission from reference [12]/John Wiley & Sons).

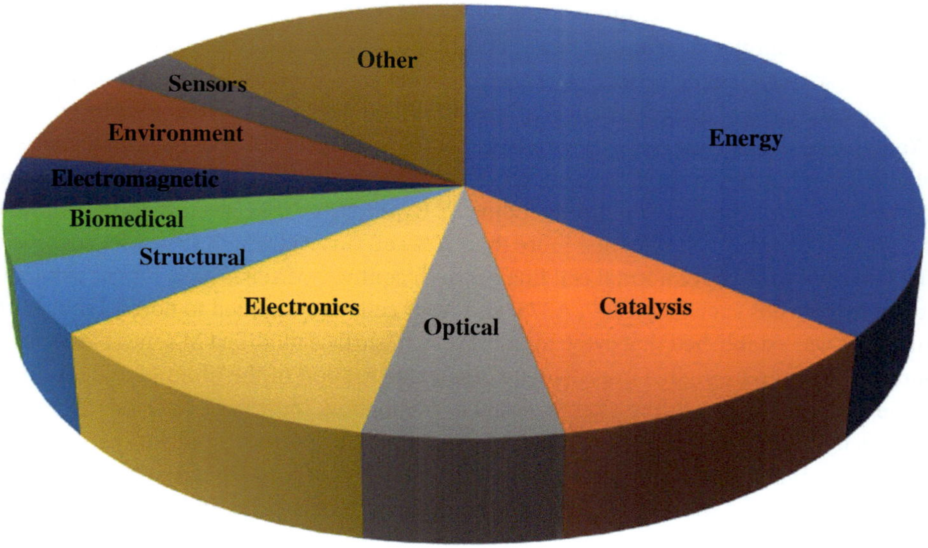

Figure 4.3 Applications of MXenes (Reproduced with permission from reference [49] Bhat A et al., 2021/Springer Nature/CC BY 4.0.

the physical properties and their role played in different applications. Depending on the composition and surface terminations, the electrical conductivity of the MXenes varies. Through continuous optimization of the synthesis of MXene, higher values of electrical conductivity can be achieved. 3D MXenes offers high moduli of electricity (~350 MPa) than 2D graphene-based systems synthesized via exfoliation technique [51]. The interplay of electrical and mechanical properties along with surface functionalities makes MXenes a far better alternative over other nano-sized materials for utilization in energy storage, and in development of better functional composites of polymers, ceramics, and metals [52, 53].

4.3 Properties of MXenes

4.3.1 Biomedical Properties of MXenes

Not surprisingly, MXenes show excellent performance in biomedical application due to their physical (large surface area), electrical properties, and good biocompatibility [54]. Over the years, efficient biosensors are synthesized via 2D materials and are vital for determining the pathogenic, toxicity, or pharmacological activity in biological systems [55].

The efficient biosensors must be capable enough to selectively detect a trace amount of analyte within the system. Owing to its large surface area and numerous surface active moities, MXenes act as an excellent bio-sensor for external stimuli [56]. In this context, Yao et al. synthesized a superoxide biosensor of Ti_3C_2MXene nanocomposite impregnated with gold platinum that possess low limit of detection (LOD) for super-oxides [57]. Liu et al. synthesized nitrite biosensor with high detection range by immobilizing haemoglobin in the biosensor (Nafion/Hb/$Ti_3C_2T_{x}$/GCE) [58]. Biosensor can specifically detect nitrite in the wide range where maximum reaches up to 11800 mM and the lowest till 0.12 mM. In another work, similar matrix was taken, and titanium oxide nanoparticles were incorporated, that shows wider detection range from 0.1 to 380 mM. Besides biosensors, MXenes are widely used in other biomedical applications, where the toxicological/cytotoxicity effect of MXene-based composites is the important factor for their practical application. It has been validated that MXenes have very low cytotoxicity, which is a favorable quality for the clinical applications. For instance, MXenes including Ti_3C_2, Ti_2N,and Ti_2C even at concentrations as high as 500 mg ml^{-1} maintained more than 70% cell survival which shows its negligible effect on the viabilities of normal cell lines [59]. In contrast, MXenes have higher cytotoxicity on tumor cells such as MCF-7, A375, and A549 cells as compared to normal cells [60]. Further, it was established that cytotoxicity of several surface modified MXenes nanosheets found to have the negligible impact, even at a concentration of the order of 160 mg ml^{-1}, on the viability of T1 cell line after 48 hours of incubation, indicating their relatively low cytotoxicity [61]. Afterwards, Chen et al. have studied the cytotoxicity of various common MAX phases including Ti_3AlC_2 and Ti_3SiC_2 on the survival of preosteoclasts (CRL-2593 and MC_3T_3-E1 cells) and they found relatively no cytotoxic effects [62]. In particular, there is lot of work available highlighting the role of physicochemical characteristics such as chemical compositions, dispersibility, surface, moieties, and surface properties, structure, size, and shape in determining the cytotoxicity of MXenes. Therefore, with different approaches, desired surface functionalization of MXenes can be achieved which will help in modulating the cytotoxicity of MXene based nanocomposites for the particular biomedical application.

4.3.2 Electronic and Transport Properties

Several reports are published on the theoretical study of the electronic and transport properties of MXenes, and several important findings have already been reported. Many reports, focus on determining of the band gap, density of states (DOS), and overall band structure. Experimental and theoretical both investigation revealed that functionalized MXenes show metallic behavior in nature, and further examination depicts that generally the band gap lies between 0.24 to 1.8 eV [63]. Some MXenes exhibit semiconductor behavior after functionalization, which is examined through the electronic structure of both pristine and functionalized MXene. In pristine MXene, d-orbital of M transition metal contributed all fermi level energy, however on functionalization the fermi level shifted downwards i.e. below the d-band [64]. The metallic behavior of the MXenes depends on the transition metal, for example, the metallicity decreases while going down in a group. Apart from having similar ordered layer structure, sometimes MXene shows altogether different electronic structure. For instance, in $Mo_2TiC_2T_n$, Mo present at the surface, whereas Ti takes the location in the centre. $Mo_2TiC_2T_n$ exhibit the behavior of semiconductor instead of metal despite of having similarity in structure with $Ti_3C_2T_n$ [65]. Furthermore, researchers investigated that some transition metal carbides such as W_2CO_2, Mo_2CO_2, and Cr_2CO_2 are insulating in nature as well [66]. The band gap varies due to near free electron (NFE) state which ultimately depends on the terminated group attached with the MXenes. Mostly, the outer transition metal decides the electronic properties and hence the surface terminations attached to it plays an important role. The electronegative group attached to the transition metal drags more electrons toward itself, which leads to a reduction of the DOS at the Fermi surface or a lowering of the Fermi energy, resulting in formation of the new energy band below the Fermi surface [67]. Thus, the surface terminations can be engineered to obtain different value of band gap. Naguib et al. predicted with his research that, band gap of Ti_3C_2 changed on surface functionalization, the bandgap of bare Ti_3C_2 metal is zero with certain value of DOS at the Fermi level but as soon as the surface functionalization is done with OH and F groups the bandgap increases from 0.05 to 0.1 eV [68]. Till date, various theoretical models have been reported to predict diverse the electronic properties of terminated MXenes varying from extremely conductive metallic (Cr_2N, Ti_2CF_2, Sc_2C, Zr_2CF_2) [69], half-metallic (Mn_2NF_2, Cr_2NO_2, Cr_2C) [70], semiconducting (Ti_2CO_2, Cr_2CF_2, Zr_2CO_2) [71] to highly insulating state. Stoichiometry also determines the electronic structure and this can be altered by dopants, defects, lattice symmetry, and the applied electric field. Balci et al. reported that bare Sc_2CF_2 MXene generally has a bandgap of 0.96 eV, which on doping with silicon, germanium, tin, and boron can be altered to 0.24–0.55 eV [72]. In some cases, biaxial strain causes band gap change from ~1.4 to 1.8 eV as observed in Sc_2CO_2MXene [73].

4.3.3 Optical Properties

The optical properties comprising reflection, scattering, absorption, and transmittance can be analyzed by in-silico models through the calculations involving the imaginary part of dielectric function which is directly linked to wavelength of photon. Using above method, the optical behavior e.g. reflection, energy loss function, and absorption spectrum of many carbides, nitrides of titanium (Ti_2C, Ti_3C_2, Ti_2N, and Ti_3N_2) have been inspected thoroughly. Random Phase Approximation (RPA), a popular method to examine optical behavior was used for Ti_2C, Ti_2N, Ti_3C_2, Ti_3N_2, functionalized $Ti_3C_2T_2$ [74, 75]. Owing to the conductivity of the order of metal, almost zero optical gap was found in the imaginary component of the dielectric function. The plasmon energy was found, on applying an electric field in the parallel direction to the surface, to be in the range of 10–11 eV [76]. This work suggested that for the photon of energy lesser than 1 eV the reflectivity became maximum nearly 100%, which clearly indicated that MXenes are fully capable of transmitting electromagnetic waves.

In past few years, researchers focused on the analysis of interesting optical parameters including optical transparency, plasmonic behavior, and efficient photo-thermal conversion [77]. Once again, terminations present on surface of MXenes have a very significant part in investigating the optical properties. Berdiyorov et al. published a study about the examination of the importance of surface moieties on the optical behavior of $Ti_3C_2X_2$, where X= F, O, OH terminations [74]. In this study value of refractive index and extinction coefficient was obtained using dielectric constant and it was found that terminations reduces all these parameters at lower photon energy. However, at higher photon energy, strong absorption was recorded in comparison to the bare MXene due to terminations. Halim et al. [78] reported the optical properties of both Ti_2CT_x and Nb_2CT_x films, analyzed that properties tend to change with transition metal. Some of them compared the absorption ability of MXenes with other popular 2D layered materials. Jiang et al. scrutinized the nonlinear optical behavior using excitation sources of different wavelength i.e. 800, 1064, 1550, and 1800 nm [79]. They recorded the absorption behavior of $Ti_3C_2T_x$ better than graphene by order of 10^2.The optical properties of MXenes including linear and non-linear depends on energy band gap and energy structure. Theoretical investigations in range of 200–700 nm revealed that –F, –OH functionalized Ti_2C displayed lesser in-plane absorption coefficient than O terminated Ti_2C [80]. MXenes quantum dots offers more photoluminescence (PL) due to enlarged quantum confinement and edge effects. A handful of MXene-derived QDs including Ti_3C_2 and V_2C based QDs have been reported so far consisting of the favorable PL characteristics [81, 82]. Study on the insights of optics, opens the door of non-linearity, which is the fundamental core of many communication systems including sensors, lasers, and photonic devices. The Ti3C_2T_x was analyzed for broadband nonlinear optical response, and the outcome explicitly indicated its utility in the industrial application for developing femtosecond laser using Ti3C_2T_x [83].

4.3.4 Magnetic Properties

Interestingly, some immaculate MXenes, such as Ti_2C, Ti_2N, Cr_2C, Cr_2N, and Mn_2C are expected to have magnetic properties. However, majority of 2D MXenes are non-magnetic in nature, thus MXenes exhibit variation in magnetic behavior. Some of them, predicted to be ferromagnetic such as Cr_2C, Cr_2N, Ta_3C_2, $Cr_3C_{2,}$ Fe_2C are, while other 2D Ti_3C_2 and $Ti_3N_{2,}$ Zr_2N are antiferromagnetic [84, 85]. In particular, magnetic MXenes with half -metallicity, have recently been explored vigorously for their utility in spintronics. Cr_2C, a combination of metallic spin and insulating spin, was reported to be the first half metal from the MXene family [86]. Subsequently, the half metallicity of other MXenes Ti_2C and Ti_2N, Ni_2N was evaluated [87, 88]. The magnetic behaviors of pristine MXenes are because of unpaired electrons available with transition metal, but surface functionalization affects the magnetic property of most of the MXenes. Exceptionally, the magnetism of Cr - and Mn-containing MXenes is not suppressed after surface functionalization [89]. The magnetic properties of MXene could be tuned for desired application by modification in the chemical ratio of two M elements and surface functionalizations present it as potential candidate for future applications. For illustration, functionalization of Cr_2C by X groups (e.g. H, F, ClorOH) results into its conversion from a ferromagnetic half metal into an antiferromagnetic semiconductor [90]. Mn-based MXenes such as bare Mn_2C have antiferromagnetic property, while Mn_2CCl_2 shows behavior of antiferromagnetic insulator apart from this Mn_2CF_2 can be transformed into half metal of ferromagnetic nature having a high curie temperature [91]. In this way, MXenes shows variation in magnetic behavior on surface passivation and attracts researcher's attention worldwide. Not only symmetrical but also non symmetrical surface terminations have been intensively studied for analyzing the influence of functionalization on MXenes. It is observed that Cr_2CXX' (X, X' = H, F, Cl, Br, OH) are all antiferromagnetic semiconductors [16]. Surprisingly, in Cr_2CFCl,it shows similar behavior to bipolar magnetic semiconductors, which widen their potential to be as entrants for spintronic applications [92]. Lambrechtet al. thoroughly inspected spin structure and magnetic behavior in different M_2NT_x nitrides, where M = Ti, Cr, Mn. They revealed that the spin symmetry/magnetic properties can be tuned for particular applications by modifying the chemical compositions or changing the transition metal of MXenes. It has been found that different magnetic interactions could be obtained by controlling spin symmetries using spin–orbit coupling mechanism [93, 94]. A recent study on surface-terminated $Ti_2MnC_2T_x$ monolayers suggests strong ferromagnetism and high

Curie temperature (495–1133 K) which is highest among 2D ferromagnetic materials [95]. Functionalized MXene 2D Ti_2NO_2, convert to a ferromagnetic half-metal from an antiferromagnetic semiconductor, resulting from crystal strains that enhance crystal splitting [96]. Semiconductor-to-metal transitions have recently been found in $Hf_2MnC_2O_2$ and $Hf_2VC_2O_2$MXenes, with high sensitivity of curie temperatures depending on applied strain [97]. The ferrimagnetism due to the interactions of inherent d electron between different Mo layers was found in molybdenum-based MXenes $Mo_3N_2F_2$ via external strain that induces half-metallicity (E_g = 0.47 eV) with a Curie temperature of 237 K [98]. Subsequently, $Hf_2VC_2F_2$ was studied, where in-plane non-collinear Y-type antiferromagnetic pattern has been demonstrated which leads to ferroelectric polarization [99].

4.3.5 Topological Properties

Although most MXenes exhibit conductivity of metal or semiconductors, however MXenes with functionality shows the behavior of 2D topological insulators [100]. In this type, the electrons propagate along the edge states resulting in less dissipation or transport of electrons. Many of the studied topological insulating MXenes contain bigger metals of 4d, 5d- series such as Mo, W, Ti, Zr, and Hf atoms. For instance, W_2CO_2 with O-functionalization was found with large state of charge-induced bandgaps, calculated using gradient approximation. An optimized surface chemistry of MXenes could be a key parameter to obtain MXenes of semiconducting or topological insulating property. Theoretically, some $Mo_2M''C_2O_2$, where M″ can be Ti, Zr, Hf transition metals are even projected to have spin Hall effect at RT, as they contain considerable large topological gaps ~0.1–0.2 eV [101]. In a work, Khazaei and his co-workers reported the presence of 2D topological insulating phase in MXenes with two transition metal elements $W_2MC_2O_2$ (M = Ti, Zr, or Hf), subject to different M atoms and bandgap reaches to 0.285 eV [102]. Some of the MXenes changes into topological insulators under electric field due to charge transfer as observed in $Sc_2C(OH)_2$ [103]. This condition occurs due to optimum surface transport channels for charge mobility. This field lacks experimental studies and hence dedicated studies are required to know the transformation of metal MXenes to topological insulators. It is further concluded that single terminations are considered better candidate than mixed terminated MXenes, in order to predict their topological states.

4.3.6 Vibrational Properties

The phonon spectra describe the quantity of phonon modes, ultimately predicting the vibrational properties of MXenes. This can be determined by inelastic neutron scattering method [104]. The phonon spectra of various bare and terminated M_2XT_2 (M = 3d, 4d, 5d transitional metal atoms, X=C, N, T=F, O, OH) systems has been discussed in the current literature [105, 106] approving some specific stabilities. Due to the similarity in bonding, chemical and crystal structure of M_2C systems, several features including vibrational modes are generally common to all [106].

The phonon dispersions exhibit three acoustic modes; out of which two are in-plane rigid-movement, whereas, the third one corresponds to out-of-the-plane vibration (have quadratic dispersion and exhibit lesser energy). Lifshitz et al. first explained the quadratic dependence in MXene, which is similar to 2D graphene and MoS_2 layered structure [107]. Basically, 3N phonon modes are detected in a system of Natoms, where three are acoustic modes and 3N−3 are optical modes. The optical modes in general are contributed by the surface functionalities, for instance, addition of 2 atoms in the crystal involves inclusion of 6 optical modes. Out of which Raman active and IR active modes can be analyzed by using group theory. Phonon modes showing E_g and A_{1g} symmetry are established to be Raman-active, while that of E_u and A_{2u} are IR-active. In the case of bare M_2X, more specifically, V_2C, the Raman-active modes relate to the low-frequency optical modes (at 224 and 359 cm^{-1}), and IR-active modes link to the upper frequency in the phonon spectrum (at 610 and 701 cm^{-1}). The Raman and IR active mode frequencies have been stated for several MXene systems. The peak position along with atomic vibrations are of prime relevance and insights of the variation with composition and surface terminations need more rigorous study.

4.3.7 Electrochemical Properties

Due to their 2D layered structure, large surface areas, easy surface modifications and highly defined structure, MXenes are auspicious candidates for energy storage devices such as batteries, supercapacitors MXenes can house various ions having different dimensions between their 2D layers of $M_{n+1}X_nT_x$. They can be spontaneously intercalated by polar organic molecules and cations such as Li^+, Na^+, and K^+ [108–110] occupying electrochemically active sites. These intercalated materials can act as an electrode material and enhanced electrochemical performance of energy storage in terms of energy density, power density can be achieved. Insight to the MXene based capacitors, a Ti_3C2T_x thin electrode demonstrated excellent capacitance of ~350 Fcm^{-3} at 20 mV sin^{-1} KOH electrolyte at charge discharge cycles up to 10,000 at 1 A g^{-1} [111]. Ling et al. stated that $Ti_3C_2T_x$/polymer composite film, synthesized by mixing $Ti_3C_2T_x$ and polyvinyl alcohol (PVA) [112]. The electrochemical performance was investigated in KOH electrolyte and found superior than many other 2D materials. The composite film displayed an impressive capacitance as high as 528 F cm^{-3} at 2 A g^{-1} with slight decrease in capacitance after 10,000 cycles. D. Agnese et al. modified the $Ti_3C_2T_x$ by intercalating it with DMSO, leading to the delamination of $Ti_3C_2T_x$ [113]. Further, $Ti_3C_2T_x$ so obtained displayed a high capacitance of 415 F cm^{-3} at 5 A g^{-1} in H_2SO_4 electrolyte and stability over 10,000 cycles. Thus, it is being realized that enhanced capacitance can be achieved with surface modification that contain more oxygen-containing (–OH,–COOH) functional groups. Interestingly, the $Ti_3C_2T_x$ produced by a HCl-LiF mixture possesses predominantly oxygen-containing functionalities with outstanding capacitances up to 900 F cm^{-3} [38].

The tuning of chemical structure and surface chemistry through modulation in synthesis has a great impact on working potential as anode and cathode in batteries (Figure 4.4). The intercalated electropositive metal ions between MXene layers exhibit high rate performance. In a study, $Ti_3C_2T_x$ were used as backbone for conductive sulfur molecules in Li–S batteries, resulting in improved cyclability and stability [115]. This is attributed to the strong interaction of polysulfide species with MXene functional groups. Similarly, encapsulation

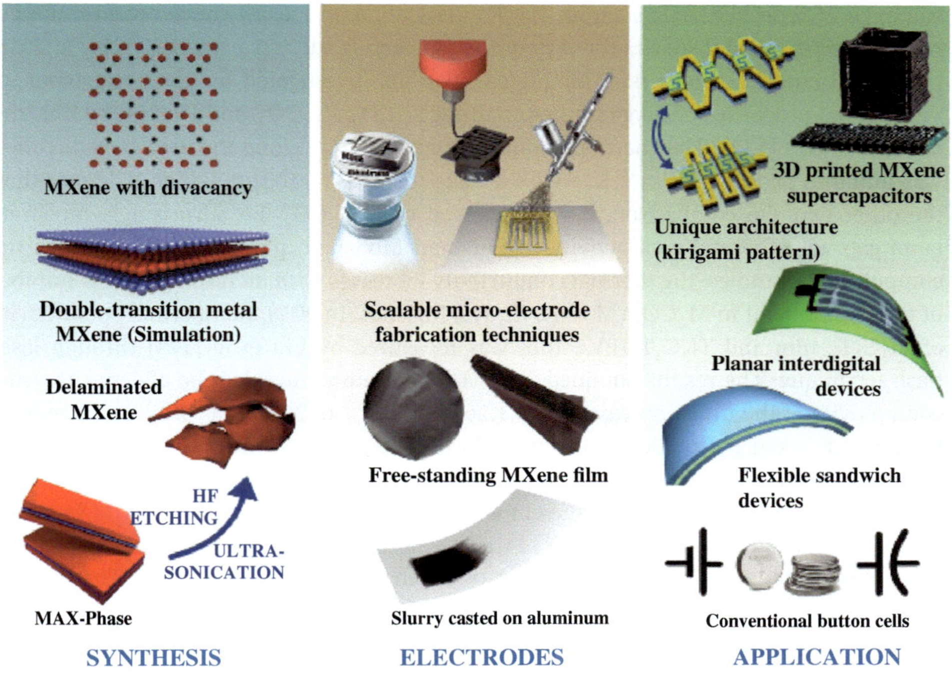

Figure 4.4 Synthesis of electrode material for energy storage devices (Reproduced with permission from reference [114] Das P et al., 2020/IOP Publishing/CC BY 4.0.

of tin NPs between layers of $Ti_3C_2T_x$ results in stable performance with a superior capacity ~ 2000 mAh g^{-1} [116].The mechanism of high capacitance of MXene was not clear as no pronounced redox peaks and no resemblance to carbon-based double layer capacitors was observed. However, later it was found that charge storage is, and it involves changes in the oxidations state of the transition metal [117]. However, the redox reaction is electrolyte dependent as well, as in sulphuric acid electrolyte; the $Ti_3C_2T_x$MXene is predominantly pseudo-capacitive in nature [112].

Also, the density of the electrode serves as a determining factor on evaluating capacitance. As carbides are much denser than carbon, freestanding bare MXene electrodes typically possess densities from ~4 g cm^{-3}, whereas MXene-based composite electrodes with polymer binders and conductive additives have lower densities ranging from 1 to 2.5 g $cm-^{3}$, which is quite large. Thus, MXene aerogel forms and thin films are better alternatives than bulk bare form. MXene-based hybrid materials can provide enhanced electrochemical and mechanical performance with the integration of conducting fillers i.e. carbon, nanotubes, graphene as they prevent restacking of the 2D layers [118, 119].

4.3.8 Thermal Properties

In many electronic devices, good thermal properties have its own significance as high thermal conductivity contributes to better heat dissipation, which ensure safety and prevents damage from excess heat [120]. Although theoretical thermal conductivity values of MXene materials such as Ti_2CO_2, $Ti_3C_2T_{x,}$ and Hf_2CO_2 have been examined, however experimental studies for various new MXene is still less explored [121]. In 2015, theoretical studies using first-principle calculations investigated for the comparison of the thermal properties of oxygen functionalized Ti_2CO_2, $Zr_2CO_{2,}$ and Hf_2CO_2 by Zha and his co-workers [122]. The thermal conductivity of Hf_2CO_2 along the armchair direction was found to be ~131.2 Wm^{-1} K^{-1} with a flake thickness ~100 μm which is higher than the other two. On comparison, it was observed that Hf_2CO_2 has surprisingly moderate band gap with thermal conductivity higher than MoS_2 and phosphorene. Based on the computational studies, the thermal conductivity increases with increasing atomic number of transition metal in M_2CO_2 (M = Ti, Zr, Hf) MXenes. In 2018, the thermal conductivity of $Ti_3C_2T_x$ film and $Ti_3C_2T_x$/PVA film was measured by Liu et al [123] through laser flash technique. The results obtained using T-type method revealed the effective thermal conductivity of the films increased from 1.26 W m^{-1} K^{-1} to 2.84 W m^{-1} K^{-1} on increasing temperature from 80 to 290 K. However, these films were synthesized using laminated multilayered $Ti_3C_2T_x$ instead of delaminated MAX phase. Later on, enhanced thermal conductivity was achieved where novel nanocomposite of palmitic acid/Ti_3C_2MXene was synthesized using two step method [124]. The nanocomposite showed thermal conductivity enhancement by 14.45% on addition of MXene in palamatic acid (PA). The thermal conductivity of pure PA was found to be 0.18 W m^{-1} K^{-1} and increased to 0.21 W m^{-1} K^{-1}in PA/0.1%MXene. The thermal conductivity of the composite is improved by doping MXene nano-flakes into PA to make it suitable material for nano-electronics including solar cell application. In a work, Liu and Li synthesized $Ti_3C_2T_x$/poly (vinyl alcohol) composites and investigated the effects of the thermal properties of MXene using temperature-dependent Raman spectroscopy [125]. By reducing the thermal coefficient of the $E_g{}^1$ mode, the composite significantly improves the thermal stability from −0.062 to −0.033 cm^{-1} K^{-1}, due to the strong Ti–O linkages between the MXene and PVA polymer. Meanwhile, the thermal conductivities of two, i.e. $Ti_3C_2T_x$ and $Ti_3C_2T_x$/PVA composites reaches nearly 55 and 48 W m^{-1} K^{-1} respectively.

4.4 Conclusion and future Perspectives

After the exploration of graphene, among 2D materials, MXene, exfoliated from layered MAX phases, has received rapid interest in past decade. The fascinating properties of MXene make them promising candidates as alternatives to beat other 2D materials. This chapter summarizes the latest progress, particularly focusing on the impact of material design on physicochemical and biological properties. The multi-functionality provide a strong reason for exploring new MXene-based materials. The chapter highlights that versatile MXenes can be utilized for biomedical applications as well. The major challenges includes the experimental/practical measurement of their magnetic and optical properties. Also, in-depth study of vibrational properties is missing in literature. Overall, properties determined using theoretical calculations are found easily in current literature. This chapter is useful for the researchers to link the effect of surface tuning on the properties of MXenes.

References

1 Al-Hamadani, Y.A.J., Jun, B.M., Yoon, M. et al. (2020). Applications of MXene-based membranes in water purification: a review. *Chemosphere* 2: 3.

2 Sun, Y., Xu, Z., Zhuang, Y. et al. (2021). A review on MXene: synthesis, properties and applications on alkali metal ion batteries. *IOP Conf. Ser.: Earth Environ. Sci* 714 (4): 042030. https://doi.org/10.1088/1755-1315/714/4/042030.

3 Lei, J.C., Zhang, X., and Zhou, Z. (2015). Recent advances in MXene: preparation, properties, and applications. *Front. Phys.* 10 (3): 276–286. https://doi.org/10.1007/s11467-015-0493-x.

4 Shuck, C.E., Sarycheva, A., Anayee, M. et al. (2020). Scalable synthesis of $Ti_3C_2T_x$ MXene. *Adv. Eng. Mater.* 22: 1901241.

5 Zhan, X., Si, C., Zhou, J., and Sun, Z. (2020). MXene and MXene-based composites: synthesis, properties and environment-related applications. *Nanoscale Horiz.* 5 (2): 235–258. https://pubs.rsc.org/en/content/articlehtml/2020/nh/c9nh00571d.

6 Papadopoulou, K.A., Chroneos, A., Parfitt, D., and Christopoulos, S.R.G. (2020). A perspective on MXenes: their synthesis, properties, and recent applications. *J. Appl. Phys.* 128 (17): 170902. https://doi.org/10.1063/5.0021485.

7 Huang, J., Li, Z., Mao, Y., and Li, Z. (2021). Progress and biomedical applications of MXenes. *Nano Select* 2 (8): 1480–1508. https://doi.org/10.1002/nano.202000309.

8 Chen, J., Huang, Q., Huang, H. et al. (2020). Recent progress and advances in the environmental applications of MXene related materials. *Nanoscale* 12 (6): 3574–3592.

9 Kumar, J.A., Prakash, P., Krithiga, T. et al. (2022). Methods of synthesis, characteristics, and environmental applications of MXene: a comprehensive review. *Chemosphere* (286): 131607.

10 Maleski, K. and Alhabeb, M. (2019). Top-down MXene synthesis (selective etching). In: *2D Metal Carbides and Nitrides (MXenes): Structure, Properties and Applications*, 69–87. https://doi.org/10.1007/978-3-030-19026-2_5.

11 Naguib, M., Kurtoglu, M., Presser, V. et al. (2011). Two-dimensional nanocrystals produced by exfoliation of Ti_3AlC_2. *Adv. Mater.* 23 (37): 4248–4253. https://doi.org/10.1002/adma.201102306.

12 Wei, Y., Zhang, P., Soomro, R.A. et al. (2021). Advances in the synthesis of 2D MXenes. *Adv. Mater.* 33 (39): 2103148. https://doi.org/10.1002/adma.202103148.

13 Gogotsi, Y. and Anasori, B. (2019). The rise of MXenes. *ACS Nano* 13 (8): 8491–8494. https://doi.org/10.1021/acsnano.9b06394.

14 Zhang, X., Shao, J., Yan, C. et al. (2021). A review on optoelectronic device applications of 2D transition metal carbides and nitrides. *Mater. Des.* (200): 109452.

15 Li, X., Huang, Z., Shuck, C.E. et al. (2022). MXene chemistry, electrochemistry and energy storage applications. *Nat. Rev. Chem.* 6 (6): 389–404. https://www.nature.com/articles/s41570-022-00384-8.

16 Ibragimova, R., Erhart, P., Rinke, P., and Komsa, H.P. (2021). Surface functionalization of 2D MXenes: trends in distribution, composition, and electronic properties. *J. Phys. Chem. Lett.* 12 (9): 2377–2384. https://doi.org/10.1021/acs.jpclett.0c03710.

17 Rasool, K., Helal, M., Ali, A. et al. (2016). Antibacterial activity of $Ti_3C_2T_x$ MXene. *ACS Nano* 10 (3): 3674–3684. https://doi.org/10.1021/acsnano.6b00181.

18 Tahir, K., Miran, W., Jang, J. et al. (2020). A novel MXene-coated biocathode for enhanced microbial electrosynthesis performance. *Chem. Eng. J.* 381: 122687.

19 Lim, G.P., Soon, C.F., Ma, N.L. et al. (2021). Cytotoxicity of MXene-based nanomaterials for biomedical applications: a mini review. *Environ. Res.* 201: 111592.

20 Yu, M. and Feng, X. (2020). Scalable manufacturing of MXene films: moving toward industrialization. *Matter.* 3 (2): 335–336.

21 SiavashIravani, S. and Varma, R. (2022). Bioinspired and biomimetic MXene-based structures with fascinating properties: recent advances. *Mater. Adv.* https://pubs.rsc.org/en/content/articlehtml/2022/ma/d2ma00151a.

22 Naguib, M., Barsoum, M.W., and Gogotsi, Y. (2021). Ten years of progress in the synthesis and development of MXenes. *Adv. Mater.* 33 (39): 2103393. https://doi.org/10.1002/adma.202103393.

23 Alhabeb, M., Maleski, K., Anasori, B. et al. (2017). Guidelines for synthesis and processing of two-dimensional titanium carbide ($Ti_3C_2T_x$ MXene). *Chem. Mater.* 29 (18): 7633–7644. https://doi.org/10.1021/acs.chemmater.7b02847.

24 Anasori, B., Lukatskaya, M.R., and Gogotsi, Y. (2017). 2D metal carbides and nitrides (MXenes) for energy storage. *Nat. Rev. Mater.* 2 (2): 1–17. https://www.nature.com/articles/natrevmats201698.

25 Persson, I., el Ghazaly, A., Tao, Q. et al. (2018). Tailoring structure, composition, and energy storage properties of MXenes from selective etching of in-plane, chemically ordered MAX phases. *Small* 14 (17): 1703676. https://doi.org/10.1002/smll.201703676.

26 Natu, V., Pai, R., Sokol, M. et al. (2020). 2D Ti3C2Tz MXene synthesized by water-free etching of Ti_3AlC_2 in polar organic solvents. *Chem* 6 (3): 616–630. https://doi.org/10.1021/nn204153h.

27 Naguib, M., Mashtalir, O., Carle, J. et al. (2012). Two-dimensional transition metal carbides. *ACS Nano* 6 (2): 1322–1331.

28 Lu, C., Liu, H., Xu, L. et al. (2022). Two-dimensional vanadium carbide for simultaneously tailoring the hydrogen sorption thermodynamics and kinetics of magnesium hydride. *J. Magnesium Alloys* 10 (4): 1051–1065. 10.1149/2.0142003JES.

29 Yu, T. and Breslin, C.B. (2020). Review—two-dimensional titanium carbide MXenes and their emerging applications as electrochemical sensors. *J. Electrochem. Soc.* 167 (3): 037514. https://doi.org/10.1007/s42864-020-00048-4.

30 Zhao, L. and Li, B. (2020). Synthesis and recent applications of MXenes with Mo, V or Nb transition metals: a review. *Tungsten* 2 (2): 176–193.

31 Soundiraraju, B. and George, B.K. (2017). Two-dimensional titanium nitride (Ti_2N) MXene: synthesis, characterization, and potential application as surface-enhanced Raman scattering substrate. *ACS Nano* 11 (9): 8892–8900. https://doi.org/10.1021/acsnano.7b03129.

32 Pogorielov, M., Smyrnova, K., Kyrylenko, S. et al. (2021). Mxenes—a new class of two-dimensional materials: structure, properties and potential applications. *Nanomaterials* 11 (12).

33 Sun, W., Wang, H.W., Vlcek, L. et al. (2020). Multiscale and multimodal characterization of two-dimensional titanium carbonitride MXene. *Adv. Mater. Interfaces* 7: 1902207.

34 Syamsai, R. and Grace, A.N. (2019). Ta4C3 MXene as supercapacitor electrodes. *J. Alloys Compd.* 792: 1230–1238. https://doi.org/10.1021/acs.inorgchem.9b03625.

35 Wang, D., Si, J., Lin, S. et al. (2020). Achieving macroscopic $V_4C_3T_x$ MXene by selectively etching Al from V_4AlC_3 single crystals. *Inorg. Chem.* 59 (5): 3239–3248.

36 Hu, F., Wang, X., Bao, S. et al. (2022). Tailoring electromagnetic responses of delaminated $Mo_2TiC_2T_x$ MXene through the decoration of Ni particles of different morphologies. *Chem. Eng. J.* 440: 135855.

37 Hantanasirisakul, K., Anasori, B., Nemsak, S. et al. (2020). Evidence of a magnetic transition in atomically thin Cr_2 TiC_2T_x MXene. *Nanoscale Horiz.* 5: 1557–1565.

38 Zhang, T., Pan, L., Tang, H. et al. (2017). Synthesis of two-dimensional Ti3C2TxMXene using HCl+LiF etchant: enhanced exfoliation and delamination. *J. Alloys Compd.* 695: 818–826.

39 Xu, C., Wang, L., Liu, Z. et al. (2015). Large-area high-quality 2D ultrathin Mo2C superconducting crystals. *Nat. Mater.* 14 (11): 1135–1141.

40 Xu, C., Chen, L., Liu, Z. et al. (2019). Bottom-up synthesis of 2D transition metal carbides and nitrides. 2D metal carbides and nitrides (MXenes): structure. *Prop. Appl.* 89–109.

41 Xiao, X., Song, H., Lin, S. et al. (2016). Scalable salt-templated synthesis of two-dimensional transition metal oxides. *Nat. Commun.* 7: 11296.

42 Kamysbayev, V., Filatov, A.S., Hu, H. et al. (2020). Covalent surface modifications and superconductivity of two-dimensional metal carbide MXenes. *Science* 369 (6506): 979–983.

43 Brady, A., Wang, H.W., Naguib, M. et al. (2021). Pre-sodiated $Ti_3C_2T_x$ MXene structure and behavior as electrode for sodium-ion capacitors. *ACS Nano* 15 (2): 2994–3003.

44 Khazaei, M., Ranjbar, A., Ghorbani-Asl, M. et al. (2016). Nearly free electron states in MXenes. *Phys. Rev. B.* 93 (20): 205125.

45 Anasori, B., Dahlqvist, M., Halim, J. et al. (2015). Experimental and theoretical characterization of ordered MAX phases Mo_2TiAlC_2 and $Mo_2Ti_2AlC_3$. *J. Appl. Phys.* 118 (9): 094304.

46 Dahlqvist, M., Alling, B., and Rosén, J. (2010 Jun 23). Stability trends of MAX phases from first principles. *Phys. Rev. B: Condens. Matter* 81 (22): 220102.

47 Deshmukh, K., Muzaffar, A., Kovářík, T. et al. (2022). Introduction to 2D MXenes: fundamental aspects, MAX phases and MXene derivatives, current challenges, and future prospects. *Mxenes Compos.* 1–47.

48 Li, N., Peng, J., Ong, W.J. et al. (2021). MXenes: an emerging platform for wearable electronics and looking beyond. *Matter* 4 (2): 377–407.

49 Bhat, A., Anwer, S., Bhat, K.S. et al. (2021). Prospects challenges and stability of 2D MXenes for clean energy conversion and storage applications. *npj 2D Mater. Appl.* 5 (1): 61.

50 Shukla, V. (2020). The tunable electric and magnetic properties of 2D MXenes and their potential applications. *Mater. Adv.* 1 (9): 3104–3121.

51 Mostafavi, E. and Iravani, S. (2022). MXene-graphene composites: a perspective on biomedical potentials. *Nano Micro Lett.* 14 (1): 130. www.advancedscience.com.

52 Hn, A. (2019). MXetronics: electronic and photonic applications of MXenes. *Nano Energy* 60: 179–197.

53 Xu, B., Zhi, C., and Shi, P. (2020). Latest advances in MXene biosensors. *J. Phys. Mater.* 3 (3): 031001. https://doi.org/10.1088/2515-7639/ab8f78.

54 Lin, H., Chen, Y., and Shi, J. (2018, 2018). Insights into 2D MXenes for versatile biomedical applications: current advances and challenges ahead. *Adv. Sci.* 5: 1800518.

55 Kalambate, P.K., Gadhari, N.S., Li, X. et al. (2019). Recent advances in MXene–based electrochemical sensors and biosensors. *TrAC, Trends Anal. Chem.* 120: 115643.

56 Xu, B., Zhi, C., and Shi, P. (2020). Latest advances in MXene biosensors. *J. Phys.: Mater.* 3 (3): 031001.

57 Rakhi, R.B., Nayuk, P., Xia, C., and Alshareef, H.N. (2016). Novel amperometric glucose biosensor based on MXene nanocomposite. *Sci. Rep.* 6, 38465.

58 Liu, H., Duan, C., Yang, C. et al. (2015). A novel nitrite biosensor based on the direct electrochemistry of hemoglobin immobilized on MXene-Ti_3C_2. *Sens. Actuators, B* 218: 60–66.

59 Jastrzębska, A.M., Szuplewska, A., Wojciechowski, T. et al. (2017). In vitro studies on cytotoxicity of delaminated Ti_3C_2 MXene. *J. Hazard. Mater.* 339: 1–8.

60 Das, S., Das, J., Samadder, A. et al. (2012). Apigenin-induced apoptosis in A375 and A549 cells through selective action and dysfunction of mitochondria. *Exp. Biol. Med.* 237 (12): 1433–1448.

61 Rashid, B., Anwar, A., Shahabuddin, S. et al. (2021). A comparative study of cytotoxicity of PPG and PEG surface-modified 2-D Ti_3C_2 MXene flakes on human cancer cells and their Photothermal response. *Materials* 14 (16): 4370.

62 Chen, K., Qiu, N., Deng, Q. et al. (2017). Cytocompatibility of Ti_3AlC_2, Ti_3SiC_2, and Ti_2AlN: in vitro tests and first-principles calculations. *ACS Biomater. Sci. Eng.* 3 (10): 2293–2301.

63 Zhang, Y., Xia, W., Wu, Y., and Zhang, P. (2019). Prediction of MXene based 2D tunable band gap semiconductors: GW quasiparticle calculations. *Nanoscale* 11 (9): 3993–4000.

64 Limbu, Y., Kaphle, G.C., Karn, A.L. et al. (2022). Electronic structure and magnetism of pristine, defected, and strained Ti_2N MXene. *J. Magn. Magn. Mater.* 563: 169895.

65 Wang, Y., Xu, Y., Hu, M. et al. (2020). MXenes: focus on optical and electronic properties and corresponding applications. *Nanophotonics* 9 (7): 1601–1620.

66 Cheng, Y.W., Dai, J.H., Zhang, Y.M., and Song, Y. (2018). Two-dimensional, ordered, double transition metal carbides (MXenes): a new family of promising catalysts for the hydrogen evolution reaction. *J. Phys. Chem. C* 122 (49): 28113–28122.

67 Lee, H.S. (2016). Chapter 11 Density of states, fermi energy and energy bands. In: *Thermoelectrics: Design and Materials: Design and Materials.* https://doi.org/10.1002/9781118848944.ch11.

68 Sun W., Xie Y., Kent PRC. Journal a me Double transition metal MXenes with wide band gaps.

69 Hart, J.L., Hantanasirisakul, K., Lang, A.C. et al. (2019). Control of MXenes' electronic properties through termination and intercalation. *Nat. Commun.* 10 (1): 522.

70 Hantanasirisakul, K. and Gogotsi, Y. (2018). Electronic and optical properties of 2D transition metal carbides and nitrides (MXenes). *Adv. Mater.* 30 (52): e1804779.

71 Khazaei, M., Ranjbar, A., Arai, M. et al. (2017). Electronic properties and applications of MXenes: a theoretical review. *J. Mater. Chem. C* 5 (10): 2488–2503.

72 Balci, E., Akkuş, Ü.Ö., and Berber, S. (2017). Band gap modification in doped MXene: Sc_2CF_2. *J. Mater. Chem. C* 5 (24): 5956–5961.

73 Frisenda, R., Drüppel, M., Schmidt, R. et al. (2017). Biaxial strain tuning of the optical properties of single-layer transition metal dichalcogenides. *npj 2D Mater. Appl.* 1 (1): 1–7.

74 Berdiyorov, G.R. (2016). Optical properties of functionalized $Ti_3C_2T_2$ (T = F, O, OH) MXene: first-principles calculations. *AIP Adv.* 6 (5): 055105.

75 Chen, G.P., Voora, V.K., Agee, M.M. et al. (2017). Random-phase approximation methods. *Annu. Rev. Phys. Chem.* 68: 421–445.

76 Kyaw, H.H., Boonruang, S., Mohammed, W.S., and Dutta, J. (2015). Design of electric-field assisted surface plasmon resonance system for the detection of heavy metal ions in water. *AIP Adv.* 5 (10): 107226.

77 Liu, Z., Nie, Y., Yuan, W. et al. (2015). Optical cavity-assisted broadband optical transparency of a plasmonic metal film. *Nanotechnology* 26 (18): 1–9.

78 Halim, J., Persson, I., Moon, E.J. et al. (2019). Electronic and optical characterization of 2D Ti_2C and Nb2C (MXene) thin films. *J. Phys.: Condens. Matter* 31 (16): 165301.

79 Jiang, X., Liu, S., Liang, W. et al. (2018). Broadband nonlinear photonics in few-layer MXene $Ti_3C_2T_x$ (T = F, O, or OH). *Laser Photonics Rev.* 12 (2): 1700229.

80 Huang, H., Wang, W., and Zhang, S. (2021). Theoretical assessment of Raman spectra on MXene Ti_2C: from monolayer to bilayer. *Phys. Chem. Chem. Phys.* 23 (35): 19884–19891.

81 Lu, S., Sui, L., Liu, Y. et al. (2019). White photoluminescent Ti_3C_2 MXene quantum dots with two-photon fluorescence. *Adv. Sci.* 6 (9): 1801470.

82 Jiang, X., Kuklin, A.V., Baev, A. et al. (2020). Two-dimensional MXenes: From morphological to optical, electric, and magnetic properties and applications. *Phys. Rep.* 848: 1–58.

83 Li, J., Zhang, Z., Du, L. et al. (2019). Highly stable femtosecond pulse generation from a MXene$Ti_3C_2T_x$ (T = F, O, or OH) mode-locked fiber laser. *Photonics Res.* 7 (3): 260.

84 Yue, Y., Wang, B., Miao, N. et al. (2021). Tuning the magnetic properties of Zr2N MXene by biaxial strain. *Ceram. Int.* 47 (2): 2367–2373.

85 Yue, Y. (2017). Fe_2C monolayer: an intrinsic ferromagnetic MXene. *J. Magn. Magn. Mater.* 434: 164–168.

86 Coey, J.M.D. and Sanvito, S. (2004). Magnetic semiconductors and half-metals. *J. Phys. D: Appl. Phys.* 37 (7): 988–993.

87 He, J., Lyu, P., and Nachtigall, P. (2016). New two-dimensional Mn-based MXenes with room-temperature ferromagnetism and half-metallicity. *J. Mater. Chem. C* 4 (47): 11143–11149.

88 Wang, G. and Liao, Y. (2017). Theoretical prediction of robust and intrinsic half-metallicity in Ni_2N MXene with different types of surface terminations. *Appl. Surf. Sci.* 426: 804–811.

89 Mozafari, M. and Soroush, M. (2021). Surface functionalization of MXenes. *Mater. Adv.* 2 (22): 7277–7307.

90 Si, C., Zhou, J., and Sun, Z. (2015). Half-metallic ferromagnetism and surface functionalization-induced metal-insulator transition in graphene-like two-dimensional Cr_2C crystals. *ACS Appl. Mater. Interfaces* 7 (31): 17510–17515.

91 Hu, L., Wu, X., and Yang, J. (2016). Mn_2C monolayer: a 2D antiferromagnetic metal with high Néel temperature and large spin–orbit coupling. *Nanoscale* 8 (26): 12939–12945.

92 Li, X., Wu, X., Li, Z. et al. (2012). Bipolar magnetic semiconductors: a new class of spintronics materials. *Nanoscale* 4 (18): 5680–5685.

93 Lambrecht, W.R.L., Miao, M.S., and Lukashev, P. (2005). Magnetic properties of transition-metal nitrides. *J. Appl. Phys.* 97 (10): 10D306.

94 Frey, N.C., Kumar, H., Anasori, B. et al. (2018). Tuning noncollinear spin structure and anisotropy in ferromagnetic nitride MXenes. *ACS Nano* 12 (6): 6319–6325.

95 Jiang, Z., Wang, P., Xing, J. et al. (2018). Screening and design of Novel 2D ferromagnetic materials with high curie temperature above room temperature. *ACS Appl. Mater. Interfaces* 10 (45): 39032–39039.

96 Jiang, X., Kuklin, A.V., Baev, A. et al. (2020). Two-dimensional MXenes: from morphological to optical, electric, and magnetic properties and applications. *Phys. Rep.* 848: 1–58.

97 Adler, D. and Brooks, H. (1967). Theory of semiconductor-to-metal transitions. *Phys. Rev.* 155 (3): 826–840.

98 Li, S.S., Hu, S.J., Ji, W.X. et al. (2017). Emergence of ferrimagnetic half-metallicity in two-dimensional MXene $Mo_3N_2F_2$. *Appl. Phys. Lett.* 111 (20): 20240.

99 Gross, I., Akhtar, W., Garcia, V. et al. (2017). Real-space imaging of non-collinear antiferromagnetic order with a single-spin magnetometer. *Nature* 549 (7671): 252–256.

100 Khazaei, M., Mishra, A., Venkataramanan, N.S. et al. (2019). Recent advances in MXenes: from fundamentals to applications. *Curr. Opin. Solid State Mater. Sci.* 23 (3): 164–178.

101 Khazaei, M., Ranjbar, A., Arai, M., and Yunoki, S. (2016). Topological insulators in the ordered double transition metals $M_2{}'$ M'' C_2 MXenes (M′= Mo, W; M″=Ti, Zr, Hf). *Phys. Rev. B* 94 (12): 125152.

102 Liang, Y., Khazaei, M., Ranjbar, A. et al. (2017). Theoretical prediction of two-dimensional functionalized MXene nitrides as topological insulators. *Phys. Rev. B* 96 (19): 195414.

103 Balcı, E., ÖzdenAkkuş, Ü., and Berber, S. (2018). Controlling topological electronic structure of multifunctional MXene layer. *Appl. Phys. Lett.* 113 (8): 083107.

104 Wu, H., Gu, J., Li, Z. et al. (2022). Characterization of phonon thermal transport of $Ti_3C_2T_x$ MXene thin film. *J. Phys.: Condens. Matter* 34 (15): 155704.

105 Yorulmaz, U., Özden, A., Perkgöz, N.K. et al. (2016). Vibrational and mechanical properties of single layer MXene structures: a first-principles investigation. *Nanotechnology* 27 (33): 335702.

106 Champagne, A., Shi, L., Ouisse, T. et al. (2018). Electronic and vibrational properties of V_2C -based MXenes: from experiments to first-principles modeling. *Phys. Rev. B* 97 (11): 1600180.

107 Zhao, Z.Y. and Liu, Q.L. (2018). Study of the layer-dependent properties of MoS_2 nanosheets with different crystal structures by DFT calculations. *Catal. Sci. Technol.* 8 (7): 1867–1879.

108 Kajiyama, S., Szabova, L., Sodeyama, K. et al. (2016). Sodium-ion intercalation mechanism in MXeneNanosheets. *ACS Nano* 10 (3): 3334–3341.

109 Placke, T., Gonzalez-Julian, J., Bärmann, P. et al. (2021). Solvent co-intercalation into few-layered $Ti_3C_2T_x$ mxenes in lithium ion batteries induced by acidic or basic post-treatment. *ACS Nano* 15 (2): 3295–3308.

110 Ming, F., Liang, H., Zhang, W. et al. (2019). Porous MXenes enable high performance potassium ion capacitors. *Nano Energy* 62: 853–860.

111 Fu, Q., Wen, J., Zhang, N. et al. (2017). Free-standing $Ti_3C_2T_x$ electrode with ultrahigh volumetric capacitance. *RSC Adv.* 7 (20): 11998–12005.

112 Zang, X., Wang, J., Qin, Y. et al. (2020). Enhancing capacitance performance of $Ti_3C_2T_x$ MXene as electrode materials of supercapacitor: from controlled preparation to composite structure construction. *Nano Micro Lett.* 12 (1): 1–24.

113 Lv, G., Wang, J., Shi, Z., and Fan, L. (2018). Intercalation and delamination of two-dimensional MXene ($Ti_3C_2T_x$) and application in sodium-ion batteries. *Mater. Lett.* 219: 45–50.

114 Das, P. and Wu, Z.S. (2020). MXene for energy storage: present status and future perspectives. *J. Phys.: Energy* 2 (3): 032004.

115 Meng, Q., Jin, Q., Wang, X. et al. (2020). 3D $Ti_3C_2T_x$ aerogel-modified separators for high-performance li–S batteries. *J. Alloys Compd.* 816.

116 Xiong, D., Li, X., Bai, Z., and Lu, S. (2018). Recent advances in layered Ti3C_2*Tx*MXene for electrochemical energy storage. *Small* 14 (17): 1703419.

117 Hu, M., Li, Z., Hu, T. et al. (2016). High-capacitance mechanism for Ti_3C_2Tx MXene by in situ electrochemical Raman spectroscopy investigation. *ACS Nano* 10 (12): 11344–11350.

118 Li, L., Zhang, D., Deng, J. et al. (2021). Application of MXene-based materials in hybrid capacitors. *Sustainable Energy Fuels* 5 (13): 3278–3291.

119 Jyoti, J., Singh, B.P., Sandhu, M., and Tripathi, S.K. (2022). New insights on MXene and its advanced hybrid materials for lithium-ion batteries. *Sustainable Energy Fuels* 6 (4): 971–1013.

120 Jamil, F., Ali, H.M., and Janjua, M.M. (2021). MXene based advanced materials for thermal energy storage: a recent review. *J. Energy Storage* 35: 102322.

121 Wang, M., Liu, Y., Zhang, H. et al. (2022). Thermal conductivities of $Ti_3C_2T_x$ MXenes and their interfacial thermal performance in MXene/epoxy composites – a molecular dynamics simulation. *Int. J. Heat Mass Transfer* 194: 123027.

122 Zha, X.H., Huang, Q., He, J. et al. (2016). The thermal and electrical properties of the promising semiconductor MXene Hf_2CO_2. *Sci. Rep.* 6.

123 Chen, L., Shi, X., Yu, N. et al. (2018). Measurement and analysis of thermal conductivity of Ti_3C_2Tx MXene films. *Materials* 11 (9): 1701.
124 Kang, R., Zhang, Z., Guo, L. et al. (2019). Enhanced thermal conductivity of epoxy composites filled with 2D transition metal carbides (MXenes) with ultralow loading. *Sci. Rep.* 9 (1).
125 Liu, R. and Li, W. (2018). High-thermal-stability and high-thermal-conductivity $Ti_3C_2T_x$ MXene/poly(vinyl alcohol) (PVA) composites. *ACS Omega* 3 (3): 2609–2617.

5

Processing and Characterization of MXenes and Their Nanocomposites

Suji Mary Zachariah[1,2], *Yves Grohens*[2], *and Sabu Thomas*[1,3,4]

[1] *International and Inter University Centre for Nanoscience and Nanotechnology (IIUCNN), Mahatma Gandhi University, Kottayam, Kerala, India*
[2] *University Bretagne Compositic, Sud, Lorient, France*
[3] *School of Energy Materials, Mahatma Gandhi University, Kottayam, Kerala, India*
[4] *Department of Chemical Sciences, University of Johannesburg, Doornfontein, Johannesburg, South Africa*

5.1 Introduction

Two research teams led by Yury Gogotsi and Michel Barsoum in 2011 made the discovery of MXenes, which are nothing but a class of two-dimensional (2D) transition metal carbides, carbonitrides, and nitrides. Due to their exceptional qualities, including metallic conductivity, abundant surface chemistry, configurable terminations, excellent processability, and others, they have made a significant advancement in a variety of applications. Integrating them into polymers, ceramics, and metals has expanded their applications [1]. These 2D layered materials are known as "MXenes" because they were created by etching A layer from the MAX (M stands for any transition metal, A stands for element from A-group of perodic elements, X for carbon or nitrogen) phase. The suffix "ene" is added to emphasize their resemblance to graphene. MAX phases are a large family of around 60+ members with the composition $M^{n+1}AX^{n}$, where M stands for an early transition metal such as Ti, V, Cr, and Nb, A stands for a group A element such as Al, Si, Sn, and In, X stands for carbon and/or nitrogen, and $n = 1$, 2, or 3. Multiple etching procedures have been proposed to synthesize MXenes, and one of these simple etching processes involves basically immersing a MAX phase in hydrofluoric acid (HF) at ambient temperature and selectively etching the A elemental layer. The MAX phase powder is mixed in aqueous HF at a particular concentration for a fixed period of time, subsequently subjected to centrifuging and/or filtration of the mixture to separate the solid from the supernatant, and then washed with deionized water (DI) until the pH of the suspension reaches 4 to 6 values [2, 3]. As a result, the MAX particles have a structure similar to exfoliated graphite. When these structures are loosely packed, they are termed as multilayer, or ML–MXenes, and when the number of stacked layers is less than 5, they are referred to as few-layer MXenes (FL–MXene). MXenes find applications in diversified areas due to their inherent properties. These areas include energy storage [4], Hg adsorption [5], photocatalysis [6], supercapacitors [7], epidermal sensors [8, 9], CO_2 reduction [10–12], and electromagnetic interference [13–15].

MXenes: Next-Generation 2D Materials: Fundamentals and Applications, First Edition.
Edited by Jay Singh, Kshitij RB Singh, Ravindra Pratap Singh, and Charles Oluwaseun Adetunji.

5.2 Processing Techniques

5.2.1 Solution Blending

This technique is an easy processing technique (general scheme shown in Figure 5.1), and hence it is the prior option to synthesize MXene-based polymer nanocomposites. This technique utilizes the superior hydrophilicity of MXene, improves dispersion between MXene and polymer, and flexibly modulates the ratio of the components in the composites. Generally, this technique uses hydrolyzable polymers such as Polyvinyl alcohol (PVA), Polylactic acid (PLA), cellulose, and chitin, along with polar dispersants such as water, Dimethyl sulphoxide (DMSO), and Dimethyl formamide (DMF). The dispersant selected should solubilize the polymer and disperse the MXene effectively. Otherwise, surface modification of MXene can also be done for efficient dispersion. Zi et al. [16] in their study applied cationic modifiers, including decyltrimethylammonium bromide (DTAB), octadecyl trimethylammonium bromide (OTAB), and di-dodecyl dimethylammonium bromide (DDAB), for altering MXene nanosheets in polystyrene (PS) matrix for high thermal stability and flame-retardant properties. Based on the findings, the modification boosted the distance between the MXene nanosheets' interlayers, enhanced their dispersion, and raised their interface interactions within the PS matrix, which improved their fireproof properties and lowered the emission of toxic fumes. However, Wu et al. [17] functionalized MXene using cetyltrimethylammonium bromide (CTAB) for better dispersion in poly(vinylidene fluoride) (PVDF) matrix. Similarly, Aakyiir et al. [18] demonstrated modification of MXene using allylamine, which endowed the nanocomposite with excellent thermal, mechanical, and electrical properties.

However, some issues, such as waste production, the challenge of evaporating the solvent, environmental contamination, uneven film thickness, wrinkling during casting, and low mechanical properties resulting from the solution blending method, prevent their practical implementation [19].

5.2.2 In Situ Polymerization Technique

This technique is a wet method and a simple technique for the synthesis of polymers. In this process (refer to Figure 5.2), monomers, initiators, or curing agents are thoroughly incorporated with the MXene nanosheets by properly maintaining the polymerization conditions.

Figure 5.1 General scheme for solution blending process. Source: Figure self-drawn.

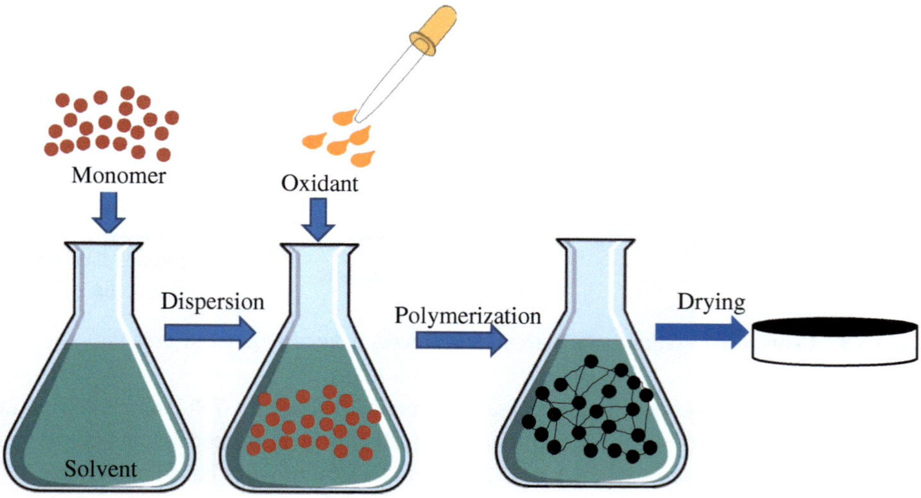

Figure 5.2 General scheme for in situ polymerization process. Source: Figure self-drawn.

This process offers the advantage of effective dispersion of MXene in the polymer matrix. Several works based on in situ polymerization of conducting polymers such as polyaniline (PANI) [20, 21], polypyrrole (PPy) [22, 23], and Polythiophene (PTh) [24, 25] on MXene have been reported. McDaniel et al. [26] covalently modified $Ti_3C_2T_z$ MXene using isocyanate for their effective dispersion in the thiourethane matrix. This modification offered a great advantage in terms of the thiol–isocyanate reaction that favored the in situ polymerization fully under ambient conditions. Wang et al. [27] reported in their work in situ polymerization of aniline on the surface of MXene. Studies showed that high reactivity and electronegative sites favored polymerization and significantly improved their performance in ammonia sensing and supercapacitor applications. Lastly, the majority of the findings point to the possibility that in situ polymerization could improve MXene dispersion in a polymer matrix and improve the thermal, electrical, and mechanical capabilities of the polymers.

5.2.3 Melt Blending

Melt blending is the most preferred mixing technique (Figure 5.3) for thermoplastics using extruder or any internal mixers. Initially, the polymer is fed to the mixer for viscosity reduction after which MXene or other fillers are fed for mixing. In other words, MXene is incorporated into the polymer matrix after the polymer has reached its melting point. If refined or optimized, this method will be very useful for the large-scale production of MXene-based nanocomposites.

Melt processing has been reported in the fabrication of MXene-based nanocomposites with EPDM, [28] ultrahigh-molecular-weight polyethylene, [29] nylon-6, [30], etc. Some of these works report modification of MXene for better dispersion in a polymer host. Sheng et al. [31] reported homogeneous dispersion of Ti_3C_2 MXene nanosheets in a thermoplastic polyurethane matrix. MXene was initially pretreated with polyethylene glycol and then mixed with Thermoplasticpolyurethane (TPU) via melt blending, which effectively

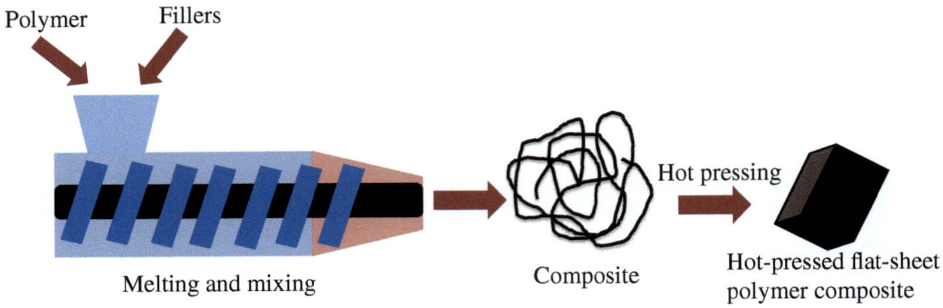

Figure 5.3 General representation of melt mixing process. Source: Figure self-drawn.

exfoliated MXene nanosheets and dispersed them effectively in the TPU matrix. This strategy also proved to be promising for manufacturing effectively reinforced MXene-based composites at an industrial scale. Gao et al. [32] reported a similar work based on MXene/TPU via melt blending, where they reported better mechanical, thermal, and rheological properties. The main disadvantage of this process is that it is limited only to thermoplastic polymers, and the use of very high temperatures can lead to oxidation or degradation of MXene.

5.2.4 Electrospinning

It is an easy, flexible, and cost-effective technique that can create continuous microfilaments in the nano- or micrometer range. Figure 5.4 depicts a fundamental setup. It has been suggested by Mayerberger et al. [33] that commonly utilized polymers such as poly(acrylic acid), poly(vinyl alcohol), and alginate can be used to electrospin suspensions of single-layer $Ti_3C_2T_x$. Studies show that MXene flakes were effectively dispersed during

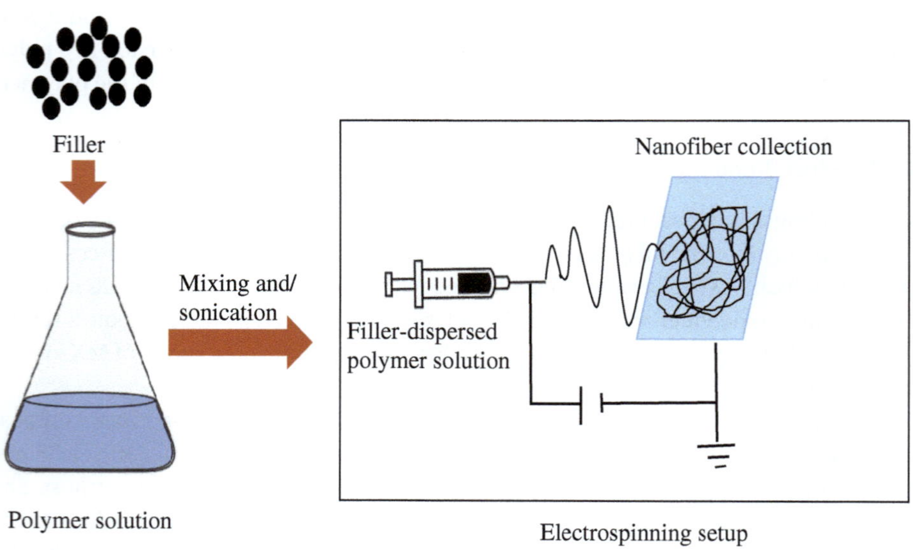

Figure 5.4 General scheme for electrospinning setup. Source: Figure self-drawn.

the electrospinning process even at 1% (w/w), affecting the solution properties of the polymer and having a larger impact on the final fiber diameter.

Levitt et al. [34] in their work used the electrospinning technique to incorporate $Ti_3C_2T_x$ within carbon nanofibers to be used as electrodes in supercapacitors. The electrospinning process initially involves making a homogeneous solution of MXene/PAN in DMF by continuous stirring. This is later fed to a syringe and a particular voltage is applied due to which fiber gets deposited onto the opposite surface resulting in the formation of a nonwoven mat of MXene/PAN composite fibers. Results suggest that this technique exhibits a higher areal capacitance (of up to 244 mF cm^{-2}) in comparison to pure carbon nanofibers and is a promising method for incorporating MXene flakes into nanoscale fibers for wearable energy storage devices.

They have a significant advantage in producing fibers with different architectures such as core–shell, bi-lobal, and composites. In this method, fiber morphology, diameter, and stacked density can be controlled by adjusting process parameters and environmental conditions [35]. However, the most difficult challenge of spinning composite solutions, especially those containing 2D materials, is agglomeration or restacking of sheets, which results in bead formation and inconsistencies across the electrospun fiber mat, which is critical for large-scale device fabrication, such as supercapacitors.

5.2.5 Vacuum-Assisted Filtration (VAF) Method

The vacuum-assisted filtration (VAF) method (Figure 5.5) is widely used to separate 2D material flakes from fluids to make various 2D material-based macroscopic structures usually in the form of papers using a very simple setup. For instance, freestanding MXene film can be fabricated using this method, which under the effect of suction using a vacuum pump delaminates MXene flakes present in MXene colloidal dispersion. This results in highly conductive, stretched MXene flakes being tightly deposited onto the filter surface in a face to face orientation.

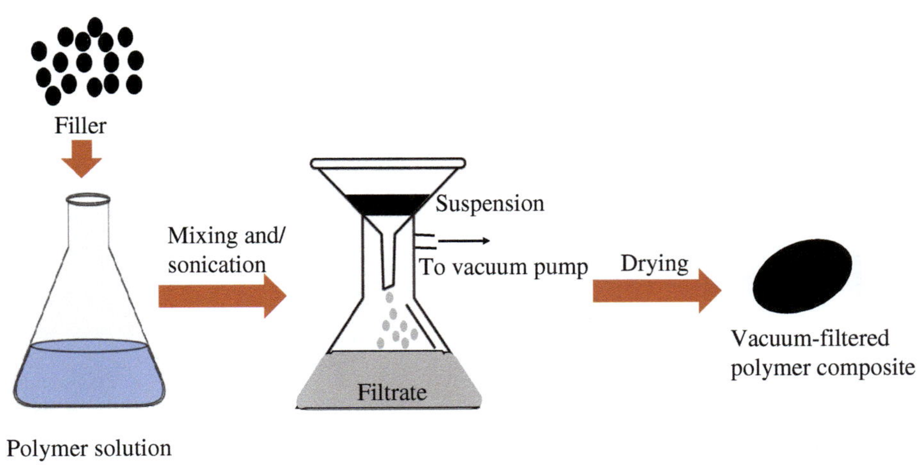

Figure 5.5 General scheme for vacuum filtration setup. Source: Figure self-drawn.

Liu et al. [36] fabricated MXene/chitosan films for electromagnetic interference shielding applications using the VAF technique. The results suggest that this technique endowed the film with a well-aligned MXene structure, resulting in better conductivity and shielding ability. Briefly, a particular amount of MXene solution was dispersed into a chitosan solution by stirring. This mixture was then filtered using a sand core funnel connected with the vacuum pump and dried overnight to obtain the composite films. Zhu et al. [37] developed an H_2 sensor using a composite of $Ti_3C_2T_x$ and palladium colloidal nanoclusters (Pd CNC) using the VAF technique. The film exhibited excellent flexibility, compactness, and low density and detected a wide range of H_2 without any additional processing.

However, this technique has certain limitations. It requires a large amount of time and energy, particularly for fabricating thick MXene films. This is due to the fact that once the filtration starts, MXene rapidly covers the surface pores, forming a very tight film, and the remaining MXene stacks over the previously formed film. Also, the high aspect ratio and barrier effect of 2D materials restrict the passage of water, resulting in very slow filtration. This eventually causes more contact time for MXene with oxygen, resulting in oxidation. However, applying high pressure to films made from water-soluble polymers results in a loss of their original weight, making it challenging to understand the connection between performance and sample composition.

5.2.6 Spin Coating

Spin-coating is a popular coating method (Figure 5.6) that is quick and effective for creating smooth, homogenous MXene-based films, typically with thicknesses between microns and nanometers. First, a homogenous solution of MXene and pertinent additions is produced by

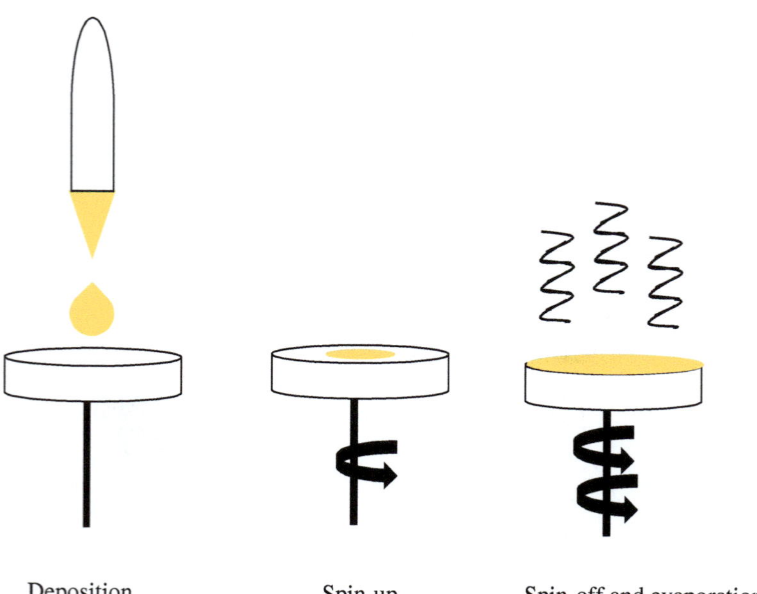

Figure 5.6 General scheme for spin-coating technique. Source: Figure self-drawn.

vigorous stirring or ultrasound treatment. Later, this is spread out using centrifugal force on a surface to allow the solvent to evaporate at high temperatures or in a vacuum. The result will be a free-standing MXene composite film. The final film thickness is determined by the angular speed, solution viscosity, and spinning time.

Zhang et al. [38] adopted the spray coating technique for synthesizing highly uniform Ti_3C_2 film for application in memristive devices. Tuning the spinning speeds revealed that higher Ti_3C_2 concentrations were noted at lower rotating speeds. The uniformity was later confirmed using various microscopic and diffraction techniques. Wu et al. [39] developed highly oriented MXene-derived metal–organic frameworks using the spin coating technique. XRD results supported the results from simulations, indicating highly oriented stacking of the nanosheets. Liu et al. [40] adopted a layer-by-layer spin coating technique to fabricate electromagnetic interference shielding material. The fabricated MXene/PDMS/BN (MXene/polydimethylsiloxane/boron nitride) film with its ordered alternate structure allowed multiple reflections and absorption of electromagnetic waves between the layers, thereby achieving a high shielding value of 32 dB.

Even though this technique the advantage of producing very fine, thin, and uniform coating, it also has certain disadvantages. This technique is generally used for depositing films on small areas on a lab scale and hence is only suitable for optimizing ink formulations. Also, the efficacy of this process is reported to be very low because only 2 to 5% of the material can be deposited, while the rest is usually flung off during the process.

5.3 Characterization Techniques

5.3.1 X-Ray Diffraction (XRD)

The simplest way to confirm the synthesis of MXene is to use X-ray diffraction (XRD). This has been depicted in each work, showing patterns for both MAX and MXene, but in many cases, there is no need for further analysis or consideration other than stating that MAX has been converted to MXene. While XRDs are useful, they can provide additional information leading to a deeper study of MXenes that would otherwise not be available. They additionally offer details on the existence of incorporated species, but they cannot provide confirmation of their bonding.

For instance, Aakyir et al. [41] in their work on Acrylonitrile butadiene rubber (NBR)/MXene composite showed intercalated lithium ions and elastomer molecules in MXene sheets. The diffraction peak at 7.5° disappeared at 2 and 7.5 vol% (low MXene content), implying MXene exfoliation. At 19.6 vol%, meanwhile, the signal returned at 6.1°, indicating an increased interlayer space, which suggests elastomer molecules intercalated into MXene sheets. Due to the intercalation in MXene sheets, the widened peak at 6.1° is noted and suggests that MXene maintains its ordered layered structure in the nanocomposite. The c-lattice parameter's effect on relative humidity was further demonstrated using XRD. According to Ghidiu et al. [42], cation exchange of Li^+ with Na^+, K^+, Rb^+, Mg^{2+}, and Ca^{2+} caused structural changes that were quite different in response to RH, as indicated by XRD, involving a noncontinuous structural widening in the direction of normal to the basal plane as H_2O molecules intercalated.

5.3.2 Scanning Electron Microscopy (SEM) and Transmission Electron Microscopy

SEM makes it easier to represent the MXene structure. However, not all etched samples are exactly alike. The first MXenes were created by immersing them in concentrated HF, but the "accordion" architecture is thought to be a sign of a successful synthesis. This structure, on the other hand, is essentially equivalent to MXene and does not adequately characterize all multilayered MXenes. SEM is used with energy-dispersive X-ray spectroscopy (EDS) in several investigations to acquire qualitative data concerning M elements in MXenes. However, if carbonitride is analyzed, this approach may not be employed to evaluate surface terminations and X ratios [43].

TEM is a very important technique used to analyze the nature of materials, and usually not used to analyze the MAX phase or its precursors [44]. In some works, TEM is coupled with EDS to get insights into material composition. For studying new phases, using XRD and TEM in parallel gives excellent microscopic and macroscopic insights about structure than either technique could give individually [45].

5.3.3 X-Ray Absorption Spectroscopy (XAS)

X-ray absorption spectroscopy (XAS) is an additional effective method for analyzing MXene structure. Most of the time, synchrotron radiation is used to create an intense adequate X-ray source. This approach is divided into two categories: X-ray absorption near-edge structure (XANES), which detects an atom's oxidation status, and the extended X-ray absorption fine structure (EXAFS), which provides information on bond length and coordination number.

Lukatskaya et al. [46] coupled in situ XAS with an electrochemical study, which revealed the capacitive mechanism of MXenes. They determined a quantity equivalent to that predicted from the gravimetric capacitance of the sample assuming redox-only capacitance. Also, this study provided important insight into electrolyte decomposition processes and concluded that the electrochemical behavior of $Ti_3C_2T_x$ in sulfuric acid is predominantly pseudocapacitive. Similarly, Temimy et al. [47] studied the enhancement of Ti_3C_2 pseudocapacitance after urea intercalation using the XAS technique. By using XAS, it was possible to examine the oxidation state and oxygen bonding of individual $Ti_3C_2T_x$ flakes prior to and subsequent to urea intercalation. After intercalation, it was shown that the entire flake had a higher Ti oxidation state than pure $Ti_3C_2T_x$. In situ XAS of intercalated $Ti_3C_2T_x$ aqueous dispersions also showed a stronger Ti oxidation, similar to dry samples; however, for virgin $Ti_3C_2T_x$, the Ti atoms are dramatically decreased in water as compared to dry samples. Djire et al. [48] used the XAS technique to elucidate the hydrogen evolution reaction (HER) in $Ti_4N_3T_x$ MXenes. XANES revealed the presence of multiple oxidation states in the MXenes during HER, while EXAFS revealed that oxygen vacancies in the basal planes are the reason for HER catalysis.

5.3.4 X-Ray Photoelectron Spectroscopy (XPS)

To determine the average material composition, XPS is more typically utilized. This technique has acquired appeal for surface examination because of its low penetration depth, thereby surface sensitivity, and ability to obtain information about chemical composition and elemental oxidation states. Even though this technique is powerful, it has certain limitations that are not particular to MXenes. Inability to analyze the charging state of the specimen causes peak offsets which is a major drawback. However, this is often neglected in conducting samples as this is crucial for insulating samples only. Due to many reasons, this technique is not trivial for MXenes. XPS analysis of transition metals and their compounds is very challenging because of their numerous oxidation states, complex peak splittings, and asymmetric peaks.

Halim et al. [49] investigated MXene surface chemistry and quantified surface groups. They also reported the average sum of the terminations' negative charges for the MXenes. This information is extremely useful when designing/fabricating 2D materials for a variety of applications.

5.3.5 Atomic Force Microscopy (AFM)

Atomic force microscopy (AFM) is increasingly beneficial for assessing 2D materials since it provides evidence of both lateral flake size and thickness. Despite this, there are constraints in measuring the thickness of 2D monolayers, which can vary considerably with respect to the existence of surface adsorbates and entrapped interfacial particles.

Mariano et al. [50] in their work on titanium carbide MXene films probed AFM for analyzing the thickness. They reported the thickness of the exfoliated single layer to be 1.5 ± 0.2 nm. Guo et al. [51] in their study on nanomechanical properties of Ti_3C_2 MXenes revealed through AFM analysis that at higher pressures, MXene showed better adhesion and friction. At the same time at higher temperatures, it showed reduced adhesion and friction. They used XPS to explain this phenomenon of temperature effect on chemical bonds and chemical structure.

5.3.6 Nuclear Magnetic Resonance

The quantitative evaluation of -F terminations in fluoride byproducts and the accurate calculation of the ratio of oxygen groups (=O) to hydroxyl groups (–OH) are two of the most challenging parts of MXene chemistry. According to a number of nuclear magnetic resonance (NMR) investigations, the $Ti_3C_2T_x$ surface contains far fewer –OH terminations than –F or –O terminations, and surface terminations are highly dependent on the synthesis method used.

Hope et al. [52] used NMR to investigate the surface modification of MXenes in their work. Because of the high sensitivity of ^{19}F and ^{1}H nuclei, it is also useful for analyzing fluorine and hydrogen-containing materials. ^{13}C NMR spectra showed a nonquantitative conversion of Ti_3AlC_2 to $Ti_3C_2T_x$. 2D ^{1}H–^{13}C heteronuclear correlation (HETCOR) spectrum recorded the connectivity of the –OH groups to the Ti_3C_2 sheets while the ^{1}H–^{19}F HETCOR spectra explored the arrangement of the different functional groups in space. Kobayashi et al. [53] used the SSNMR technique along with DFT calculations to identify and assign multiple types of –OH groups, residing on the external and internal surfaces of MXene. Harris et al. [54] directly measured the surface terminal groups and connectivity in MXene V_2CT_x. By measuring interactions with the MXene surface, 1H NMR spectra provide direct proof of OH moieties in the surface layer and reveal a significant amount of water hydrogen-bonded to the surface hydroxide layer. Fluoride moieties were found bonded to the MXene surface, with spectra caused by strong interactions with the metallic/semiconducting MXene.

5.3.7 Raman Spectroscopy

Vibrational spectroscopic techniques are commonly used to predict the structure of 2D materials because they are used to assess the fingerprint region of molecules. In the MXene field, Raman spectroscopy is used more than IR spectroscopy due to the water molecules present between the layers which could affect the IR spectrum. It is capable of identifying amorphous compounds as well as examining transition metal oxides, which often occur in MXenes, and it can detect phonons (lattice vibrations), which may yield details regarding bonding in MXenes. Sarycheva et al. [55] investigated the surface chemistry, stacking, and quality of $Ti_3C_2T_x$ using Raman spectroscopy. They discovered that the Raman features altered according to the synthesis route, flake size, and environment surrounding the MXene flakes. Confocal Raman measurements were used to investigate the composition of Ti_2CT_x [56] and the phase stability of $Ti_3C_2T_x$ at various annealing conditions [57].

5.4 Conclusion

In this chapter, we concentrated on the synthesis and characterization of MXene-based nanocomposites, as well as their applications. EMI shielding, liquid and gas separation, batteries, supercapacitors, and other practical applications have been described. Nonetheless, MXene nanocomposite development is still in its early stages, and there are many pressing issues that must be addressed before these materials can be used in practice. Researchers have made several efforts to develop new fabrication strategies due to their distinctive properties and deep applications. MXene synthesis is still difficult due to its instability after exfoliation. Greener methods for producing MXenes with high yield and quality must also be considered. When it comes to the synthesis of MXene nanocomposites, their dispersion in the host matrix has proven to be a major issue because they play a critical part in tuning the internal structure and final properties. Many studies have shown that effective surface modification can increase the number of active sites and, as a result, improve dispersibility. Better MXene dispersion methodologies must be developed in order for MXene nanocomposites to be widely used. Another issue that has been raised is the tendency of MXene sheets to stack. This self-stacking mechanism is critical in electrochemical applications where ion transmission decreases due to self-stacking. Therefore, an in-depth study is essential so that this wonder material meets all the requirements for industrial application including its cost-effectiveness and greener synthesis. Despite the few limitations discussed previously, MXene nanocomposites have proven to be valuable materials for applications ranging from energy storage to biomedical.

References

1 Gogotsi, Y. and Anasori, B. (2019). The rise of MXenes. *ACS Nano* 13 (8): 8491–8494.

2 Naguib, M., Mochalin, V.N., Barsoum, M.W., and Gogotsi, Y. (2014). 25th anniversary article: MXenes: a new family of two-dimensional materials. *Adv. Mater.* 26 (7): 992–1005.

3 Naguib, M., Mashtalir, O., Carle, J. et al. (2012). Two-dimensional transition metal carbides. *ACS Nano* 6 (2): 1322–1331.

4 Anasori, B., Lukatskaya, M.R., and Gogotsi, Y. (2017). 2D metal carbides and nitrides (MXenes) for energy storage. *Nat. Rev. Mater.* 2 (2): 1–17.

5 Shahzad, A., Jang, J., Lim, S.R., and Lee, D.S. (2020). Unique selectivity and rapid uptake of molybdenum-disulfide-functionalized MXene nanocomposite for mercury adsorption. *Environ. Res.* 182: 109005.

6 Sun, Y., Meng, X., Dall'Agnese, Y. et al. (2019). 2D MXenes as co-catalysts in photocatalysis: synthetic methods. *Nano Micro Lett.* 11 (1): 1–22.

7 Hu, M., Zhang, H., Hu, T. et al. (2020). Emerging 2D MXenes for supercapacitors: status, challenges and prospects. *Chem. Soc. Rev.* 49 (18): 6666–6693.

8 Li, X., He, L., Li, Y. et al. (2021). Healable, degradable, and conductive MXene nanocomposite hydrogel for multifunctional epidermal sensors. *ACS Nano* 15 (4): 7765–7773.

9 Wu, X., Liao, H., Ma, D. et al. (2020). A wearable, self-adhesive, long-lastingly moist and healable epidermal sensor assembled from conductive MXene nanocomposites. *J. Mater. Chem. C* 8 (5): 1788–1795.

10 Handoko, A.D., Chen, H., Lum, Y. et al. (2020). Two-dimensional titanium and molybdenum carbide MXenes as electrocatalysts for CO_2 reduction. *Iscience* 23 (6): 101181.

11 Li, N., Chen, X., Ong, W.J. et al. (2017). Understanding of electrochemical mechanisms for CO_2 capture and conversion into hydrocarbon fuels in transition-metal carbides (MXenes). *ACS Nano* 11 (11): 10825–10833.

12 Guo, Z., Li, Y., Sa, B. et al. (2020). M_2C-type MXenes: promising catalysts for CO_2 capture and reduction. *Appl. Surf. Sci.* 521: 146436.

13 Shahzad, F., Alhabeb, M., Hatter, C.B. et al. (2016). Electromagnetic interference shielding with 2D transition metal carbides (MXenes). *Science* 353 (6304): 1137–1140.

14 Cao, M.S., Cai, Y.Z., He, P. et al. (2019). 2D MXenes: electromagnetic property for microwave absorption and electromagnetic interference shielding. *Chem. Eng. J.* 359: 1265–1302.

15 Han, M., Shuck, C.E., Rakhmanov, R. et al. (2020). Beyond $Ti_3C_2T_x$: MXenes for electromagnetic interference shielding. *ACS Nano* 14 (4): 5008–5016.

16 Si, J.Y., Tawiah, B., Sun, W.L. et al. (2019). Functionalization of MXene nanosheets for polystyrene towards high thermal stability and flame retardant properties. *Polymers* 11 (6): 976.

17 Wu, W., Zhao, W., Sun, Q. et al. (2021). Surface treatment of two dimensional MXene for poly(vinylidene fluoride) nanocomposites with tunable dielectric permittivity. *Compos. Commun.* 23: 100562.

18 Aakyiir, M., Yu, H., Araby, S. et al. (2020). Electrically and thermally conductive elastomer by using MXene nanosheets with interface modification. *Chem. Eng. J.* 397: 125439.

19 Chen, X., Zhao, Y., Li, L. et al. (2021). MXene/polymer nanocomposites: preparation, properties, and applications. *Polym. Rev.* 61 (1): 80–115.

20 Zhao, L., Wang, K., Wei, W. et al. (2019). High-performance flexible sensing devices based on polyaniline/MXene nanocomposites. *InfoMat* 1 (3): 407–416.

21 VahidMohammadi, A., Moncada, J., Chen, H. et al. (2018). Thick and freestanding MXene/PANI pseudocapacitive electrodes with ultrahigh specific capacitance. *J. Mater. Chem. A* 6 (44): 22123–22133.

22 Wei, D., Wu, W., Zhu, J. et al. (2020). A facile strategy of polypyrrole nanospheres grown on Ti_3C_2-MXene nanosheets as advanced supercapacitor electrodes. *J. Electroanal. Chem.* 877: 114538.

23 Wang, Q.W., Zhang, H.B., Liu, J. et al. (2019). Multifunctional and water-resistant MXene-decorated polyester textiles with outstanding electromagnetic interference shielding and joule heating performances. *Adv. Funct. Mater.* 29 (7): 1806819.

24 Peymanfar, R., Mohammadi, A., and Javanshir, S. (2020). Preparation of graphite-like carbon nitride/polythiophene nanocomposite and investigation of its optical and microwave absorbing characteristics. *Compos. Commun.* 21: 100421.

25 Ren, Y., Zhu, J., Wang, L. et al. (2018). Synthesis of polyaniline nanoparticles deposited on two-dimensional titanium carbide for high-performance supercapacitors. *Mater. Lett.* 214: 84–87.

26 McDaniel, R.M., Carey, M.S., Wilson, O.R. et al. (2021). Well-dispersed nanocomposites using covalently modified, multilayer, 2D titanium carbide (MXene) and in-situ "Click" polymerization. *Chem. Mater.* 33 (5): 1648–1656.

27 Wang, X., Zhang, D., Zhang, H. et al. (2021). In situ polymerized polyaniline/MXene (V_2C) as building blocks of supercapacitor and ammonia sensor self-powered by electromagnetic-triboelectric hybrid generator. *Nano Energy* 88: 106242.

28 Wu, H., Hu, X., Li, X. et al. (2022). Large-scale fabrication of flexible EPDM/MXene/PW phase change composites with excellent light-to-thermal conversion efficiency via water-assisted melt blending. *Composites, Part A* 152: 106713.

29 Zhang, H., Wang, L., Chen, Q. et al. (2016). Preparation, mechanical and anti-friction performance of MXene/polymer composites. *Mater. Des.* 92: 682–689.

30 Carey, M., Hinton, Z., Sokol, M. et al. (2019). Nylon-6/$Ti_3C_2T_z$ MXene nanocomposites synthesized by in situ ring opening polymerization of ε-caprolactam and their water transport properties. *ACS Appl. Mater. Interfaces* 11 (22): 20425–20436.

31 Sheng, X., Zhao, Y., Zhang, L., and Lu, X. (2019). Properties of two-dimensional Ti_3C_2 MXene/thermoplastic polyurethane nanocomposites with effective reinforcement via melt blending. *Compos. Sci. Technol.* 181: 107710.

32 Gao, Q., Feng, M., Li, E. et al. (2020). Mechanical, thermal, and rheological properties of $Ti_3C_2T_x$ MXene/thermoplastic polyurethane nanocomposites. *Macromol. Mater. Eng.* 305 (10): 2000343.

33 Mayerberger, E.A., Urbanek, O., McDaniel, R.M. et al. (2017). Preparation and characterization of polymer-$Ti_3C_2T_x$ (MXene) composite nanofibers produced via electrospinning. *J. Appl. Polym. Sci.* 134 (37): 45295.

34 Levitt, A.S., Alhabeb, M., Hatter, C.B. et al. (2019). Electrospun MXene/carbon nanofibers as supercapacitor electrodes. *J. Mater. Chem. A* 7 (1): 269–277.

35 Hou, C., Jiao, T., Xing, R. et al. (2017). Preparation of TiO_2 nanoparticles modified electrospun nanocomposite membranes toward efficient dye degradation for wastewater treatment. *J. Taiwan Inst. Chem. Eng.* 78: 118–126.

36 Liu, F., Li, Y., Hao, S. et al. (2020). Well-aligned MXene/chitosan films with humidity response for high-performance elecomagnetic interference shielding. *Carbohydr. Polym.* 243: 116467.

37 Zhu, Z., Liu, C., Jiang, F. et al. (2020). Flexible and lightweight $Ti_3C_2T_x$ MXene@ Pd colloidal nanoclusters paper film as novel H_2 sensor. *J. Hazard. Mater.* 399: 123054.

38 Zhang, M., Wang, Y., Gao, F. et al. (2019). Formation of new MXene film using spinning coating method with DMSO solution and its application in advanced memristive device. *Ceram. Int.* 45 (15): 19467–19472.

39 Wu, H., Almalki, M., Xu, X. et al. (2019). MXene derived metal–organic frameworks. *J. Am. Chem. Soc.* 141 (51): 20037–20042.

40 Liu, H., Fu, R., Su, X. et al. (2021). Electrical insulating MXene/PDMS/BN composite with enhanced thermal conductivity for electromagnetic shielding application. *Compos. Commun.* 23: 100593.

41 Aakyiir, M., Araby, S., Michelmore, A. et al. (2020). Elastomer nanocomposites containing MXene for mechanical robustness and electrical and thermal conductivity. *Nanotechnology* 31 (31): 315715.

42 Ghidiu, M., Halim, J., Kota, S. et al. (2016). Ion-exchange and cation solvation reactions in Ti_3C_2 MXene. *Chem. Mater.* 28 (10): 3507–3514.

43 Shekhirev, M., Shuck, C.E., Sarycheva, A., and Gogotsi, Y. (2021). Characterization of MXenes at every step, from their precursors to single flakes and assembled films. *Prog. Mater Sci.* 120: 100757.

44 Srimuk, P., Halim, J., Lee, J. et al. (2018). Two-dimensional molybdenum carbide (MXene) with divacancy ordering for brackish and seawater desalination via cation and anion intercalation. *ACS Sustainable Chem. Eng.* 6: 3739–3747.

45 Anasori, B., Shi, C., Moon, E.J. et al. (2016). Control of electronic properties of 2D carbides (MXenes) by manipulating their transition metal layers. *Nanoscale Horiz.* 1: 227–234.

46 Lukatskaya, M.R., Bak, S.M., Yu, X. et al. (2015). Probing the mechanism of high capacitance in 2D titanium carbide using in situ X-ray absorption spectroscopy. *Adv. Energy Mater.* 5 (15): 1500589.

47 Al-Temimy, A., Anasori, B., Mazzio, K.A. et al. (2020). Enhancement of Ti_3C_2 MXene pseudocapacitance after urea intercalation studied by soft X-ray absorption spectroscopy. *J. Phys. Chem. C* 124 (9): 5079–5086.

48 Djire, A., Zhang, H., Reinhart, B.J. et al. (2021). Mechanisms of hydrogen evolution reaction in two-dimensional nitride MXenes using in situ X-ray absorption spectroelectrochemistry. *ACS Catal.* 11 (5): 3128–3136.

49 Halim, J., Cook, K.M., Naguib, M. et al. (2016). X-ray photoelectron spectroscopy of select multi-layered transition metal carbides (MXenes). *Appl. Surf. Sci.* 362: 406–417.

50 Mariano, M., Mashtalir, O., Antonio, F.Q. et al. (2016). Solution-processed titanium carbide MXene films examined as highly transparent conductors. *Nanoscale* 8 (36): 16371–16378.

51 Guo, Y., Zhou, X., Wang, D. et al. (2019). Nanomechanical properties of Ti_3C_2 mxene. *Langmuir* 35 (45): 14481–14485.

52 Hope, M.A., Forse, A.C., Griffith, K.J. et al. (2016). NMR reveals the surface functionalisation of Ti_3C_2 MXene. *Phys. Chem. Chem. Phys.* 18 (7): 5099–5102.

53 Kobayashi, T., Sun, Y., Prenger, K. et al. (2020). Nature of terminating hydroxyl groups and intercalating water in $Ti_3C_2T_x$ MXenes: a study by ^1H solid-state NMR and DFT calculations. *J. Phys. Chem. C* 124 (25): 13649–13655.

54 Harris, K.J., Bugnet, M., Naguib, M. et al. (2015). Direct measurement of surface termination groups and their connectivity in the 2D MXene V_2CT_x using NMR spectroscopy. *J. Phys. Chem. C* 119 (24): 13713–13720.

55 Sarycheva, A. and Gogotsi, Y. (2020). Raman spectroscopy analysis of the structure and surface chemistry of $Ti_3C_2T_x$ MXene. *Chem. Mater.* 32 (8): 3480–3488.

56 Lai, S., Jeon, J., Jang, S.K. et al. (2015). Surface group modification and carrier transport properties of layered transition metal carbides (Ti_2CT_x, T: -OH, -F and -O). *Nanoscale* 7 (46): 19390–19396.

57 Han, M., Yin, X., Wu, H. et al. (2016). Ti_3C_2 MXenes with modified surface for high-performance electromagnetic absorption and shielding in the X-band. *ACS Appl. Mater. Interfaces* 8 (32): 21011–21019.

6

Progressive Approach Toward MXenes Hydrogel

Abbas Mohammadi, Ghazal Amini, and Saman Abrishamkar

Department of Chemistry, University of Isfahan, Isfahan, Iran

6.1 Hydrogels

In 1960, the first synthetic biocompatible hydrogel with a high affinity for water was prepared based on polyhydroxy ethyl methacrylate (PHEMA), which was utilized in the manufacturing of contact lenses [1]. The historical progression of hydrogel research can be summarized as follows: The initial phase of hydrogel development witnessed a diverse range of experimental methods, including various random cross-linking approaches and the utilization of initiators for chemical modifications. These attempts played a crucial role in advancing the understanding of hydrogel materials and paved the way for further innovations in this field. The main purpose was to synthesize hydrogel substances that possess both excellent mechanical properties and a high degree of swelling. Subsequent efforts focused on the tailored design of hydrogels to demonstrate responsiveness to specific stimuli, such as variations in pH, temperature, pressure, and solvent concentration. Finally, the synthesis of the third generation of hydrogels comprised of stereo-complexed materials has been successfully achieved [2–5].

Hydrogels with three-dimensional polymeric structures have a remarkable ability to absorb biofluids or water. These compounds can absorb water up to more than 1000 times the mass of the polymer without dissolving [6, 7]. Hydrogels possess soft and rubbery surfaces, structures, and physicochemical properties that make them more akin to living tissues than other synthetic materials. Therefore, the technology of using hydrogels in biomedicine and medicine has been developed [8, 9].

Hydrogels can undergo cross-linking through the application of either chemical or physical methods [10, 11]. The unique properties of hydrogels are influenced by various factors, most notably hydrophilicity and the extent of cross-linking in the polymer chains [12]. Hydrophilic functional groups, such as carboxylic, amine, hydroxyl, amide, and sulfonate groups, embedded within the polymer chains are primarily responsible for the exceptional water adsorption capacity exhibited by hydrogels, while their solvent resistance is because of the presence of cross-links between polymer chains [13].

MXenes: Next-Generation 2D Materials: Fundamentals and Applications, First Edition.
Edited by Jay Singh, Kshitij RB Singh, Ravindra Pratap Singh, and Charles Oluwaseun Adetunji.

Recently, hydrogels have gained significant interest because of their indispensable and practical properties. These polymers are prepared in different physical forms such as microparticles, nanoparticles, coatings, and films. These hydrogels can be used in medicine, tissue engineering, etc. [14–16].

6.1.1 Hydrogels Classification

There are multiple classifications for hydrogels. The cause of this classification is the differences in source, synthesis method, ionic charge, composition, and type of cross-links.

Hydrogels are categorized into natural and synthetic types, depending on the source of the material. Natural hydrogels are derived from proteins and polysaccharides such as alginate and chitosan [17, 18]. Over the past twenty years, synthetic hydrogels have been increasingly favored over natural structure hydrogels. This is due to their extended durability, substantial water absorption capabilities, and superior gel strength. Moreover, synthetic hydrogels usually have a specific structure and can be modified to obtain functionality and biodegradability. Synthetic hydrogels can be easily prepared by reacting monomers via classical chemical methods [3, 19, 20]. Hybrid hydrogels can also be produced by combining natural and synthetic polymers, resulting in unique material properties [21–23].

Based on the composition, hydrogels can be divided into homopolymer, copolymer, and interpenetrating polymer network (IPN) hydrogels [24]. Homopolymer hydrogels consist of a single hydrophilic monomer, polymerized to form the network structure. Factors such as the monomer's nature, the polymerization technique, and the type of cross-links determine the overall properties of these hydrogels. A pertinent example of homopolymer hydrogel is the poly(ethylene glycol)-based hydrogel [25]. On the other hand, copolymer hydrogels result from the polymerization of two distinct monomers, wherein at least one monomer is hydrophilic, responsible for hydrogel swelling. These hydrogels are synthesized by polymerizing monomers with an initiator and a cross-linking agent [25, 26]. IPN hydrogels are composed of two polymer networks that are interconnected without any covalent bonds [27]. The hydrogels obtained by the IPN strategy are dense hydrogels with controllable physical properties and good mechanical properties [28].

Hydrogels can be grouped into four categories based on the electric charge of their network. The first group is neutral hydrogels, which are devoid of any charge on their main chain or side groups. The second group is ionic hydrogels, which can be either cationic or anionic. Cationic hydrogels have positively charged groups, such as amines, and show high swelling values at low pH. On the other hand, anionic hydrogels have negatively charged groups, such as carboxylic acid, and are highly swellable at high pH values. The third group is amphoteric or ampholytic hydrogels, which possess both acidic and basic groups. Lastly, zwitterionic or polybetaines hydrogels have either anionic or cationic groups in their repeating units [25, 29, 30].

Hydrogels can be categorized into three types based on pore size: nonporous, microporous, and macroporous. Depending on the preparation method, they can take on a variety of morphologies, including particles, powders, membranes, films, matrices, or emulsions. The products obtained from polymerization can be different depending on the technique used [31].

6.1.2 Properties of Hydrogels

The most important properties of hydrogels are their water storage ability and permeability. The cross-linked polymer hydrogels swell when immersed in water or solvent but do not dissolve. The swelling property, generally used to describe hydrogels with the degree of swelling, depends on multiple characteristics including network density, solvent–polymer interactions, and solvent properties. By increasing the number of hydrophilic groups and decreasing the cross-linking density, the water-holding capacity of the hydrogel can be enhanced [6, 8, 20]. The swelling rate of hydrogels is a crucial parameter that is directly associated with the amount of water that enters the hydrogel. Another essential factor in many fields, including pharmaceutical and biomedical, is the mechanical properties of hydrogels [31–35]. Finally, in addition to high water absorption capacity, an ideal hydrogel should have other characteristics such as stability, non-toxicity, pH neutrality, suitable porosity, re-wetting capability, long service life, and minimal monomer residual in the solution [36, 37]. Hydrogels have been found to be highly stable and durable in swelling environments, which has made them valuable in a range of applications including separation, drug delivery, and tissue engineering. The development of stimulus-responsive hydrogels has garnered significant interest due to their potential for targeted and controlled release of drugs and other substances. This particular type of hydrogel is highly responsive to a range of environmental factors, including temperature, pH levels, solvent composition, salt concentration, pressure, ionic strength, and electric or magnetic fields [38–41].

6.2 MXene-Based Hydrogels

Recently, hydrogels have attracted much attention for various applications. MXenes with several unique properties can be used as composite reinforcements. Research has shown that MXene-containing hydrogels perform better in many applications than MXens or hydrogels alone. By introducing MXenes into hydrogel systems, it becomes possible to create soft materials based on MXenes that possess distinct and customizable features. This offers a range of options for designing MXene-based soft materials with specific and adjustable properties. MXene nanosheets can be effectively utilized in a hydrogel substrate, which serves as a highly suitable platform [42, 43]. The resulting MXene-based hydrogels possess unique properties that make them suitable for energy storage, catalysis, tissue engineering, etc. Hence, the MXene-based hydrogels exhibit immense potential for forthcoming applications [44, 45].

The hydrophilic surface groups (–OH, –O, –Cl, and –F) of MXene enable it to exhibit excellent dispersibility in aqueous solvents. However, in aqueous media, MXene tends to agglomerate due to the influence of Van der Waals force and hydrogen bonding. This poses a challenge during the fabrication of MXene-based hydrogels, as nanosheet aggregation can affect the uniformity of the hydrogel. Furthermore, MXene nanosheets have a tendency to oxidize in prepared hydrogels, which can have a significant impact on the final performance of the hydrogel. For this reason, another component in the hydrogel matrix is often necessary to solve these problems, which can balance the hydrophilicity of MXenes [40, 46, 47].

The incorporation of MXene into hydrogels has been found to significantly enhance the stability of MXene, which has been a limiting factor in many MXene-based applications.

Furthermore, simple modifications can lead to the synthesis of MXene hydrogel derivatives, such as aerogels, thereby augmenting their versatility [47–49]. This chapter will delve into the synthesis mechanism, properties, and applications of MXene hydrogels.

6.2.1 Applications of MXene Hydrogels

Biocompatibility is an important attribute of hydrogels, rendering them highly advantageous for various biomedical applications. This characteristic ensures that they are safe for use in biological systems without causing any adverse reactions. Their exceptional water absorption capacity provides countless applications in separation technology, agriculture, and medicine [50]. In particular, due to their porosity and high-water absorption capacity, hydrogels are ideal for use as adsorbents in separation technology [51, 52].

Many applications have been identified for MXenes. These applications include energy storage systems, which can benefit from the high surface area and excellent electrical conductivity of MXenes. Furthermore, their superior electromagnetic interference shielding performance surpasses all synthetic materials with the same thickness, thus making them ideal for electronic devices and communication systems. Another area where MXenes have shown promise is in composite reinforcement, where their high strength and stiffness can improve the mechanical properties of materials. Additionally, MXenes have been explored for water treatment systems, where they can efficiently remove pollutants and contaminants from water [53–55]. In addition, MXenes have excellent antibacterial properties, which can be even more effective than graphene oxide in reducing and killing live bacterial cells. MXenes are also potentially attractive for catalytic applications due to their diversity in surface chemistry, the presence of transition metal oxides, and their high specific surface area [56, 57].

6.2.2 Mechanisms of Synthesis and Gelation of MXene Hydrogels

According to studies, most of the MXene-based hydrogels are made of $Ti_3C_2T_x$. In a typical MXene-based hydrogel, the continuous phase (e.g. water) is limited by a cross-linker in a three-dimensional structural network. To date, different cross-linkers have been reported for MXene-based hydrogels, including MXenes themselves, inorganic ions, polymers, reduced graphene oxide (rGO), and/or hybrids of any of them [58, 59]. In the following, various synthesis techniques used for MXene-based hydrogels, which are classified according to the type of cross-linker, are discussed.

6.2.2.1 All-MXene Hydrogels

The formation of all-MXene hydrogels poses a challenge due to the inherent hydrophilicity of MXene nanosheets, which causes agglomeration through Van der Waals forces, making it difficult for them to form a gel on their own. To overcome this challenge, a cross-linker or another component is often utilized in the hydrogel to balance the hydrophilicity of MXene surfaces and aid in hydrogel formation [60].

Lin et al. proposed an innovative method for preparing all-MXene hydrogels using vacuum-assisted filtration of delaminated $Ti_3C_2T_x$ suspensions. In this procedure, the vacuum is disconnected once the non-colloidal solution remains on the porous membranes.

Hydrogel film is then dipped into the acetone to make it easier to peel off. The dynamic physical cross-linking between MXene nanosheets is significantly influenced by the presence of unremoved water molecules, resulting in the formation of interconnections and a gel-like texture [45]. However, it is important to note that this physical cross-linking cannot overcome the Van der Waals interlayer forces in MXene nanolayers. Therefore, two strategies have been proposed to control the restacking of MXene nanosheets and prevent hydrogel collapsing.

In the first method, developed by Lin et al. the undried hydrogel film was immersed in heat-stable ionic liquid (IL); for example, 1-ethyl-3-methylimidazolium bis (trifluoromethyl sulfonyl) imide (EMI-TFSI) to provide a solvent exchange. Then the film was placed in a vacuum at a temperature of 81 °C to dry. After drying the film, it was observed that the IL was not removed from the MXene film. The remaining ionic liquid between MXene nanoplates prevented the restacking of the layers and formed a $Ti_3C_2T_x$ ionogel instead of a hydrogel. Figure 6.1 shows the filtration process and SEM images of the $Ti_3C_2T_x$ ionogel film [45].

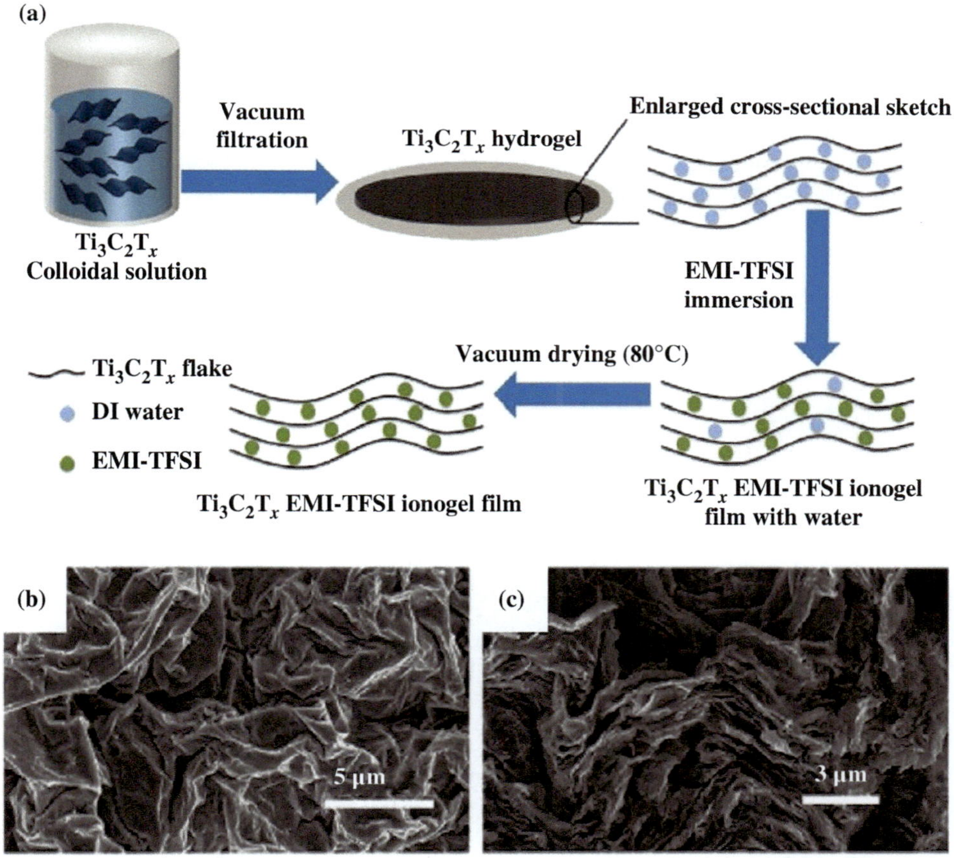

Figure 6.1 (a) Schematic illustration of $Ti_3C_2T_x$ EMI-TFSI ionogel film formation. (b, c) SEM images of $Ti_3C_2T_x$ hydrogel film: (b) top view; (c) cross-sectional view. Source: Reproduced with permission from Ref. [45]. 2016, Elsevier.

In the second method, performed by Lukatskaya et al. to prevent the agglomeration of MXene nanosheets, the MXene hydrogel was placed in the H_2SO_4 electrolyte for three days [59].

6.2.2.2 MXene-GO Nanocomposite Hydrogels

As already discussed, the presence of cross-linkers is essential for the gelation process in MXene-based hydrogels. Moreover, because of the limited cross-linking sites at MXene surfaces, it is essential to choose a suitable cross-linker [61]. In this regard, it was found that using a cross-linker based on a two-dimensional (2D) material such as GO enables surface interactions with MXene nanosheets [59, 62].

Chen et al. discovered that interfacial cross-linking can occur between $Ti_3C_2T_x$ and rGO nanosheets. This process involves $Ti_3C_2T_x$ being added to a GO solution, where the Ti acts as a reducing agent that converts the hydrophilic GO surface into the more hydrophobic rGO (Figure 6.2a). As the gel forms, the reduced electrostatic repulsion between $Ti_3C_2T_x$ and rGO decreases the self-convergence of MXene nanosheets into the rGO structure, which is caused by hydrogen bonds between the $Ti_3C_2T_x$ and rGO [63].

In order to better enhance the available surface in this reaction (gelation of MXene-GO hydrogel), Shang et al. used ethylenediamine (EDA). EDA creates a distance between the MXene and rGO plates, improving the chemical bonds between these nanoplates. These chemical bonds between the MXene sheets and the rGO layers result from the opening of the epoxy rings on the GO sheets (Figure 6.3). The MXene-rGO nanocomposite hydrogels exhibit a greater thickness and flexibility compared to the GO hydrogels [64].

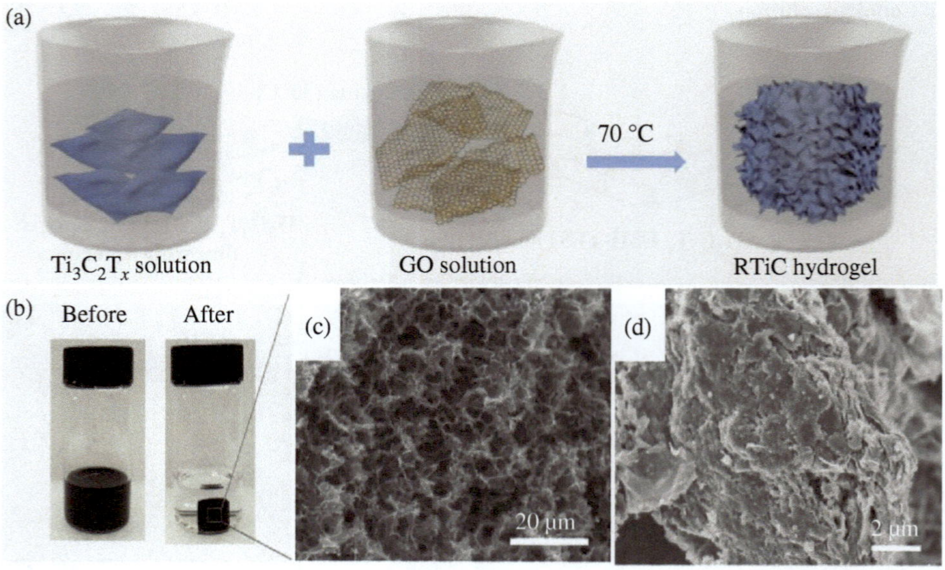

Figure 6.2 (a) Schematic illustrations of synthesis process for $Ti_3C_2T_x$–rGO (RTiC) hydrogel. (b) Digital photos of a homogeneous GO/$Ti_3C_2T_x$ colloid before and after reaction at 70 °C. (c) SEM image of RTiC hydrogel, and (d) SEM image of RTiC powder. Source: Reproduced with permission from Ref. [63]. 2018, American Chemical Society.

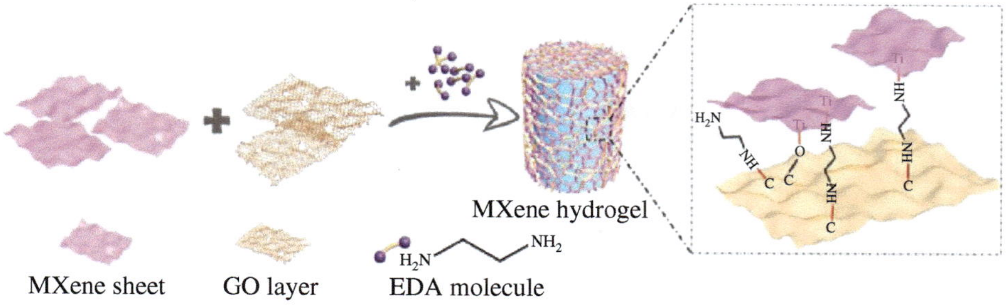

Figure 6.3 Schematic illustrations of gelation process for an EDA-initiated $Ti_3C_2T_x$–rGO hydrogel. Source: Reproduced with permission from Ref. [64]. 2019, John Wiley and Sons.

6.2.2.3 MXene-polymer Nanocomposite Hydrogels

The incorporation of hydrophilic MXene nanosheets into polymeric hydrogel networks, which have a high capacity to absorb water, offers exceptional versatility with multiple functions. Additionally, the high viscosity of polymers can prevent the restacking of MXene nanosheets, thus improving their dispersion stability. However, evenly distributing MXene nanosheets between polymer networks is challenging due to the strong bonding effect between the polymer chains [54].

The preparation of MXene-polymer nanocomposite hydrogels requires a monomer, a cross-linker agent, and an initiator. In the polymerization step, if one type of monomer is used, the synthesized three-dimensional network is called MXene-homopolymeric nanocomposite hydrogel. Various MXene-homopolymeric nanocomposite hydrogels can be prepared using different polymers, for example, cellulose, chitosan, polyvinyl alcohol (PVA), and acrylate polymers. A copolymeric hydrogel network is formed when two or more different types of monomers are utilized. If this happens in the presence of MXene nanosheets, MXene-copolymeric nanocomposite hydrogels are formed. For example, in a study by Liao et al. acrylamide (AAm) and PVA monomers were mixed in water containing $Ti_3C_2T_x$ nanosheets, resulting in MXene-containing copolymer hydrogels (Figure 6.4). Among the PVA chains, a cross-linker such as sodium tetraborate decahydrate (borax) can be used, and the –OH groups are connected with borate ions ($B(OH)_4^{-1}$). Finally, a polyacrylamide network (PAAm) was obtained by polymerizing AAm monomers at 60–65 °C. In this reaction, the entanglements of polymer chains accommodate MXene nanosheets (hydrogel network). In addition, OH groups present in $Ti_3C_2T_x$ also establish a chemical bond with borate ions [65].

6.2.2.4 MXene-metal Hybrid Nanocomposite Hydrogels

One of the problems that rarely happens during the previous methods is oxidation, which destroys the properties of the hydrogel. Therefore, a rapid gelation process can be used to eliminate or reduce oxidation. This method increases the speed of MXenes phase separation from water and prevents the re-accumulation of nanosheets. For this purpose, divalent metal ions can be used as cross-linkers to form MXene-metal hybrid hydrogels. The divalent ions (such as Fe^{2+}) strongly interact with the surface hydroxyl groups of $Ti_3C_2T_x$ nanosheets and cause cross-linking. For instance, by adding a metal salt solution of $FeCl_2{\cdot}4H_2O$ to a suspension of MXene, a strong bond is formed between the Fe^{2+} and OH groups. This

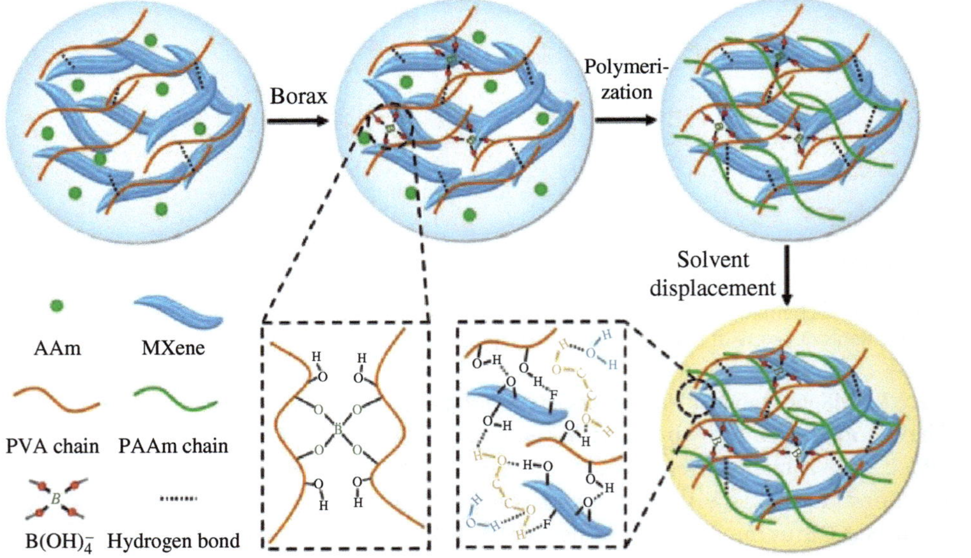

Figure 6.4 Schematic illustration of MXene–PAAm–PVA nanocomposite hydrogel preparation. Source: Reproduced with permission from Ref. [65]. 2019, John Wiley and Sons.

reduces the hydrophilicity of the negatively charged MXene sheets and increases the phase separation within them, resulting in effective prevention of the oxidation of MXenes. This fast-gelation process that utilizes divalent metal ions takes only a few minutes to complete. It is worth noting that other divalent metal ions (such as Mg^{2+}, Co^{2+}, and Ni^{2+}) and trivalent metal ions (such as Al^{3+}) can also be used with this method. However, monovalent metal ions are not suitable for this process as they possess weaker hydration energy and lead to the aggregation of MXene nanosheets [66]. As a result, the careful selection of metal ions plays a crucial role in controlling the gelation process and achieving the desired dispersion and stability of MXene nanosheets within the hydrogel matrix.

6.2.3 Properties of MXene-Based Hydrogels

When MXenes are incorporated into hydrogels, they further improve the behavior and properties of the hydrogels. These effects include rheological and electrochemical properties. It should be noted that the improvement in hydrogel properties in the presence of MXene strongly depends on the MXene suspension properties, such as the size, thickness, and even concentration of nanosheets.

According to the tests conducted on hydrogels, some characteristics such as mechanical stability, electrical conductivity, and rheological properties are increased in hydrogels prepared from thinner MXene nanosheets with a large specific surface area. In addition, in MXene hydrogels, the hydrogel network obstacles the aggregation of MXene nanosheets, leading to the higher penetration of electrolytes, resulting in good electrochemical properties. MXenes have several unique properties, including very thin structures, large specific surface area, photothermal properties, antibacterial properties, and mechanical strength.

Among 2D materials, the $Ti_3C_2T_x$ monolayer has the highest young modulus. Therefore, the presence of the MXene nanosheets in hydrogels improves the mechanical properties of hydrogels. In this regard, gelatin and PVA-based hydrogels incorporated with MXene nanosheets showed higher mechanical strength than pure hydrogels [42, 43, 67]. Moreover, MXenes with a negatively charged hydrophilic surface have a uniform distribution within the hydrogel. Therefore, the incorporation of MXene nanosheets into the hydrogels causes an increase in the swelling capacity of the prepared hydrogels. The 3D network structure, good swelling capacity, polar terminal groups, and great anchoring surface sites allow multiple drugs to be easily attached to the MXene surfaces with higher adsorption and release capacity. Besides, MXene-based hydrogels have more biocompatibility than polymer hydrogels, which are attractive for biomedical applications [47, 68]. In conclusion, it can be said that MXene-based hydrogels have a bright future for various applications.

6.2.4 Applications of MXene-Based Hydrogels

MXene-containing hydrogels have multiple applications, including energy storage, biomedicine, catalysts, and sensors.

6.2.4.1 Energy Storage

The most critical application of MXene hydrogels in the field of energy storage is in the electrochemical capacitors (ECs) and batteries. In the first synthesized MXene-only hydrogel as a supercapacitor electrode prepared by Lin et al. all $Ti_3C_2T_x$ hydrogels had a capacitance of 70 Fg^{-1} with a vast potential window of 3 volts. However, when a colloidal solution was filtered from MXene fragments to prepare the MXene electrode, they reassembled, in other words, the distance between their layers became less, and this method was irreversible. The reassembly of MXene parts in the filtration reaction and their small interlayer distance limit the access to electrolyte ions, mainly the large ions of ILs, and limit the full use of the surface. As a result of these events, we see poor behavior in electrochemical properties. Carbon nanotubes and graphene materials are used to solve this problem, that is, to increase the capacity of MXene-based capacitors [45].

Solvent-free salt liquid at room temperature is called IL. They have low vapor pressure, electrochemical cal, and nonflammable properties and can also work in high-voltage energy storage devices. However, their disadvantages include high viscosity and low mobility of ions. These factors make these electrolytes unsuitable for electrodes with high porosity or multilayer materials with low interlayer distance. To solve this problem, to control the agglomeration of nanoplates, we must use the first synthesis method described previously, vacuum filtration. For example, to synthesize a $Ti_3C_2T_x$ hydrogel film, the $Ti_3C_2T_x$ dried film and hydrogel film were immersed in an IL electrolyte (i.e.1-ethyl-3-methylimidazolium bis(trifluoromethylsulfonyl) imide [EMI-TFSI]). After drying in a vacuum, IL remained in the $Ti_3C_2T_x$ film, which increased the interlayer space of MXene nanosheets and prevented restacking.

The electrochemical performance of $Ti_3C_2T_x$ hydrogel film was compared with a vacuum-dried $Ti_3C_2T_x$ film. Unlike the dried $Ti_3C_2T_x$ film, which showed a very low capacitance of 1 Fg^{-1}, the $Ti_3C_2T_x$ hydrogel film showed a capacitance of up to 70 Fg^{-1} [46, 67]. It is found that the electrochemical properties of MXens depend on their lateral dimensions. The synthesis of controlled-dimension MXens makes it possible to design and fabricate electrodes with high electronic and ionic conductivity.

Electron accessibility and ion exchange capacity can be optimized by controlling the size of nanosheets to improve electrochemical performance. More extensive MXene nanosheets generally increase electrical conductivity because of their low connection resistance, as smaller nanosheets show shorter ion diffusion paths. Therefore, a higher volume capacity is obtained until the hydrogel electrode consists of a mixture of large (about 15 mm) and small (about 1–2 mm) MXene nanosheets with a weight ratio of 1:1.

6.2.4.2 Biomedical Applications

Soft hydrogel matrices, similar to host tissues, can be used as a cover for lost host tissue after an injury as a substrate for stem cell transfer to support tissue regeneration. The 2D planar shape of MXene nanosheets provides a very large surface for cells to attach and allow them to grow. After an injury, infection often endangers patients' lives or even leads to organ failure. Therefore, there is a great need to build intelligent multifunctional composites that increase regeneration and improve issues like putrefaction and/or inflammation. Therefore, numerous studies were conducted to fabricate bio-hydrogels with electrical conductivity, biocompatibility, and antibacterial and anti-inflammatory properties for tissue engineering [69, 70].

In this regard, a composite hydrogel was made by combining titanium carbide MXene nanosheets (Ti_3C_2) and honey in a clinically approved chitosan hydrogel. The first reason to use honey is that it is an old remedy for nutrient deficiencies and can precipitate wound recovery. In addition, due to the presence of flavonoids and polyphenols in honey, we can expect anti-inflammatory, antibacterial, and antioxidant properties from them. Honey also provides the conditions for cell growth because of its sugar content (mono-, di-, and oligo-saccharides) and vitamins, amino acids, and proteins [69, 71]. As mentioned before, MXenes are biocompatible materials with antibacterial properties [72, 73]. Recently, Ti_3C_2 MXene-based biomaterials have been reported to have inherent anti-inflammatory and immunomodulatory properties. For instance, it is found that MXene-chitosan nanocomposite hydrogel has properties such as biocompatibility with a high swelling ratio, porous structure as well as high ability to load and release the drug [74–76].

MXene hydrogels can be used in biological applications, like cancer treatment and drug release. Compared to hydrogels based on other 2D nanomaterials like graphene, MXene hydrogels have strong hydrophilicity, leading to better dispersion in biological environments. It also makes it easier to attach drugs to the polar end groups of MXene. For example, MXene-cellulose nanocomposite hydrogel treats tumors and controls drug (doxorubicin, DOX) release. MXenes can absorb near-infrared (NIR) and convert it into thermal energy. Therefore, due to the good photothermal properties of $Ti_3C_2T_x$, the killing of cancer cells was reached by synergistic photothermal therapy (PTT) and chemotherapy [62].

The development of cancer treatments is an ongoing process that requires advanced and effective therapies with minimal side effects. Topical chemotherapy using smart hydrogels has gained popularity due to its ability to be activated by various stimuli, such as pH, enzymes, hypoxia, and oxidation. Moreover, the use of external stimuli such as magnetic field, ultrasound, temperature, and light has shown promising results in adapting to hydrogels. In particular, NIR light has emerged as a highly beneficial tool for biomedical treatment due to its significant penetration, low optical toxicity, and high space–time remote control ability. By using NIR-responsive hydrogels, doctors and researchers can achieve precise and accurate spatial and temporal control of drug dosage, making it a valuable addition to the field of topical drug delivery for clinical applications [77, 78].

When it comes to enhancing drug release, photothermal agents that generate heat after absorbing NIR light can prove to be quite useful. However, most such agents have not been able to meet the necessary criteria for increased uptake in the biological NIR region. This is where MXenes come into the picture. These agents are capable of absorbing NIR and converting it into thermal energy, all while maintaining biocompatibility. As a result, MXene hydrogel has emerged as a promising drug carrier. These hydrogels respond to NIR by undergoing a phase change or disruption of their structure, which leads to the release of the trapped therapeutic agents (drugs) [79, 80].

Agarose is a type of polysaccharide. Agar has characteristics such as good mechanical properties and high biological activity. This water-soluble polysaccharide shows high chemical reactivity for modification or functionalization. Agarose hydrogels that contain drugs can release the drug by increasing the temperature [81, 82]. A recent study revealed the development of an intelligent drug delivery system that employs MXene and agarose nanosheets to carry the potent anticancer drug, doxorubicin (DOX). This novel approach enables easy regulation of drug release by way of 808 nm NIR light irradiation, as shown in Figure 6.5. When exposed to NIR radiation, the encapsulated drug is swiftly released, thereby effectively eliminating cancer cells. The kinetics of drug release can be fine-tuned by adjusting the agarose concentration, MXene concentration, radiation intensity, and exposure time [83].

In another work conducted by Changyu Yang, MXene ($Ti_2C_3T_x$) was included in PNIPAM, and as a result, MXene/PNIPAM composite hydrogel was prepared. This hydrogel showed unique properties and characteristics. For example, this hydrogel shows good photothermal properties under NIR laser irradiation. This feature can be used for medical applications and drug delivery [84].

6.2.4.3 Catalysts

To date, several MXenes exhibited excellent catalytic performance for different reactions. Moreover, their excellent electrical conductivity and high hydrophilicity are often used

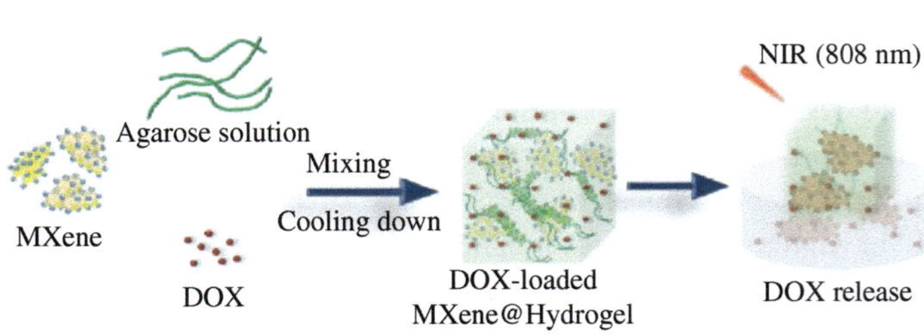

Figure 6.5 Schematic preparation of DOX-loaded MXene hydrogel for NIR-controlled drug release. Source: Reproduced with permission from Ref. [83]. 2021, Elsevier.

as catalyst supports because of their excellent electrical conductivity. When MXenes are incorporated into the hydrogels, the active surface is enhanced, leading to improved catalytic activity. For example, $Ti_3C_2T_x$-based hydrogels have a porous and interconnected structure with high potential as a promising platform for photo-redox catalysis. When Eosin Y (EY) photosensitizer was incorporated into the hydrogel, the resulting hydrogel showed more photoactivity and was easier to operate than its powder counterpart. This improvement was attributed to the porous hydrogel network, which enhanced active surface sites and promoted ion diffusion. In addition to the improved catalytic activity, $Ti_3C_2T_x$-rGO/EY hydrogel exhibited more mechanical properties and recyclability than its powder sample, which allows it to be reused without significant performance reduction [85].

6.2.4.4 Sensors

Recently, conductive hydrogels have received much attention. We can mention electronic devices, soft robotics, and prosthetics among their applications. In order to use these hydrogels, susceptible sensors are needed, and they must also have the ability to stick to different surfaces such as human skin [86]. Hydrogels can react to various external stimuli by mechanisms inducing measurable changes in electrical, geometrical, and optical properties.

Unfortunately, in the case of electromechanical sensing, hydrogels generally show relatively low sensitivity, and their electrical response to mechanical stimuli exhibits hysteresis and oscillation, which is because of the viscoelastic properties of hydrogels [87].

Recently, particular types of conductive nanofillers have been used to increase the sensitivity and response rate of hydrogels, whose geometry and conductivity can be altered by shape-changing.

In this regard, MXene hydrogels have shown great potential as sensors. These hydrogels show high sensing ability to different movement patterns. The resistance of MXene hydrogels increases when an object moves on its surface parallel to the flow direction (positive signal), and it decreases when the object moves in a vertical direction (negative signal). MXene nanosheets can significantly improve the sensitivity of the hydrogel. For instance, MXene hydrogels can be used to distinguish between two different types of handwriting or to identify signatures that vary from person to person in strength, speed, and order of writing. In contrast, new hydrogels have failed to detect these differences [88, 89]. Moreover, it is found that MXene-based hydrogels show even higher sensitivity under compressive strain than tensile strain. In addition, an extraordinary tensile sensitivity was observed in these hydrogels, which is ten times higher than the new hydrogel. According to studies, MXene hydrogels have instant self-healing ability, excellent compatibility, and good adhesion to various surfaces, including human skin [87, 90]. One of the works done by Zhang, K.H. Lee, D in 2018 was the production of PVA hydrogel containing MXene. This hydrogel has a very high sensory capability, which can be used for signature recognition and handwriting recognition. That is, this hydrogel could distinguish between “Hi” written by several people [91, 92]. When an object moves on the surface of hydrogels parallel to the flow direction, the signal increases, and if the object moves in the perpendicular direction, the signal decreases [93].

6.3 Conclusions

Hydrogels are 3D polymeric networks with numerous hydrophilic groups, which cause them to have a large capacity to absorb water or biofluids. Thanks to their biocompatibility and high-water absorption capacity, hydrogels are used for various applications. However, in order to improve the properties of the hydrogels and also introduce new exciting properties, MXene nanosheets can be incorporated into the hydrogel structure. MXenes are a new family of 2D metal carbides, carbonitrides, and nitrides with excellent properties such as layered structure, high hydrophilicity, large specific surface area, and good mechanical strength, which make them suitable as excellent reinforcements for hydrogels. Despite the high hydrophilicity of MXenes, they tend to agglomerate in aqueous media under the effect of Van der Waals force and hydrogen bonding. Thus, nanosheet aggregation is likely to occur during the MXene-based hydrogel fabrication process. To overcome this problem, it is necessary to have another component in the hydrogel matrix that can act as a cross-linker, including MXenes themselves, inorganic ions, polymers, rGO, or combinations of any of them. The resulting MXene-based hydrogels with highly improved properties have great potential for a wide range of applications.

References

1. Campbell, T.D. (2007). *Synthesis and Physical Characterization of Biocompatible Hydrogels*. The Florida State University.
2. Seow, W.Y. and Hauser, C.A. (2014). Short to ultrashort peptide hydrogels for biomedical uses. *Mater. Today* 17 (8): 381–388.
3. Wichterle, O. and Lim, D. (1960). Hydrophilic gels for biological use. *Nature* 185 (4706): 117–118.
4. Buwalda, S.J. et al. (2014). Hydrogels in a historical perspective: from simple networks to smart materials. *J. Controlled Release* 190: 254–273.
5. Abebe, D.G. and Fujiwara, T. (2012). Controlled thermoresponsive hydrogels by stereocomplexed PLA-PEG-PLA prepared via hybrid micelles of pre-mixed copolymers with different PEG lengths. *Biomacromolecules* 13 (6): 1828–1836.
6. Nguyen, Q.V., Park, J.H., and Lee, D.S. (2015). Injectable polymeric hydrogels for the delivery of therapeutic agents: a review. *Eur. Polym. J.* 72: 602–619.
7. Radhakrishnan, J., Krishnan, U.M., and Sethuraman, S. (2014). Hydrogel based injectable scaffolds for cardiac tissue regeneration. *Biotechnol. Adv.* 32 (2): 449–461.
8. Thambi, T., Li, Y., and Lee, D.S. (2017). Injectable hydrogels for sustained release of therapeutic agents. *J. Controlled Release* 267: 57–66.
9. Riederer, M.S., Requist, B.D., Payne, K.A. et al. (2016). Injectable and microporous scaffold of densely-packed, growth factor-encapsulating chitosan microgels. *Carbohydr. Polym.* 152: 792–801.
10. Pacelli, S., Acosta, F., Chakravarti, A.R. et al. (2017). Nanodiamond-based injectable hydrogel for sustained growth factor release: preparation, characterization and in vitro analysis. *Acta Biomater.* 58: 479–491.

11 Drury, J.L. and Mooney, D.J. (2003). Hydrogels for tissue engineering: scaffold design variables and applications. *Biomaterials* 24 (24): 4337–4351.

12 Luo, Y., Kobler, J.B., Heaton, J.T. et al. (2010). Injectable hyaluronic acid-dextran hydrogels and effects of implantation in ferret vocal fold. *J. Biomed. Mater. Res. Part B* 93 (2): 386–393.

13 Hoffman, A.S. (2012). Hydrogels for biomedical applications. *Adv. Drug Delivery Rev.* 64: 18–23.

14 Rehab, A., Akelah, A., Issa, R. et al. (1991). Controlled release of herbicides supported on polysaccharide based hydrogels. *J. Bioact. Compat. Polym.* 6 (1): 52–63.

15 Pulapura, S. and Kohn, J. (1992). Trends in the development of bioresorbable polymers for medical applications. *J. Biomater. Appl.* 6 (3): 216–250.

16 Mittal, H., Ray, S.S., and Okamoto, M. (2016). Recent progress on the design and applications of polysaccharide-based graft copolymer hydrogels as adsorbents for wastewater purification. *Macromol. Mater. Eng.* 301 (5): 496–522.

17 Silva, S.S., Mano, J.F., and Reis, R.L. (2010). Potential applications of natural origin polymer-based systems in soft tissue regeneration. *Crit. Rev. Biotechnol.* 30 (3): 200–221.

18 Li, F., Li, S., El Ghzaoui, A. et al. (2007). Synthesis and gelation properties of PEG–PLA– PEG triblock copolymers obtained by coupling monohydroxylated PEG– PLA with adipoyl chloride. *Langmuir* 23 (5): 2778–2783.

19 El-Rehim, H.A., Hegazy, E.S.A., and El-Mohdy, H.A. (2004). Radiation synthesis of hydrogels to enhance sandy soils water retention and increase plant performance. *J. Appl. Polym. Sci.* 93 (3): 1360–1371.

20 Ahmed, E.M. (2015). Hydrogel: preparation, characterization, and applications: a review. *J. Adv. Res.* 6 (2): 105–121.

21 Thakur, S., Thakur, V.K., and Arotiba, O.A. (2018). History, classification, properties and application of hydrogels: an overview. In: *Hydrogels: Recent Advances*, 29–50.

22 Barba, B.J.D., Tranquilan-Aranilla, C., and Abad, L.V. (2016). Hemostatic potential of natural/synthetic polymer based hydrogels crosslinked by gamma radiation. *Radiat. Phys. Chem.* 118: 111–113.

23 Nur, M. and Vasiljevic, T. (2017). Can natural polymers assist in delivering insulin orally? *Int. J. Biol. Macromol.* 103: 889–901.

24 Lim, J., Chouai, A., Lo, S.T. et al. (2009). Design, synthesis, characterization, and biological evaluation of triazine dendrimers bearing paclitaxel using ester and ester/disulfide linkages. *Bioconjugate Chem.* 20 (11): 2154–2161.

25 Singhal, R. and Gupta, K. (2016). A review: tailor-made hydrogel structures (classifications and synthesis parameters). *Polym. Plast. Technol. Eng.* 55 (1): 54–70.

26 Mitsumata, T., Suemitsu, Y., Fujii, K. et al. (2003). pH-response of chitosan, κ-carrageenan, carboxymethyl cellulose sodium salt complex hydrogels. *Polymer* 44 (23): 7103–7111.

27 Jaiswal, M. and Koul, V. (2013). Assessment of multicomponent hydrogel scaffolds of poly (acrylic acid-2-hydroxy ethyl methacrylate)/gelatin for tissue engineering applications. *J. Biomater. Appl.* 27 (7): 848–861.

28 Miyata, T. (2002). Gels and interpenetrating polymer networks. In: *Supramolecular Design for Biological Applications* (ed. S. Seiffert), 109–150. CRC Press.

29 Zhang, Y., Wu, F., Li, M. et al. (2005). pH switching on-off semi-IPN hydrogel based on cross-linked poly (acrylamide-co-acrylic acid) and linear polyallyamine. *Polymer* 46 (18): 7695–7700.

30 Cheng, W.M., Hu, X.M., Zhao, Y.Y. et al. (2017). Preparation and swelling properties of poly (acrylic acid-co-acrylamide) composite hydrogels. *e-Polymers* 17 (1): 95–106.

31 Maitra, J. and Shukla, V.K. (2014). Cross-linking in hydrogels-a review. *Am. J. Polym. Sci.* 4 (2): 25–31.

32 Schurz, J. (1991). Rheology of polymer solutions of the network type. *Prog. Polym. Sci.* 16 (1): 1–53.

33 Okumura, Y. and Ito, K. (2001). The polyrotaxane gel: a topological gel by figure-of-eight cross-links. *Adv. Mater.* 13 (7): 485–487.

34 Zohourian, M.M. and Kabiri, K. (2008). Superabsorbent polymer materials: a review. *Iran. Polym. J.* 17 (6): 451–477.

35 Shibayama, M. (2012). Structure-mechanical property relationship of tough hydrogels. *Soft Matter* 8 (31): 8030–8038.

36 Richter, A. (2010). Hydrogels for actuators. In: *Hydrogel Sensors and Actuators: Engineering and Technology*, 221–248.

37 Laftah, W.A., Hashim, S., and Ibrahim, A.N. (2011). Polymer hydrogels: a review. *Polym. Plast. Technol. Eng.* 50 (14): 1475–1486.

38 Ullah, F., Othman, M.B.H., Javed, F. et al. (2015). Classification, processing and application of hydrogels: a review. *Mater. Sci. Eng., C* 57: 414–433.

39 Khan, A., Othman, M.B.H., Razak, K.A. et al. (2013). Synthesis and physicochemical investigation of chitosan-PMAA-based dual-responsive hydrogels. *J. Polym. Res.* 20: 1–8.

40 Geever, L.M., Cooney, C.C., Lyons, J.G. et al. (2008). Characterisation and controlled drug release from novel drug-loaded hydrogels. *Eur. J. Pharm. Biopharm.* 69 (3): 1147–1159.

41 Bawa, P., Pillay, V., Choonara, Y.E. et al. (2009). Stimuli-responsive polymers and their applications in drug delivery. *Biomed. Mater.* 4 (2): 022001.

42 Zhang, Y.Z., Wang, Y., Jiang, Q. et al. (2020). MXene printing and patterned coating for device applications. *Adv. Mater.* 32 (21): 1908486.

43 Alhabeb, M., Maleski, K., Anasori, B. et al. (2017). Guidelines for synthesis and processing of two-dimensional titanium carbide (Ti3C2T x MXene). *Chem. Mater.* 29 (18): 7633–7644.

44 Ye, G., Wen, Z., Wen, F. et al. (2020). Mussel-inspired conductive Ti2C-cryogel promotes functional maturation of cardiomyocytes and enhances repair of myocardial infarction. *Theranostics* 10 (5): 2047.

45 Lin, Z., Barbara, D., Taberna, P.L. et al. (2016). Capacitance of $Ti_3C_2T_x$ MXene in ionic liquid electrolyte. *J. Power Sources* 326: 575–579.

46 Kayali, E., VahidMohammadi, A., Orangi, J. et al. (2018). Controlling the dimensions of 2D MXenes for ultrahigh-rate pseudocapacitive energy storage. *ACS Appl. Mater. Interfaces* 10 (31): 25949–25954.

47 Xing, C., Chen, S., Liang, X. et al. (2018). Two-dimensional MXene (Ti_3C_2)-integrated cellulose hydrogels: toward smart three-dimensional network nanoplatforms exhibiting light-induced swelling and bimodal photothermal/chemotherapy anticancer activity. *ACS Appl. Mater. Interfaces* 10 (33): 27631–27643.

48 Buchholz, F. (1998). Applications of superabsorbent polymers. In: *Modern Superabsorbent Polymer Technology* (ed. F. Buchholz and A. Graham), xvii+279. New York: Wiley.

49 Saraydin, D., Karadağ, E., Öztop, H.N. et al. (1994). Adsorption of bovine serum albumin onto acrylamid—maleic acid hydrogels. *Biomaterials* 15 (11): 917–920.

50 Karadağ, E., Saraydin, D., and Güven, O. (1997). Interaction of nicotine and its pharmaceutical derivatives with acrylamide/itaconic acid hydrogels. *J. Appl. Polym. Sci.* 66 (4): 733–739.

51 Karadağ, E., Saraydin, D., and Güven, O. (1995). Behaviors of acrylamide/itaconic acid hydrogels in uptake of uranyl ions from aqueous solutions. *Sep. Sci. Technol.* 30 (20): 3747–3760.

52 Kloxin, A.M., Kloxin, C.J., Bowman, C.N. et al. (2010). Mechanical properties of cellularly responsive hydrogels and their experimental determination. *Adv. Mater.* 22 (31): 3484–3494.

53 Zhong, Y., Xia, X., Shi, F. et al. (2016). Transition metal carbides and nitrides in energy storage and conversion. *Adv. Sci.* 3 (5): 1500286.

54 Gogotsi, Y.G. and Andrievski, R.A. (2012). *Materials Science of Carbides, Nitrides and Borides*, vol. 68. Springer Science & Business Media.

55 Naguib, M., Mashtalir, O., Carle, J. et al. (2012). Two-dimensional transition metal carbides. *ACS Nano* 6: 1322–1331.

56 Pang, J., Mendes, R.G., Bachmatiuk, A. et al. (2019). Applications of 2D MXenes in energy conversion and storage systems. *Chem. Soc. Rev.* 48 (1): 72–133.

57 Fu, Z., Wang, N., Legut, D. et al. (2019). Rational design of flexible two-dimensional MXenes with multiple functionalities. *Chem. Rev.* 119 (23): 11980–12031.

58 Anasori, B., Lukatskaya, M., and Gogotsi, Y. (2017). 2D Metal carbides and nitrides (MXenes) for energy storage. *Nat. Rev. Mater.* 2: 16098.

59 Lee, K.H., Zhang, Y.Z., Jiang, Q. et al. (2020). Ultrasound-driven two-dimensional Ti(3)C(2)T(x) MXene hydrogel generator. *ACS Nano* 14 (3): 3199–3207.

60 Chen, Y., Xie, X., Xin, X. et al. (2019). $Ti_3C_2T_x$-based three-dimensional hydrogel by a graphene oxide-assisted self-convergence process for enhanced photoredox catalysis. *ACS Nano* 13 (1): 295–304.

61 Shang, T., Lin, Z., Qi, C. et al. (2019). 3D Macroscopic architectures from self-assembled MXene hydrogels. *Adv. Funct. Mater.* 29.

62 Liao, H., Guo, X., Wan, P. et al. (2019). Conductive MXene nanocomposite organohydrogel for flexible, healable, low-temperature tolerant strain sensors. *Adv. Funct. Mater.* 29: 1904507.

63 Deng, Y., Shang, T., Wu, Z. et al. (2019). Fast gelation of $Ti_3C_2T_x$ MXene initiated by metal ions. *Adv. Mater.* 31 (43): 1902432.

64 Lukatskaya, M.R., Kota, S., Lin, Z. et al. (2017). Ultra-high-rate pseudocapacitive energy storage in two-dimensional transition metal carbides. *Nat. Energy* 2 (8): 1–6.

65 Zhou, Z., Liu, J., Zhang, X. et al. (2019). Ultrathin MXene/calcium alginate aerogel film for high-performance electromagnetic interference shielding. *Adv. Mater. Interfaces* 6 (6): 1802040.

66 Minden-Birkenmaier, B.A., Meadows, M.B., Cherukuri, K. et al. (2019). The effect of Manuka honey on dHL-60 cytokine, chemokine, and matrix-degrading enzyme release under inflammatory conditions. *Med One* 4 (2): e190005.

67 Rafieerad, A., Yan, W., Sequiera, G.L. et al. (2019). Application of Ti_3 C_2 MXene quantum dots for immunomodulation and regenerative medicine. *Adv. Healthcare Mater.* 8 (16): e1900569.

68 Saravanan, S., Sareen, N., Abu-El-Rub, E. et al. (2018). Graphene oxide-gold nanosheets containing chitosan scaffold improves ventricular contractility and function after implantation into infarcted heart. *Sci. Rep.* 8 (1): 15069.

69 Soleymaniha, M., Shahbazi, M.A., Rafieerad, A.R. et al. (2019). Promoting role of MXene nanosheets in biomedical sciences: therapeutic and biosensing innovations. *Adv. Healthcare Mater.* 8 (1): e1801137.

70 Wang, H. and Roman, M. (2023). Effects of chitosan molecular weight and degree of deacetylation on chitosan−cellulose nanocrystal complexes and their formation. *Molecules* 28 (3): 1361.

71 Kivity, S., Agmon-Levin, N., Blank, M. et al. (2009). Infections and autoimmunity--friends or foes? *Trends Immunol.* 30 (8): 409–414.

72 Li, R., Zhang, L., Shi, L. et al. (2017). MXene Ti_3C_2: an effective 2D light-to-heat conversion material. *ACS Nano* 11 (4): 3752–3759.

73 Lipatov, A., Lu, H., Alhabeb, M. et al. (2018). Elastic properties of 2D $Ti_3C_2T_x$ MXene monolayers and bilayers. *Sci. Adv.* 4 (6): eaat0491.

74 Shen, K.H., Lu, C.H., Kuo, C.Y. et al. (2021). Smart near infrared-responsive nanocomposite hydrogels for therapeutics and diagnostics. *J. Mater. Chem. B* 9 (35): 7100–7116.

75 Sanchez-Casanova, S., Martin-Saavedra, F.M., Escudero-Duch, C. et al. (2020). Local delivery of bone morphogenetic protein-2 from near infrared-responsive hydrogels for bone tissue regeneration. *Biomaterials* 241: 119909.

76 Yang, X., Zhang, C., Deng, D. et al. (2022). Multiple stimuli-responsive MXene-based hydrogel as intelligent drug delivery carriers for deep chronic wound healing. *Small* 18 (5): e2104368.

77 Zhang, H., Fan, T., Chen, W. et al. (2020). Recent advances of two-dimensional materials in smart drug delivery nano-systems. *Bioact. Mater.* 5: 1071–1086.

78 Khodadadi Yazdi, M., Taghizadeh, A., Taghizadeh, M. et al. (2020). Agarose-based biomaterials for advanced drug delivery. *J. Controlled Release* 326: 523–543.

79 Kim, C., Jeong, D., Kim, S. et al. (2019). Cyclodextrin functionalized agarose gel with low gelling temperature for controlled drug delivery systems. *Carbohydr. Polym.* 222: 115011.

80 Dong, Y., Li, S., Li, X. et al. (2021). Smart MXene/agarose hydrogel with photothermal property for controlled drug release. *Int. J. Biol. Macromol.* 190: 693–699.

81 Yang, C., Xu, D., Peng, W. et al. (2018). $Ti_2C_3T_x$ nanosheets as photothermal agents for near-infrared responsive hydrogels. *Nanoscale* 10 (32): 15387–15392.

82 Liu, X., Wei, Y., and Qiu, Y. (2021). Advanced flexible skin-like pressure and strain sensors for human health monitoring. *Micromachines* 12 (6): 695.

83 Zhang, Y.Z., El-Demellawi, J.K., Jiang, Q. et al. (2020). MXene hydrogels: fundamentals and applications. *Chem. Soc. Rev.* 49 (20): 7229–7251.

84 Ge, G., Lu, Y., Qu, X. et al. (2019). Muscle-inspired self-healing hydrogels for strain and temperature sensor. *ACS Nano* 14 (1): 218–228.

85 Wang, C., Zhao, J., Ma, C. et al. (2017). Detection of non-joint areas tiny strain and anti-interference voice recognition by micro-cracked metal thin film. *Nano Energy* 34: 578–585.

86 Ge, G., Yuan, W., Zhao, W. et al. (2019). Highly stretchable and autonomously healable epidermal sensor based on multi-functional hydrogel frameworks. *J. Mater. Chem. A* 7 (11): 5949–5956.

87 Wang, X., Gu, Y., Xiong, Z. et al. (2014). Silk-molded flexible, ultrasensitive, and highly stable electronic skin for monitoring human physiological signals. *Adv. Mater.* 26 (9): 1336–1342.

88 Hu, T., Wang, J., Zhang, H. et al. (2015). Vibrational properties of Ti_3C_2 and $Ti_3C_2T_2$ (T = O, F, OH) monosheets by first-principles calculations: a comparative study. *Phys. Chem. Chem. Phys.* 17 (15): 9997–10003.

89 Zhang, Y.Z., Lee, K.H., Anjum, D.H. et al. (2018). MXenes stretch hydrogel sensor performance to new limits. *Sci. Adv.* 4: eaat0098.

90 Yang, C., Xu, D., Peng, W. et al. (2018). Ti_2C_3Tx nanosheets as photothermal agents for near-infrared responsive hydrogels. *Nanoscale* 10 (32): 15387–15392.

91 Zhang, Y.-Z., El-Demellawi, J.K., Jiang, Q. et al. (2020). MXene hydrogels: fundamentals and applications. *Chem. Soc. Rev.* 49 (20): 7229–7251.

92 Ge, G., Yuan, W., Zhao, W. et al. (2019). Highly stretchable and autonomously healable epidermal sensor based on multi-functional hydrogel frameworks. *J. Mater. Chem. A* 7 (11): 5949–5956.

93 Ge, G., Lu, Y., Qu, X. et al. (2019). Muscle-inspired self-healing hydrogels for strain and temperature sensor. *ACS Nano* 14 (1): 218–228.

7

Comparison of MXenes with Other 2D Materials

Yoshiyuki Sato, Daisuke Nakane, and Takashiro Akitsu

Department of Chemistry, Faculty of Science Division II, Tokyo University of Science, Tokyo, Japan

7.1 Introduction of MXenes

In recent years, the demand for renewable energy has been rising globally due to increasing environmental concerns, and novel materials for metal-ion batteries with higher efficiency have been actively developed. According to the operating principles of electrochemical energy conversion, surfaces and/or interfaces play a crucial role in intercalation of (metal) ions, physical and/or chemical adsorption, and redox chemical reactions [1]. Among them, two-dimensional (2D) materials, such as (sheet-shaped carbon) graphene, boron nitride, and transition metal dichalcogenide (TMDC), which possess a quite large surface area and good electrical conductivity, are considered to be preferred materials for chemical energy conversion [2–5]. Besides electrochemical energy conversion or storage, 2D materials offer the following three main advantages for reusable metal-ion batteries: (i) the crystallized internal structure facilitates interlayer transport of electrons at the atomic level; (ii) the layered internal structure creates a number of requisite pathways for the diffusion of metal ions and creates a favorable environment for it; and (iii) the 2D structure provides an elastic space that covers the volume change when electrochemical reactions occur and maintains excellent contact between the layers [3, 6, 7].

Among 2D materials, new 2D metal carbides and/or 2D metal nitrides (so-called MXenes) exhibiting characteristic surface electrochemical performance, tunable structural features, and physical and/or chemical properties have attracted great interest for the use of purposes of energy conversion and/or storage. Since the oldest and first report of study on $Ti_3C_2T_x$ in 2011, a number of compositions have been rapidly developed, and approximately 30 MXene-based compositions have been reported to date [8]. This newly synthesized material, MXene, prepared during etching of the MAX ($M_{n+1}AX_n$) phase, is now widely considered to be a highly promising material for supercapacitors [9]. MXene also has been expected for applications in several types of metal-ion batteries, such as lithium, sodium, and potassium, due to its oxygen-terminated and/or fluorine functionalized surface [10–15]. However, most literature on MXene has focused on its properties, and few studies have addressed modification operations of the surface of MXene. These, however, are critical for their application to high-performance storage in technology of energy devices.

MXenes: Next-Generation 2D Materials: Fundamentals and Applications, First Edition.
Edited by Jay Singh, Kshitij RB Singh, Ravindra Pratap Singh, and Charles Oluwaseun Adetunji.

MXenes generally refer to M_nX_{n-1} (n = 2, 3 and 4) layers of metal-containing ceramic called MAX ($M_{n+1}AX_n$) phase. They have interlayer A atoms removed (M, A, and X denote early transition metals, 3 or 4 group elements, and carbon or nitrogen elements, respectively) [16]. Since the MX hexagonal lattice can be maintained after the A atoms are removed, the M_nX_{n-1} layer is prepared after extracting the elements. The MXene films are generally aligned horizontally as prior to MAX separation. Since most MXenes exhibit desired mechanical properties, they are expected to be broadly applied to nanodevices with mechanical requirements. There are several types of etching methods of MXenes.

MXene can be divided into several types according to the separation method that occurs during etching, and the separation method differs depending on the characteristics of the surface portion of the MXene to be separated. Among the several etching methods that exist, the most traditional or conventional etching technique is using hydrofluoric (HF) acid presented in 2011 [17] in order for stratifying the Ti_3AlC_2 MAX phase. The easy substitution reaction of HF to produce H_2 facilitated the loss of the Al layer from the phase of Ti_3AlC_2 MAX. The solution of deionized water and HF achieved $Ti_3C_2T_x$ (where T denotes –OH, –F, or =O components) by causing a reaction that produced Ti_3C_2. Moreover, by using the HF etching method, Ti_2AlC, Ta_4AlC_3, $(Ti_{0.5}Nb_{0.5})_2AlC$, $V_0(Ti_{0.5}Nb_{0.5})_2AlC$, $(V_{0.5}Cr_{0.5})_3AlC_2$, Ti_3AlCN, Nb_2AlC, Ti_3SiC_2, Mo_2Ga_2C, and $Zr_3Al_3C_5$ were exfoliated to MXene [18–22]. Indeed, such HF etching continues to be the most widely used preparation of MXenes. Time, temperature, and density of F^- ion during etching play an essential role in the formation of high-quality MXene layers. Etching with relatively stronger acids, however, may result in larger defects of the surface and reduced MXenes output [23]. Alhabeb et al. [24] confirmed that $Ti_3C_2T_x$ from conc. HF gave rise to ideally layered morphology in contrast to other acids. The MXene from etching with HF exhibited the original surface morphology with functional groups as already mentioned.

The addition of two types of transition metals into MXene creates a unique structure of MXene. Specifically, the addition of two transition metals gives rise to so-called solid solutions like (Ti, Nb) CT_x. Additionally, such transition metals may also yield a sandwich of transition metal atoms, for example $Mo_2TiC_2T_x$, or 2D flake with an in-plane ordered $(Mo_{2/3}Y_{1/3})_2CT_x$. Since 2017, numerous ordered MAX phases have been developed by using two metals. Moreover, their magnetic or other features have been rigorously exploited [25–27]. Recently, a number of possible MXene frameworks have also been inferred from calculations of MXene and its precursor derivatives [27–31]. In fact, many MXenes of nonstoichiometric compositions can be synthesized with finely tuned properties and two metals and carbon nitrides of (M and X sites) solid solutions [32].

The stability, electronic states, and various properties of MXenes are greatly affected by the terminal groups and structures present on their surfaces. Both thermodynamically and kinetic aspects by insertion or diffusion in ion batteries are of particular interest for MXenes. Theoretical and experimental studies sometimes differed from each other in proposed mechanisms of MXenes including surface groups. Theoretical calculations proposed the importance of surface groups, namely, –OH enhanced capacitance of MXenes. The capacitance (and energy density) depends not only on functional groups but also on the number of layers inside of MXenes.

The first application of MXene that was considered was energy storage. Recently, the use of MXene in the biomedical field has emerged as a new and especially active research area, and its applications in photo-therapy of cancer, biosensors, and so on have attracted substantial attention [33, 34]. Other areas of research are also underway, including electromagnetic applications, such as interference in electromagnetic shielding and printed antennas [35], and applications in electronic components and structures. However, many of the predicted functions of MXene (e.g. topological insulators and ferromagnetism) have never been experimentally examined. These applications in nanoelectronics, however, indicate that MXene can be considered as one of the important 2D materials in industry. In general, MXene may be relatively more predominant in the field of energy materials [36].

Mo_2CT_x also helps to assist electron transfer in hydrogen evolution reaction (HER) catalysis [37, 38] from the viewpoint of electron conductivity or transfer. Preparation and structural studies of MoS_2 encapsulated Mo_2CT_x were carried out by Yu et al. [39]. The $MoS_2@Mo_2CT_x$ hybrids synthesized by the hydrothermal method resulted in improved HER performance under alkaline conditions, showing a low overpotential and a very low transfer resistance at a current density of $10\,mA\,cm^{-2}$. This synthetic strategy and design approach demonstrates that the properties of MXene can be tailored to the desired application by preparing a combination of organic-like molybdenum carbide and MoS_2 nanoflowers. Furthermore, Kuznetsov et al. reported that substitution of transition metals in the 3 or 4 groups in the host MXene matrix leads to a relatively stable MXene with a significant increase in catalytic efficiency [40, 41]. MXene has also been applied to clean energy technology. The ternary carbides (e.g. Mo_2Ga_2C or Ti_2AlC) produce MXene, for example Ga and Al free Mo_2CT_x and Ti_2CT_x by HF etching [42]. The electrochemical activity was investigated by a common electrochemical cell, in which MXene was drop-cast on a glassy carbon substrate. As a result, the highest (283 mV) potential HER material with an overpotential was Mo_2CT_x, yielding a current density of approximately $10\,mA\,cm^{-2}$. In contrast, Ti_2CT_x showed relatively lower data with an overpotential to attain a current density of $10\,mA\,cm^{-2}$.

The oxygen reduction reaction (ORR) constitutes a key phenomenon in energy conversion systems. Liu et al. [43] studied the influence of surface-terminating groups of MXene critically affecting ORR data theoretically. F-terminated surfaces are expected to show better activity, although they are less stable than O-terminated surfaces. Yang et al. developed a non-precious metal-containing composite (Fe–N–C@ $Ti_3C_2T_x$) catalyst using a separated pyrolysis method and demonstrated that $Ti_3C_2T_x$ not only acts as an excellent conductive material but also inhibits the aggregation and Fe–N–C's breakdown after carbonization. In this way, enhancing the ORR activity and stability of the Fe–N–C catalyst was realized [32]. Fe–N–C@MXene has a half-wave potential of 0.887 V and a limiting current density of $6.4\,mA\,cm^{-2}$, indicating excellent electrocatalytic ORR activity (similar level to Pt/C available commercially). Purified samples of MXene did not show ORR activity at all due to the lack of active sites. Interestingly, Fe–N–C@MXene achieved a tremendous enhancement of ORR activity, with 0.777 V of a half-wave potential and a limiting current density under acidic conditions. At present, predicting the needed properties of MXene materials with desired performance for superior OER and ORR may be still basic research due to their slow rate. Thus, the present report will establish an appropriate strategy to encourage interest in pursuing the engineering of MXene for OER and/or ORR applications in energy fields. Such studies will validate the usage of MXene as a special 2D material that can be combined with other active catalytic materials having heterostructures. The development of MXene hybrids incorporating inexpensive metal nanoparticles with good redox activity, efficiency, and stability remains an important and difficult ongoing task.

7.2 MXenes vs. Carbon Materials

Among 2D materials, graphene exhibits the unique properties of being naturally thin in atomic level and mechanically flexible. It also possesses excellent thermal properties, which have attracted great interest for thermal management. These properties have the potential to solve increasingly serious heat dissipation problems in (opto)electronic devices. As a consequence, research on thermal interface materials based on 2D materials has been widely conducted [44].

When compared to reduced graphene oxide (rGO), experiments showed that $Ti_3C_2T_x$ coated on cellulose exhibited superior light absorption properties. In addition, temperature changes on the surface of both bulk water molecules and two photothermal films were measured by means of IR thermography. It was observed that the temperature on photothermal $Ti_3C_2T_x$/cellulose film's surface increased more effectively than that of rGO after 10 minutes. This phenomenon indicates that $Ti_3C_2T_x$ possesses better photo-absorption data and light-to-heat conversion ratios than rGO and can promote better water transfer to the surface of photothermal films. Therefore, $Ti_3C_2T_x$ MXene seems to be especially promising for solar desalination [45].

In the fabrication of MXene, materials combining metal–organic frameworks (MOFs) have also been extensively studied. Here, we introduce a composite material of MXene, Ti_3C_2, and a Co-containing ZIF-67 (a famous zeolite imidazole MOF). The conversion of nitrogen gases to ammonia molecules is a highly important mechanism in ecosystems. Redox preparation of ammonia from nitrogen (with water as a byproduct) is also an environmentally friendly process with great application prospects. Therefore, many attempts have been devoted to enhance activity and/or selectivity as catalysts. Indeed, because of MOFs' high porosity and large surface area in active and the excellent conductivity of Ti_3C_2, such a composite can efficiently produce ammonia electrochemically. In particular, the ZIF-67@Ti_3C_2 catalyst showed an excellent amount of product of ammonia. This is markedly larger than the values shown by Ti_3C_2 and ZIF-67 alone, resulting in good efficiency [46]. This report constituted a major step to extend the application of the MXene materials in nitrogen reduction by electrochemical reactions. In addition, the combination of MOFs used as a composite, taking into account their types and dimensionality, may lead to realization of the development of materials that exhibit higher activity.

7.3 MXenes vs. 2D-chalcogenide/Carbide/Nitride

One of the most well-known examples of the application of the series of processes that transform multilayered three-dimensional graphite into 2D graphene to other materials is the 2D TMDC. Generally, 2D materials have advantage to tune or realize characteristic crystal or electronic structures different from 3D materials (namely, bulk crystalline materials). Monolayer TMDCs have a crystal (bulk solid-state 3D) structure that is very similar to the

2D one of graphene, with the difference being that unlike graphene, they have a gap in the Dirac cone that makes them semiconductors. This band gap enables field-effect transistor operation with high switching performance [47].

The advantages of 2D MXene as a nanomaterial indicate great potential for the realization of new secondary batteries. However, as with other 2D materials, their practical value has been compromised by severe self-aggregation, small capacity, and unsatisfactory durability when used for large metal ions, such as sodium and potassium ions. Here, we describe a new ternary heterostructure exhibiting ultrafast ion transport properties, self-assembled from selenides with transition metals (MSe and M = Cu, Ni, Co, respectively), MXene nanostructured sheets and N-rich carbon nanoribbons, for application in sodium and potassium ion batteries. Ternary heterostructures with ultrafast ion transport properties will be discussed. Positively charged MSe is electrostatically adsorbed on the MXene's surface with a negatively charged –OH (hydroxyl) functional group, whereas N-rich carbon nanoribbons are bonded to the opposite side of MXene by bridged -NH_2 groups and hydrogen bondings. This characteristic of MXene's heterostructure made it impossible to re-stack such 2D materials with increasing conductivity. Interestingly, it provides an ultrafast ion transport between faces pathway and additional sites for storage at the surface and interface to enhance the high-rate storage for applications such as sodium and potassium ion batteries. Computational study with DFT reveals that the interfacial transport of ions may be higher than the data of MXenes alone. This significantly improves the storage capacity of Na^+ and K^+, resulting in good stability in long-term cycling. As with graphene and time, the introduction of TMDCs within the MXene structure facilitated augmented functionality. These studies should provide novel and valuable insights into the field of 2D materials engineering [48].

Many investigations exist that focus on MXene and TMDC complexes of hetero-component nano-scaled structures for energy applications. So far, many MXene and hierarchical MXene/TMDC hetero-component structures can be prepared by selective etching and exfoliation of MAX phase in the layered structures. Generally speaking, when synthesizing MXene from the MAX phase, F-containing materials may be used as etching agents, resulting in MXene's surface being modified with abundant or common terminating groups such as –O, –F, and –OH, which reduces the conductivity of MXene and degrades its electrochemical performance. MXenes, however, can be hybridized with other active materials, such as TMDCs (MoS_2, CoS, $MoSe_2$, etc.). This provides an effective method to control electronic states and features for MXenes, due to their terminating functional groups [49]. The good natures of MXene and hierarchical MXene/TMDC hetero-component structures guarantee excellent properties as sodium and potassium ion batteries. Moreover, when MXene and hierarchical MXene/TMDC hetero-component structures are employed in supercapacitors, their lamellar structure, abundant active sites, and polar active materials can improve the data as supercapacitors. MXene is promising as a flexible electrode for rechargeable electrochemical devices due to its 2D structure and/or conductivity. Moreover, the hierarchical MXene/TMDC hetero-component structure exhibits desirable catalytic properties due to its conductivity as metal, tunable electronic structure, many terminating groups, and outside metal sites. Therefore, MXene and hierarchical MXene/TMDC hetero-component structures have exhibited good energy materials, but several challenges remain to be overcome for practical use. The first good synthetic route to high-quality MXene was with high (good) yield, not expensive, and regulated termination of surface. In particular, the preparation of MXenes with terminal functional groups is urgently needed to elucidate the interrelationship between the termination groups and MXenes' data. The subsequent step is to fabricate pure monolayer MXenes without surface terminations by chemical vapor deposition (CVD) and further investigate their several physical properties to obtain additional research data. Other goals include defining the intercalation mechanism, charge accumulation mechanism, electronic structure, and other physicochemical properties of TMDC-hybridized MXene. MXene/TMDC heterostructures can also be combined with high-performance electrocatalysts and tunable bandgap semiconductors to improve electrochemical performance, especially in rechargeable batteries and superconducting devices, and to tune catalytic reactions. As a consequence, an urgent need exists to further understand their heterointerfaces. Furthermore, MXene and hierarchical MXene/TMDC hetero-component structures are anticipated to be useful in other important applications, for example, water purification, gas separation, and analytical sensors.

7.4 MXenes vs. 2D Metal–Organic Frameworks

MOFs are composed of coordination compounds with bridging organic ligands. MOFs are attracting increasing attention as multifunctional materials because they exhibit a broad range of functions by combining the various elements that make up the MOF. Among the many studies on MOFs, the use of MOFs in 2D membrane structures has been actively researched, and efforts are underway to apply the unique functions generated between the oriented layers of 2D MOFs to electronic materials.

As already mentioned, numerous investigations exist on hybrid materials combining 2D MOFs and MXene. In this section, we introduce an approach to the application of MOFs that is different from the abovementioned examples. Using the organometallic structure-based approach, an example of combining hierarchical porous mixed Ni and Co sulfides (abbreviated as NiCoS) on $Ti_3C_2T_x$ MXene has been reported [50]. $Ti_3C_2T_x$ (where T_x denotes surface end groups, for example, $-F_x$ and $-(OH)_x$) MXene MOF-derived/2D MXene hybrid materials were prepared by forming ZIF-67 on the sheet (hereinafter referred to as NiCoS/ $Ti_3C_2T_x$) and converting it into hierarchical porous mixed Ni and Co sulfides, whose zeta potentials were also reported with the relationship of terminal groups. Co^{2+} immobilized by electrostatic interaction due to negative charges on $Ti_3C_2T_x$ surface coordinates with 2-methylimidazole ligands to uniformly yield ZIF-67/$Ti_3C_2T_x$ composite material. Using $Ni(NO_3)_2$ reacted here, Ni^{2+} ions etched ZIF-67 and Co^{2+}/Co^{3+} ions co-precipitated with Ni^{2+} ions to give rise to layered mixed Ni and Co sulfides (namely, NiCo-LDH/ $Ti_3C_2T_x$).

The ZIF series can be said to be an excellent example of MOF with various applications. Let's look at the superiority of MXene or materials above over MOFs with a catalyst application example. Molecular nitrogen adsorption or desorption were also tested to verify the adsorption properties of $Ti_3C_2T_x$, NiCoS, and NiCoS/$Ti_3C_2T_x$ in their native state. It was found that they all showed similar adsorption/desorption curves, suggesting the existence of mesopores in the structure. The BET surface area of NiCoS/$Ti_3C_2T_x$ was higher than those of $Ti_3C_2T_x$ and NiCoS. Furthermore, distribution curve of pore size confirmed that NiCoS/$Ti_3C_2T_x$ has hierarchical pores that simultaneously contain mesopores about 3 nm and macropores about 70 nm. Therefore, the hierarchical porous structure of NiCoS/$Ti_3C_2T_x$ with high surface area and high pore volume allows for large amounts of surface-active sites and easy access to the electrolyte, which may be advantage points as a catalyst.

Next, their electrochemical characterization was performed in a common method. The results of linear sweep voltammetry (LSV) curves revealed that $Ti_3C_2T_x$ alone did not exhibit OER activity. On the other hand, NiCo-LDH and NiCoS showed high catalytic activity with high current density and low onset potential for both samples of NiCo-LDH/$Ti_3C_2T_x$ and NiCoS/$Ti_3C_2T_x$ composite materials, suggesting that the $Ti_3C_2T_x$ matrix has a positive effect on OER behavior. In this test, NiCoS/$Ti_3C_2T_x$ exhibited the best data, with a small overpotential to obtain a standard current density. The results also suggested a weak synergistic effect of the physical mixture of these two materials.

ICP-AES measurement showed that the contents of Ni and Co in NiCoS/$Ti_3C_2T_x$ were 44.59 wt% and 21.22 wt%, respectively. Therefore, the turnover frequency (TOF) data evaluated suggested that all Ni and Co sites in the catalyst are activated. These values are markedly higher than those of RuO_2. Under a constant current, the researchers investigated the durability of the catalyst and revealed that chronopotentiometry curve kept the potential of NiCoS/$Ti_3C_2T_x$ for six hours. As expected in analogy with OER performance, NiCoS/$Ti_3C_2T_x$ exhibited the best stability among commercial RuO_2, NiCoS, NiCo-LDH/$Ti_3C_2T_x$, and NiCoLDH.

With the aid of correlation of capacitance, XPS measurement was also carried out for NiCoS/$Ti_3C_2T_x$ to examine OER catalytic data electrochemically. For evaluation of electrochemical double-layer capacitance, cyclic voltammetry (CV) curves were used for these calculations of active materials such as NiCoS/$Ti_3C_2T_x$, NiCoS, NiCo-LDH/$Ti_3C_2T_x$, NiCo-LDH, and RuO_2. As for electrochemical impedance spectroscopy, Nyquist plot of the electro-catalysts revealed a characteristic semicircle being ascribed to interfacial charge transfer resistance. Bode modulus plot also provides similar results, that is the lowest resistance of NiCoS/$Ti_3C_2T_x$. The water oxidation performance of NiCoS/$Ti_3C_2T_x$ may be supported by its high conductivity, which resulted in effective tuning of NiCo-LDH by sulfidation treatment and binding of NiCo-LDH.

Comparison of XPS before/after OER tests was also used for structure and activity investigation for NiCoS/$Ti_3C_2T_x$ by proving Ni 2p 853.28 and 870.55 eV peaks and their shifts by OER procedures (or decreasing by Ni–O formation) ascribed to Ni-S bonds. In the case of Co 2p peaks associated with Co–S bond formation, Co of NiCoS/$Ti_3C_2T_x$ was oxidized by OER reactions, which was also supported by splitting O1s peaks of NiCoLDH/$Ti_3C_2T_x$ by H_2O adsorption. For NiCoS/$Ti_3C_2T_x$, oxide/hydroxide of Ni or Co after OER processes resulted in decreasing S peaks because of appearance of OH^-, O^{2-}, and H_2O species. In this way, water oxidizing catalyst NiCoS/$Ti_3C_2T_x$ not only oxide/hydroxide of Ni or Co but also irreversible changes of structures after OER could be observed by means of XPS.

Feasibility of NiCoS/$Ti_3C_2T_x$ as rechargeable battery materials of zinc-air type (assembled anode: Zn; cathode: NiCoS/$Ti_3C_2T_x$–NiCoS/$Ti_3C_2T_x$–Pt/C) can be discussed based on the results of OER tests. NiCoS/$Ti_3C_2T_x$ + Pt/C exhibited better data of open-circuit voltage and charge–discharge curve than RuO_2 + Pt/C and worse data than noble metals.

Stability of charge–discharge curves was exhibited for NiCoS/$Ti_3C_2T_x$ + Pt/C, which was similar to RuO_2 + Pt/C, which was also confirmed keeping red light-emitting diode on for five minutes.

That NiCoS and $Ti_3C_2T_x$ sheets exhibited good charge transfer conductivity, and OER was attributed to their structures and interfacial interactions. For example, OER process changes NiCoS into NiCoOOH–NiCoS, which was confirmed by intrinsically active species for water oxidation on NiCoOOH's surface. Such hybrid materials can be used as air cathode showing overpotential and stability. Of course, water oxidation may be an important energy-conversion technology to use solar cells of low costs.

In this way, the comparison with MOF reveals the structural superiority of MXene to such purposes mentioned in this section. MOF is a solid-state material, but there are organic ligands that are essentially insulators between transition metal ions. Therefore, a 2D structure can be created only when a suitable ligand is used. In addition, at present, a 2D structure cannot be obtained unless a special synthesis method (on the interface, etc.) is used. Furthermore, in order to intentionally produce a low-dimensional conductive material, it is often an advantage that a periodic crystal structure can be constructed in as small a unit as possible. Organic ligands are vulnerable to redox and MOF's ordered structures are easily destroyed. Furthermore, although nano-scaled "void space" is also a feature of MOFs, it also reduces the number of active sites per surface area and regularly provides "defects" in conductive solids. In this respect, restrictions appear in the purposeful material design of the structure of the MOF having a relatively large organic ligand between the metal ions.

7.5 Summary

Although extensive research on 2D materials has been carried out in the past decades, numerous challenges remain in the practical application of MXenes nanocomposites. Indeed, extant literature on MXenes is still at an early stage compared to other types of 2D nanomaterials. Future efforts in the synthesis of MXenes include the topic of how to fabricate 2D materials as thinly as feasible using active exfoliation techniques. Most of the MXene developed has been achieved by selectively etching. Some reagents are highly toxic, but etching of element A has been successfully achieved by using more safe etchants. Moreover, as an alternative to conventional acids, a new etchant has been developed. Further efforts are needed to develop fluorine-free and environmentally friendly etching agents as alternatives to conventional acids.

MXene materials' reactivity and surface are improved by reducing their size. The surface functional groups should have an important influence on the electrochemical properties of MXene. However, morphology increasing the number of active sites should be improved. For instance, to apply for capacitance, pore size and its volume should be well designed. The pore size of MXene is much larger than that of other materials; however, it is still challenging to tune the porosity of MXene. Additionally, unlike graphene, which possesses certain constraints, growing large monolayer MXene sheets is important but still challenging.

Functionalization through covalent bonding may also be important for 2D chemical substances. From a materials engineering perspective, the interface and shape design of MXene-based nanocomposites are critical in order to achieve high catalytic efficiency and long-term stability. MXenes have been studied in clean energy conversion applications, with the exception of HER catalysis. However, limited research has been carried out so far.

For new energy storage and power generation systems, it is expected that the nature of van der Waals interlayer interactions and the variety of 2D materials will be used to build 2D structures of other components. MXene composites can also be used as high performance, low cost, and safe electrodes. However, durability and chemical stability need to be further improved. Thus, solving the abovementioned problems facilitates the search for 2D MXene nanomaterials in energy storage and energy transformation devices.

MXene is said to be highly competitive because it has a variety of applications centered on energy and catalytic (electrochemical) materials, can meet a wide range of technical needs of users, and is economically cost effective. However, at present, it is in the basic research stage, and there are issues to be overcome such as instability (more seriously with acid), orientation, and controllability.

There are effective ways to store MXene suspensions for extended periods of time, but the disadvantage is that they require a frozen environment. It is also difficult to develop such a long-term stable MXene-based 2D layered material. Currently, research is concentrated on the first compound, that is, $Ti_3C_2T_x$. However, in order to adjust the bandgap and conductivity, the combination of element substitutions and the study of surface functional groups are also being conducted. Comprehensive theoretical and experimental strategies will continue to be needed for further utilization.

References

1 Pang, J.B., Mendes, R.G., Bachmatiuk, A. et al. (2019). Applications of 2D MXenes in energy conversion and storage systems. *Chem. Soc. Rev.* 48: 72–133.

2 Tan, C.L. and Zhang, H. (2015). Two-dimensional transition metaldichalcogenide nanosheet-based composites. *Chem. Soc. Rev.* 44: 2713–2731.

3 Sun, Y.F., Gao, S., and Xie, Y. (2014). Atomically-thick two-dimensional crystals: electronic structure regulation and energy device construction. *Chem. Soc. Rev.* 43: 530–546.

4 Pacile, D., Meyer, J., Girit, Ç., and Zettl, A. (2008). The two-dimensional phase of boron nitride: few-atomic-layer sheets and suspended membranes. *Appl. Phys. Lett.* 92: 133107.

5 Wang, Q.H., Kalantar-Zadeh, K., Kis, A. et al. (2012). Electronics and optoelectronics of two-dimensional transition metal dichalcogenides. *Nat. Nanotechnol.* 7: 699.

6 Xiong, D.B., Li, X.F., Bai, Z.M. et al. (2018). Recent advances in layered $Ti_3C_2T_x$ MXene for electrochemical energy storage. *Small* 14: 1703419.

7 Liu, Y.T., Zhu, X.D., and Pan, L. (2018). Hybrid architectures based on 2D MXenes and low-dimensional inorganic nanostructures: methods, synergies, and energy-related applications. *Small* 14: 1803632.

8 Anasori, B., Xie, Y., Beidaghi, M. et al. (2015). Two-dimensional, ordered, double transition metals carbides (MXenes). *ACS Nano* 9: 9507–9516.

9 Mendoza-Sanchez, B. and Gogotsi, Y. (2016). Synthesis of two-dimensional materials for capacitive energy storage. *Adv. Mater.* 28: 6104–6135.

10 Zhang, X., Zhang, Z.H., and Zhou, Z. (2018). MXene-based materials for electrochemical energy storage. *J. Energy Chem.* 27: 73–85.

11 Anasori, B., Lukatskaya, M.R., and Gogotsi, Y. (2017). 2D metal carbides and nitrides (MXenes) for energy storage. *Nat. Rev. Mater.* 2: 16098.

12 Zheng, W., Li, Y., Zhang, P. et al. (2018). Energy storage and application for 2D nano-material MXenes. *Mater. Rev.* 32: 2513–2537.

13 Xie, Y., Naguib, M., Mochalin, V.N. et al. (2014). Role of surface structure on li-ion energy storage capacity of two-dimensional transition-metal carbides. *J. Am. Chem. Soc.* 136: 6385–6394.

14 Qi, X., Chen, X., Peng, S.K. et al. (2019). Research progress on two-dimensional nanomaterials MXenes and their application for lithium-ion batteries. *J. Mater. Eng.* 47: 10–20.

15 Liang, X., Garsuch, A., and Nazar, L.F. (2015). Sulfur cathodes based on conductive MXene nanosheets for high-performance lithium-sulfur batteries. *Angew. Chem. Int. Ed.* 54: 3907–3911.

16 Barsoum, M.W. (2013). *MAX Phases: Properties of Machinable Ternary Carbides and Nitrides*. Wiley.

17 Naguib, M., Kurtoglu, M., Presser, V. et al. (2011). Two-dimensional nanocrystals produced by exfoliation of Ti_3AlC_2. *Adv. Mater.* 23: 4248–4253.

18 Naguib, M., Mashtalir, O., Carle, J. et al. (2012). Two-dimensional transition metal carbides. *ACS Nano* 6: 1322–1331.

19 Mashtalir, O., Lukatskaya, M.R., Zhao, M.Q. et al. (2015). Amine-assisted delamination of Nb_2C MXene for Li-ion energy storage devices. *Adv. Mater.* 27: 3501–3506.

20 Halim, J., Kota, S., Lukatskaya, M.R. et al. (2016). Synthesis and characterization of 2D molybdenum carbide (MXene). *Adv. Funct. Mater.* 26: 3118–3127.

21 Zhou, J., Zha, X.H., Chen, F.Y. et al. (2016). A two-dimensional zirconium carbide by selective etching of Al_3C_3 from nanolaminated $Zr_3Al_3C_5$. *Angew. Chem. Int. Ed.* 55: 5008–5013.

22 Alhabeb, M., Maleski, K., Mathis, T.S. et al. (2018). Selective etching of silicon from Ti_3SiC_2 (MAX) to obtain 2D titanium carbide (MXene). *Angew. Chem. Int. Ed.* 57: 5444–5448.

23 Sang, X., Xie, Y., Lin, M.W. et al. (2016). Atomic defects in monolayer titanium carbide ($Ti_3C_2T_x$) MXene. *ACS Nano* 10: 9193–9200.

24 Alhabeb, M., Maleski, K., Anasori, B. et al. (2017). Guidelines for synthesis and processing of two-dimensional titanium-carbide ($Ti_3C_2T_x$ MXene). *Chem. Mater.* 29: 7633–7644.

25 Petruhins, A., Lu, J., Hultman, L., and Rosen, J. (2019). Synthesis of atomically layered and chemically ordered rare-earth (RE) i-MAX phases; $(Mo_{2/3}RE_{1/3})_2GaC$ with RE= Gd, Tb, Dy, Ho, Er, Tm, Yb, and Lu. *Mater. Res. Lett.* 7: 446–452.

26 Dahlqvist, M., Lu, J., Meshkian, R. et al. (2017). Prediction and synthesis of a family of atomic laminate phases with Kagomé-like and in-plane chemical ordering. *Sci. Adv.* 3: e1700642.

27 Ashton, M., Hennig, R.G., Broderick, S.R. et al. (2016). Computational discovery of stable M_2AX phases. *Phys. Rev. B* 94: 054116.

28 Dahlqvist, M., Petruhins, A., Lu, J. et al. (2018). Origin of chemically ordered atomic laminates (i-MAX): expanding the elemental space by a theoretical/experimental approach. *ACS Nano* 12: 7761–7770.

29 Rajan, A.C., Mishra, A., Satsangi, S. et al. (2018). Machine-learning-assisted accurate band gap predictions of functionalized MXene. *Chem. Mater.* 30: 4031–4038.

30 Hu, T., Li, Z., Hu, M. et al. (2017). Chemical origin of termination-functionalized MXenes: $Ti_3C_2T_2$ as a case study. *J. Phys. Chem. C* 121: 19254–19261.

31 Zha, X.H., Luo, K., Li, Q. et al. (2015). Role of the surface effect on the structural, electronic and mechanical properties of the carbide MXenes. *EPL (Europhys. Lett.)* 111: 26007.

32 Wang, W.T., Batool, N., Zhang, T.H. et al. (2021). When MOFs meet MXenes: superior ORR performance in both alkaline and acidic solutions. *J. Mater. Chem. A* 9: 3952–3960.

33 Anasori, B. and Gogotsi, Y. (2019). *2D Metal Carbides and Nitrides (MXenes): Structure, Properties and Applications*. Springer.

34 Cheng, L., Wang, X., Gong, F. et al. (2020). 2D nanomaterials for cancer theranostic applications. *Adv. Mater.* 32: 1902333.

35 Sarycheva, A., Polemi, A., Liu, Y. et al. (2018). 2D titanium carbide (MXene) for wireless communication. *Sci. Adv.* 4: eaau0920.

36 Sun, S., Liao, C., Hafez, A.M. et al. (2018). Two-dimensional MXenes for energy storage. *Chem. Eng. J.* 338: 27–45.

37 Lim, K.R.G., Handoko, A.D., Nemani, S.K. et al. (2020). Rational design of two-dimensional transition metal carbide/nitride (MXene) hybrids and nanocomposites for catalytic energy storage and conversion. *ACS Nano* 14: 10834–10864.

38 Liang, J., Ding, C., Liu, J. et al. (2019). Heterostructure engineering of Co-doped MoS_2 coupled with Mo_2CT_x MXene for enhanced hydrogen evolution in alkaline media. *Nanoscale* 11: 10992–11000.

39 Ren, J., Zong, H., Sun, Y. et al. (2020). 2D organ-like molybdenum carbide (MXene) coupled with MoS_2 nanoflowers enhances the catalytic activity in the hydrogen evolution reaction. *CrystEngComm* 22: 1395–1403.

40 Li, S., Sun, J., and Guan, J. (2021). Strategies to improve electrocatalytic and photocatalytic performance of two-dimensional materials for hydrogen evolution reaction. *Chin. J. Catal.* 42: 511–556.

41 Kuznetsov, D.A., Chen, Z., Kumar, P.V. et al. (2019). Single site cobalt substitution in 2D molybdenum carbide (MXene) enhances catalytic activity in the hydrogen evolution reaction. *JACS* 141: 17809–17816.

42 Seh, Z.W., Fredrickson, K.D., Anasori, B. et al. (2016). Two-dimensional molybdenum carbide (MXene) as an efficient electrocatalyst for hydrogen evolution. *ACS Energy Lett.* 1: 589–594.

43 Liu, C.Y. and Li, E.Y. (2018). Termination effects of Pt/v-$Ti_{n+1}C_nT_2$ MXene surfaces for oxygen reduction reaction catalysis. *ACS Appl. Mater. Interfaces* 11: 1638–1644.

44 Jiang, X., Kuklin, A.V., Baev, A. et al. (2020). *Phys. Rep.* 848.

45 Zha, X.J., Zhao, X., Pu, J.H. et al. (2019). *ACS Appl. Mater. Interfaces* 11: 36589.

46 Liang, X., Ren, X., Yang, Q. et al. (2021). A two-dimensional MXene-supported metal–organic framework for highly selective ambient electrocatalytic nitrogen reduction. *Nanoscale* 13: 2843–2848.

47 Zhang, Y. and Iwasa, Y. (2014). Transition metal chalcogenide two-dimensional crystals. *J. Crystallogr. Soc. Jpn.* 69: 771–776.

48 Cao, J., Li, J., Li, D. et al. (2021). Strongly Coupled 2D transition metal chalcogenide-MXene-carbonaceous nanoribbon heterostructures with ultrafast ion transport for boosting sodium/potassium ions storage. *Nano Micro Lett.* 13: 113.

49 Jin, J., Xiao, T., Zhang, Y. et al. (2013). Hierarchical MXene/transition metal chalcogenide heterostructures for electrochemical energy storage and conversion. *Nanoscale* 13: 19740–19770.

50 Zou, H., He, B., Kuang, P. et al. (2018). Metal-organic framework-derived nickel-cobalt sulfide on ultrathin mxene nanosheets for electrocatalytic oxygen evolution. *ACS Appl. Mater. Interfaces* 10: 22311–22319.

8

Newly Emerging 2D MXenes for Hydrogen Storage

Sneha Lavate, Sankhula Lokesh, and Rohit Srivastava

Catalysis & Hydrogen Research Lab, Department of Petroleum Engineering, School of Energy Technology, Pandit Deendayal Energy University, Gandhinagar, Gujarat, India

8.1 Introduction

Two-dimensional (2D) materials possess antique optical, electronic, and mechanical properties leading to the extensive exploration, study, fabrication, and application [1] in multiple sectors displayed in Figure 8.1. The various benefits of 2D materials, including their high aspect ratio, outstanding mechanical rigidity, and superior thermal conductivity, make them viable choices for H_2 storage applications. Since the discovery of graphene and its wide applications, a new range of 2D materials reached beyond because of their astonishing thermal, electronic, optical, and mechanical properties. Several 2D materials have been widely investigated, which include MXenes, transition metals dichalcogenides (TMDs), phosphorene, layered double hydroxides (LDHs), and their derivatives [2]. Graphene has many exceptional properties for H-sorption including light weight, good mechanical strength, chemical stability, and environmental stability. Phosphorene-based nanomaterials also perform better in H_2 capturing at room temperature in a reversible manner due to their existence of periodic channels. These materials are effective for storing H_2 under atmospheric circumstances due to their high dissociation energy, significant H_2 physisorption on phosphorene surface, and internal diffusion. The capacity to store H_2 through chemisorption and dissociation of H_2 molecules on the LDH surface, followed by diffusion of H_2, can be another feature of LDH materials. Among these 2D nanostructures, MXene has been considered as an exceptional catalyst because of its highest capacity of H_2 adsorption of 8.6 wt%, which is far greater than metal-based complex hydrides. The reversibility of H_2 storage property under low temperatures due to the incorporation of MXene-based materials also enhances the H_2 sorption parameter, so these materials have gained major attention of researchers across the globe. In 2011, 2D layered carbide powder (Ti_3C_3) was invented as the first member of the MXene family. The timeline of MXene is shown in Figure 8.2. The family of MXene includes nitrides, transition metal carbides, and carbonitrides [3].

MXene possesses a few unique characteristics such as great biocompatibility, obvious security capability, ultra-large interlayer spacing, and environmental benignity except some MXenes [4].

MXenes: Next-Generation 2D Materials: Fundamentals and Applications, First Edition.
Edited by Jay Singh, Kshitij RB Singh, Ravindra Pratap Singh, and Charles Oluwaseun Adetunji.

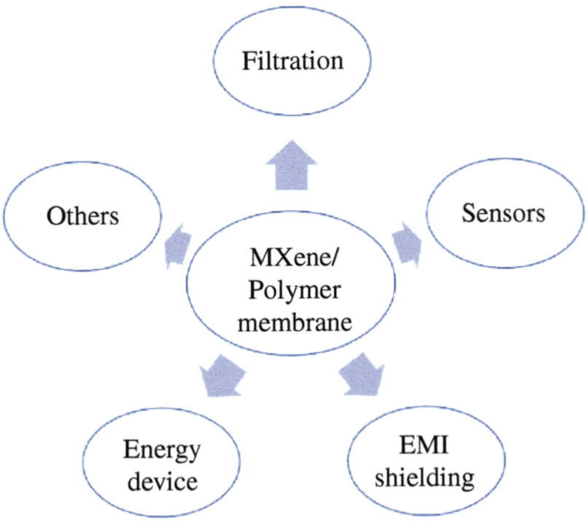

Figure 8.1 Applications of MXene in various sectors.

The unique properties of MXene compared to other 2D materials are represented in Table 8.1. Table 8.1 gives a concrete information about the characteristics of 2D materials like graphene, phosphorene, and LDH.

Since MXene has a wide surface area and several sites for H_2 adsorption and desorption, it has recently been employed as an effective catalyst material for energy storage. H_2 being a "green fuel" is acknowledged as the demandable energy source taking the place of conventional fossil fuels due to its remarkable properties such as zero carbon emissions, high energy output, and low mass density [9]. There are three different ways to store H_2: solid-state storage, cryogenic storage, and high-pressure compression at over 70 MPa. Among all of the three types, solid-state H_2 storage is the most convenient way as it does not require high-pressure compression and high-temperature treatment [10]. Among other storage methods of H_2, solid H_2 storage is more promising and will be a game changer toward the H_2 economy as shown in Table 8.2.

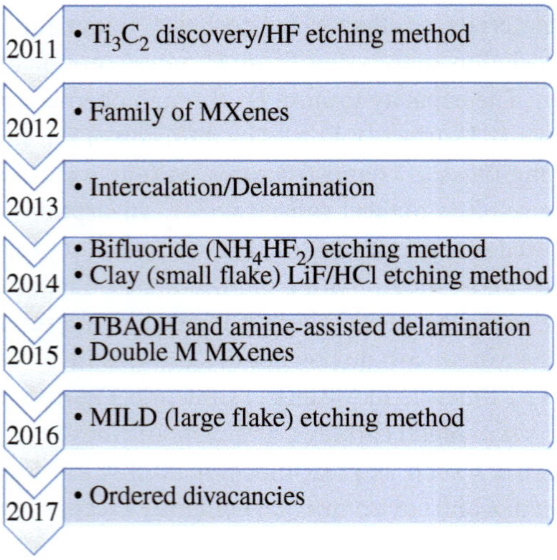

Figure 8.2 MXenes: from Ti_3C_2 discovery to ordered divacancies.

Table 8.1 Properties of different 2D materials for H_2 storage applications.

2D material	Parameters responsible for H_2 storage	Challenges	Refs.
Graphene	High surface area, chemical inertness, good mechanical strength, light weight, low cost	Limited reversibility of hydrogen molecules, higher desorption temperature, and pressure	[5]
Phosphorene	Hydrogen adsorption in room temperature	Low desorption rate	[6]
Layered double hydroxide	High surface area, easy structural tenability, good conductivity, better hydrogen adsorption rate, good stability	High desorption temperature, less regeneration of adsorbed molecules	[7]
MXene	Good mechanical strength, lower desorption temperatures, high storage capability, high surface area, good reversibility of hydrogen atoms	Still not considered as conventional material	[8]

Solid state H_2 storage is a major challenge due to limited storage capacity, high energy consumption during adsorption and desorption process. MXenes help in improving the H_2 adsorbing/desorbing capability of different metal and complex hydrides by a significant amount, thus 2D MXenes are of major interest for many researchers to develop a novel composite material for enhancing H_2 desorption capability. There are several techniques available for H_2 production, reaction of ammonia borane with water molecules is one prominent method for meeting global energy needs [18–21]. Although H_2 has been identified as the fuel of the future, because of its low boiling point, high flammability, and high gravimetric energy density. Its present storage and transportation techniques are of

Table 8.2 H_2 storage capability of different MXene materials.

Nanomaterial	E_{ad} (eV/H or eV/H_2)	H_2 adsorption capacity	Refs.
MgH_2-10 wt% $(Ti_{0.5}V_{0.5})_3C_2$	NIL	6.5	[11]
Ti_2C	5.027	8.6	[12]
Cr_2C	0.96	7.6	[13]
$Ti_2N(OH)_2$	0.153	5.31	[14]
Ti_2NO_2	0.142	5.38	[14]
Ti_2N	3.63	8.55	[14]
Ti_2NF_2	0.137	5.176	[14]
MgH_2-5 wt% $Nb_4C_3T_x$	Nil	3.50	[15]
NaH/Al-8 wt% Ti_3C_2	Nil	4.9	[16]
Sc_2C	4.703	9.0	[17]

great concern to experts all around the world. The primary issues with H_2 storage should be safer, simpler technology that is reversible and has great gravimetric/volumetric capabilities [22–24]. Interestingly, ammonia borane hydrolysis reaction has shown better results of good stability, higher H_2 density, and relative ease of hydrolysis [25–31].

$$NH_3BH_3 + 4H_2O \rightarrow NH_4^+ + B(OH)_4^- + 3H_2 \left(\frac{Pt}{Ti_3C_2T_x}, 20 - 85°C, 45 \text{ min} \right) \quad (8.1)$$

The $NaAlH_4$ decomposition reaction was classified into three stages for liberating 7.5 wt% H_2. There are some practical limitations for this decomposition reaction as high decomposition reaction temperature for the third reaction step, so only 5.6 wt% H_2 (first two steps) can be evolved from $NaAlH_4$ practically below 200 °C [16].

$$NaAlH_4 \rightarrow \frac{1}{3}Na_3AlH_6 + \frac{2}{3}Al + H_2 \quad (8.2)$$

$$Na_3AlH_6 \rightarrow 3NaH + Al + \frac{3}{2}H_2 \quad (8.3)$$

$$NaH \rightarrow Na + \frac{1}{2}H_2 \quad (8.4)$$

8.2 Structural Properties of MXene

Typically, $M_{n+1}X_nT_x$ ($n = 1, 2, 3$) is used to represent MXene, where M stands for an early transition metal, X stands for C or N, and T_x represents the surface functional group [2]. MAX phases are used as precursors for synthesis of MXenes [32].

The structure of MXene resembles its MAX phase precursor; M atoms are positioned at the octahedral interstitial sites, while X atoms are placed in densely packed arrays. One X layer is placed between two M layers to form the MXene monolayer. MXene's overall structure is a packed hexagonal structure to achieve stability, the ordering of M may shift [33].

8.3 Synthesis Techniques

There are several other methods available rather than conventional methods as mentioned in Figure 8.3 for the synthesis of high-yield and good-quality MXenes such as pyrolysis, chemical vapor deposition, and wet etching followed by a stripping process. By using the chemical vapor deposition process, Xu et al. [34] created a layered structure with a high specific surface area called α-Mo2C. Several investigations have been conducted on the basis of the structural morphological requirements in order to establish an effective synthesis process for MXene [35–39].

Due to their outstanding qualities, including their large specific surface area, tunable surface chemistry, and transition metal oxide availability, MXenes have a wide range of uses.

These nanoparticles may be applied to conductive coatings, sensors and chemical noses, triboelectric nanogenerators, desalination processes, batteries, and pseudo-capacitors. By absorbing solar light, the Ti_3C_2 MXene-based nanostructure may cleanse water using evaporation processes with an energy efficiency of 84%. For its energy storage applications,

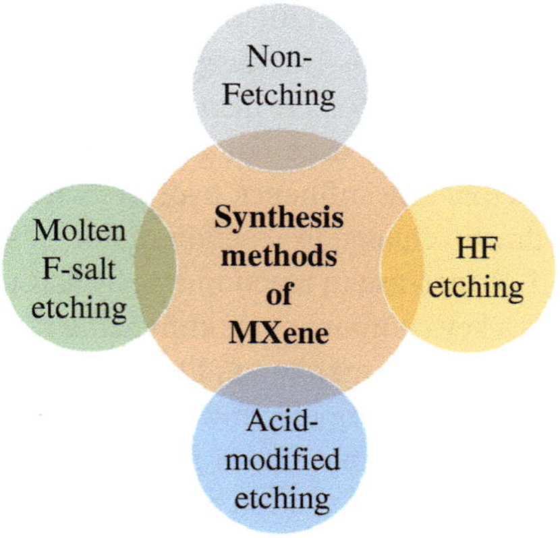

Figure 8.3 Synthesis methods of MXene.

they are able to exfoliate stacked layers into a sheet-like structure with unique properties including electrical conductivity and flexibility because of the intercalation of dimethyl sulfoxide (DMSO) into the MXene structure. Because MXenes can receive electrons through the inclusion of polymeric composites, they have the extraordinary potential to convert muscle motions into electrical power. In order to protect polymeric materials against mechanical stretching or bending, MXene layered structures may also be used as highly electrically conductive coatings. The MXene structures have remarkable properties in that they can detect low gas concentrations even while the molecules of gas are passing across the surface. The chemical makeup and porous structure of MXene facilitate the tracing of various gas molecules that are undetectable when using other materials.

8.4 H_2 Storage Reaction Mechanisms

The solid H_2 storage is more challenging compared to liquid and gaseous H_2 storage. The mechanism of solid H_2 storage using MXene as a novel 2D material has been mentioned as follows.

8.4.1 Adsorption

Adsorption is classified into two categories, namely, physisorption and chemisorption. Physisorption is primarily governed by weaker van der Walls forces of attraction between adsorbate and adsorbent molecules, whereas chemisorption process involves the formation of reaction coordinates and formation of covalent bonds with the solid material surface. Physisorption is generally considered reversible in nature, but chemisorption involves strong bonds between molecules, thus requires more energy for regeneration of adsorbed molecules. Generally, for practical cases, MXenes partial metallic parameter provides impetus for molecular H_2 adsorption having a binding capacity of 10–20 kJ/mol. Chen et al. studied that reaction of ammonia borane with water molecule follows Langmuir–Hinshelwood reaction kinetic mechanism with O—H bond cleavage by Pt-bound NH_3BH_2 is slow reaction step that determines the rate of reaction (where * is a free active site) [40]. Several researchers explain that rate-determining step for hydrolysis reaction is O—H cleavage [41–47].

Since B—N bond dissociation is not considered as crucial reaction for composite, hydrolysis reaction does not get affected by the NH_4^+ concentration increase. There are many factors that influence hydrolysis reaction of ammonia borane including solution pH and drastic change in reaction rate in basic and acidic mediums [29, 48]. We have found that 30% increase in reaction rate by addition of NaOH to the reaction mixture states that OH^- is a key component for enhancing reaction kinetics. Water and basic solution dissociates in a reversible manner on Pt substrate surface according to Eqs. (8.5) and (8.6) [49]. Platinum surface having electron deficient properties, that helps in increasing OH^* concentration and also enables B—H bond dissociation, thus influences rate determining step and also improves the reaction rate of Pt/MXene-H_2O and Pt/MXene-O_3 catalyst composite.

$$H_2O + * \leftrightarrow OH^* + H^+ + e^- \tag{8.5}$$

$$OH^- + * \leftrightarrow OH^* + e^- \tag{8.6}$$

We should have a thorough research on the reaction kinetics developed by Chen et al., focusing on subsequent $NH_3BH_2^*$ reactions and O—H bond cleavage. The adsorption of ammonia borane on Pt surface leads to the B—H bond breakage and forms $NH_3BH_2^*$ and H^*, whereas water molecule dissociates into OH^*, H^+, and e^- after adsorbing on catalyst surface, thus H^+/e^- pairs with H^* to produce H_2. The $NH_3BH_2^*$ can be formed by the recombination of OH^* intermediate and $NH_3BH_2(OH)^*$, thus desorbs into the bulk medium. The gravimetric methods were used to analyze the catalyst samples for the dehydrogenation parameters using a Sieverts' kind of instrument. The 100 mg samples were used for experimental tests and analysis after being dried at a ramp rate of 2 °C min^{-1} to 350 °C. Using a device with an online mass spectrometer, temperature-programmed desorption-mass spectroscopy measurement (TPD-MS) studies were carried out (MS, Hiden HPR-20). The sample (40 mg) was heated to 500 °C under an Ar flow stream at a rate of 5 °C min^{-1}. The performance of catalyst samples during non-isothermal dehydration was examined using a differential scanning calorimeter (DSC, TA Q2000) equipment heated to 400 °C while being surrounded by an argon gas stream.

$$Mg(BH_4)_2 + 2Ti_3C_2F_X \rightarrow MgF_2 + 2B + 4H_2 + 2Ti_3C_2 \tag{8.7}$$

$$Mg(BH_4)_2 \rightarrow Mg + MgB_2 + 2B + 8H_2 \tag{8.8}$$

$$2Mg(BH_4)_2 + 2Ti_3C_2F_X \rightarrow MgF_2 + MgB_2 + 2B + 8H_2 + 2Ti_3C_2 \tag{8.9}$$

Two stages were identified in the dehydrogenation process of the $Mg(BH_4)_2$-40 wt% Ti_3C_2 compound. The first one is a low-temperature process that matches Eq. (8.7) and has an activation energy of 151.3 kJ mol^{-1}. The second mechanism is reaction Eq. (8.8), whose activation energy is 178.0 kJ mol^{-1}. In contrast to pure $Mg(BH_4)_2$, the activation energies of the reaction stages are lower in the second mechanism (451.6 kJ mol^{-1}). The fluoride anion will be accommodated in the hydride, which will boost the H_2 sorption process. In comparison to pure $Mg(BH_4)_2$, the Ti_3C_2 MXene and $Mg(BH_4)_2$ composite has a greater surface area and provides impetus for desorption, which makes the sorption mechanism easier. In comparison to pure $Mg(BH_4)_2$, the start temperature for the dehydrogenation process of $Mg(BH_4)$ 240 wt% Ti_3C_2 is reduced by 128 °C, and it liberates 7.5 wt% H_2 at 260 °C.

The low reaction temperature caused by the fluorine molecules adsorbed on the surface of Ti_3C_2 MXene results in the formation of MgF_2, which is essential for the creation of H_2. Ball milling also aids in the average size reduction of magnesium borohydride and shows more flaws, which can boost the rate of H_2 gaseous molecule desorption. The 2D structure of Ti_3C_2 MXene will have more H_2 diffusion patterns. The calculated activation energy of the two-step dehydrogenation reaction pathway is much less than that of pure $Mg(BH_4)_2$. Density functional theory (DFT) calculations using CASTEP code 46 were used to determine the total energy pseudopotential. Vanderbilt ultra-soft pseudopotentials were examined in a plane wave basis set having an energy cutoff of 480 eV. The BFGS minimization method was used for analyzing structural parameters and atomic positions for a convergence tolerance limit of force <0.01 eV/Å, energy $<5.0\times10^{-6}$ eV, displacement $<5.0\times10^{-4}$ Å, and stress <0.02 GPa. The H_2 binding energy (E_b) with the Ti_2C substrate was examined by following equation:

$$E_b = (E_{host} + nE_{H2} - E_{host+nH2})/n \tag{8.10}$$

where E_{host} is the host materials' total energy, E_{H2} is the free H_2 molecule's energy, $E_{host+nH2}$ is the energy accumulated with new H_2 molecules adsorbed on host materials, and n is the total adsorbed H_2 molecules [12].

8.4.2 Kinetics and Thermodynamics

8.4.2.1 Kinetic Models

Through several scientific interventions, including ball milling, metal alloying, thin film architectures, and catalytic mechanisms, the problems of energy storage kinetics were examined. However, in terms of capabilities and energy usage, no material has outperformed others for real-world energy storage applications. To address the storage issue, a number of hybrid materials have been researched, including transition metal hydrides, LDHs, and graphene nanocomposites. The intake of H_2 and H_2 molecules regeneration in H_2 systems is mainly due to adsorption, surface diffusion, internal movement, and chemical reaction [50]. Figure 8.4 explains the detailed steps involved for these processes: geometrical contraction and nucleation growth are considered as most probable reaction mechanisms for hydrogenation and dehydrogenation, which was also revealed by several experimental procedures [51]. These models were developed on the basis of isothermal and non-isothermal fitting experimental methods [52]. These techniques help for better understanding of kinetic parameters, reaction modules, and reaction mechanisms for enhancing H_2 storage systems.

8.4.2.2 Geometrical Contraction

This model assumes that reaction steps progress consequently from solid surface to outer medium, and the catalyst particles are assumed to be of uniform structure [52]. The relation between particle sphericity and reaction fraction is found by:

$$\varepsilon = 1 - \left(\frac{r}{r_o}\right)^d \tag{8.11}$$

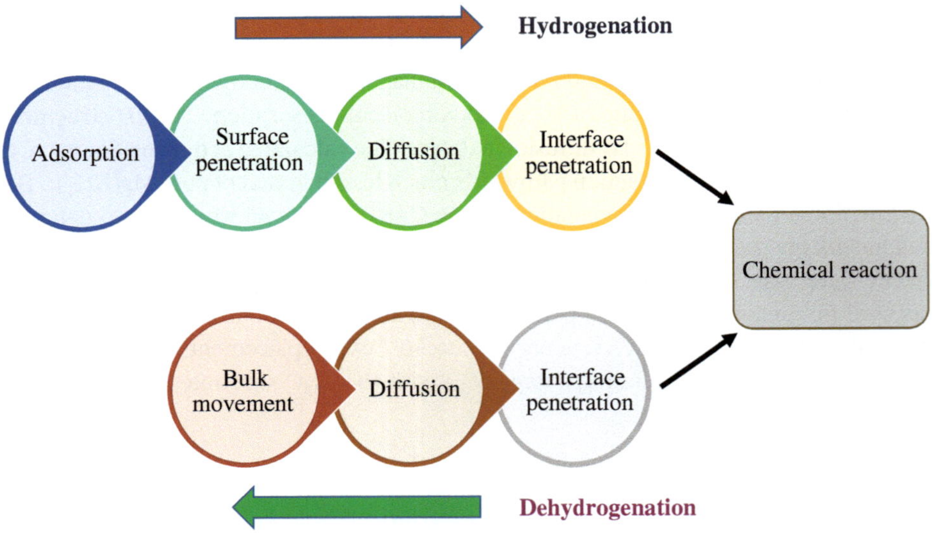

Figure 8.4 Hydrogenation and dehydrogenation steps involved for storing in solid form.

where d = dimensional parameter, r = unconsumed particle length, r_o = length of overall particles, and ε = fraction of reaction progress [53–57]. These are some of the geometrical contraction models.

8.4.2.3 Contracting Volume Model

$$\frac{d_r}{d_t} = -K_{int} \tag{8.12}$$

$$1 - (1 - \varepsilon)^{1/d} = \frac{K_{int}}{r_o} t = kt \tag{8.13}$$

where k_{int} = interface-controlled reaction rate constant and k = rate constant generalized.

8.4.2.4 Jander Model

$$\rho \frac{d_r}{d_t} = \frac{D\Delta C}{r_o - r} \tag{8.14}$$

$$\left[1 - (1 - \varepsilon)^{\frac{1}{d}}\right]^2 = \frac{2D\Delta C}{\rho {r_o}^2} t = kt \tag{8.15}$$

where ΔC = difference in concentration, D = diffusivity, and ρ = density.

8.4.2.5 Ginstling–Brounshtein Model

$$\rho \frac{d_r}{d_t} = \frac{D\Delta C}{r \log \frac{r_o}{r}} \tag{8.16}$$

$$1 - \varepsilon(\log(1 - \varepsilon)) + \varepsilon = \frac{4D\Delta C}{\rho {r_o}^2} = kt \tag{8.17}$$

For 2D cylindrical shape:

$$\rho\frac{d_r}{d_t} = \frac{D\Delta C\, r_o}{(r_o - r)r} \tag{8.18}$$

$$1 - \frac{2}{3}\varepsilon - (1-\varepsilon)^{\frac{2}{3}} = \frac{2D\Delta C}{\rho {r_o}^2}t = kt \tag{8.19}$$

For 3D cylindrical shape:

8.4.2.6 Valensi–Carter Model

$$\rho\frac{d_r}{d_t} = \frac{D\Delta C}{r - \frac{r^2}{(Z{r_o}^3 + r^3(1-z))^{1/3}}} \tag{8.20}$$

$$\frac{Z - (1+(z-1)\varepsilon)^{\frac{2}{3}} - (z-1)(1-\varepsilon)^{\frac{2}{3}}}{Z-1} = \frac{2D\Delta C}{\rho {r_o}^2}t = kt \tag{8.21}$$

where z = volume ratio (product to reactant). This model says that hydrogenation rate is limited by the surface interface process. The dimensional parameters and generalized constants will be found through isothermal curve fitting and analysis [52–54]. Bösenberg et al. explained that $LiHMgB_2$ with Vanadium and Titanium deposits follow the CV model as it is based on interfacial control [58]. The Jander Model is mainly analyzed for finding penetration rates through the material. This model has two major assumptions including (a) composite catalyst material volume utilized for H_2 storage process remains constant for entire sorption process. (b) Interfacial area also remains the same during two- and three-dimensional diffusion processes [53, 59]. However, there are certain limitations for this model assumption because volume may not be the same for the entire adsorption and desorption process. Moreover, Shao et al. revealed that Mg doped with Ti at 1 MPa of H_2 has similar relations with this model [60]. Chaudhary et al. explained that synthesized MgH_2-Si by different operations like milling and ultra-sonicating process produces desorption curves that follows this model [61]. The kinetic model is undoubtedly the most precise one in the geometric contraction model but still not exhibited for H_2 sorption systems [62–64].

8.4.2.7 Nucleation-Growth Impingement Models

This model is referred as the Johnson–Mehl–Avrami–Kolomogorov (JMAK) that defines entire reaction mechanisms into three steps: particle nucleation, crystal growth, and impingement [65–68]. This model is described by the following definite integral:

$$\varepsilon = F\int_0^t I(r)V(r)d_r \tag{8.22}$$

where I = nucleation function, r = rate of nucleation, F = impingement mechanism gives the relation between fraction of real and extended reaction, and V = growth parameter. The following definite integral describes the interfacial growth control and diffusion control given by classical JMAK (C-JMAK) equation as

$$V_{(r)} = \left[Go\int_r^t \exp\left(\frac{-\Delta Eg}{RT}\right)d\eta\right]^{\frac{d}{m}}. \tag{8.23}$$

where ΔE_g = nucleation growth activation energy, G_0 = rate of intrinsic impregnation, m = crystal nucleation mode parameter, and $\frac{d}{m}$ = growth factor. The C-JMAK model briefly describes the process of adsorption and desorption phenomena of nucleation, growth, and impingement. Generally, nucleation is mainly defined as saturation of solution, growth of particles from interface and diffusion sites, and isotropic nuclei growth. The Avrami exponent of the desorption kinetics for $NaAlH_4$ doped with Titanium was found to be three using this model, which explains that 3D interface process is the slowest step. However, the brief description of kinetic mechanisms for H_2 sorption-capable materials based on the literature survey is quite difficult because the analysis methods have certain drawbacks of reaction mechanism misconception [69].

8.5 Factors Influencing H_2 Storage

Adsorption and desorption temperatures, surface morphology, binding energy of H_2 molecules to the particle surface, and system pressure are some of the factors that may affect the H_2 storage application. There are different techniques available for enhancing the H_2 sorption capability of MgH_2 by the incorporation of ML-Ti_3C_2 [70]. The structural morphology, surface interface, and elemental composition range were further characterized by using various physical experimental methods such as XRD, SEM, EDS, TEM, and XPS. Outstanding H_2 sorption performance implies that 2D layered structure helps for high specific surface area and more active sites, and thus facilitating the H_2 diffusion into the system.

8.6 Recent Advances in MXene-Based Compounds for H_2 Storage

The large-scale synthesis of MXenes is still a challenge to researchers as it involves toxic etching agents. To overcome this drawback and enhance the production of MXene at a larger scale, Jun Mei and co-workers proposed a new synthesis strategy shown in Figure 8.5. The typical procedure involves the removal of lightly bonded sulfur from MAX phases by thermal annealing, leaving behind graphene-like Ti_2C nanosheets [71].

Effective adsorption and desorption kinetics of H_2 were obtained by adding a small amount of Ti_3C_2 MXene to $NaAlH_4$ [72]. Jiaguang Zheng et al. fabricated a unique "cake-like" structure of Ti_3C_2–$Mg(BH_4)_2$ with consistently distributed Ti_3C_2 MXene over $Mg(BH_4)_2$, reducing the particle size for better dehydrogenation activity of the hybrid [73]. Aihu Feng et al. replaced the HF with bifluoride for etching Ti_3AlC_2. The material's SEM images (Figure 8.6) show a fine layered structure concluding the successful synthesis of Ti_3AlC_2 [74]. Another synthesis of 2D $Nb_4C_3T_x$ was carried out to advocate the catalytic activity of $Nb_4C_3T_x$ on H_2 storage of MgH_2. The synthesized material showed excellent catalytic activity for H2ation-deH2ation switching at certain temperatures on MgH_2 [15]. Kaicheng et al. developed a reactive hydride system composed of a $2LiH + MgB_2$ mixture with Ti_3C_2 embedded in the mix by ball milling. Table 8.3 represents the list of additives with their respective parameters for hydrogenation and dehydrogenation [75].

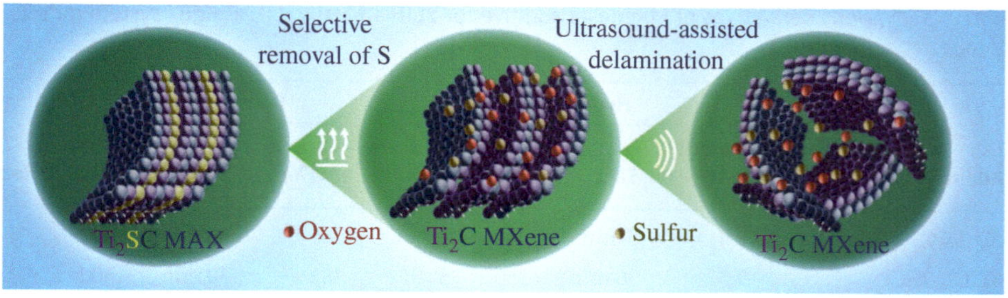

Figure 8.5 Schematic of MAX to MXene by thermal reduction. Source: Reproduced with permission from [71]. Copyright 2022, Elsevier.

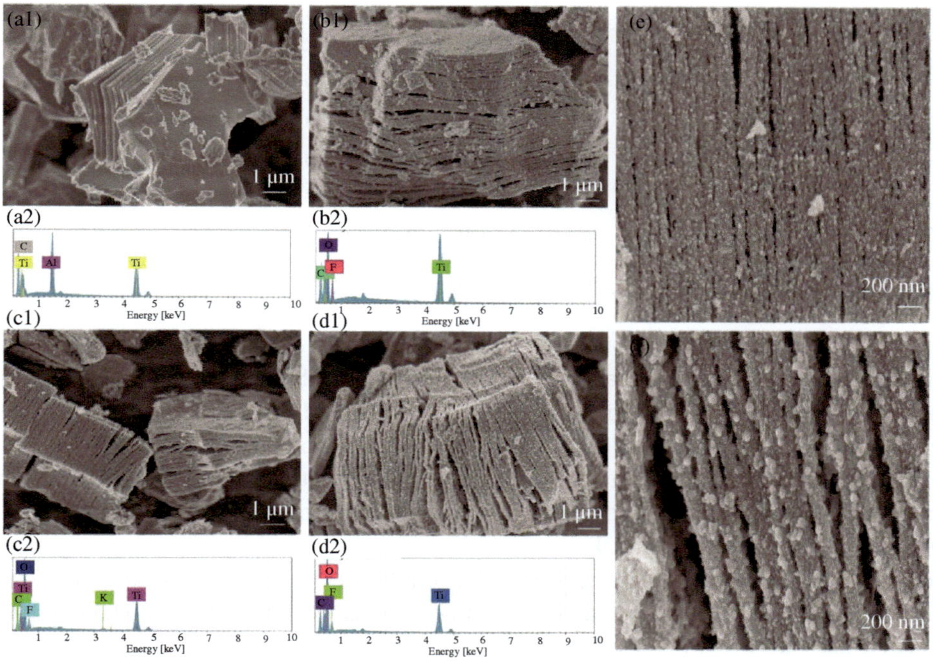

Figure 8.6 SEM and EDS images of Ti_3AlC_2 and Ti_3C_2 (a1, a2) SEM and EDS images Ti_3AlC_2 (b1, b2) SEM and EDS images of samples exfoliated by $NaHF_2$, (c1, c2) SEM and EDS images of samples exfoliated by NH_4HF_2, (d1, d2) SEM and EDS images of samples exfoliated by NH_4HF_2, (e, f) High magnification SEM images of Ti_3C_2 obtained by exfoliation of Ti_3AlC_2 with KHF_2 or NH_4HF_2. Source: Reproduced with permission from [74]. Copyright 2022, Elsevier.

Xin Zhang et al. successfully prepared a novel $TiVO_{3.5}$ catalyst for H_2 storage on MgH_2 by oxidation of solid-solution $(Ti_{0.5}V_{0.5})_3C_2$, which showed structural properties similar to its precursor MXene. The addition of this composite to MgH_2 reduced the activation energy of desorption from 153.5 to 62.4 kJ/mol [85]. Yanping Fang and co-workers prepared a 2D MXene/A-TiO_2 composite to enhance the H_2 storage capacity of $NaAlH_4$. Researchers first fabricated Ti_3AlC_2 powders by ball milling of elemental Ti, Al, and graphite powders with a ratio of 3:1.1:2 followed by cold-press under 1 GPa pressure and finally heated at 1450 °C

Table 8.3 Various derivatives of MXene as additive for $2LiH + MgH_2/2LiH + MgB_2$ system for H_2 storage.

	Hydrogenation				Dehydrogenation			
Additive	**Capacity (wt%)**	**Time**	**Temp (°C)**	**Pressure (MPa)**	**Capacity (wt%)**	**Time**	**Temp (°C)**	**Refs.**
2D Ti_3C_2	9.4	5 min	350	10	9.5	30 min	400	[75]
$3TiCl_3{\cdot}AlCl_3$	9.5	30 min	350	5	9.5	30 min	400	[76]
TiC	7.8	16 h	350	5	7.8	20 h	380	[77]
TiN	8.2	16 h	350	5	8.2	4 h	380	[77]
TiF_3	7.6	4 h	350	7.5	8.0	32.5 min	400	[78]
TiO_2	7.6	1.8 h	350	5	8.1	2.7 h	380	[77]
Nano-TiO_2	~10.0	25 min	400	5	~10.0	50 min	400	[79]
Ni-B	9.4	70 min	400	10	9.4	70 min	400	[77]
Activated carbon	9.0	420 min	400	10	9.0	240 min	400	[80]
MWCNTs-TiO_2	6.8	8 h	400	8	6.8	8 h	400	[81]
MoS_2	—	—	—	—	4.13	133 min	400	[82]
Ti isopropoxide	8	2.5 h	350	5	8.5	4 h	400	[83]
CoNiB	—	—	—	—	10.2	360 min	400	[84]

in a tube furnace [86]. The researchers performed a trial on the addition of Ti_3C_2 MXene as a catalyst to improve the H_2 storage kinetics of $LiBH_4$. The $LiBH_4 + 40$ wt% Ti_3C_2 mixture exhibited activation energy of 70.27 kJ/mol, lower than the pristine sample [87].

DFT has been used to investigate the H_2 storage of functional material HF_2CF_2 decorated with Li. Calculations of formation enthalpy and phonon dispersion curves indicated good stability of the composite. First-principles density functional calculations were done to check nH_2 adsorption characteristics. The results concluded that this kind of MXene shows superior ionic surface and charge transfer capability and structural stability [88]. Yadav and co-workers investigated Cr_2C for H_2 storage capacity with the help of DFT. Sites of adsorption on the surface of Cr_2C were explored to check sorption of H_2 molecule. Some of the sites were found to be capable of Kubas-type interaction which explains that the material can exhibit better performance for reversible H_2 storage at ambient atmosphere [13]. Tianping Huang et al. derived Ni nanoparticles from MOFs and dispersed them into exfoliated MXene monolayer as a functional carrier for H_2 storage on MgH_2. First principles calculations were carried out to explore the effect of Ti^0 on $Ni/Mg_2Ni/Mg_2NiH_4$ as "H_2 pump" behavior of Mg_2NiH_4 can be observed [89]. CeF_3 nanoparticles incorporated Ti_3C_2 nanosheets exhibited good cyclability, enhanced catalytic activity, better dehydrogenation, and reversibility. The structural configuration that includes Ti—F—Ce bonding was responsible for better H_2 storage performance [90].

In addition to Ti_2C_3, Wen Zhu et al. incorporated Ni into Ti_2C_3, forming the composite $MgH_2 + Ni@Ti\text{-}MX$ with improved H_2 sorption kinetics at low operating temperatures shown in Figure 8.7 [91].

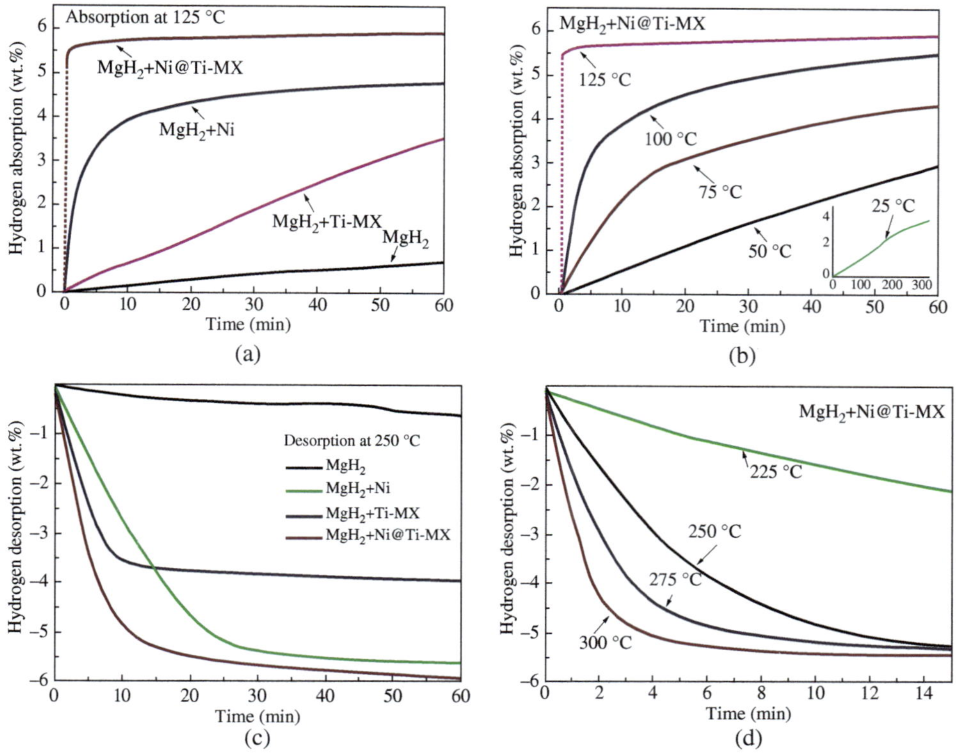

Figure 8.7 (a) MgH_2 + Ni@Ti-MX, MgH_2 + Ni, MgH_2 + Ti-MX composites and MgH_2 isothermal hydriding curves at 125 °C, 3 MPa H-pressure, (b) different temperature dynamic hydriding profiles of MgH_2 + Ni@Ti-MX. (c) MgH_2 + Ni@Ti-MX, MgH_2 + Ni, MgH_2 + Ti-MX composites and MgH_2 isothermal dehydrogenation curves at 250 °C, 0.002 MPa H-pressure, (d) different temperature dehydrogenation profiles of MgH_2 + Ni@Ti-MX. Source: Adapted from Zeyi Wang et al. (2019)

Zeyi Wang et al. first synthesized NbTiAlC solid MAX phase compound by sintering the mixture of Nb, Ti, Al, and C at 1500 °C for two hours; as-synthesized MAX phase was etched with 48% HF solution to obtain NbTiC MXene. DFT study interpreted a process of charge transfer in NbTi cluster from Ti to Nb atoms with less adsorption energy value of H_2 on NbTi, explaining the advantage to both Mg—H bond breaking and detaching H_2 from the surface of NbTi, which leads to better catalytic activity [92].

Ruicheng Jiang et al. applied the MILD method (fluoride and HCl salt instead of HF) to synthesize Ti_3C_2 followed by ball milling Al, NaH, and Ti_3C_2. Ti_3C_2-doped NaH/Al composite exhibited excellent cyclic stability, reduction in activation energy, and increased rate of H_2 desorption and adsorption [16]. Qianqian Kong et al. reported a novel hamamelis-like structure composed of $K_2Ti_6O_{13}$ obtained by reacting Ti_3C_2 with a solution mixture of H_2O_2 and KOH shown in Figure 8.8. The enhanced H_2 adsorption/desorption properties of MgH_2-(5 wt%) $K_2Ti_6O_{13}$ were due to mutual effect among Ti^{2+}, $KMgH_3$, and Ti metal [93].

Recently, Zang et al. [94] have developed lithium borohydride on Ti_3C_2 supports (LBH@$x Ti_3C_2$) to have composite hybrids under room temperature by facile impregnation technique and demonstrated the importance of 2D MXene nanoparticles for enhancing

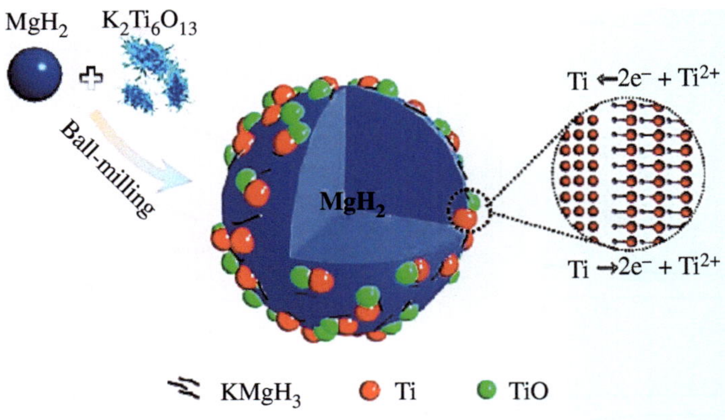

Figure 8.8 Schematic for fabrication of MgH_2-$K_2Ti_6O_{13}$. Source: Adapted from Zeyi Wang et al. (2019).

the adsorption/desorption parameters of $LiBH_4$. Liu et al. [95] investigated the sorption properties of MgH_2 and stated that catalyst properties can be improved by adding Ti_3C_2 MXene. The H_2 binding capability of $Mg(BH_4)_2$ would be increased by several folds due to key confinement and catalytic mechanism of MXenes. Zheng et al. [73] performed several experiments and described the importance of incorporating 2D MXene structures for achieving lower $Mg(BH_4)_2$ desorption temperatures and thus enhancing desorption rate. These experiments also explained that remarkable catalytic activity upgrading of $Mg(BH_4)_2$-Ti_3C_2 in H_2 storage applications is mainly due to Ti_3C_2, which provides Ti metallic active sites to increase the rate of $Mg(BH_4)_2$ decomposition. Optimum requirement of 2D MXenes for enhancing the rate of desorption can be found by incorporating Ti_3C_2 MXene as additives for $Mg(BH_4)_2$-x wt% Ti_3C_2 composite synthesis. The $Mg(BH_4)_2$ catalyst particles are distributed on the Ti_3C_2 substrate filled in the multilayered structure thus Ti_3C_2 can have more number of active sites. The synthesis of 2D materials by several techniques, dehydrogenation reaction products obtained at different temperatures and to find the fluorine as vital component during sorption process. Ti_3C_2 MXene and $Mg(BH_4)_2$ mixture obtained by milling exhibits a low onset decomposition temperature and diffusion mechanism. There are several complex hydrides like $NaAlH_4$ that can be considered as facile H_2 storage systems [96].

Recently, Liu et al. [97] revealed the ultra-high storage capability (8.8 wt%) of Ti_2CT_x MXene at standard conditions. For a deeper comprehension of the storage process at work and to thoroughly examine the H-storage structure, the composition of the samples has to be characterized and examined. To achieve 8.8 wt% capacity, three layers of H_2 molecules must be held in the interlayer region of MXene with the structure of $Ti_2CF_2H_{14}$. At normal temperature, the H_2 layers with a 2D structure resembling graphene exist in a solid/liquid state, which is important for investigating novel materials with unexpected features.

$Ti_3C_2T_x$-based nanomaterials act as support for Pt nanoparticles and examine the catalytic performance in ammonia borane hydrolysis reaction, which is considered as prominent H_2 carrier technique. Adsorbed molecules will be regenerated from ammonia borane by catalytic hydrolysis phenomenon, where metal and the support interactions play a vital role. These results imply that the platinum electronic environment was manipulated by MXene oxidizing, thus implementing a new technique for developing novel catalyst composites, water and ozone oxidants are used for this purpose [98]. In addition, Cu nanoparticles (NPs) of ~5 nm sizes were distributed uniformly on 2D titanium carbide MXene (Ti_3C_2). XPS studies and DFT calculations give the information of transfer of electrons from copper nanoparticles to MXene structure upon hybridization. Interfacial charge transfer is based on the work function difference between Ti_3C_2 (5.78 eV) and Cu NPs (4.63 eV). The formation of Cu/Ti_3C_2 due to fermi energy reference is mainly by d-band centre (ε_d) of Cu shifting from −2.38 to −2.27 eV. The adsorption energy of H_2 was greater for Cu/Ti_3C_2 (−2.57 eV) compared to Cu particles (−2.45 eV). The composite material Cu/Ti_3C_2 exhibits an excellent catalyst performance for the conversion of 4-nitrophenol into 4-aminophenol having rate of reaction constant is 43.1 min^{-1} mg^{-1} was 6.3 times greater compared to Cu NPs. These type of enhanced catalysts leads to favorable H_2 molecules adsorption and better hydrophilicity of monohybrid structures [99]. The $Mg(BH_4)_2$ dehydrogenation performance was also examined by introducing various weight percentage (10, 20, 40, 60 wt%) of Ti_3C_2 MXene. The $Mg(BH_4)_2$-40 wt% Ti_3C_2 catalyst composite liberates 7.5 wt% H_2 at 260 °C, but pristine $Mg(BH_4)_2$ releases only 2.9 wt% H_2 at standard reaction parameters and desorption temperature was reduced from 210 to 82 °C. The 2D layered structural morphology of Ti_3C_2 MXene and fluorine is important in dehydrogenation reaction kinetics especially for reduction in temperatures below 200 °C [100].

The doping of MXenes to $NaAlH_4$ significantly improves H_2 sorption performances and stability. The Ti_3C_2-doped NaH/Al composite material has H_2 adsorption rate of 4.2 wt% at the temperature of 110 °C in a short span of 4.5 minutes. The onset temperature for dehydrogenation of Ti_3C_2-doped NaH/Al material is 76 °C. This composite catalyst has low activation energy (92.5 kJ/mol for the first reaction and 58.1 kJ/mol for the second reaction), compared to as-milled $NaAlH_4$. Ti_3C_2-doped NaH/Al has an excellent cycling H_2 sorption properties, 99% of initial capability can be observed for 10 cycles [16]. H_2 desorption properties of $LiBH_4$ can be modified by introduction of 2D MXenes. The temperature-programmed dehydrogenation (TPD) and isotherm dehydrogenation (TD) results describe that introducing Ti_3C_2 structures ascribes to enhancement in H_2 sorption capability of $LiBH_4$. Reduction in onset temperature for dehydrogenation reaction and enhancement in the rate of dehydrogenation occurs with increasing the Ti_3C_2 composition. The onset temperature for dehydrogenation process of composite $LiBH_4$+40 wt% Ti_3C_2 starts from 120 °C and approximately liberates 5.37 wt% H_2 in a span of 1 h at a temperature of 350 °C. Furthermore, $LiBH_4$+ 40 wt% Ti_3C_2 exhibits an activation energy reduction to 70.3 kJ/mol, which is much lower compared to pure $LiBH_4$. In order to enhance the kinetics of hydrogenation reaction for the composite $LiBH_4$ doped with 40 wt% Ti_3C_2, the apparent activation energy (E_a) related to dehydrogenation of $LiBH_4$+ 40 wt% Ti_3C_2 is given by Kissinger's peak shift method.

$$\text{In}\left(\frac{\beta}{T_p^2}\right) = -\frac{E_a}{RTp} + A \tag{8.24}$$

where β (K/min) is the ramp rate, T_p (K) is the maximum temperature, E_a is energy of activation, A is the exponential factor, and R is the universal gas constant (8.314 kJ/mol).

Here, T_p was found by the MS profiles with various ramp rates. The literature survey revealed that the activation energy of pure $LiBH_4$ was approximately 187 ± 24 kJ/mol. The E_a value for $LiBH_4$ + 40 wt% MXene is decreased to 70.3 kJ/mol, which is less compared to pristine $LiBH_4$. This explains that 40 wt% MXene incorporation to $LiBH_4$ exhibits an improvement in catalytic performance by reducing the activation energy barrier related to pure $LiBH_4$. The catalytic behavior of Ti_3C_2 for $LiBH_4$ desorption phenomena obtained by ball-milled and dehydrogenated samples was analyzed by X-ray diffraction techniques and Fourier transform infrared spectroscopy for a better understanding of reaction kinetics. The IR bands of B—H bond stretching and bending at 2384, 2218, and 1114 cm^{-1} are observed and indicate that $LiBH_4$ is the main component in the ball-milled samples, and two bands at 2307 and 2279 cm^{-1} are determined. The temperature rise to 150 °C can be observed by the XRD pattern peaks, which are attributed to LiF and LiH, and the $LiBH_3F$ peaks in the FT-IR spectra could not be found. This explains that the $LiBH_3F$ decomposition liberates H_2 during its initial stage. The XRD pattern of $LiBH_4$ becomes weak, and the new peaks at $2\theta = 34.2°$ and 44.5° can be found, indicating TiB_2 formation in the composites with the increase in temperature to 250 °C. As the temperature increases to 350 °C, TiB_2 peak intensity gradually strengthens, and the intensity of $LiBH_4$ peaks decreases revealing that the $LiBH_4$ and MXene reaction occurs in 250–350 °C temperature range, which is also confirmed by TPD results. The $LiBH_4$ peaks in X-ray diffraction technique and B—H bond stretching in the Fourier transform of infrared spectroscopy could not be detected at 400 °C, which implies the decomposition of $LiBH_4$ [101].

The MgH_2+ML-Ti_3C_2 multilayered composite was also developed by ball milling technique. These experiments reveal that onset temperature for desorption process of MgH_2-6 wt% ML-Ti_3C_2 composite is decreased to 142 °C with a H_2 storage capability of 6.56 wt%. The H_2 desorption process activation energy for MgH_2-6 wt% ML-Ti_3C_2 composite system is 99 kJ/mol, which is 35.3% less compared to pristine MgH_2. The H_2 sorption kinetics of ML-Ti_3C_2 was mainly due to two synergistic effects: one is that better regeneration of H_2 molecules can be facilitated by Ti molecules, while the other is easier H_2 conversion is mainly due to electron transfer generated by multivalent Ti. The activation energy reduction by 35.3% from 153 kJ/mol of pristine MgH_2 to 99 kJ/mol of MgH_2-6 wt% ML-Ti_3C_2. DSC results show that the incorporation of ML-Ti_3C_2 doesn't enhance the key properties but greatly improves the desorption kinetic parameters. This composite also exhibits good adsorption reversibility by adsorbing H_2 molecules after dehydrogenation of about 6.3 wt% at room temperature, which is 40 °C less compared to pristine MgH_2. Density function theory studies are helpful for analyzing H_2 storage capacity of 2D MXene Cr_2C. All these studies on Cr_2C surface adsorption sites reveal that they are suitable for physisorption and chemisorption of H_2 and Kubas-type interaction, which might be useful for H_2 storage reversible systems under standard conditions. A better understanding of interactions between Cr_2C and adsorbed H_2 molecules was done by electronic structure calculations and charge transfer analysis. The following H_2 storage capability of Cr_2C is about 7.6 wt%, out of which 1.2 wt% is attributed to chemisorption, 3.2 wt% for Kubas interactions, and remaining for weak electrostatic interactions having binding energy of 0.26 eV/H_2 and charge transfer of 0.09 e− to H atom from Cr atom [13].

8.7 Conclusions

Storing H_2 is the major concern because of its low boiling point and highest gravimetric energy density, hence H_2 is mostly stored in gaseous and liquid form. The solid H_2 storage by using metal hydrides and complex hydrides has been developed, and major research was performed to enhance the H_2 sorption capability. Different metal and nonmetal-based hydrides utilized for H_2 storage have some limitations during H_2 molecule regeneration. 2D MXenes plays an important role in enhancing the H_2 storage capacity of metal hydrides either by forming composite or itself can also be used as H_2 storage material. The composites of metal hydrides and MXenes have lower desorption temperature along with increased storage capacity, which implies that energy consumption can be reduced multiple times by the use of 2D structures. This book chapter gives brief description about the role of MXenes in modifying parameters for effective H_2 storage. This chapter also provides an insightful fundamental understanding about MXenes 2D material and its application in hydrogen storage and it may be also helpful to a wide spectrum of researchers working in the field of Green Hydrogen and play important role toward hydrogen economy.

8.8 Future Perspectives and Challenges

The MXene layered structures have been considered as promising catalyst for storing H_2 due to their light weight, thermodynamic stability, and reversibility of adsorbed H_2 molecules. These nanomaterials possess higher H_2 Storage capacity observed by using density function theory. MXenes have exceptional properties of high surface area, hydrophilicity, good surface reactivity, and high adsorption capability compared to other nanomaterials, their incorporation into metal-complex hydrides helps to reduce the desorption temperature and the regenerative storage process will be easier. Utilizing the MXene for wide range of applications in various sectors can help to enhance the efficiency of systems. As hydrogen is a green regenerative fuel, it can be used in defense sectors and aircrafts to replace conventional fuels.

Acknowledgment

R.S. would like to thank the United States-India Science & Technology Endowment Fund (USISTEF) (Grant No.: USISTEF/IG-STAGE-I/019/2022) for providing financial support.

References

1 Anasori, B., Lukatskaya, M.R., and Gogotsi, Y. (2017). 2D metal carbides and nitrides (MXenes) for energy storage. *Nat. Rev. Mater.* 2 (2): 1–17. http://doi.org/10.1038/natrevmats.2016.98.

2 Liu, J., Peng, W., Li, Y. et al. (2020). 2D MXene-based materials for electrocatalysis. *Trans. Tianjin Univ.* (0123456789). https://doi.org/10.1007/s12209-020-00235-x.

3 Kannan K, Sadasivuni KK, Abdullah AM, Kumar B. Current trends in MXene-based nanomaterials for energy storage and conversion system: a mini review. 2020;1–28. https://doi.org/10.3390/catal10050495.

4 Sun, S., Liao, C., Hafez, A.M. et al. (2018). Two-dimensional MXenes for energy storage. *Chem. Eng. J.* 338 (December 2017): 27–45. https://doi.org/10.1016/j.cej.2017.12.155.

5 Jastrzębski K. and Kula P. (2021). Emerging technology for a green, sustainable energy-promising materials for hydrogen storage, from nanotubes to graphene—a review. *Materials (Basel)* 14 (10): 2499. https://doi.org/10.3390/ma14102499.

6 Garara, M., Benzidi, H., Lakhal, M. et al. (2019). Phosphorene: a promising candidate for H_2 storage at room temperature. *Int. J. Hydrogen Energy* 44 (45): 194. https://doi.org/10.1016/j.ijhydene.2019.07.194.

7 Ensafi, A.A., Jafari-Asl, M., Nabiyan, A. et al. (2016). Hydrogen storage in hybrid of layered double hydroxides/reduced graphene oxide using spillover mechanism. *Energy*. 99: 103-114. https://doi.org/10.1016/j.energy.2016.01.042.

8 Kumar, P., Singh, S., Hashmi, S.A.R., and Kim, K. (2021). MXenes: emerging 2D materials for hydrogen storage. *Nano Energy* 85 (March): 105989. https://doi.org/10.1016/j.nanoen.2021.105989.

9 Davoodabadi, A., Mahmoudi, A., and Ghasemi, H. (2021). The potential of hydrogen hydrate as a future hydrogen storage medium. *iScience* 24 (1): 101907. https://doi.org/10.1016/j.isci.2020.101907.

10 Rusman, N.A.A. and Dahari, M. (2016). A review on the current progress of metal hydrides material for solid-state hydrogen storage applications. *Int. J. Hydrogen Energy* 41 (28): 12108–12126. https://doi.org/10.1016/j.ijhydene.2016.05.244.

11 Kumar, P., Singh, S., Hashmi, S.A., and Kim, K.H. (2021). MXenes: Emerging 2D materials for hydrogen storage. *Nano Energy*. 85: 105989. https://doi.org/10.1016/j.nanoen.2021.105989.

12 Hu, Q., Sun, D., Wu, Q. et al. (2013). MXene: a new family of promising hydrogen storage medium. *J. Phys. Chem. A* 117 (51): 14253–14260. https://doi.org/10.1021/jp409585v.

13 Yadav, A., Dashora, A., Patel, N. et al. (2016). Study of 2D MXene Cr_2C material for hydrogen storage using density functional theory. *Appl. Surf. Sci.* https://doi.org/10.1016/j.apsusc.2016.07.083.

14 Li, Y., Guo, Y., Chen, W. et al. (2019). Reversible hydrogen storage behaviors of Ti_2N MXenes predicted by first-principles calculations. *J. Mater. Sci.* 54 (1): 493–505. https://doi.org/10.1007/s10853-018-2854-7.

15 Liu, Y., Gao, H., Zhu, Y. et al. (2019). Applied surface science excellent catalytic activity of a two-dimensional Nb_4 C_3T_x (MXene) on hydrogen storage of MgH_2. *Appl. Surf. Sci.* 493 (July): 431–440. https://doi.org/10.1016/j.apsusc.2019.07.037.

16 Jiang, R., Xiao, X., Zheng, J. et al. (2020, 2019). Remarkable hydrogen absorption/desorption behaviors and mechanism of sodium alanates in-situ doped with Ti-based 2D MXene, 122529. *Mater. Chem. Phys.* 242: https://doi.org/10.1016/j.matchemphys.2019.122529.

17 Lea, T.N.-M. and Cheng-chau Chiu, J.-L.K. (2020). From the perspective of DFT calculations, thermodynamic modeling and kinetic Monte carlo simulations: the interactin

between hydrogen and Sc_2C monolayers. *Phys. Chem. Chem. Phys.* 19 (22): 4387–4401. https://doi.org/10.1039/C9CP05796J.

18 Cheng, H., Kamegawa, T., Mori, K., and Yamashita, H. (2014). Surfactant-free non-aqueous synthesis of plasmonic molybdenum oxide nanosheets with enhanced catalytic activity for hydrogen generation from ammonia borane under visible light. *Angew Chem. Int. Ed.* 53 (11): 2910–2914. https://doi.org/10.1002/anie.201309759.

19 Yamashita, H., Mori, K., and Kuwahara, Y. (2021). Design of plasmonic catalysts utilizing nanostructures. *J. Jpn. Pet. Inst.* 64 (4): 155–165. https://doi.org/10.1627/jpi.64.155.

20 Eberle, U., Felderhoff, M., and Schüth, F. (2009). Chemical and physical solutions for hydrogen storage. *Angew Chem. Int. Ed.* 48 (36): 6608–6630. https://doi.org/10.1002/anie.200806293.

21 Leitao, E.M., Jurca, T., and Manners, I. (2013). Catalysis in service of main group chemistry offers a versatile approach to p-block molecules and materials. *Nat. Chem.* 5 (10): 817–829. https://doi.org/10.1038/nchem.1749.

22 Grochala, W. and Edwards, P.P. (2004). Thermal decomposition of the non-interstitial hydrides for the storage and production of hydrogen. *Chem. Rev.* 104 (3): 1283–1316. https://doi.org/10.1021/cr030691s.

23 Sewell, L.J., Lloyd-Jones, G.C., and Weller, A.S. (2012). Development of a generic mechanism for the dehydrocoupling of amine-boranes: a stoichiometric, catalytic, and kinetic study of H3B·NMe 2H using the $[Rh(PCy_3)_2]$+fragment. *J. Am. Chem. Soc.* 134 (7): 3598–3610. https://doi.org/10.1021/ja2112965.

24 Metin, O. and Özkar, S. (2009). Hydrogen generation from the hydrolysis of ammonia-borane and sodium borohydride using water-soluble polymer-stabilized cobalt(0) nanoclusters catalyst. *Energy Fuels* 23 (7): 3517–3526.

25 Zhang, N., Jalil, A., Wu, D. et al. (2018). Refining defect states in W18 O49 by Mo doping: a strategy for tuning N2 activation towards solar-driven nitrogen fixation. *J. Am. Chem. Soc.* 140 (30): 9434–9443.

26 Akbayrak, S., Tonbul, Y., and Özkar, S. (2020). Magnetically separable Rh0/Co_3 O_4 nanocatalyst provides over a million turnovers in hydrogen release from ammonia borane. *ACS Sustain. Chem. Eng.* 8 (10): 4216–4224.

27 Yao, Q., Ding, Y., and Lu, Z.-H. (2020). Noble-metal-free nanocatalysts for hydrogen generation from boron- and nitrogen-based hydrides. *Inorg. Chem. Front.* 7 (20): 3837–3874.

28 Wen, Z., Wu, J., and Fan, G. (2020). Facile fabrication of rhodium/nanodiamond hybrid as advanced catalyst toward hydrogen production from ammonia–borane. *Catalysts* 10 (9): 1037.

29 Slot, T.K., Yue, F., Xu, H. et al. (2020). Surface oxidation of $Ti_3C_2T_x$ enhances the catalytic activity of supported platinum nanoparticles in ammonia borane hydrolysis. *2D Mater.* 8 (1): 015001.

30 Palmer, C., Saadi, F., and McFarland, E.W. (2018). Technoeconomics of commodity chemical production using sunlight. *ACS Sustain. Chem. Eng.* 6 (5): 7003–7009.

31 Kumar, R., Karkamkar, A., Bowden, M., and Autrey, T. (2019). Solid-state hydrogen rich boron–nitrogen compounds for energy storage. *Chem. Soc. Rev.* 48 (21): 5350–5380.

32 Zhan, X., Si, C., Zhou, J., and Sun, Z. (2020). MXene and MXene-based composites: synthesis, properties and environment-related applications. *Nanoscale Horiz.* 5 (2): 235–58. https://doi.org/10.1039/C9NH00571D.

33 Gao, L., Li, C., Huang, W. et al. (2020). MXene/polymer membranes: synthesis, properties, and emerging applications. *Chem. Mater.* 32 (5): 1703–1747.

34 Xu, C., Wang, L., Liu, Z. et al. (2015). Large-area high quality 2D ultrathin Mo2C superconducting crystals. *Nat. Mater.* 14 (11): 1135–1141.

35 Gogotsi, Y. and Barsoum, M.W. (2012). Two-dimensional transition metal carbides. *ACS Nano* 6 (2): 1322–1331.

36 Mashtalir O, Lukatskaya MR, Zhao M, Barsoum MW. Amine-assisted delamination of Nb 2 C MXene for Li-ion energy storage devices. 2015;1–6.

37 Halim, J., Kota, S., Lukatskaya, M.R. et al. (2016). Synthesis and characterization of 2D molybdenum carbide (MXene). *Adv. Funct. Mater.* 26 (18): 3118–3127.

38 Zhou, J., Zha, X., Chen, F.Y. et al. (2016). A two-dimensional zirconium carbide by selective etching of Al_3 C_3 from nanolaminated Zr_3 Al_3 C_5. *Angew. Chem. Int. Ed.* 55 (16): 5008–5013.

39 Naguib, M., Kurtoglu, M., Presser, V. et al. (2011). Two-dimensional nanocrystals produced by exfoliation of Ti_3 AlC_2. *Adv. Mater.* 23: 4248–4253.

40 Chen, W., Li, D., Wang, Z. et al. (2017). Reaction mechanism and kinetics for hydrolytic dehydrogenation of ammonia borane on a Pt/CNT catalyst. *AlChE J.* 63 (1): 60–65.

41 Feng, K., Zhong, J., Zhao, B. et al. (2016). Cu_x Co_{1-x} O nanoparticles on graphene oxide as a synergistic catalyst for high-efficiency hydrolysis of ammonia-borane. *Angew Chem. Int. Ed.* 55 (39): 11950–11954.

42 Xu, F. and Liu, X. (2021). Magnetic Co-Pd/C nanocomposites for hydrogen evolution upon the hydrolytic dehydrogenation of NH_3 BH_3, $NaBH_4$, and Me_2 $NHBH_3$. *ACS Appl. Nano Mater.* 4 (7): 7479–7485.

43 Fu, F., Wang, C., Wang, Q. et al. (2018). Highly selective and sharp volcano-type synergistic Ni_2 Pt@ZIF-8-catalyzed hydrogen evolution from ammonia borane hydrolysis. *J. Am. Chem. Soc.* 140 (31): 10034–10042.

44 Li, H., Wang, M., Luo, L., and Zeng, J. (2019). Static regulation and dynamic evolution of single-atom catalysts in thermal catalytic reactions. *Adv. Sci.* 6 (3): 1801471.

45 Wu, H., Luo, Q., Zhang, R. et al. (2018). Single Pt atoms supported on oxidized graphene as a promising catalyst for hydrolysis of ammonia borane. *Chin. J. Chem. Phys.* 31 (5): 641–648.

46 Banu, T., Debnath, T., Ash, T., and Das, A.K. (2015). Hydrolysis of ammonia borane and metal amidoboranes: a comparative study. *J. Chem. Phys.* 143 (19): 194305.

47 Long, L.-L., Liu, X.-Y., Chen, J.-J. et al. (2018). Ultrafine and well-dispersed nickel nanoparticles with hierarchical structure for catalytically breaking a boron—hydrogen bond. *ACS Appl. Nano Mater.* 1 (12): 6800–6807.

48 Yao, Q., Lu, Z.-H., Yang, Y. et al. (2018). Facile synthesis of graphene-supported Ni-CeO_x nanocomposites as highly efficient catalysts for hydrolytic dehydrogenation of ammonia borane. *Nano Res.* 11 (8): 4412–4422.

49 Qi, H., Huang, Z., Wang, M. et al. (2018). Bifunctional ligands for Pd-catalyzed selective alkoxycarbonylation of alkynes. *J. Catal.* 363: 63–68.

50 Abd Khalim Khafidz, N.Z., Yaakob, Z., Lim, K.L., and Timmiati, S.N. (2016). The kinetics of lightweight solid-state hydrogen storage materials: a review. *Int. J. Hydrogen Energy* 41 (30): 13131–13151.

51 Chou, K.-C. and Xu, K. (2007). A new model for hydriding and dehydriding reactions in intermetallics. *Intermetallics* 15 (5–6): 767–777.

52 Pang, Y. and Li, Q. (2016). A review on kinetic models and corresponding analysis methods for hydrogen storage materials. *Int. J. Hydrogen Energy* 41 (40): 18072–18087.

53 Booth, F. (1948) A note on the theory of surface diffusion reactions. Trans. Faraday Soc. 44: 796–801. https://doi.org/10.1039/TF9484400796.

54 Carstensen, J.T. (1974). Stability of solids and solid dosage forms. *J. Pharm. Sci.* 63 (1): 1–14.

55 Khawam, A. and Flanagan, D.R. (2006). Solid-state kinetic models: basics and mathematical fundamentals. *J. Phys. Chem. B* 110 (35): 17315–17328.

56 Koga, N. and Criado, J.M. (1997). Influence of the particle size distribution on the CRTA curves for the solid-state reactions of interface shrinkage type. *J. Therm. Anal.* 49 (3): 1477–1484.

57 Koga, N. and Criado, J.M. (2005). Kinetic analyses of solid-state reactions with a particle-size distribution. *J. Am. Ceram. Soc.* 81 (11): 2901–2909.

58 Bösenberg, U., Kim, J.W., Gosslar, D. et al. (2010). Role of additives in $LiBH_4$–MgH_2 reactive hydride composites for sorption kinetics. *Acta Mater.* 58 (9): 3381–3389.

59 Jander, W. (1927). Reactions in the solid state at higher temperatures. Reaction rates of endothermic reactions. *J. Inorg. Gen. Chem.* 163 (1): 1–30. https://doi.org/10.1002/zaac.19271630102.

60 Shao, H., Xin, G., Zheng, J. et al. (2012). Nanotechnology in Mg-based materials for hydrogen storage. *Nano Energy* 1 (4): 590–601.

61 Ngqalakwezi, A. and Nkazi, D.B. (2020). Hydrogen storage: materials, kinetics and thermodynamics. In: *Advanced Applications of Hydrogen and Engineering Systems in the Automotive Industry* (2 December 2020), 137. IntechOpen.

62 Carter, R.E. (1961). Kinetic model for solid-state reactions. *J. Chem. Phys.* 34 (6): 2010–2015.

63 Shimada, E., Yamashita, H., Matsumoto, S. et al. (1999). Oxidation kinetics of spherically shaped palladium powder. *J. Mater. Sci.* 34 (16): 4011–4015.

64 Gray, R.D. (1969). Kinetics of oxidation of copper(I) by molecular oxygen in perchloric acid-acetonitrile solutions. *J. Am. Chem. Soc.* 91 (1): 56–62.

65 Avrami, M. (1941). Granulation, phase change, and microstructure kinetics of phase change. III. *J. Chem. Phys.* 9 (2): 177–184.

66 Avrami, M. (1940). Kinetics of phase change. II Transformation-time relations for random distribution of nuclei. *J. Chem. Phys.* 8 (2): 212–224.

67 Avrami, M. (1939). Kinetics of phase change. I General theory. *J. Chem. Phys.* 7 (12): 1103–1112.

68 Pang, Y., Sun, D., Gu, Q. et al. (2016). Comprehensive determination of kinetic parameters in solid-state phase transitions: an extended Jonhson–Mehl–Avrami–Kolomogorov model with analytical solutions. *Cryst. Growth Des.* 16 (4): 2404–2415.

69 Fanfoni, M. and Tomellini, M. (1998) The Johnson-Mehl- Avrami-Kohnogorov model: A brief review. *Il Nuovo Cimento D.* 20: 1171–1182 . https://doi.org/10.1007/BF03185527.

70 Wu, Z., Fang, J., Liu, N. et al. (2021). The improvement in hydrogen storage performance of MgH_2 enabled by multilayer Ti_3C. 2: 1–14.

71 Mei, J., Ayoko, G.A., and Hu, C. (2020). Thermal reduction of sulfur-containing MAX phase for MXene production. *Chem. Eng. J.* 395 (April): 125111. https://doi.org/10.1016/j.cej.2020.125111.

72 Wu, R., Du, H., Wang, Z. et al. (2016). Remarkably improved hydrogen storage properties of NaA_lH_4 doped with 2D titanium carbide. *J. Power Sources* 327: 519–525. https://doi.org/10.1016/j.jpowsour.2016.07.095.

73 Zheng, J., Cheng, H., Xiao, X. et al. (2019). ScienceDirect enhanced low temperature hydrogen desorption properties and mechanism of Mg (BH_4) 2 composited with 2D MXene. *Int. J. Hydrogen Energy* 44 (44): 24292–24300. https://doi.org/10.1016/j.ijhydene.2019.07.177.

74 Feng, A., Yu, Y. et al. (2017). Two-dimensional MXene Ti_3C_2 produced by exfoliation of Ti_3AlC_2. *Mater. Des.* 114: 161–166. https://doi.org/10.1016/j.matdes.2016.10.053.

75 Xian, K., Gao, M., Li, Z. et al. (2019). Superior kinetic and cyclic performance of a 2D titanium carbide incorporated $2LiH^+$ MgB_2 composite toward highly reversible hydrogen storage. *ACS Appl. Energy Mater.* 2: 4853–4864.

76 Le, T.T., Pistidda, C., Puszkiel, J. et al. (2018). Design of a nanometric AlTi additive for MgB_2-based reactive hydride composites with superior kinetic properties. *J. Phys. Chem. C* 122 (14): 7642–7655.

77 Saldan, I., Campesi, R., Zavorotynska, O. et al. (2012). Enhanced hydrogen uptake/release in 2LiH-MgB_2 composite with titanium additives. *Int. J. Hydrogen Energy* 37 (2): 1604–1612. https://doi.org/10.1016/j.ijhydene.2011.10.047.

78 Wang, P., Wang, P., Ma, L. et al. (2009). Improved hydrogen storage property of Li-Mg-B-H system by milling with titanium trifluoride. *Energy Environ. Sci.* 2 (1): 120–123.

79 Puszkiel, J.A., Castro Riglos, M.V., Ramallo-López, J.M. et al. (2017). A novel catalytic route for hydrogenation-dehydrogenation of $2LiH + MgB_2$: via in situ formed core-shell $LixTiO_2$ nanoparticles. *J. Mater. Chem. A* 5 (25): 12922–12933.

80 Wang, K., Kang, X., Zhong, Y. et al. (2014). Improved reversible dehydrogenation properties of $2LiBH_4$-MgH_2 composite by milling with graphitic carbon nitride. *Int. J. Hydrogen Energy* 39 (25): 13369–13374. https://doi.org/10.1016/j.ijhydene.2014.04.079.

81 Plerdsranoy, P., Chanthee, S., and Utke, R. (2017). Compaction of $LiBH_4$-MgH_2 doped with MWCNTs-TiO_2 for reversible hydrogen storage. *Int. J. Hydrogen Energy* 42 (2): 978–986.

82 Wang, J., Han, S., Zhang, W. et al. (2013). Effects of MoS_2 addition on the hydrogen storage properties of $2LiBH_4$-MgH_2 systems. *Int. J. Hydrogen Energy* 38 (34): 14631–14637.

83 Bösenberg, U., Doppiu, S., Mosegaard, L. et al. (2007). Hydrogen sorption properties of MgH_2-$LiBH_4$ composites. *Acta Mater.* 55 (11): 3951–3958.

84 Zhao, Y., Ding, L., Zhong, T. et al. (2014). Hydrogen storage behavior of $2LiBH_4/MgH_2$ composites improved by the catalysis of CoNiB nanoparticles. *Int. J. Hydrogen Energy* 39 (21): 11055–11060. https://doi.org/10.1016/j.ijhydene.2014.05.063.

85 Zhang, X., Shen, Z., Jian, N. et al. (2018). ScienceDirect A novel complex oxide $TiVO_3$ 5 as a highly active catalytic precursor for improving the hydrogen storage properties of MgH_2. *Int. J. Hydrogen Energy*: https://doi.org/10.1016/j.ijhydene.2018.10.216.

86 Fan, Y., Yuan, Z., Zou, G. et al. (2017). Two-dimensional MXene/A-TiO_2 composite with unprecedented catalytic activation for sodium alanate. *Catal. Today* 0–1. https://doi.org/10.1016/j.cattod.2017.11.018.

87 Fan, Y., Chen, D., Liu, X. et al. (2019). ScienceDirect improving the hydrogen storage performance of lithium borohydride by Ti_3C_2 MXene. *Int. J. Hydrogen Energy* 44 (55): 29297–29303. https://doi.org/10.1016/j.ijhydene.2019.01.011.

88 Gencer, A., Aydin, S., Surucu, O. et al. (2021). Applied surface science enhanced hydrogen storage of a functional material: Hf2 CF2 MXene with Li decoration. *Appl. Surf. Sci.* 551: 149484. https://doi.org/10.1016/j.apsusc.2021.149484.

89 Huang, T., Huang, X., Hu, C. et al. (2020). MOFs derived Ni nanoparticles dispersed on monolayer MXene functional carrier: an effective double transition metal-based catalyst to improve hydrogen storage properties of MgH_2. *Chem. Eng. J.* 127851. https://doi.org/10.1016/j.cej.2020.127851.

90 Yuan, Z., Zhang, D., Fan, G. et al. (2021). Synergistic effect of CeF_3 nanoparticles supported on Ti_3C_2 MXene for catalyzing hydrogen storage of NaA_lH_4. *ACS Appl. Energy Mater.* 4 (3): 2820–2827 https://doi.org/10.1021/acsaem.1c00122.

91 Zhu, W., Panda, S., Lu, C. et al. (2020). Using a self-assembled two-dimensional MXene-based catalyst (2D-Ni@ Ti_3C_2) to enhance hydrogen storage properties of MgH2. *ACS Appl. Mater. Interfaces.* 12 (45): 50333–50343. https://doi.org/10.1021/acsami.0c12767.

92 Wang, Z., Zhang, X., Ren, Z. et al. (2019). In situ formed ultra fi ne NbTi nanocrystals from a NbTiC solid-solution MXene for hydrogen storage. 14244–52.

93 Xian, K., Gao*, M., Li, Z. et al. (2019). Superior kinetic and cyclic performance of a 2D titanium carbide incorporated $2LiH + MgB_2$ composite toward highly reversible hydrogen storage. *ACS Appl. Energy Mater.* 2 (7): 4853–4864.

94 Zang, L., Sun, W., Liu, S. et al. (2018). Enhanced hydrogen storage properties and reversibility of $LiBH_4$ confined in two-dimensional Ti_3C_2. *ACS Appl. Mater. Interfaces* 10 (23): 19598–19604.

95 Liu, Y., Du, H., Zhang, X. et al. (2016). Superior catalytic activity derived from a two-dimensional Ti_3C_2 precursor towards the hydrogen storage reaction of magnesium hydride. *Chem. Commun.* 52 (4): 705–708.

96 Bogdanović, B. and Schwickardi, M. (1997). Ti-doped alkali metal aluminium hydrides as potential novel reversible hydrogen storage materials. *J. Alloys Compd.* 253–254: 1–9.

97 Liu, S., Liu, J., Liu, X. et al. (2021) Hydrogen storage in incompletely etched multilayer Ti_2CT_x at room temperature. *Nat. Nanotechnol.* 16: 331–336 .

98 Rasool, K., Helal, M., Ali, A. et al. (2016). Antibacterial $Ti_3C_2T_X$ MXene nanosheets: towards advanced wastewater treatment membranes. In: *Abstracts of Papers of the*

American Chemical Society (20 August 2016), Volume 252, 1155. Washington, DC USA: American Chemical Society.

99 Liu, L., Zhao, Q., Liu, R., and Zhu, L. (2019). Hydrogen adsorption-induced catalytic enhancement over Cu nanoparticles immobilized by layered Ti_3C_2 MXene. *Appl. Catal. B* 252 (April): 198–204. https://doi.org/10.1016/j.apcatb.2019.04.026.

100 Feng, X., Yuan, J., Lv, Y. et al. (2020). Improvement of desorption performance of $Mg(BH_4)_2$ by two-dimensional Ti_3C_2 MXene addition. *Int. J. Hydrogen Energy* 45 (33): 16654–16662. https://doi.org/10.1016/j.ijhydene.2020.04.132.

101 Fan, Y., Chen, D., Liu, X. et al. (2019). Improving the hydrogen storage performance of lithium borohydride by Ti_3C_2 MXene. *Int. J. Hydrogen Energy* 44 (55): 29297–29303. https://doi.org/10.1016/j.ijhydene.2019.01.011.

9

MXenes for Supercapacitor Applications

Samaneh Shahsavarifar[1], Hamidreza Parsimehr[2], and Amir Ershad-Langroudi[3]

[1]*Department of Metrology and Optoelectronics, Faculty of Electronics, Telecommunications, and Informatics, Gdansk University of Technology, Gdansk, Poland*
[2]*Department of Chemistry, University of New Brunswick, Fredericton, Canada*
[3]*Color and Surface Coatings Group, Polymer Processing Department, Iran Polymer and Petrochemical Institute (IPPI), Tehran, Iran*

9.1 Introduction

Storage for heavy items technologies, such as electric cars [1], static battery systems [2], and massive power usage equipment in energy production, are in a growing market nowadays, in line with associated technological advancements. Due to its high-power applications that are difficult to meet using batteries, supercapacitor research is gaining traction. Although it is possible to store a lot of energy with the battery, it has a longer charge and discharge time. It should be replaced with a supercapacitor, which charges considerably more quickly than batteries. The fundamental drawback of supercapacitors in practical applications is their increased self-discharge rate. Furthermore, maintaining a steady output voltage as the charge drops is problematic. Supercapacitor research is still ongoing to enhance the voltage window and improve cyclability [3]. Current lithium-ion batteries (LIBs) [4], and supercapacitors [5] frequently it was impossible to achieve [1] high power density and energy for one-of-a-kind products at the same time, [2] restricted availability of resources like lithium source, and [3] costly of existing rechargeable batteries. As a result, this field of study is looking for innovative and cost-effective ideas.

Many substances were exfoliated at the beginning of two-dimensional (2D) matter applications, including hexagonal BN [6], MoS_2 [7], hydroxides [8], graphenes [6], and 2D material research on transition metal oxides [8] in its early stages. MXenes, a type of 2D material with better electrochemical and mechanical characteristics, generate considerable interest. MXenes were found in 2011 [9]. These are transition metal nitrides and carbides, containing MAX (or $M_{n+1}AX_n$) phases, where $n = 1$–4. M stands for beginning transition metal (Ti, Mo, Sc, Zr, Cr, V, Nb, Hf, Ta), A stands for a component (usually Al, Ti, In, S, P, Cd, Si, Ga, As, Pb, Sn, Ge), and X stands for nitrogen or carbon. $M_{n+1}X_nT_x$ 2D layers called MXenes are formed when the "A" layer is selectively eliminated by etching. The letters T stand for $^-$F, $^-$O, and $^-$OH [10]. The MAX phases have multilayer crystalline phases with the p63/mmc

MXenes: Next-Generation 2D Materials: Fundamentals and Applications, First Edition.
Edited by Jay Singh, Kshitij RB Singh, Ravindra Pratap Singh, and Charles Oluwaseun Adetunji.

space group [11]. It has an anisotropic structure with a layered structure [10]. Each unit cell will have two formula units. Typically, net elemental layers of A are interspersed with nearly compact M layers, with X-atoms covering the octahedral spaces between them [9].

MAX phases were promising for energy storage before MXenes, which are bonded with each other by very strong covalent or metallic interactions [12]. Ti_3SiC_2 was the first MAX phase that attracted significant interest. Silicon anodes have been popular in energy studies for a while. Gogotsi et al. introduced the use and etching of MAX-phase titanium silicon carbide (Ti_3SiC_2) in energy research. These compounds can be a potential anode substance as ion batteries due to their unique properties such as huge lithium-ion storage capacity, electrical conductivity, layer structure favorable configuration for anode materials, fast release of lithium, and low operating voltage [13].

MXenes have an $[MX]_n$ M configuration, wherein the $n+1$ layers of M encapsulate the n number of X layers. If there are many initial transition metals, MXenes can exist as either a solid solution or an organized phase. In the case of a solid solution, the two transition metals may be arbitrarily stacked. These include $(Ti, Nb)_2C$, $(Ti, V)_2C$, $(Ti, V)_3C_2$, $(Ti, Nb)_4C_3$, $(Cr, V)_3C_2$, and $(Nb, Zr)_4C_3$. Single to triple layers of transition metal are layered inside the other in a two-dimensional carbide or nitride arrangement in an ordered MXene.

MXenes have a strong metallic conductivity and hydrophilic surface [3]. The existence of moieties on the surfaces of MXenes, such as oxygen, hydroxyl, or fluorine, confers hydrophilicity [14]. These substances can provide high energy storage densities as well as quick electrochemical charge storage [15–17]. MXenes are interesting options for supercapacitors [18, 19], LIBs [20–23], lithium-sulfur batteries [23], a wide spectrum of sensors [24, 25], water treatment [26], and other applications. Although several MXenes materials exist, supercapacitor research is focusing on titanium- and vanadium-based MXenes. The majority of them make use of composite for electrodes, as well as careful electrolyte selection to boost capacitance [27, 28]. Researchers were leading the way with irregular capacitor layouts to see how they performed [29]. The electrochemical characteristics of MXenes are constantly influenced by the fabrication methods utilized, how electrodes are made, and how the MXene is arranged [30].

9.2 Two-dimensional MXenes Structure

$M_{n+1}X_nT_x$ phases, where $n = 1–3$, are the representation of MXene, where M stands as a transition metal (e.g., Ti, Sc, Nb, Hf, Ta, Zr, Mo, V, and others), X refers for nitrogen or carbon, T_x is for surface endpoints (e.g., oxygen, fluorine, or hydroxyl), and x refers for the number of surface areas. MXenes may be frequently generated by specific scratching with hydrofluoric (HF) corrosive as an etchant to remove the A layers from their comparable MAX substrates. Aside from the conventional $M_{n+1}AX_n$ (n = 1–3) synthesis, MAXs have a place with a group of multilayered ternary carbides or nitrides. M refers to quick-change metals, which typically consist of 13–16 elements, and X is for nitrogen or carbon, as in Ti_3AlC_2 [31]. MXene appears to be the most promising rival as a cathode material in a study of the above clarity due to the spacing between MXenes can be adjusted [32].

9.3 MXenes' Characteristics

MXene, as a new material, has certain distinct features when compared to previous compounds. In this part, their structure as well as their electrical, optical, and magnetic characteristics are considered.

9.3.1 Characteristics of the Structure

The hexagonal-packed layering structure of MXene is comparable to that of the MAX phase compound. The component X atoms are placed in octahedral orientations, while the transition metal (M) atoms are arranged in a nearly tightly packed pattern. Van der Waals energies are about similar in other 2D substances, linking adjacent layers. The majority of MXenes are synthesized from the aqueous phase. As a result, the MXene surface is made up of a combination of $^{-}$OH, $^{-}$O, and $^{-}$F endings [33–37]. In MXene, three surface termination locations are possible: (i) just on top of the transition metal atom, (ii) the vacant location between the upper metal atoms, and (iii) the void site between the next layer of X atoms. Even though many theoretical simulations have been undertaken to investigate the impact of a single surface terminal, MXene cannot be obtained with or without surface endings, and the terminal positions are assigned at random. To achieve a stable state, two electrons are required for oxygen termination (–O), whereas only one electron is required for $^{-}$OH and $^{-}$F termination. As a result, in interstitial bonding with two metal atoms, the $^{-}$O end site is more likely to fill unoccupied sites, causing the interstitial bonding to be stabilized. The oxygen-functionalized Ti_3C_2 has greater thermodynamic stability and strength than its equivalents due to a considerable amount of electron transfer from the inside bond to the outside surface of the material [38].

MXene structures generated by diverse synthesis methods, like HF and LiF-HCl etching, are similar [39]. XRD test indicates that the reflected peak (0002) of the LiF-HCl etched specimen shifts to a lower angle (according to an ac parameter of ~25 Å) as compared to the MXene etched by HF, which moves to approximately 19–20 Å [39]. X-ray photoelectron spectrometry (XPS) is a critical tool that may be utilized to investigate the chemical situations of the atoms in MXene as well as quantify the surface ending groups [40]. The chemical elements found in D-Mo_2CT_z, namely, the Mo 3d, C 1s, O 1s, and F 1s high-resolution XPS spectrum peak matching curves [41]. $Ti_3C_2T_z$ is considered as a chemical exfoliation substance produced by targeted corrosion of "Al" from Ti_3AlC_2's MAX phase. T_z is used to indicate moiety such as –F, –OH, and –O. Heating altered the structure of $Ti_3C_2T_z$, either by providing additional TiC layers to generate $Ti_4C_3T_z$ and $Ti_5C_4T_z$ or by converting $Ti_3C_2T_z$ to TiC [42]. $Ti_3C_2T_z$ structure may appear as a Ti–C–Ti–C–Ti five-element layered shape, that is ended through T_z at the surface [42]. The optimized $Ti_{n+3}C_{n+2}$ crystal structure by using DFT results in a theoretical STEM picture that can be simulated. Only the experimental STEM pictures can detail the manner to specify the number of h-TiC layers.

9.3.2 Electronic Characteristics

In comparison to other 2D materials, MXenes exhibit the most unique electrical characteristics. MXenes' widespread use in the storage of energy as well as electrochemistry is due to their superior electronic properties [42]. The parameters impacting the electrical characteristics of MXene, such as various M, X, and chemical bonding on their surfaces, have recently been investigated using theoretical calculations [43, 44]. MXenes' electronic characteristics can vary from metal to semiconductor because they include a variety of transition metals. Some MXenes, however, have been anticipated to be geometrical insulators because they contain heavy transition metals (e.g., Cr, Mo, and W) [45–48]. The surface ending will also alter the electrical characteristics of bare MXene. Investigations in MXene are used to determine the electrical characteristics of Ti_2C [48], Ti_3C_2, Mo_2C, Mo_2TiC_2, and $Mo_2Ti_2C_3$ [40, 49, 50]. The electrical characteristics are impacted by changes in the external M layer, according to these investigations. Even though $Ti_3C_2T_z$ is a metal, MXenes containing Mo exhibited semiconductor-like properties, including positive magnetic properties in 10 K [49, 51]. MXene's transport capabilities may be altered by altering the posttreatment at the termination of the surface [52, 53]. As a result, the F− and OH− groups can only receive 1 e^-, and their electronic structures are identical; however, the O^- group can accept 2 e^-, hence its electronic structure is distinct [33, 54]. The surface endings, various levels of surface imperfections, and the MXene synthesis technique may all have an impact on its electrical characteristics. The more faults there are, the poorer the conductivity is owing to the disintegration of the organized structure and the free electrical current. Doping is a method for generating defects into MXene. Charge imbalance can be caused by the interchange of particular ions with varying electric densities. MXene's conductivity is also influenced by its surface ending [47, 55, 56].

9.3.3 Optical Characteristics

MXene possesses optical features including light emission and absorption, which makes it helpful in a variety of functions and biomedical applications. MXenes' linear and nonlinear optical characteristics (e.g., photoluminescence (PL), nonlinear refractive index, and saturation absorption) are intimately linked to the energy structure outlined abovementioned (such as bandgap, topological barriers, and direct/indirect bandgaps) [45, 46]. The linear/nonlinear dielectric scattering, or refractive index, specifies optical characteristics. To study the nonlinear optical characteristics of $Ti_3C_2T_z$ thin films, the open aperture Z-scan technique is commonly utilized [57]. Theoretical study identifies major aspects that will influence MXene's optical properties. The M_2C bandgap (where M = Ti, Zr, Hf) is 0.92 eV to 1.75 eV depending on the composition of M, in the visible light absorption range (VIS) [58–61]. By adjusting the surface ending, the bandgap may be changed to fit the ultraviolet (UV) band [62, 63]. MXene exhibits significant absorption characteristics and a spectrum that extends UV to NIR (up to 1350 nm), making it ideal for photothermal treatment in biological environments. MXenes are also the desirable material for improving Raman scattering patterns. MXenes have significant local surface plasmon resonance effects because of their high metal conductivity, making them an attractive material for surface-enhanced spectrometry [64, 65]. In addition, the hydrophilic and flexible surface of the Raman label can give stability and close contact. Both of these characteristics help to boost the Raman scattering intensity. Many optical features of MXene, such as luminous performance, emission color, and mechanism, have to be investigated extensively. MXene can exhibit layer thickness-related properties in the UV–VIS–NIR region, as evidenced by the transmittance of the spin-cast $Ti_3C_2T_z$ coating [66].

9.3.4 Magnetic Characteristics

The magnetic characteristics of the MXene class's 2D layers are significant. 2D Cr_2C, Cr_2N, Ta_3C_2, and Cr_3C_2 are ferromagnets that can be removed from their MAX state in MXenes, but 2D Ti_3C_2 and Ti_3N_2 are antiferromagnets [67–69]. In Ti_2C and Ti_2N, near-half-metal ferromagnetism was already observed [65]. The antiferromagnetic Mn_2C monolayer has a significant Neel temperature of 720 K, following theoretical estimates. Surface modification (OH, F, and Cl) converts that to the ferromagnetic system with a high Curie temperature (520 K) [68]. Several MXenes, such as Cr_3C_2 and Mn_2C, seem based on theoretical predictions and have not been chemically synthesized. Crystal magnetism is significantly influenced by crystal tension, according to theoretical investigations. Ion doping and void introducing are two key strategies for controlling its magnetic characteristics that have been proposed [51].

9.4 MXenes as a Source of Energy Storage

9.4.1 Supercapacitor Energy Storage Mechanism

Because MXenes have a lot of binding sites, between the interface/surface and between the layers, the active groups are thought to be the key energy storage sites. Energy is normally stored in energy storage via adsorption–desorption; however, the electrical double layer or thin sites on the particle surface exhibit a function in supercapacitors [58]. Electrical double-layer capacitors and pseudocapacitors are the two categories that supercapacitors fall under. While the latter relies on quick redox processes, the former maintains charge by creating an electrical double layer at the interface of the electrode–electrolyte. This discrepancy, which is independent of electrode surface area, encourages research into pseudo capacitors [15]. The non-Faradaic charge storage is followed by the double-layer capacitance, whereas the faradaic chemical process is followed by the pseudocapacitance charge storage [70]. The Faradaic process involves the transport of ions or electrons from the electrolyte to the bulk phase across the electrode. The ions or electrons that enter the electrode must exit the electrode following an electrode reaction, which is a key criterion. In addition, the resultant varieties will enter an exterior bulk phase. This procedure can be sustained if the external bulk phase remains constant and does not change the composition. Species entering the electrode, on the other hand, will not exit the electrode in a non-Faradaic process. The charge is held here, along with the current flow's progress. In most cases, there is no electrode response. Otherwise, the electrode reaction is limited to the atoms of the electrode. In this case, oxidized or reduced reacting species will be trapped inside the electrode. The electrode's capacity (C/g) and capacitance (F/g) may be calculated using the materials' output of charge vs. voltage. As a result, the capacitive process is named after the non-Faradaic electrode method [71]. The form of cyclic voltammograms can be utilized to detect the mechanism of charge storage in various substances [72]. When it comes to energy storage devices, by combining materials with various methods of energy storage, researchers are working to achieve high energy density and stability. Until recently, MXene-based hybrid designs have been designed for various types of energy storage systems. The charge storage strategy of MXene supercapacitors, however, is affected by a variety of circumstances. Understanding these contributing aspects will aid in the development of a supercapacitor that meets our needs.

9.4.2 Morphology's Effect on MXenes' Energy Storage

Morphologies have a significant effect on the electrochemical MXenes' characteristics. MXenes stacks have 2D slits after layer MAX phase A is etched away [72]. A significant increase in capacitance is achievable if water molecules can be contained within these openings [73]. Because these water molecules' dipolar polarization may show an outside electric field and so enhance the capacitance of a double-layer electric capacitor with a negative dielectric constant, they have a negative dielectric constant. Ion diffusion and capacitance can be improved by reducing the lateral dimension of layers [74]. MXene is a 2D material that can be subjected to different morphologies. MXenes based on Nb exhibited a hybrid surface morphology that changed with temperature, ranging from featherlike to barrel-shaped. When $Ti_3C_2T_x$ was created by rapid oxidation, it produced bi-pyramidal and nanocuboids. Even after 30 cycles, they still had a reversible capacity of 220 mAh/g at a C/18 cycling rate. As a result, fine-tuning the oxidation process may contribute to achieving specific morphologies and increase supercapacitors. This opens up a lot of possibilities for future study into optimizing the temperature conditions, mechanisms, and factors that are related to desirable morphologies.

9.4.3 MXene Functional Group Reactivity and Supercapacitors

The reasons for MXenes' instability on aqueous colloids are a popular study topic in MXene chemistry. The oxidation of MXenes is a broad theory that describes MXene instability in aqueous solutions. MXene stored in an aqueous solution with argon has a strong anatase peak, but MXene protected in propane-2-ol, which contains a lot of oxygen, indicates any oxidation. As a result, it is suggested that the instabilities are caused by the transition of titanium carbide MXenes into titania in aqueous conditions, rather than oxidation [75]. MXene oxidative breakdown in aqueous solutions has shown a variety of effects. Some researchers identified TiO_2 and CO_2 [76] as carbide end products, whereas others identified carbon with titanium dioxide [77–80]. When it comes to MXene reactivity, its environmental effect must also be considered. The use of MXenes in wastewater treatment. Water intercalation as an applied potential will have a major impact on the chemistry of MXene surfaces. Even without applying an electric current, oxygen will be used to functionalize a multilayer MXene. A functionalization shift from oxygen to hydrogen has been detected when a negative potential is applied. This will determine the electrical characteristic. As a result, a deeper knowledge of functionalization and applied potential is required for the electrochemical use of MXenes.

According to DFT calculations, the theoretical capacity of Ti_3C_2 is 320 mAh g^{-1} due to the ideal Li storage locations at the top of the carbon atoms, which are ideal for Li storage [81].

O-terminated MXenes have a higher theoretical capacity due to the action of functional groups [82]. During Li cycling, however, a hydroxyl group is rather irreversible [83]. For interlayer-expanded bare Ti_3C_2 MXenes, the theoretical capacity of storage was increased to 413 mAh g^{-1}. OH terminations are not very good for MXenes when it comes to metal ion adsorption [35]. OH terminations are not ideal for MXenes in terms of metal ion adsorption [35]. An exothermic and thermodynamically viable method of extracting hydrogen from OH was discovered by adding Li or Pb. The technique was reproduced with O-terminated MXene nanostructure by the same team. All of the other adatoms, with the exception of Al, have negative adsorption energy, suggesting that the adsorption layer has a good probability of developing. Finally, when an atom can transfer additional electrons, just a small number of them are required to complete the oxygen bands. As a result, the repulsion of future adatom addition is increased. The layer produced by other adatoms (i.e., monovalent or divalent) will be more stable than the layer produced by an adsorbed aluminum atom [84].

9.4.4 Electrolytes' Role in the Storage Technology

Electrolytes are classified into several types: ionic, aqueous, organic, and solid electrolytes. All of them have a variety of implications on supercapacitor quality. Conducting salts such as $TEABF_4$ are dissolved in organic solvents such as acetonitrile to form organic electrolytes. Organic electrolytes, as opposed to aqueous electrolytes, can expand the supercapacitor's voltage window [85, 86]. Their practical usefulness has been limited because of their extreme volatility, flammability, cost, and toxicity. Poor conductivity and low specific capacitance are two further problems that put them at a disadvantage. Even though the fact that organic solvents may hold more salt than other solvents, acetonitriles are environmentally hazardous. Propylene carbonate, on the other hand, is ecologically beneficial. Most significantly, it offers a large voltage range over a wide range of operating temperatures. To create an optimized specific capacity, the electrode material's pore size ought to be equivalent to or bigger than the ions' sizes in the electrolyte, regardless of the electrolyte utilized [87]. Otherwise, the ions will be unable to access the reaction sites. The electrolyte solution concentration must be increased to achieve the capacitor's maximum potential voltage output. Because even a small amount of water, such as 3–5 ppm, can significantly affect the voltage. An alkaline, acidic, or neutral aqueous electrolyte can be used. When compared to organic electrolytes, they have a larger ionic concentration and, as a result, a higher conductivity. However, their potential window is relatively small, ranging about 1.2 V [87]. The significance of functional groupings has previously been highlighted. Therefore, combining the proper functional group on a carbon surface with the suitable electrolyte can only result in the best supercapacitor effectiveness [18, 87, 88]. The bulk process influences a low scan rate for acidic or alkaline electrolytes, whereas the surface processes control a high scan rate [89]. In an acidic electrolyte, $Ti_3C_2T_x$'s pseudocapacitive nature plays a critical function in water molecules. H_2O molecules can activate Ti redox activity, which compensates for the charge by diffusing protons. The ideal number of OH functional groups allows for this water to intercalate between layers, as explained in the section on the impact of functional groups. The incorporation of this to MXenes such as $V_2C_2T_x$, $Nb_2C_3T_x$, and others might considerably enhance energy storage [90].

It is used as a new method to evaluate the capacitance of intercalated $Ti_3C_2T_x$ with dipicolinic acid in a balanced electrolyte with a water base [91]. Because of the intercalation of H_3O^+ ion from dipicolinic acid hydrolysis with layers of MXene and the resulting Faraday capacitance, they outperformed acidic electrolytes. This greater capacitance may be due to the adsorption of opposite charges of electrolytes during the intercalation processes [91]. Even though the fact that there are limited experimental experiments, nitrides MXenes are quite more advantageous than carbides. In $MgSO_4$ aqueous electrolyte, the areal capacitance of Ti_2NT_x is 1350 mF/cm^2. They also validate the theoretical prediction of $Mg^{2+} > Al^{3+} > H^+ > Li^+ > Na^+ > K^+$ for capacitance superiority related to cation [92].

9.5 Supercapacitor Systems of MXene and Hybrid

9.5.1 MXene in Their Original State

In 1 M H_2SO_4, the supercapacitors mechanism in $Ti_3C_2T_x$ is pseudocapacitive. This indicates that alterations in the titanium oxidation state and electrode potential have a linear relationship [15]. They showed that by adopting correct electrode design methodologies, MXene capacitance could be improved to its theoretical value. The horizontal alignment of MXene layers is generated after vacuum filtering. Although this can produce a high capacitance, it can also hinder ion mobility and charge transfer. To make open-structure electrodes, they adapted PMMA microspheres as a template. For a pseudocapacitive substance, these macroporous electrodes have the highest rate handling capabilities. A gravimetric capacitance of 310 F g^{-1} is recorded at 10 mV s^{-1} scan rate [15]. This research focused on improving ion accessibility through electrode design. They used hydrogels in addition to macroporous structures. They believe that by using unique electrode architecture, they will be able to improve the rate performance of all types of existing MXenes. In this study, they were able to demonstrate it with micro-range thick electrodes. When sulfuric acid is employed as the electrolyte, the voltage window of $Mo_{1.33}C$ is limited. In acid electrolytes, a maximum of 1.3 V was achievable even for an asymmetric system. In an acid electrolyte, a maximum voltage of 1.3 V was possible even for an asymmetric device. However, in a LiCl electrolyte, $Mo_{1.33}C$ provides a voltage window in the range of −1.2 to + 0.3 with a volume capacity of 815 F/cm^3 [93].

9.5.2 MXene Heterostructures

The perfect mix of high electrical conductivity and pseudo capacitance is found in extensible and fiber-like rGo/MXene conductors containing the intercalated nanostructure (see Figure 9.1) [94].

This finding creates a way for further research into the synergistic interactions of MXene with graphene. By just using 5% wt% MXene, a bulk capacitance of roughly 345.2 F/cm^3 and a gravimetric rate 195 F/g at 0.1 A/g was produced (see Figure 9.2). This is greater than previously reported graphene or CNT-based electrodes. It has a capacitance retention of 124% after 7500 cycles at 0.5 A/g. Work represents the significance of the amounts

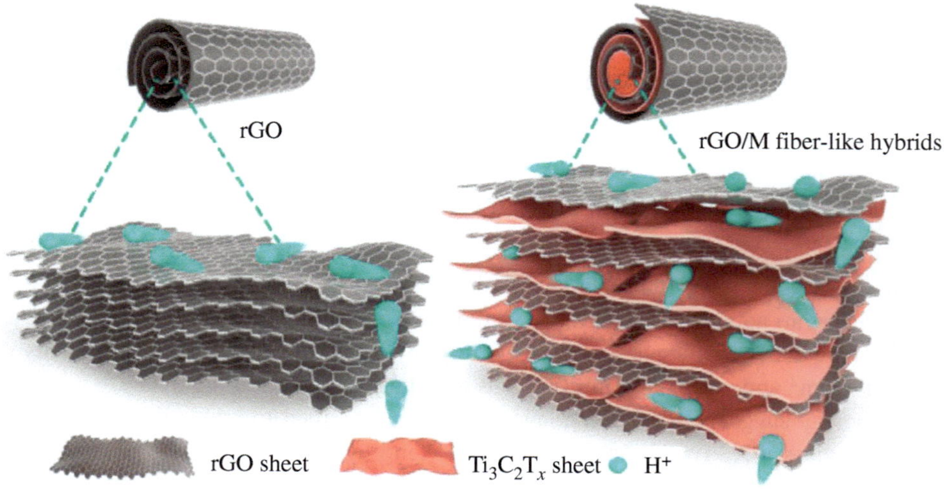

Figure 9.1 A Scheme of H^+ circulation in rGo/Ti_3C_2 flexible fiber hybrid. Source: Reproduced from Wang et al. [94] Copyright (2021), with permission from Elsevier.

of MXene utilized to fabricate electrodes. Clustering of MXene sheets happens when the weight percent of MXene increases, which has a detrimental influence on the pseudocapacitor's performance. To generate a good connection and the resulting synergistic interaction between layers, a proper weight ratio of reduced graphene oxide and MXene is required. Adriana et al. studied heterostructures of reduced graphene oxides and MXenes [95]. They discovered a voltage window extending up to 0.7 V along with increased capacitance. The amine attached on reduced graphene oxide is responsible to shift from negative to positive order in the surface charge to generate a stable binding of reduced graphene oxides and MXene. As a result, the heterostructure of rGO and Ti_3C_2 should be correctly formed. At the rGO:Ti_3C_2 weight ratio of 1:99, the electrodes get a capacitance of 254 F/g at 2 mV/s.

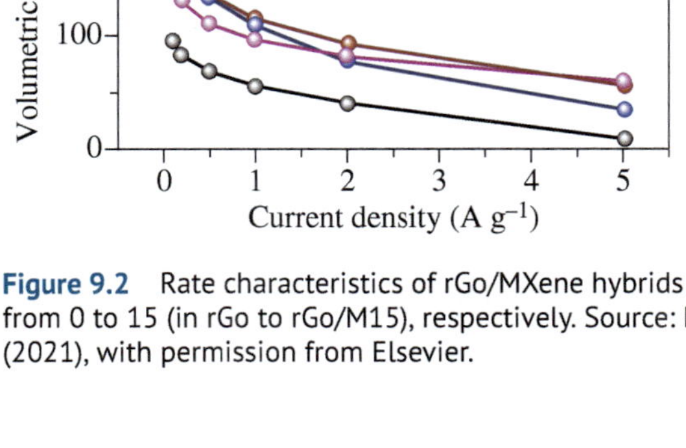

Figure 9.2 Rate characteristics of rGo/MXene hybrids as the weight percent of MXenes increases from 0 to 15 (in rGo to rGo/M15), respectively. Source: Reproduced from Wang et al. [94] Copyright (2021), with permission from Elsevier.

The voltage window can be expanded by using appropriate cathode materials that function at greater potentials than the reference electrode [96]. Expandable electrodes can be made by combining rGO with an optimal value of MXene composites that aren't titanium based.

9.5.3 Hybrids of Transition Metal Oxides in MXene

Flexibility is an advantage of an all-solid-state energy storage system. Aside from that, because transition metal oxide is layered between the two, it adds pseudocapacitance by providing an additional near-surface redox site. It has capacitance of 46.4 mF/cm^2 in Li_2SO_4 aqueous electrolyte at a current density of 0.5 mA/cm^2. With a power density of 0.176 mW/cm^2 and strong cyclability, it has an energy density of roughly 0.970 Wh/cm^2. This topology suggests that MXene layers might be modified to create thin-film energy storage devices [97]. Transition metal oxide MXene hybrids have high gravimetric and volumetric capacities [98], but when it comes to an industrial-scale need, many of those mentioned fall short. A determinant metal oxide loading within MXene can increase the pseudocapacitance to some scope, as discussed previously. To fulfill industrial requirements, the same approach may be used for different transition metal oxide MXene anode architectures [99].

9.5.4 Hierarchical Anode Structure

The multilayered structure made up of Ti_3C_2 and polyaniline nanotubes is a new arrangement that can improve the diffusion of electrolyte ions [100]. At 0.1 $A\,g^{-1}$, it has a capacitance of 596.6 $F\,g^{-1}$. It was able to preserve 94.7% of its original capacitance after 5000 cycles. Ti_3C_2 layers can operate as a substrate framework for the polyaniline hollow tubular channels. This distinct shape results in high-speed electrical transport tube lines, which improve electrochemical efficiency. The active regions on the PANI outer surface that interact with electrolytes facilitate ion access to electrodes. Furthermore, a symmetric supercapacitor built of the same material has a high-power density (1610.8 W/kg at 13.2 $Wh\,kg^{-1}$) and an energy density (25.6 $Wh\,kg^{-1}$ at 153.2 $Wh\,kg^{-1}$). After 4000 cycles, it has a great cycling durability of 81.1%. The benefit of the PANI hollow structure with Ti_3C_2 in increasing electrochemical efficiency has already been explored [100]. Their hierarchical arrangement creates an interdependence between PANI and MXene, which prevents the Ti_3C_2 sheets from being stacked again.

The hydrothermal technique by adding nitrogen and NH^{4+} to the electrolyte in nitrogen-doped intercalation improved the interaction between ions and MXene active sites. TiO_2/TiN assists in the prevention of MXene layers self-assembling [101]. At 1 $A\,g^{-1}$, the gravimetric capacitance was 361 $F\,g^{-1}$. The enhancement in electrochemical capacitance was attributed to physical confinement into the electrode structure. *In situ* polymerization of polyaniline and MXene increased the charge storage via pseudocapacitive means [16]. They succeeded in improving the electrochemical properties of thin and thick hybrid materials. In addition to having a large volumetric capacitance, the electrode's voltage window has to be expanded.

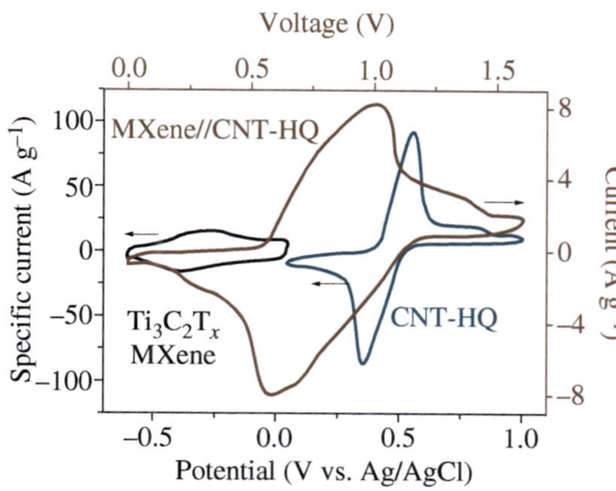

Figure 9.3 In hydrogen-ion rocking chair hybrid supercapacitors, the voltage window has been widened. Source: Reproduced with permission from Hu et al. [102] Copyright (2021) American Chemical Society.

9.5.5 Appropriate Positive Electrode Design

In the selection of MXene materials, the high energy density and a range of voltage windows may be required in addition to a high volumetric capacitance. Asymmetric supercapacitors can be a good option for positive electrodes. In this related, hydroquinone and carbon nanotubes combination were designed supercapacitors based on hydrogen-ion rocking chairs with working voltage of 1.6 V, in which the transfer of a proton between the negative and positive electrodes gives a 62 Wh kg^{-1} energy density (see Figure 9.3) [102].

9.5.6 Microsupercapacitors

Advances in wearable electronics that are wireless and portable have necessitated the use of a high energy density and a steady power source [103, 104]. Although micro-batteries can be an alternative solution, the use of micro-capacitors can be a better solution, especially in cases where a fast power supply is required and the battery cannot be replaced. However, the instability of $Ti_3C_2T_x$ in aqueous conditions limits the storage time of micro-supercapacitors. The charge was stored for 80 days in an open-air environment at ambient temperature using sodium ascorbate caped $Ti_3C_2T_x$ dispersion.

9.6 Prospects

The structure of MXenes and the several varieties of MXenes are investigated. Many of them have so far been studied only in theory. Studies and experiments have shown that titanium-based MXene is suitable for energy storage systems. In addition, the electrode architecture can be used to improve the use of supercapacitors. Also, the reliability of the MXene hybrid anode design needs further investigation. Pseudocapacitance advantages should also be generalized to MXene composites without titanium. By building a symmetrical supercapacitor, its potential can be used to the fullest. Whereas, in an asymmetric supercapacitor, the voltage window expansion may be accomplished by applying the optimal cathode selection. Capacitive capacitance can be altered by synthesis processes, functional groups, formed defects, and impurities used as bonding in hybrids. The use of carbon nanotubes and carbon compounds can improve flexibility and electrochemical characteristics. The importance of electrolytes choosing in the energy system of supercapacitors cannot be overstated. Supercapacitors of the solid type with gel electrolytes have more flexibility and storage capacity. In addition, future wearable energy storage applications will also call for the most modern type of micro-supercapacitors.

9.7 Conclusion

Supercapacitors based on 2D inorganic compounds have been widely used in energy storage systems, which has led to extensive research on the design and construction of supercapacitors by increasing the voltage window and improving cycle abilities. In them, single, double, and triple layers of transition metals are arranged inside each other in a 2D MXenes carbide or nitride arrangement. MXenes have strong metallic conductivity as well as hydrophilic surfaces. The research of supercapacitors on the electrochemical characteristics of MXenes is influenced by the manufacturing method and the way of arrangement, which uses composites for electrodes and the precise selection of electrolytes to increase the capacity.

Most MXenes are synthesized from the aqueous phase and their surface is a combination of F^-, O^-, and OH^- ends. They are often made by etching to remove the A layer from the substrate and have multiple layers of carbides or ternary nitrides. In addition, they have electrical, optical, and magnetic properties, so surface end groups and surface defects, and synthesis techniques all affect their electric properties. In addition, MXenes indicate remarkable absorption properties that make them suitable candidates for photothermal treatment in biological environments.

MXenes usually store energy through absorption–desorption, and the shape of cyclic voltammograms can be used to detect the mechanism of charge storage in different materials, and researchers are trying to achieve energy density and stability.

The morphology of MXenes has a significant effect on the electrochemical properties. The reactivity of the functional groups and the instability of MXenes are important for their potential application in electrochemistry. In addition, the type of electrolyte and electrode design affects the quality of supercapacitors and energy storage technology, and in heterogeneous structures, the voltage window can be expanded by using suitable cathode materials that have a higher potential compared to the reference electrode.

References

1 Hoque, M., Hannan, M., Mohamed, A., and Ayob, A. (2017). Battery charge equalization controller in electric vehicle applications: a review. *Renew. Sustain. Energy Rev.* 75: 1363–1385.

2 Mekhilef, S., Saidur, R., and Kamalisarvestani, M. (2012). Effect of dust, humidity and air velocity on efficiency of photovoltaic cells. *Renew. Sustain. Energy Rev.* 16 (5): 2920–2925.

3 Huang, S., Zhu, X., Sarkar, S., and Zhao, Y. (2019). Challenges and opportunities for supercapacitors. *APL Mater.* 7 (10): 100901.

4 Lv, X., Wei, W., Huang, B., and Dai, Y. (2019). Achieving high energy density for lithium-ion battery anodes by Si/C nanostructure design. *J. Mater. Chem. A* 7 (5): 2165–2171.

5 Zheng, S., Xue, H., and Pang, H. (2018). Supercapacitors based on metal coordination materials. *Coord. Chem. Rev.* 373: 2–21.

6 Huang, J.Y., Ding, F., Yakobson, B.I. et al. (2009). In situ observation of graphene sublimation and multi-layer edge reconstructions. *Proc. Natl. Acad. Sci. U. S. A.* 106 (25): 10103–10108.

7 (a) Liu, C. and Singh, O. (1984). *Thin Solid Films* 113: 165–172. Google Scholar There is no corresponding record for this reference. (b) Joensen, P., Frindt, R.F., and Morrison, S.R. (1986). *Mater. Res. Bull.* 21: 457–461.

8 Balendhran, S., Walia, S., Nili, H. et al. (2013). Two-dimensional molybdenum trioxide and dichalcogenides. *Adv. Funct. Mater.* 23 (32): 3952–3970.

9 Coleman, J.N., Lotya, M., O'Neill, A. et al. (2011). Two-dimensional nanosheets produced by liquid exfoliation of layered materials. *Science* 331 (6017): 568–571.

10 Shuck, C., Deysher, G., Frey, N. et al. (ed.) (2020). Synthesis of Mo 4 VAlC 4 MAX Phase and Two-Dimensional Mo 4 VC 4 Mxene with Five Atomic Layers of Transition Metals. In: *2020 Virtual AIChE Annual Meeting*. AIChE.

11 Barsoum, M. (2000). A new class of solids: thermodynamically stable nanolaminates. *Prog. Solid State Chem.* 28: 201.

12 Anasori, B., Lukatskaya, M., and Yu, G. (2017). 2D metal carbides and nitrides (MXenes) for energy storage. *Nat. Rev. Mater.* 2 (2): 16098.

13 Alhabeb, M., Maleski, K., Mathis, T.S. et al. (2018). Selective etching of silicon from Ti_3SiC_2 (MAX) to obtain 2D titanium carbide (MXene). *Angew. Chem.* 130 (19): 5542–5546.

14 Lukatskaya, M.R., Mashtalir, O., Ren, C.E. et al. (2013). Cation intercalation and high volumetric capacitance of two-dimensional titanium carbide. *Forensic Sci.* 341 (6153): 1502–1505.

15 Lukatskaya, M.R., Kota, S., Lin, Z. et al. (2017). Ultra-high-rate pseudocapacitive energy storage in two-dimensional transition metal carbides. *Nat. Energy* 2 (8): 1–6.

16 VahidMohammadi, A., Moncada, J., Chen, H. et al. (2018). Thick and freestanding MXene/PANI pseudocapacitive electrodes with ultrahigh specific capacitance. *J. Mater. Chem. A* 6 (44): 22123–22133.

17 Kayali, E., VahidMohammadi, A., Orangi, J., and Beidaghi, M. (2018). Controlling the dimensions of 2D MXenes for ultrahigh-rate pseudocapacitive energy storage. *ACS Appl. Mater. Interfaces* 10 (31): 25949–25954.

18 Agnese, Y.D., Lukatskaya, M., Cook, K. et al. (2014). *Electrochem. Commun.* 48: 118.

19 Zhao, M.Q., Ren, C.E., Ling, Z. et al. (2015). Flexible MXene/carbon nanotube composite paper with high volumetric capacitance. *Adv. Mater.* 27 (2): 339–345.

20 Come, J., Naguib, M., Rozier, P. et al. (2012). A non-aqueous asymmetric cell with a Ti2C-based two-dimensional negative electrode. *J. Electrochem. Soc.* 159 (8): A1368.

21 Naguib, M., Come, J., Dyatkin, B. et al. (2012). MXene: a promising transition metal carbide anode for lithium-ion batteries. *Electrochem. Commun.* 16 (1): 61–64.

22 Ng, V.M.H., Huang, H., Zhou, K. et al. (2017). Recent progress in layered transition metal carbides and/or nitrides (MXenes) and their composites: synthesis and applications. *J. Mater. Chem. A* 5 (7): 3039–3068.

23 Mashtalir, O., Lukatskaya, M.R., Zhao, M.Q. et al. (2015). Amine-assisted delamination of Nb2C MXene for Li-ion energy storage devices. *Adv. Mater.* 27 (23): 3501–3506.

24 Harris, P.A., Taylor, R., Minor, B.L. et al. (2019). The REDCap consortium: Building an international community of software platform partners. *J. Biomed. Inf.* 95: 103208.

25 Lorencova, L., Bertok, T., Dosekova, E. et al. (2017). Electrochemical performance of $Ti_3C_2T_x$ MXene in aqueous media: towards ultrasensitive H_2O_2 sensing. *Electrochim. Acta* 235: 471–479.

26 Wu, Y., Chen, H., Xiao, J. et al. (2015). Adsorptive separation of methanol–acetone on isostructural series of metal–organic frameworks M-BTC (M = Ti, Fe, Cu, Co, Ru, Mo): a computational study of adsorption mechanisms and metal-substitution impacts. *ACS Appl. Mater. Interfaces* 7 (48): 26930–26940.

27 Shan, Q., Mu, X., Alhabeb, M. et al. (2018). Two-dimensional vanadium carbide (V2C) MXene as electrode for supercapacitors with aqueous electrolytes. *Electrochem. Commun.* 96: 103–107.

28 Peng, C., Kuai, Z., Zeng, T. et al. (2019). WO3 Nanorods/MXene composite as high performance electrode for supercapacitors. *J. Alloys Compd.* 810: 151928.

29 Venkateshalu, S. and Grace, A.N. (2020). $Ti_3C_2T_x$ MXene and vanadium nitride/porous carbon as electrodes for asymmetric supercapacitors. *Electrochim. Acta* 341: 136035.

30 Orangi, J., Hamade, F., Davis, V.A., and Beidaghi, M. (2019). 3D printing of additive-free 2D $Ti_3C_2T_x$ (MXene) ink for fabrication of micro-supercapacitors with ultra-high energy densities. *ACS Nano* 14 (1): 640–650.

31 Naguib, M., Mochalin, V.N., Barsoum, M.W., and Gogotsi, Y. (2014). 25th anniversary article: MXenes: a new family of two-dimensional materials. *Adv. Mater.* 26 (7): 992–1005.

32 Gogotsi, Y. and Anasori, B. (2019). *The Rise of MXenes*, 8491–8494. ACS Publications.

33 Khazaei, M., Arai, M., Sasaki, T. et al. (2013). Novel electronic and magnetic properties of two-dimensional transition metal carbides and nitrides. *Adv. Funct. Mater.* 23 (17): 2185–2192.

34 Khazaei, M., Arai, M., Sasaki, T. et al. (2014). Two-dimensional molybdenum carbides: potential thermoelectric materials of the MXene family. *Phys. Chem. Chem. Phys.* 16 (17): 7841–7849.

35 Xie, Y., Naguib, M., Mochalin, V.N. et al. (2014). Role of surface structure on Li-ion energy storage capacity of two-dimensional transition-metal carbides. *J. Am. Chem. Soc.* 136 (17): 6385–6394.

36 Cc, C., Rl, R., and Collin, S. (2019). First evidence of the cysteine and glutathione conjugates of 3-sulfanylpentan-1-ol in hop (*Humulus lupulus* L.). *J. Agric. Food. Chem.* 67 (14): 4002–4010.

37 Ji, X., Xu, K., Chen, C. et al. (2016). Probing the electrochemical capacitance of MXene nanosheets for high-performance pseudocapacitors. *Phys. Chem. Chem. Phys.* 18 (6): 4460–4467.

38 Tan, R., Carrey, J., Desvaux, C. et al. (2009). Magnetoresistance and collective Coulomb blockade in superlattices of ferromagnetic CoFe nanoparticles. *Phys. Rev. B* 79 (17): 174428.

39 Hope, M.A., Forse, A.C., Griffith, K.J. et al. (2016). NMR reveals the surface functionalisation of Ti_3C_2 MXene. *Phys. Chem. Chem. Phys.* 18 (7): 5099–5102.

40 Halim, J., Kota, S., Lukatskaya, M.R. et al. (2016). Synthesis and characterization of 2D molybdenum carbide (MXene). *Adv. Funct. Mater.* 26 (18): 3118–3127.

41 Kumar, M., Tomar, R.S., Lade, H., and Paul, D. (2016). Methylotrophic bacteria in sustainable agriculture. *World J. Microbiol. Biotechnol.* 32 (7): 1–9.

42 Sang, X., Xie, Y., Yilmaz, D.E. et al. (2018). In situ atomistic insight into the growth mechanisms of single layer 2D transition metal carbides. *Nat. Commun.* 9 (1): 1–9.

43 Si, C., Zhou, J., and Sun, Z. (2015). Half-metallic ferromagnetism and surface functionalization-induced metal–insulator transition in graphene-like two-dimensional Cr_2C crystals. *ACS Appl. Mater. Interfaces* 7 (31): 17510–17515.

44 Khazaei, M., Arai, M., Sasaki, T. et al. (2015). OH-terminated two-dimensional transition metal carbides and nitrides as ultralow work function materials. *Phys. Rev. B* 92 (7): 075411.

45 Khazaei, M., Ranjbar, A., Arai, M., and Yunoki, S. (2016). Topological insulators in the ordered double transition metals M 2′ M ″C 2 MXenes (M′ = Mo, W; M″ = Ti, Zr, Hf). *Phys. Rev. B* 94 (12): 125152.

46 Liang, Y., Khazaei, M., Ranjbar, A. et al. (2017). Theoretical prediction of two-dimensional functionalized MXene nitrides as topological insulators. *Phys. Rev. B* 96 (19): 195414.

47 Si, C., Jin, K.-H., Zhou, J. et al. (2016). Large-gap quantum spin Hall state in MXenes: d-band topological order in a triangular lattice. *Nano Lett.* 16 (10): 6584–6591.

48 Halim, J., Persson, I., Moon, E.J. et al. (2019). Electronic and optical characterization of 2D Ti_2C and Nb2C (MXene) thin films. *J. Phys.: Condens. Matter* 31 (16): 165301.

49 Halim, J., Moon, E.J., Eklund, P. et al. (2018). Variable range hopping and thermally activated transport in molybdenum-based MXenes. *Phys. Rev. B* 98 (10): 104202.

50 Halim, J., Palisaitis, J., Lu, J. et al. (2018). Synthesis of two-dimensional Nb1. 33C (MXene) with randomly distributed vacancies by etching of the quaternary solid solution (Nb2/3Sc1/3) 2AlC MAX phase. *ACS Appl. Nano Mater.* 1 (6): 2455–2460.

51 Hart, J.L., Hantanasirisakul, K., Lang, A.C. et al. (2019). Control of MXenes' electronic properties through termination and intercalation. *Nat. Commun.* 10 (1): 1–10.

52 Hu, T., Zhang, H., Wang, J. et al. (2015). Anisotropic electronic conduction in stacked two-dimensional titanium carbide. *Sci. Rep.* 5 (1): 1–8.

53 Zha, X.-H., Zhou, J., Luo, K. et al. (2017). Controllable magnitude and anisotropy of the electrical conductivity of $Hf_3C_2O_2$ MXene. *J. Phys.: Condens. Matter* 29 (16): 165701.

54 Liu, J.-H., Kan, X., Amin, B. et al. (2017). Theoretical exploration of the potential applications of Sc-based MXenes. *Phys. Chem. Chem. Phys.* 19 (48): 32253–32261.

55 Weng, H., Ranjbar, A., Liang, Y. et al. (2015). Large-gap two-dimensional topological insulator in oxygen functionalized MXene. *Phys. Rev. B* 92 (7): 075436.

56 Cartwright, J.H. (2000). Emergent global oscillations in heterogeneous excitable media: the example of pancreatic β cells. *Phys. Rev. E* 62 (1): 1149.

57 Dewprashad, B. and Eisenbraun, E. (1994). Fundamentals of epoxy formulation. *J. Chem. Educ.* 71 (4): 290.

58 Dall'Agnese, C., Dall'Agnese, Y., Anasori, B. et al. (2018). Oxidized Ti_3C_2 MXene nanosheets for dye-sensitized solar cells. *New J. Chem.* 42 (20): 16446–16450.

59 Yu, Z., Feng, W., Lu, W. et al. (2019). MXenes with tunable work functions and their application as electron-and hole-transport materials in non-fullerene organic solar cells. *J. Mater. Chem. A* 7 (18): 11160–11169.

60 Gao, G., O'Mullane, A.P., and Du, A. (2017). 2D MXenes: a new family of promising catalysts for the hydrogen evolution reaction. *ACS Catal.* 7 (1): 494–500.

61 Zha, X.-H., Luo, K., Li, Q. et al. (2015). Role of the surface effect on the structural, electronic and mechanical properties of the carbide MXenes. *Europhys. Lett.* 111 (2): 26007.

62 Hong, L., Klie, R.F., and Öğüt, S. (2016). First-principles study of size-and edge-dependent properties of MXene nanoribbons. *Phys. Rev. B* 93 (11): 115412.

63 Kim, H. and Alshareef, H.N. (2019). MXetronics: MXene-enabled electronic and photonic devices. *ACS Mater. Lett.* 2 (1): 55–70.

64 Zhang, C., Anasori, B., Seral-Ascaso, A. et al. (2017). Transparent, flexible, and conductive 2D titanium carbide (MXene) films with high volumetric capacitance. *Adv. Mater.* 29 (36): 1702678.

65 Gao, G., Ding, G., Li, J. et al. (2016). Monolayer MXenes: promising half-metals and spin gapless semiconductors. *Nanoscale.* 8 (16): 8986–8994.

66 Dillon, A.D., Ghidiu, M.J., Krick, A.L. et al. (2016). Highly conductive optical quality solution-processed films of 2D titanium carbide. *Adv. Funct. Mater.* 26 (23): 4162–4168.

67 Zhang, Y. and Li, F. (2017). Robust half-metallic ferromagnetism in Cr_3C_2 MXene. *J. Magn. Magn. Mater.* 433: 222–226.

68 He, J., Lyu, P., and Nachtigall, P. (2016). New two-dimensional Mn-based MXenes with room-temperature ferromagnetism and half-metallicity. *J. Mater. Chem. C* 4 (47): 11143–11149.

69 Ingason, A.S., Dahlqvist, M., and Rosén, J. (2016). Magnetic MAX phases from theory and experiments; a review. *J. Phys.: Condens. Matter* 28 (43): 433003.

70 Conway, B. and Pell, W. (2003). Double-layer and pseudocapacitance types of electrochemical capacitors and their applications to the development of hybrid devices. *J. Solid State Electrochem.* 7 (9): 637–644.

71 Biesheuvel, P., Porada, S., and Dykstra, J. (2018). The difference between Faradaic and non-Faradaic electrode processes. *arXiv* 180902930.

72 Gogotsi, Y. and Penner, R.M. (2018). *Energy Storage in Nanomaterials–Capacitive, Pseudocapacitive, or Battery-Like?* 2081–2083. ACS Publications.

73 Naguib, M., Mashtalir, O., Carle, J. et al. (2012). *ACS Nano* 6: 1322. Crossref, ISI.

74 Sugahara, A., Ando, Y., Kajiyama, S. et al. (2019). Negative dielectric constant of water confined in nanosheets. *Nat. Commun.* 10 (1): 1–7.

75 Lipatov, A., Alhabeb, M., Lukatskaya, M.R. et al. (2016). Effect of synthesis on quality, electronic properties and environmental stability of individual monolayer Ti_3C_2 MXene flakes. *Adv. Electron. Mater.* 2 (12): 1600255.

76 Huang, S. and Mochalin, V.N. (2019). Hydrolysis of 2D transition-metal carbides (MXenes) in colloidal solutions. *Inorg. Chem.* 58 (3): 1958–1966.

77 Maleski, K., Ren, C.E., Zhao, M.-Q. et al. (2018). Size-dependent physical and electrochemical properties of two-dimensional MXene flakes. *ACS Appl. Mater. Interfaces* 10 (29): 24491–24498.

78 Naguib, M., Mashtalir, O., Lukatskaya, M.R. et al. (2014). One-step synthesis of nanocrystalline transition metal oxides on thin sheets of disordered graphitic carbon by oxidation of MXenes. *Chem. Commun.* 50 (56): 7420–7423.

79 Mashtalir, O., Cook, K.M., Mochalin, V.N. et al. (2014). Dye adsorption and decomposition on two-dimensional titanium carbide in aqueous media. *J. Mater. Chem. A* 2 (35): 14334–14338.

80 Lee, Y., Kim, S.J., Kim, Y.-J. et al. (2020). Oxidation-resistant titanium carbide MXene films. *J. Mater. Chem. A* 8 (2): 573–581.

81 Fredrickson, K.D., Anasori, B., Seh, Z.W. et al. (2016). Effects of applied potential and water intercalation on the surface chemistry of Ti_2C and Mo_2C MXenes. *J. Phys. Chem. C* 120 (50): 28432–28440.

82 Tang, Q., Zhou, Z., and Shen, P. (2012). Are MXenes promising anode materials for Li ion batteries? Computational studies on electronic properties and Li storage capability of Ti_3C_2 and $Ti_3C_2X_2$ (X=F, OH) monolayer. *J. Am. Chem. Soc.* 134 (40): 16909–16916.

83 Karlsson, L.H., Birch, J., Halim, J. et al. (2015). Atomically resolved structural and chemical investigation of single MXene sheets. *Nano Lett.* 15 (8): 4955–4960.

84 Yu, Y.-X. (2016). Prediction of mobility, enhanced storage capacity, and volume change during sodiation on interlayer-expanded functionalized Ti_3C_2 MXene anode materials for sodium-ion batteries. *J. Phys. Chem. C* 120 (10): 5288–5296.

85 Hu, T., Li, Z., Hu, M. et al. (2017). Chemical origin of termination-functionalized MXenes: Ti_3C_2 T_2 as a case study. *J. Phys. Chem. C* 121 (35): 19254–19261.

86 Wang, Z., Tammela, P., Strømme, M., and Nyholm, L. (2015). Nanocellulose coupled flexible polypyrrole@ graphene oxide composite paper electrodes with high volumetric capacitance. *Nanoscale* 7 (8): 3418–3423.

87 Jung, N., Kwon, S., Lee, D. et al. (2013). Synthesis of chemically bonded graphene/carbon nanotube composites and their application in large volumetric capacitance supercapacitors. *Adv. Mater.* 25 (47): 6854–6858.

88 Wang, Y., Song, Y., and Xia, Y. (2016). Electrochemical capacitors: mechanism, materials, systems, characterization and applications. *Chem. Soc. Rev.* 45 (21): 5925–5950.

89 Fu, J., Li, L., Lee, D. et al. (2020). Enhanced electrochemical performance of $Ti_3C_2T_x$ MXene film based supercapacitors in H_2SO_4/KI redox additive electrolyte. *Appl. Surf. Sci.* 504: 144250.

90 Shao, H., Lin, Z., Xu, K. et al. (2019). Electrochemical study of pseudocapacitive behavior of $Ti_3C_2T_x$ MXene material in aqueous electrolytes. *Energy Storage Mater.* 18: 456–461.

91 Wang, H. and Wu, X. (2020). High capacitance of dipicolinic acid-intercalated MXene in neutral water-based electrolyte. *Chem. Eng. J.* 399: 125850.

92 Djire, A., Bos, A., Liu, J. et al. (2019). Pseudocapacitive storage in nanolayered Ti_2NT_x MXene using Mg-Ion electrolyte. *ACS Appl. Nano Mater.* 2 (5): 2785–2795.

93 El Ghazaly, A., Zheng, W., Halim, J. et al. (2021). Enhanced supercapacitive performance of Mo1. 33C MXene based asymmetric supercapacitors in lithium chloride electrolyte. *Energy Storage Mater.*

94 Wang, Z., Chen, Y., Yao, M. et al. (2020). Facile fabrication of flexible rGO/MXene hybrid fiber-like electrode with high volumetric capacitance. *J. Power Sources* 448: 227398.

95 Navarro-Suárez, A.M., Maleski, K., Makaryan, T. et al. (2018). 2D titanium carbide/reduced graphene oxide heterostructures for supercapacitor applications. *Batteries Supercaps* 1 (1): 33–38.

96 Byeon, A., Glushenkov, A.M., Anasori, B. et al. (2016). Lithium-ion capacitors with 2D Nb_2CT_x (MXene)–carbon nanotube electrodes. *J. Power Sources* 326: 686–694.

97 Li, H., Liu, Y., Lin, S. et al. (2021). Laser crystallized sandwich-like MXene/Fe_3O_4/MXene thin film electrodes for flexible supercapacitors. *J. Power Sources* 497: 229882.

98 Zhang, Y., Cao, J., Yuan, Z. et al. (2021). Assembling Co3O4 Nanoparticles into MXene with Enhanced electrochemical performance for advanced asymmetric supercapacitors. *J. Colloid Interface Sci.* 599: 109–118.

99 Bo, Z., Yi, K., Yang, H. et al. (2021). More from less but precise: industry-relevant pseudocapacitance by atomically-precise mass-loading MnO2 within multifunctional MXene aerogel. *J. Power Sources* 492: 229639.

100 Wu, W., Wang, C., Zhao, C. et al. (2020). Facile strategy of hollow polyaniline nanotubes supported on Ti_3C_2-MXene nanosheets for High-performance symmetric supercapacitors. *J. Colloid Interface Sci.* 580: 601–613.

101 Yang, B., She, Y., Zhang, C. et al. (2020). Nitrogen doped intercalation TiO_2/TiN/$Ti_3C_2T_x$ nanocomposite electrodes with enhanced pseudocapacitance. *Nanomaterials* 10 (2): 345.

102 Hu, M., Cui, C., Shi, C. et al. (2019). High-energy-density hydrogen-ion-rocking-chair hybrid supercapacitors based on $Ti_3C_2T_x$ MXene and carbon nanotubes mediated by redox active molecule. *ACS Nano* 13 (6): 6899–6905.

103 Kyeremateng, N.A., Brousse, T., and Pech, D. (2017). Microsupercapacitors as miniaturized energy-storage components for on-chip electronics. *Nat. Nanotechnol.* 12 (1): 7–15.

104 Shen, C., Xu, S., Xie, Y. et al. (2017). A review of on-chip micro supercapacitors for integrated self-powering systems. *J. Microelectromech. Syst.* 26 (5): 949–965.

10

MXenes-based Biosensors

Sushma Thapa[1,2], Kshitij RB Singh[1,2], Arunadevi Natarajan[3], Rout George Kerry[4], Jay Singh[2], Shyam S. Pandey[1], and Ravindra Pratap Singh[5]

[1]*Graduate School of Life Science and Systems Engineering, Kyushu Institute of Technology, Wakamatsu, Kitakyushu, Japan*
[2]*Department of Chemistry, Institute of Science, Banaras Hindu University, Varanasi, Uttar Pradesh, India*
[3]*Department of Chemistry, PSGR Krishnammal College for Women, Coimbatore, Tamil Nadu, India*
[4]*PG Department of Biotechnology, Utkal University, Bhubaneswar, Odisha, India*
[5]*Department of Biotechnology, Faculty of Science, Indira Gandhi National Tribal University, Amarkantak, Madhya Pradesh, India*

10.1 Introduction

In recent years, two-dimensional (2D) nanomaterials have caught considerable interest owing to their exceptional physiochemical and structural properties, high surface area, and functionalized surfaces, which have made them beneficial in a broad array of applications since the invention of graphene in 2004 [1, 2]. Currently, MXenes have gained significant research interest among the other 2D nanomaterials and have become one of the most prominent families of 2D materials due to their outstanding mechanical, electronic, and physicochemical properties, large surface area, and rich surface functional groups [3–5]. An MXene is basically a 2D inorganic compound comprising of transition metal carbide, nitride, or carbonitride initially fabricated by Gogotsi and coworkers in 2011 at Drexel University. MXenes could be usually characterized by a fundamental formula $M_{n+1}X_nT_x$ where M is an early transition metal such as titanium, vanadium, chromium, molybdenum, tantalum, or vanadium, whereas X represents nitrogen or carbon, and T_x denotes the surface, functional groups, for example, –O, –OH, and –F [6–8]. 2D MXene is developed from their corresponding parent MAX phase by selective etching of A elements and further delamination. $M_{n+1}AX_n$ is the general formula of the MAX phase where n varies from 1 to 3, and A represents the 13 and 14 elements of the periodic table. Up to now, over 20 different compositions of MXenes family have been developed while 70 diverse MAX phases have been reported, making it one of the rapidly growing 2D material families, and the product yield, particle size, surface properties, defect rate, and conductivity rely on the optimization of the reaction condition.

Moreover, the 2D MXene exhibits three atomic structures, which are M_2X, M_3X_2, and M_4X_3 [9–11]. Wet chemical etching in hydrofluoric (HF)-forming or HF acid is the most promising synthesis path of MXene, which is based on three simple steps etching,

MXenes: Next-Generation 2D Materials: Fundamentals and Applications, First Edition.
Edited by Jay Singh, Kshitij RB Singh, Ravindra Pratap Singh, and Charles Oluwaseun Adetunji.

delamination, and intercalation. Filtration is carried out to isolate the solid particles after the etching step. Subsequently, the as-prepared multilayered MXene could be delaminated into separated sheets using the sonication method [12]. The exfoliation and etching methods lead to the surface functionalization of MXenes with OH, O, and F groups, resulting in their hydrophilic nature. Notwithstanding, HF processes inhibit its use in scaled-up procedures because it employs hazardous chemistry solutions. Also, the acidic solution of HF is harmful to humankind and the environment as well as it decreases the performance of MXenes. Other etchants could be employed to avoid the pernicious impact of HF, such as HCl/LiF since it offers innocuous and much milder etching conditions. Numerous parameters were likely to affect MXenes etching, for example, composition and particle size, intercalation agent, MAX phase quality, temperature, time, and etching solution [3, 13, 14].

As MXenes are produced from the carbides and nitrides of the transition metals, it exhibits excellent volume capacitors and good conductivity contrary to the other 2D materials. MXene has been primarily used in polymer nanocomposite fillers, transparent optical conductive coatings, energy storage, electronic devices, sensors, and water purifications. MXene displayed incredible competitiveness in the biosensing field among all the applications [15, 16]. Other 2D nanomaterials like graphene confine their integration in sensor design at an industrial scale because of several challenges, such as low biocompatibility, difficulty in surface functionalization, low electrical conductivity, and high hydrophobicity. The surface functional group is present in 2D MXenes, which has outstanding potential for designing sensors in terms of electrical conductivity, hydrophilicity, and biocompatibility. Moreover, it has a high intercalation capacity, which is absent in other 2D nanomaterials [17, 18]. Therefore, MXene has been frequently utilized to produce numerous improved biosensors for biomedical, environmental remediation, and agriculture safety applications.

Furthermore, MXenes are the most appropriate alternative for biosensor applications because of their high selectivity and low detection limit (LOD) [19, 20]. The recent COVID-19 pandemic has shown that biosensor has significant application in alleviating the transmission of diseases and facilitating therapeutics. Hence, the biosensor is a chemical sensing analytical device that converts a biological response into an electrical signal by a transducer that an instrument can read and primarily consists of the bioreceptor, transducer, and amplifier. The bioreceptor is essentially the biomolecule that binds and catalyzes a particular reaction for an analyte to generate a measurable signal, and the transducer changes the observed biochemical signal into an electrical signal that can be stored, analyzed, and displayed. The amplifier amplifies the converted signal into an electrical signal [21–23]. This book chapter highlights the advanced biosensor developed based on the exceptional properties of 2D MXene for various applications such as biomedical, agriculture safety, and environmental remediation. Also, this book chapter provides the challenges, limitations, and prospects of MXene.

10.2 Biosensing Application

10.2.1 Biomedical

Due to the substantial growth in the field of biomedicine, numerous researchers have focused on the biomedical application of 2D nanomaterials because of their multifaceted properties, excellent performance, and ultra-thin structure [3, 24, 25]. Currently, over other

2D materials, the family of MXenes has fascinated more research interest in biomedical applications such as therapeutic methods, biosensors, tissue engineering, drug delivery, and bioimaging owing to their multitudinous advantages comprising high specific surface areas, hydrophilicity, light-heat conversion performance, and excellent physiochemical properties [26–28]. MXene-based materials also possess several functional groups on their surface and have remarkable biocompatibility, which could be employed to load various drugs and functional nanoparticles for complex surface alteration before further biomedical applications. In addition, for drug delivery, the presence of hydrophilic functional groups is equally essential.

MXenes consisting of paramagnetic transition and high atomic number metals are preferable for biomedical imaging as they have good X-ray attenuation for computed tomography (CT) scans. Therefore, MXene-based materials possess a high potential for biomedical applications and pave the way for novel solutions for highly sensitive diagnosis [29–31]. For instance, MXenes can be used in photothermal therapy (PTT) for tumor ablation because MXene has transformation capability in near-infrared (NIR) and intense light absorption. 2D MXenes show strong absorption in the NIR region, due to which it provides the possibility for PTT and in vivo photoacoustic imaging (PAI), which is pertinent [4, 32].

Nevertheless, 2D MXenes could also be utilized as a novel biosensing system, besides the multifaceted applications of 2D MXenes for diagnostic imaging and therapeutics [11, 33, 34]. MXenes are well known for their excellent electrical conductivity, water dispersibility, large surface functional groups, and outstanding biocompatibility [35, 36]. The aforementioned properties enable MXenes to be acknowledged as a developed biosensing tool that could identify biomarkers, including big and small biomolecules. In the medical field, quantitatively identifying biomarkers is essential for early diagnosis, effective medication, and risk evaluation. Biomarkers could be detected in body fluids and tissue samples [37, 38]. Moreover, the prolonged durability of the manufactured MXenes biosensors affirms excellent data reproducibility, which could be vastly advantageous when employing such biosensors for diagnostic applications [39, 40]. Accordingly, numerous advanced MXene-based biosensors have been developed for biomedical applications.

Most diseases are related to variations in the concentration in the blood; as a result, glucose is considered the essential human biomarker. In diabetes patients, hyperglycemia is associated with decreased blood sugar levels caused by insulinemia [2]. Rakhi and her coworkers fabricated one of the earliest 2D MXene-based biosensors for the electrochemical sensing of glucose. Glucose oxidase (GO_x) was immobilized with gold nanoparticles (AuNPs), ultrathin MXene, (AuNPs), and Nafion on the glassy carbon electrode (GCE). GO_x breakdown glucose into gluconolactone and hydrogen peroxide (H_2O_2). Nafion was employed to enhance the enzyme adhesion to the GCE and biosensors' selectivity, disperse the MXene, and eradicate the interfering signals. AuNPs enable enzymes to preserve biological activity during adsorption and reduce the insulating effect for direct electron transfer (DET). The linearity and sensitivity of biosensors are augmented by 2D MXenes because it has a large surface area. The fabricated GO_x/Au/MXene/GCE biosensor exhibited exceptional repeatability, reproducibility, and stability, a wide detection range from 0.1 to 18 mM, a LOD of 5.9 μM, and high sensitivity of 4.2 μA mM^{-1} cm^{-2}. The developed enzymatic biosensor is befitting for evaluating the glucose in biological samples to identify Diabetes Mellitus [16]. The development of nonenzymatic biosensor has been rising day after day because enzymatic biosensor has some disadvantages, such as the high cost of analysis, reproducibility of enzyme immobilization, and risk of enzyme inactivation. The rising focus on wearable electronic devices has increased research toward developing noninvasive biomarker monitoring [41]. Currently, wearable devices are being developed that could be mated with individual skin to constantly monitor human activities comprising heart rate, bodily motion, skin temperature, and respiration rate, which would facilitate to determine a human health baseline and medical specialists of abnormalities signifying that further medical help may be necessary [42, 43]. For instance, Li et al. fabricated MXene/Prussian blue (MXene/PB) based modular multifunctional, wearable, and stretchable biosensors for sensitive detection biomarkers such as lactate and glucose in sweat (Figure 10.1). For noninvasive biomarkers monitoring, wearable electrochemical biosensors are considered to be substantial. Compared to carbon nanotubes (CNTs)/PB and graphene/PB composites, MXene/PB shows an improved electrochemical activity of the sensor due to their unique morphology and conductivities [44]. In clinical and medical industries, H_2O_2 is widely employed owing to its great reducing and oxidation properties. H_2O_2 has a significant role in the biological system since it is a byproduct of several oxidases in the body and the primary medium for phagocytic cells. Therefore,

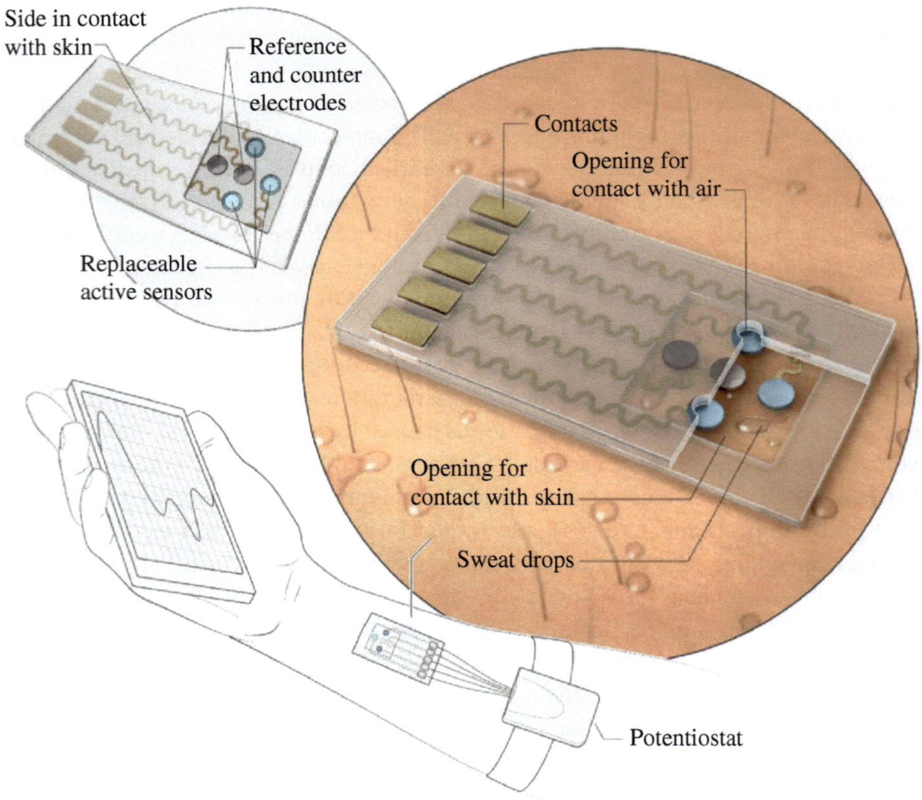

Figure 10.1 Perspiration analysis (Source: Reproduced with permission from Sawangphruk et al. [44]).

electrochemical identification of H_2O_2 at a minimal price is extremely prominent and could be determined either through nonenzymatic or enzymatic ways [45, 46].

DET occurs among the enzyme and the electrode, which is crucial for developing mediator-free biosensors. However, there is a chance of losing the bioactivity when the enzyme gets adsorbed on the plain electrode surface, which is still a challenge. The redox-active center is entrenched deep inside the protein, which makes DET complex [47–49]. 2D MXenes have the potential to furnish a preventive environment for the immobilized enzyme keeping its bioactivity and facilitating DET because of their excellent biocompatibilities, great electronic properties, and high surface area [50]. Thereby, Wang and his coworkers constructed a biosensor for the determination of H_2O_2 by immobilizing hemoglobin (Hb) on TiO_2 improvised Ti_3C_2 and exhibited good stability and biological activity to protein. Nafion improved the biosensor selectivity and enzyme cohesion to the GCE. Moreover, it also removes interfering signals. The DET of Hb was facilitated because of the exceptional characteristics of MXene. The fabricated biosensor Nafion/Hb/TiO_2-Ti_3C_2/GCE showed a LOD of 14 nM, a wide detection range of 0.1–380 μM for H_2O_2, and sensitivity of 447.3 μA mM^{-1} cm^{-2} and also it has vast potential in the biomedical analysis [37]. Field-effect transistors (FET) based biosensors are also acknowledged as a potential candidate for direct, label-free, and fast detection of biological events [51]. For example, Xu et al. fabricated an ultrasensitive MXene-based FET biosensor for monitoring spiking activity in primary hippocampal neurons and determination of dopamine. Dopamine molecules interacted with an electron from the surface MXenes groups, resulting in greater conductance, and were estimated by the FET in real time. The developed biosensor has a relatively wide linear range and a very LOD [52]. Kumar and colleagues reported a selective and sensitive electrochemical biosensor by utilizing biofunctionalized MXene for identifying cancer biomarker carcinoembryonic antigen (CEA). People affected by breast, liver, ovarian, and colorectal cancers CEA biomarker is present. MXene was fabricated by selective etching of A elements with fluoride salts, and the surface of MXene is functionalized with (3-aminopropyl) triethoxysilane (APTES). The developed electrochemical biosensor manifests a sensitivity of 37.9 μA ng^{-1} mL cm^{-2} per decade and a linear detection range of 0.0001–2000 ng mL^{-1} [53]. Likewise, Wu et al. presented a selective and ultrasensitive surface plasmon resonance (SPR) biosensor in order to sense CEA biomarker, which consisted of a MXene-based biosensing platform and multi-walled carbon nanotube (MWCNTs)-polydopamine (PDA)-Ag nanoparticle (AgNPs) signal enhancer with a LOD of 0.07 fM. SPR biosensing method acted as an efficient tool to detect CEA markers as it provides timely detection, ease of miniaturization, fast detection, and high selectivity and sensitivity. Then MWCNTs-PDA-AgNPs nanohybrids combined with polyclonal anti-CEA antibodies (Ab_2) (MWPAg-Ab_2) with a sandwich form to give a range of 2×10^{-16} to 2×10^{-8} M for cancer biomarker detection. The constructed biosensor established high specificity and good reproducibility for CEA in human serum for treatment and monitoring of cancer [54].

10.2.2 Environmental

In the 20th century, various inorganic and organic contaminants comprising phenolic compounds (e.g. bisphenol, catechol, and resorcinol) were categorized as phenomenon products; however, in the 21st century, it resulted in consequential environmental setbacks, for example, degrading and polluting several environmental arenas that involve water, soil, and air due to the augmentation in the industrial growth [55, 56]. Scientists have dedicated considerable effort to determining a feasible resolution to mitigate environmental pollutants. Consequently, new technologies are developing for remediation, like nanomaterial-based techniques, which are outstanding substitutes for traditional approaches as these methods are economical and eco-friendly [57–59].

Hence, for environmental remediation, MXenes-based nanocomposite has been widely employed as they provide exceptional characteristics [60, 61]. Phenolic compounds are a severe threat to the environment and highly poisonous to the ecological and biological environment. The phenolic compound should be monitored cautiously, and the MXene-based platform has significant potential for identifying the phenolic compound [62–64]. For example, Huang et al. developed a promising Ti_3C_2-MWCNTs nanocomposite-based electrochemical sensor for concurrently determining two phenolic compounds, namely, hydroquinone (HQ) and catechol (CT). MWCNTs have been embodied into 2D Ti_3C_2 for better spacing. The reported electrochemical biosensor possesses good repeatability, high stability, and selectivity. The limit of detection for CT is 3.9 nM and for HQ is 6.6 nM, and a wide linear range from 2 to 150 μM for both CT and HQ was achieved (Figure 10.2). Therefore, the developed sensor demonstrates outstanding potential for sensing phenolic compound [65]. Lu et al. fabricated MXene-based tyrosinase (Tyr) biosensors for sensing phenol with the detection of a limit of 12 nmol L^{-1}. 2D MXene offers a favorable microenvironment for enzyme immobilization and retaining their bioactivity. Therefore, the MXene

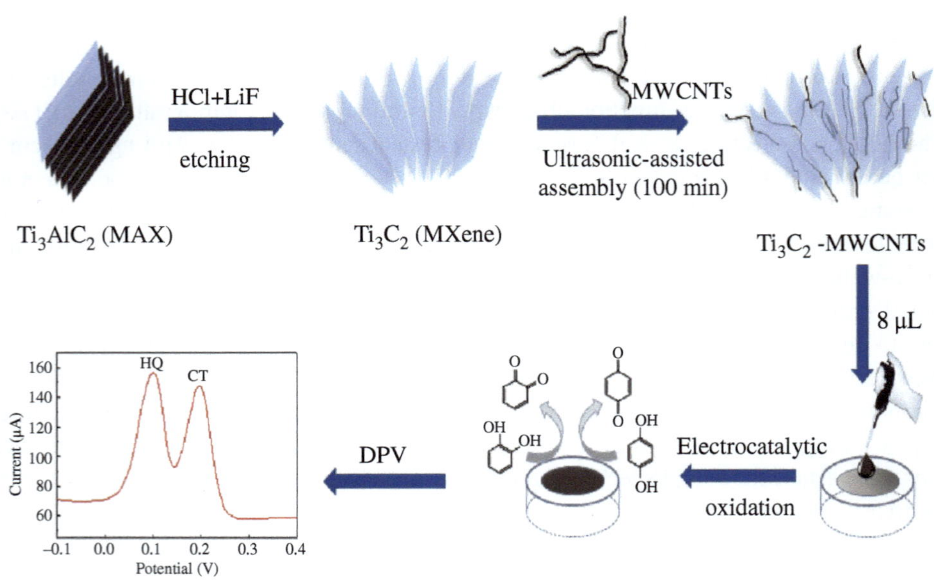

Figure 10.2 Schematic representation of the detection and preparation of Ti_3C_2-MWCNTs nanocomposite (Source: Reproduced with permission from Huang et al. [65]).

was considered a novel matrix to immobilize tyrosinase. The fabricated mediator-free biosensor shows a linear detection range of 0.05–15.5 μmol L^{-1} with a sensitivity of 414.4 mA M^{-1}, and it also has good reproducibility, high recovery, and long-term stability [66]. Likewise, nitroaromatic compounds are also harmful to environmental contamination and extensively utilized in dyes, plasticizers, pesticides, and explosives. 4-Nitrophenol (4-NP) is an extremely detrimental compound amongst other nitroaromatic compounds; moreover, consumption of 4-NP causes terrible harm to the human kidney, central nervous system, blood, and liver. Because of electron-withdrawing solid groups, the 4-NP has poor environmental degradation, and water polluted with 4-NP is arduous to clean [62, 67]. Accordingly, Jiang et al. fabricated an electrochemical sensor based on Nb_2CT_x/Zn-Co-NC composite for the identification 4-NP compound. The composite was fabricated using a simple self-assembling technique and then used as the electrode material to develop the sensor. Zn-Co-NC was decorated evenly on Nb_2CT_x to prevent restacking, and Nb_2CT_x was partly covered with Zn-Co-NC nanocage, due to which numerous open active surfaces were accessible for 4-NP electrocatalysis. The produced electrochemical sensor presented a linear detection range of 1–500 μM, a LOD of 0.070 μM, high sensitivity of 4.65 μA μM^{-1} cm^{-2}, and excellent stability and selectivity [68]. The existence of contamination in water/wastewater streams also causes a worldwide environmental concern because almost all of them are dangerous and toxic for living beings. The prime water pollutant sources are dyes and heavy metals [69, 70].

Heavy metal ions such as Cr(III), Ni(II), Cu(II), As(III), Cd(II), and Cr(VI) are the most threatening environmental pollutants because of their durable toxicities even at a small concentration, carcinogenic, and nondegradable, which make them high-risk contaminate that need prompt consideration [71]. MXene and their derivatives are excellent potential adsorbents for heavy metals as they have strong adsorption toward heavy metal ions, are eco-friendly, and are economically viable [70]. In particular, Zhu et al. reported the very first electrochemical sensor based on MXene for the sensing of heavy metals with a detection limit of 0.0041, 0.130, 0.098, and 0.041 μM for Pb(II), Hg(II), Cd(II), and Cu((II) using square wave stripping voltammetry (SWASV). 2D Alk-Ti_3C_2 was fabricated by acid etching and alkaline intercalation treatment. The Alk-Ti_3C_2 has excellent electrochemical performance and a unique structure compared to the Ti_3C_2 electrodes. Moreover, the sensor is beneficial for identifying four heavy metals with good linearity and high selectivity and sensitivity [72].

Similarly, Lv et al. developed an electrochemical sensor-based H-C_3N_4/$Ti_3C_2T_x$ composite to determine Pb(II) and Cd(II). On the surface of $Ti_3C_2T_x$, a colossal amount of H-C_3N_4 is spread out as a coating material because of the electrostatic interaction. The constructed sensor has a high sensitivity for sensing Pb(II) and Cd(II) with a detection limit of 0.6 nM and 1 nM, respectively [73]. For electrochemical identification of anionic analytes such as bromate (BrO_3), 2D MXenes have also been employed. In particular, Rasheed and coworkers fabricated GCE modified with a Nafion/Ti_3C_2 composite, which is applied as the sensor for sensitive electrochemical detection of BrO_3^-. The produced sensor exhibits unique electrocatalytic properties toward bromate reduction, and the differential pulse voltammetry DPV peak current has a linear relationship with BrO_3^- concentration over the range of 50 nM–5 μM with a limit of detection of 41 nM [74].

10.2.3 Agricultural

Researchers have apprehended enormous interest in agriculture safety due to science, technology, and quality of life advancement. MXenes also exhibited significant potential for agriculture safety besides its high sensing applicability for biomedical and environmental remediation. Pesticide and agrochemical residues are the main contaminants that menace for humankind and animals because of the illicit utilization and spread of pesticide residues in the environment. Moreover, it contaminates human food and aquatic systems [75, 76]. Accordingly, many reviews reported the noteworthy application of MXene sensors in agriculture safety for pesticide detection [74]. For instance, Zhou and coworkers reported MXene-based acetylcholinesterase (AChE) biosensors for fast-detecting organophosphate (Ops) pesticides. Chitosan (CS) was mixed with $Ti_3C_2T_x$ to form CS-$Ti_3C_2T_x$ nanocomposite, which was employed to assist AChE immobilization. The constructed AChE/CS-$Ti_3C_2T_x$/GCE biosensor presented an excellent performance for the sensing of malathion with the LOD of 0.3×10^{-14} M and the linearity in the range of 1×10^{-14}–1×10^{-8} M. Moreover, the amperometric biosensor exhibits acceptable stability, anti-interference ability, and reproducibility [77].

Similarly, Zhao et al. fabricated an electrochemical biosensor based on MXene/Au-Pd nanocomposite to identify OP pesticides. OP pesticides are commonly utilized to inhibit parasitic weeds, plant diseases, and pests and usually include phosphorus elements. Bimetallic nanoparticles Au-Pd were developed by a self-reduction on the surface of the MXene nanosheet, where ultrathin MXene nanosheet was synthesized through selective etching of the Al layers of the MAX phase. The fabricated enzymatic biosensor showed a linear detection range of 0.1–1000 $\mu g\, L^{-1}$ with a LOD of 1.75 $ng\, L^{-1}$ [78]. Song et al. constructed an electrochemical biosensor by employing a composite of MnO_2/Mn_3O_4 and MXene/AuNPs for sensing methamidophos which is an organophosphate pesticide. The fabricated AChE-Chit/MXene/AuNPs/MnO_2/Mn_3O_4/GCE sensor exhibit a linearity range from 10^{-12} to 10^{-6} M with a LOD of 1.34×10^{-13} M. A synergistic impact of perpendicularly arranged, highly structured nanosheets of MnO_2/Mn_3O_4 and Ti_3C_2/Au NPs with outstanding electrical conductivities and the high surface area produced an amplification impact and better electrochemical performance (Figure 10.3) [79].

Benzimidazole fungicides are a family of fungicides; for example, carbendazim (CBZ) has been extensively implemented in agriculture, forest, gardening, and horticulture; still, it is adverse to the individual health and atmosphere. CBZ engenders harmful consequences, including modified hormone concentration, infertility, and disruption of the endocrine system, as well as it also affects the dehydrogenase and phosphatase activities of plants. Thus, the precise and sensitive identification of CBZ is of utmost importance to humankind and protecting the environment [80–82]. By a simple green electrochemical method, Gao et al. fabricated MXene and reduced graphene (MXene/ERGO) and considered it as electrode material for the synthesis of an electrochemical sensor for the detection of CBZ pesticides. The MXene/ERGO/GCE sensor demonstrated a wide linear range of 2.0 Nm–10.0 μM with a LOD of 0.67 nM [83]. Likewise, Zhu et al. synthesized an electrochemical sensor based on phosphorene (BP)/MXene for naphthalene acetic acid (NAA) pesticide, and an extensive linear range of 0.02 μM with LOD of 1.6 nM was achieved. By ultrasonic-assisted liquid-phase exfoliation in an organic solvent, a highly

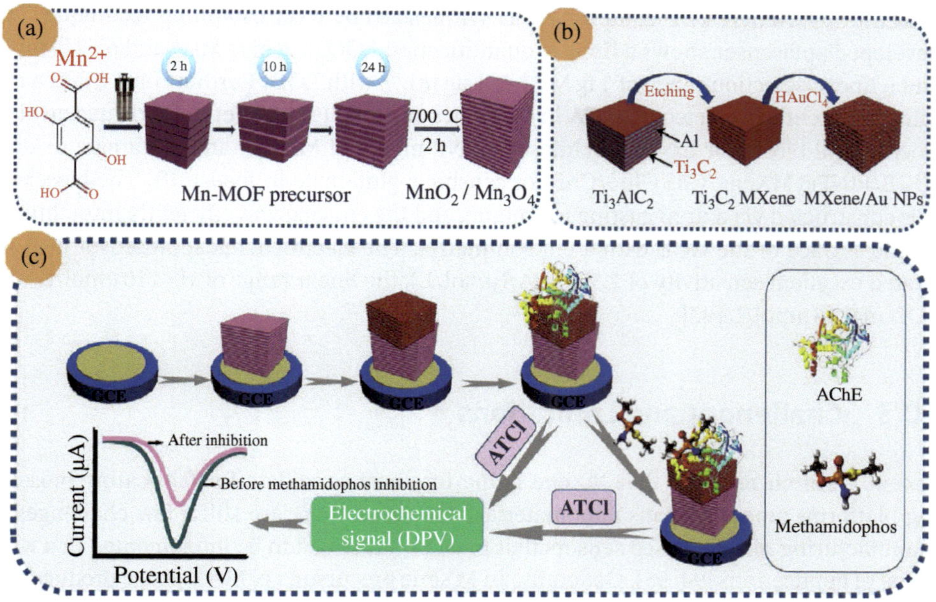

Figure 10.3 Illustration of the synthesis of AChE-Chit/MXene/Au NPs/MnO_2/Mn_3O_4/GCE biosensor for detection of methamidophos (Source: Reproduced with permission from Song et al. [79]).

ambient-stable BP nanohybrid with MXene was developed. NAA is broadly employed in crops to augment fat and protein content, but it still has a significant hazard to human health. The developed sensor exhibits reasonable practicability and excellent electrochemistry [84]. Nitrite (NO_2^-) has been widely applied in the food processing industry as a preservative and additive to prohibit growing bacteria and food spoilage. Regardless of their usage, nitrite was considered a hazardous food preservative since it interacts with secondary amines, making carcinogenic nitrosamines that lead to esophagus cancer and react with hemoglobin, causing methemoglobinemia [85–87].

Zou et al. developed an electrochemical sensor based on Au NPs/$Ti_3C_2T_x$ composite. The platform was constructed by merging functionalized MXene with Au NPs reduced in situ to detect nitrite ions, and the prepared composite has a large surface area and a high electrocatalytic activity. As MXene has a large surface area that offer even distribution of AuNPs. The fabricated biosensor showed a LOD of 0.14 μM, a linear detection range of 1.0–458.1 μM, and also it showed favorable anti-interference ability [88]. Ochratoxin A (OTA) is the most mycotoxin primarily manufactured by the genera *Penicillium* and *Aspergillus*. OTA is found in numerous food items like dairy products, coffee, cereals, spices, and wine though it produces harmful effects such as carcinogenic, genotoxic, and neurotoxic effects on mankind and animals. Moreover, it impacts the immune system, kidneys, neurons, and liver of the human system. Accordingly, there is an immediate necessity to identify OTA in agriculture and food products [89–91]. To overcome this challenge, Al-Dhahebi developed an aptasensor based on MXene/polyvinylidene (PVDF) for sensitive sensing of OTA. The loading of MXene into the PVDF for processing augments the hydrophilicity, conductivity, and electrochemical properties, which are advantageous for numerous applications, particularly biosensors. MXene/PVDF composite was synthesized by electrospinning technique. The developed aptasensor shows a limit of quantification LOQ of 2.15 fg Ml^{-1} and 6.52 fg mL^{-1} and a linear detection range of 1 fg Ml^{-1} to 1 ng mL^{-1} with LOD. Furthermore, it has a significant potential for detecting OTA in beverages and food [92]. Liu et al. fabricated an electrochemical biosensor based on chitosan, PtNP-modified MXene, and glutamate oxidase (GCE/PtNP@MXene/Chi/Gluo/Chi) for sensing L-glutamate in foodstuffs. The biosensor was constructed via a drop-casting technique, and the enzyme was efficiently immobilized on the surface of the GCE with a chitson matrix. The electrochemical biosensor demonstrated excellent sensitivity of 1.5906 nA/(μmol/L), the linear range of 10–110 μmol/L, and LOD of 0.45 μmol/L [93].

10.3 Challenges and Limitations

Recent research revealed that MXene is the foremost candidate for fabricating biosensing platforms among various nanomaterials. However, there are still a few challenges in manufacturing MXene-based sensors that must be addressed to be implemented in a wide range of applications [94, 95]. Generally, an MXene precursor, i.e. MAX phase produced by the high-temperature method of aluminum, titanium, etc., in powder form, which all are ignitable materials, and for acquiring a fine MAX powder, numerous milling methods are needed. Moreover, the existing fabrication approach employs HF as an etchant to develop MXenes, which inhibits manufacturing at an industrial scale because it is not eco-friendly, the trace residue will have an irreversible effect on individual health, and it cannot be easily handled. In place of HF, less lethal substitutes, such as electrochemical methods, molten $ZnCl_2$, and LiF/HCl, can be utilized to synthesize 2D MXene. Nevertheless, the yield of such a process is a considerable concern. Just a few MXenes have been fabricated by exfoliating the MAX phase, and a significant number of MXenes have been only anticipated abstractly. Besides, the fabrication of MXenes is costly because of the high-priced MAX precursor, product fabrication, delamination, and etching [96, 97]. The fabrication method becomes too long as the synthesis of 2D MXenes is barely shifted from the conventional chemical strategies. To enhance the synthesis procedure, microwave methods could be employed. Furthermore, profound knowledge of the terminal groups presented during the etching process is requisite because these groups significantly affect the sensor's performance and electron transfer rate.

A new approach, such as the bottom-up technique, is instantly required as it allows novel paths that offer constant surface termination, better morphology, and control over morphology because the fabrication of MXenes, with control over size, uniform surface termination, and lateral size, remains a significant challenge [98, 99]. The key challenges to be determined for constructing an MXene-based sensor are large-scale production, reproducibility, low stability of MXene in an oxygenated solvent, and cytotoxicity of MXene. Researchers have been engrossed in combining MXenes with other materials to resolve this problem [100]. The stability of 2D MXenes in aqueous and humid environments is an additional concern. Both single and multilayer MXenes are vulnerable and degrade in an aqueous, hot, and sultry environment, which confines the application of MXenes for developing into stable sensor probes. Thus, different types of MXenes have been fabricated with high stability and good conductivity to be utilized as a sensor. In addition, this deterioration could be reduced by choosing appropriate organic solvents as the MXene-based sensor enduring stability in an aqueous solution is unidentified, which needs to be thoroughly analyzed to assess the performance of these sensors in an aqueous environment. Particular emphasis must be given to the thermal and chemical stability of MXene nanosheets, particularly in a moist climate with oxygen due to the irreversible oxidation resulting in the enduring degradation of MXenes, which would be a significant challenge obstructing MXene-based sensors from various practical applications [34, 101]. The durable toxicity of MXenes has not been consistently evaluated in-depth and is unknown; therefore, the toxicity (such as genotoxicity and cytotoxicity) and the long-term biosafety, biodegradation, and biocompatibility of MXenes have to be comprehensively explored before their real-time application as sensors.

Notwithstanding, few preliminary research studies have reported that MXene's forthcoming applications, comprising sensor fabrication, would significantly rely on their biosafety [18, 102]. The majority of MXene sensors are tested under clinical detection and include complex samples with diverse environmental parameters like pH value, whereas laboratory tests involve buffers, which may result in a considerable shift in the detection result. To figure out the issue of stability and affinity of biomolecules, comprehensive knowledge of how the surface chemistry of MXenes performs could be a valuable tool as MXenes biosensor relies on good nanohybrid biocompatibility. The sensitivity and selectivity of MXene-based sensors can be bettered by surface engineering in order to give definite catalytic and adsorption activity toward a specific molecule [103, 104]. Considering the problem will further enhance the advancement of new MXene-based sensors, which are anticipated to be applied in several applications. MXene-based sensors have the substantial ability to affect many industries. Therefore, more efforts should be paid toward MXenes preparation methods, large-scale production functionalized modification, properties, cytotoxicity, and biocompatibility [105].

10.4 Conclusion and Prospects

In brief, MXene are 2D material primarily consisting of transition metal nitrides, carbides, or carbonitrides, and within a short time, it has incredible research interest. 2D MXenes or their nanocomposite are broadening their applications in sensor technology for detecting various target molecules, namely, biomolecules, environmental pollutants, pesticides, and other detrimental chemical substances, owing to their exceptional properties. Compared to the formerly established 2D nanomaterials, MXenes exhibit outstanding characteristics, particularly metallic-like conductivity, high surface area, ease of functionalization, hydrophilicity, low cytotoxicity, biocompatibility, and exciting, layered structure, which makes them highly significant for biosensing applications. As a result, ultrathin MXenes assist in the fabrication of state-of-the-art biosensors, which paves the way to novel opportunities in the sensing of various classes of target molecules and enhances the performance of biosensors; however, multiple challenges exist for industrial employment of these MXene-based biosensors. In the future, considerable emphasis and efforts should be concerned with novel synthesis methods and surface modification strategies for commercializing MXene nanocomposite to fabricate enhanced and efficient biosensors. Therefore, this book chapter provides an abridgment of the recently developed MXene-based biosensor in biomedical, environmental remediation, and agriculture safety applications.

References

1 Zhang, H., Cheng, H.-M., and Ye, P. (2018). 2D nanomaterials: beyond graphene and transition metal dichalcogenides. *Chem. Soc. Rev.* 47 (16): 6009–6012.

2 Kalambate, P.K., Gadhari, N.S., Li, X. et al. (2019). Recent advances in MXene–based electrochemical sensors and biosensors. *TrAC Trends Anal. Chem.* 120: 115643.

3 Naguib, M., Kurtoglu, M., Presser, V. et al. (2011). Two-dimensional nanocrystals produced by exfoliation of Ti3AlC2. *Adv. Mater.* 23 (37): 4248–4253.

4 Sinha, A., Dhanjai, Zhao, H. et al. (2018). MXene: an emerging material for sensing and biosensing. *TrAC Trends Anal. Chem.* 105: 424–435.

5 Lukatskaya, M.R., Mashtalir, O., Ren, C.E. et al. (2013). Cation intercalation and high volumetric capacitance of two-dimensional titanium carbide. *Science* 341 (6153): 1502–1505.

6 Hantanasirisakul, K. and Gogotsi, Y. (2018). Electronic and optical properties of 2D transition metal carbides and nitrides (MXenes). *Adv. Mater.* 30 (52): 1804779.

7 Gogotsi, Y. and Anasori, B. (2019). The rise of MXenes. *ACS Nano* 13 (8): 8491–8494.

8 Lei, J.-C., Zhang, X., and Zhou, Z. (2015). Recent advances in MXene: preparation, properties, and applications. *Front. Phys.* 10 (3): 276–286.

9 Naguib, M., Mochalin, V.N., Barsoum, M.W., and Gogotsi, Y. (2014). 25th Anniversary article: MXenes: a new family of two-dimensional materials. *Adv. Mater.* 26 (7): 992–1005.

10 Shahzad, F., Alhabeb, M., Hatter, C.B. et al. (2016). Electromagnetic interference shielding with 2D transition metal carbides (MXenes). *Science* 353 (6304): 1137–1140.

11 Soleymaniha, M., Shahbazi, M.-A., Rafieerad, A.R. et al. (2019). Promoting role of MXene nanosheets in biomedical sciences: therapeutic and biosensing innovations. *Adv. Healthc. Mater.* 8 (1): 1801137.

12 Huang, K., Li, Z., Lin, J. et al. (2018). Correction: two-dimensional transition metal carbides and nitrides (MXenes) for biomedical applications. *Chem. Soc. Rev.* 47 (17): 6889–6889.

13 Naguib, M., Mashtalir, O., Carle, J. et al. (2012). Two-dimensional transition metal carbides. *ACS Nano* 6 (2): 1322–1331.

14 Zhu, J., Ha, E., Zhao, G. et al. (2017). Recent advance in MXenes: a promising 2D material for catalysis, sensor and chemical adsorption. *Coord. Chem. Rev.* 352: 306–327.

15 An, H., Habib, T., Shah, S. et al. (2018). Surface-agnostic highly stretchable and bendable conductive MXene multilayers. *Sci. Adv.* 4 (3).

16 Rakhi, R.B., Nayak, P., Xia, C., and Alshareef, H.N. (2016). Novel amperometric glucose biosensor based on MXene nanocomposite. *Sci. Rep.* 6 (1): 36422.

17 Khazaei, M., Ranjbar, A., Arai, M. et al. (2017). Electronic properties and applications of MXenes: a theoretical review. *J. Mater. Chem. C* 5 (10): 2488–2503.

18 Khan, R. and Andreescu, S. (2020). MXenes-based bioanalytical sensors: design, characterization, and applications. *Sensors* 20 (18): 5434.
19 Feng, X., Liu, Y., Kong, Q. et al. (2010). Direct electrochemistry of myoglobin immobilized on chitosan-wrapped rod-constructed ZnO microspheres and its application to hydrogen peroxide biosensing. *J. Solid State Electrochem.* 14 (6): 923–930.
20 Kirsch, J., Siltanen, C., Zhou, Q. et al. (2013). Biosensor technology: recent advances in threat agent detection and medicine. *Chem. Soc. Rev.* 42 (22): 8733.
21 Morales-Narváez, E. and Dincer, C. (2020). The impact of biosensing in a pandemic outbreak: COVID-19. *Biosens. Bioelectron.* 163: 112274.
22 Thapa, S., Singh, K.R., Verma, R. et al. (2022). State-of-the-art smart and intelligent nanobiosensors for SARS-CoV-2 diagnosis. *Biosensors* 12 (8): 637.
23 Turner, A.P.F. (2013). Biosensors: sense and sensibility. *Chem. Soc. Rev.* 42 (8): 3184.
24 Anasori, B., Lukatskaya, M.R., and Gogotsi, Y. (2017). 2D metal carbides and nitrides (MXenes) for energy storage. *Nat. Rev. Mater.* 2 (2): 16098.
25 Lin, H., Chen, Y., and Shi, J. (2018). Insights into 2D MXenes for versatile biomedical applications: current advances and challenges ahead. *Adv. Sci.* 5 (10): 1800518.
26 Lü, J.-M., Lin, P.H., Yao, Q., and Chen, C. (2010). Chemical and molecular mechanisms of antioxidants: experimental approaches and model systems. *J. Cell. Mol. Med.* 14 (4): 840–860.
27 Vitale, F., Driscoll, N., and Murphy, B. (2019). Biomedical applications of MXenes. In: *2D Metal Carbides and Nitrides (MXenes)*, 503–524. Cham: Springer International Publishing.
28 Huang, K., Li, Z., Lin, J. et al. (2018). Two-dimensional transition metal carbides and nitrides (MXenes) for biomedical applications. *Chem. Soc. Rev.* 47 (14): 5109–5124.
29 Han, X., Huang, J., Jing, X. et al. (2018). Oxygen-deficient black titania for synergistic/enhanced sonodynamic and photoinduced cancer therapy at near infrared-II biowindow. *ACS Nano* 12 (5): 4545–4555.
30 Jastrzębska, A.M., Szuplewska, A., Wojciechowski, T. et al. (2017). In vitro studies on cytotoxicity of delaminated Ti_3C_2 MXene. *J. Hazard. Mater.* 339: 1–8.
31 Pham, V.P. and Yeom, G.Y. (2016). Recent advances in doping of molybdenum disulfide: industrial applications and future prospects. *Adv. Mater.* 28 (41): 9024–9059.
32 Lin, H., Gao, S., Dai, C. et al. (2017). A two-dimensional biodegradable niobium carbide (MXene) for photothermal tumor eradication in NIR-I and NIR-II biowindows. *J. Am. Chem. Soc.* 139 (45): 16235–16247.
33 Zhu, C., Zeng, Z., Li, H. et al. (2013). Single-layer MoS 2-based nanoprobes for homogeneous detection of biomolecules. *J. Am. Chem. Soc.* 135 (16): 5998–6001.
34 Pei, Y., Zhang, X., Hui, Z. et al. (2021). $Ti_3C_2T_x$ MXene for sensing applications: recent progress, design principles, and future perspectives. *ACS Nano* 15 (3): 3996–4017.
35 Wu, H., Zhang, X., Wei, C. et al. (2021). Modular assembly of enzyme loaded nanoparticles in 3D hollow fiber electrode for electrochemical sensing. *Chem. Eng. J.* 421: 129721.
36 Moon, J.M., Thapliyal, N., Hussain, K.K. et al. (2018). Conducting polymer-based electrochemical biosensors for neurotransmitters: a review. *Biosens. Bioelectron.*

37 Wang, F., Yang, C., Duan, M. et al. (2015). TiO2 nanoparticle modified organ-like Ti_3C_2 MXene nanocomposite encapsulating hemoglobin for a mediator-free biosensor with excellent performances. *Biosens. Bioelectron.* 74: 1022–1028.

38 Lei, Y., Cui, Y., Huang, Q. et al. (2019). Facile preparation of sulfonic groups functionalized Mxenes for efficient removal of methylene blue. *Ceram. Int.* 45 (14): 17653–17661.

39 Zhang, Q., Wang, F., Zhang, H. et al. (2018). Universal Ti_3C_2 MXenes based self-standard ratiometric fluorescence resonance energy transfer platform for highly sensitive detection of exosomes. *Anal. Chem.* 90 (21): 12737–12744.

40 Chimene, D., Alge, D.L., and Gaharwar, A.K. (2015). Two-dimensional nanomaterials for biomedical applications: emerging trends and future prospects. *Adv. Mater.* 27 (45): 7261–7284.

41 Kim, J., Campbell, A.S., and Wang, J. (2018). Wearable non-invasive epidermal glucose sensors: A review. *Talanta* 177: 163–170.

42 Khan, Y., Ostfeld, A.E., Lochner, C.M. et al. (2016). Monitoring of vital signs with flexible and wearable medical devices. *Adv. Mater.* 28 (22): 4373–4395.

43 Zhang, Y.-Z., Lee, K.H., Anjum, D.H. et al. (2018). MXenes stretch hydrogel sensor performance to new limits. *Sci. Adv.* 4 (6).

44 Lei, Y., Zhao, W., Zhang, Y. et al. (2019). A MXene-based wearable biosensor system for high-performance in vitro perspiration analysis. *Small* 15 (19): 1901190.

45 Sawangphruk, M., Sanguansak, Y., Krittayavathananon, A. et al. (2014). Silver nanodendrite modified graphene rotating disk electrode for nonenzymatic hydrogen peroxide detection. *Carbon N. Y.* 70: 287–294.

46 Xiao, L., Ding, Y., Zhai, Q. et al. (2019). Enzymatic biosensor for hydrogen peroxide based on the direct electron transfer on MWCNTs/IL/CPO-GC: the dual function of ionic liquids. *J. Electrochem. Soc.* 166 (8): G67–G74.

47 Liu, H., Duan, C., Yang, C. et al. (2015). A novel nitrite biosensor based on the direct electron transfer hemoglobin immobilized in the WO3 nanowires with high length–diameter ratio. *Mater. Sci. Eng. C* 53: 43–49.

48 Wooten, M., Karra, S., Zhang, M., and Gorski, W. (2014). On the direct electron transfer, sensing, and enzyme activity in the glucose oxidase/carbon nanotubes system. *Anal. Chem.* 86 (1): 752–757.

49 Nöll, T. and Nöll, G. (2011). Strategies for "wiring" redox-active proteins to electrodes and applications in biosensors, biofuel cells, and nanotechnology. *Chem. Soc. Rev.* 40 (7): 3564.

50 Wang, F., Yang, C., Duan, C. et al. (2015). An organ-like titanium carbide material (MXene) with multilayer structure encapsulating hemoglobin for a mediator-free biosensor. *J. Electrochem. Soc.* 162 (1): B16–B21.

51 Liu, Y., Zhang, X., Dong, S. et al. (2017). Synthesis and tribological property of $Ti_3C_2T_x$ nanosheets. *J. Mater. Sci.* 52 (4): 2200–2209.

52 Xu, B., Zhu, M., Zhang, W. et al. (2016). Ultrathin MXene-micropattern-based field-effect transistor for probing neural activity. *Adv. Mater.* 28 (17): 3333–3339.

53 Kumar, S., Lei, Y., Alshareef, N.H. et al. (2018). Biofunctionalized two-dimensional Ti_3C_2 MXenes for ultrasensitive detection of cancer biomarker. *Biosens. Bioelectron.* 121: 243–249.

54 Wu, Q., Li, N., Wang, Y. et al. (2019). A 2D transition metal carbide MXene-based SPR biosensor for ultrasensitive carcinoembryonic antigen detection. *Biosens. Bioelectron.* 144: 111697.

55 Doan, V.-D., Phan, T.L., Le, V.T. et al. (2022). Efficient and fast degradation of 4-nitrophenol and detection of Fe(III) ions by *Poria cocos* extract stabilized silver nanoparticles. *Chemosphere* 286: 131894.

56 Berkani, M., Smaali, A., Almomani, F., and Vasseghian, Y. (2022). Recent advances in MXene-based nanomaterials for desalination at water interfaces. *Environ. Res.* 203: 111845.

57 Moradi, M., Vasseghian, Y., Arabzade, H., and Mousavi Khaneghah, A. (2021). Various wastewaters treatment by sono-electrocoagulation process: a comprehensive review of operational parameters and future outlook. *Chemosphere* 263: 128314.

58 Doan, V.-D., Huynh, B.-A., Le Pham, H.A. et al. (2021). Cu2O/Fe3O4/MIL-101(Fe) nanocomposite as a highly efficient and recyclable visible-light-driven catalyst for degradation of ciprofloxacin. *Environ. Res.* 201: 111593.

59 Zhang, Y., Wang, L., Zhang, N., and Zhou, Z. (2018). Adsorptive environmental applications of MXene nanomaterials: a review. *RSC Adv.* 8 (36): 19895–19905.

60 Yoon, J., Shin, M., Lim, J. et al. (2020). Recent advances in MXene nanocomposite-based biosensors. *Biosensors* 10 (11): 185.

61 Eklund, P., Rosen, J., and Persson, P.O.Å. (2017). Layered ternary M_{n+1} AX_n phases and their 2D derivative MXene: an overview from a thin-film perspective. *J. Phys. D. Appl. Phys.* 50 (11): 113001.

62 Zhang, J., Cui, S., Ding, Y. et al. (2018). Two-dimensional mesoporous $ZnCo_2O_4$ nanosheets as a novel electrocatalyst for detection of o-nitrophenol and p-nitrophenol. *Biosens. Bioelectron.* 112: 177–185.

63 Chen, W., Mo, J., Du, X. et al. (2019). Biomimetic dynamic membrane for aquatic dye removal. *Water Res.* 151: 243–251.

64 Li, H., Kong, W., Liu, J. et al. (2015). Fluorescent N-doped carbon dots for both cellular imaging and highly-sensitive catechol detection. *Carbon N. Y.* 91: 66–75.

65 Huang, R., Chen, S., Yu, J., and Jiang, X. (2019). Self-assembled Ti_3C_2/MWCNTs nanocomposites modified glassy carbon electrode for electrochemical simultaneous detection of hydroquinone and catechol. *Ecotoxicol. Environ. Saf.* 184: 109619.

66 Wu, L., Lu, X., Dhanjai et al. (2018). 2D transition metal carbide MXene as a robust biosensing platform for enzyme immobilization and ultrasensitive detection of phenol. *Biosens. Bioelectron.* 107: 69–75.

67 Zhou, Y., Qu, Z., Zeng, Y. et al. (2014). A novel composite of graphene quantum dots and molecularly imprinted polymer for fluorescent detection of paranitrophenol. *Biosens. Bioelectron.* 52: 317–323.

68 Huang, R., Liao, D., Liu, Z. et al. (2021). Electrostatically assembling 2D hierarchical Nb2CTx and zifs-derivatives into Zn-Co-NC nanocage for the electrochemical detection of 4-nitrophenol. *Sens. Actuators B Chem.* 338: 129828.

69 Gavrilescu, M., Demnerová, K., Aamand, J. et al. (2015). Emerging pollutants in the environment: present and future challenges in biomonitoring, ecological risks and bioremediation. *N. Biotechnol.* 32 (1): 147–156.

70 Carolin, C.F., Kumar, P.S., Saravanan, A. et al. (2017). Efficient techniques for the removal of toxic heavy metals from aquatic environment: a review. *J. Environ. Chem. Eng.* 5 (3): 2782–2799.

71 Qu, X., Brame, J., Li, Q., and Alvarez, P.J.J. (2013). Nanotechnology for a safe and sustainable water supply: enabling integrated water treatment and reuse. *Acc. Chem. Res.* 46 (3): 834–843.

72 Zhu, X., Liu, B., Hou, H. et al. (2017). Alkaline intercalation of Ti_3C_2 MXene for simultaneous electrochemical detection of Cd(II), Pb(II), Cu(II) and Hg(II). *Electrochim. Acta* 248: 46–57.

73 Lv, X., Pei, F., Feng, S. et al. (2020). Facile synthesis of protonated carbon nitride/$Ti_3C_2T_x$ nanocomposite for simultaneous detection of Pb^{2+} and Cd^{2+}. *J. Electrochem. Soc.* 167 (6): 067509.

74 Rasheed, P.A., Pandey, R.P., Rasool, K., and Mahmoud, K.A. (2018). Ultra-sensitive electrocatalytic detection of bromate in drinking water based on Nafion/$Ti_3C_2T_x$ (MXene) modified glassy carbon electrode. *Sens. Actuators B Chem.* 265: 652–659.

75 Carvalho, F.P. (2017). Pesticides, environment, and food safety. *Food Energy Secur.* 6 (2): 48–60.

76 Ishag, A.E.S.A., Abdelbagi, A.O., Hammad, A.M.A. et al. (2016). Biodegradation of chlorpyrifos, malathion, and dimethoate by three strains of bacteria isolated from pesticide-polluted soils in Sudan. *J. Agric. Food Chem.* 64 (45): 8491–8498.

77 Zhou, L., Zhang, X., Ma, L. et al. (2017). Acetylcholinesterase/chitosan-transition metal carbides nanocomposites-based biosensor for the organophosphate pesticides detection. *Biochem. Eng. J.* 128: 243–249.

78 Zhao, F., Yao, Y., Jiang, C. et al. (2020). Self-reduction bimetallic nanoparticles on ultrathin MXene nanosheets as functional platform for pesticide sensing. *J. Hazard. Mater.* 384: 121358.

79 Song, D., Jiang, X., Li, Y. et al. (2019). Metal–organic frameworks-derived MnO_2/Mn_3O_4 microcuboids with hierarchically ordered nanosheets and Ti_3C_2 MXene/Au NPs composites for electrochemical pesticide detection. *J. Hazard. Mater.* 373: 367–376.

80 Suresh, I., Selvaraj, S., Nesakumar, N. et al. (2021). Nanomaterials based non-enzymatic electrochemical and optical sensors for the detection of carbendazim: a review. *Trends Environ. Anal. Chem.* 31: e00137.

81 Su, S., Wu, W., Gao, J. et al. (2012). Nanomaterials-based sensors for applications in environmental monitoring. *J. Mater. Chem.* 22 (35): 18101.

82 Grujic, S., Radisic, M., Vasiljevic, T., and Lausevic, M. (2005). Determination of carbendazim residues in fruit juices by liquid chromatography-tandem mass spectrometry. *Food Addit. Contam.* 22 (11): 1132–1137.

83 Xie, Y., Gao, F., Tu, X. et al. (2019). Facile synthesis of MXene/electrochemically reduced graphene oxide composites and their application for electrochemical sensing of carbendazim. *J. Electrochem. Soc.* 166 (16): B1673–B1680.

84 Zhu, X., Lin, L., Wu, R. et al. (2021). Portable wireless intelligent sensing of ultra-trace phytoregulator α-naphthalene acetic acid using self-assembled phosphorene/Ti_3C_2-MXene nanohybrid with high ambient stability on laser induced porous graphene as nanozyme flexible electrode. *Biosens. Bioelectron.* 179: 113062.

85 Rao, H., Liu, Y., Zhong, J. et al. (2017). Gold nanoparticle/chitosan@N,S Co-doped multiwalled carbon nanotubes sensor: fabrication, characterization, and electrochemical detection of catechol and nitrite. *ACS Sustain. Chem. Eng.* 5 (11): 10926–10939.

86 Wang, Y., Zeng, Z., Qiao, J. et al. (2021). Ultrasensitive determination of nitrite based on electrochemical platform of AuNPs deposited on PDDA-modified MXene nanosheets. *Talanta* 221: 121605.

87 Bryan, N.S., Alexander, D.D., Coughlin, J.R. et al. (2012). Ingested nitrate and nitrite and stomach cancer risk: an updated review. *Food Chem. Toxicol.* 50 (10): 3646–3665.

88 Zou, H., Zhang, F., Wang, H. et al. (2019). Au nanoparticles supported on functionalized two-dimensional titanium carbide for the sensitive detection of nitrite. *New J. Chem.* 43 (6): 2464–2470.

89 Yang, J., Gao, P., Liu, Y. et al. (2015). Label-free photoelectrochemical immunosensor for sensitive detection of Ochratoxin A. *Biosens. Bioelectron.* 64: 13–18.

90 Iqbal, S.Z., Asi, M.R., Hanif, U. et al. (2016). The presence of aflatoxins and ochratoxin A in rice and rice products; and evaluation of dietary intake. *Food Chem.* 210: 135–140.

91 Alhamoud, Y., Yang, D., Fiati Kenston, S.S. et al. (2019). Advances in biosensors for the detection of ochratoxin A: bio-receptors, nanomaterials, and their applications. *Biosens. Bioelectron.* 141: 111418.

92 Al-Dhahebi, A.M., Jose, R., Mustapha, M., and Saheed, M.S.M. (2022). Ultrasensitive aptasensor using electrospun MXene/polyvinylidene fluoride nanofiber composite for Ochratoxin A detection. *Food Chem.* 390: 133105.

93 Liu, J., Fan, Y., Chen, G., and Liu, Y. (2021). Highly sensitive glutamate biosensor based on platinum nanoparticles decorated MXene-$Ti_3C_2T_x$ for l-glutamate determination in foodstuffs. *LWT* 148: 111748.

94 Xu, B., Zhi, C., and Shi, P. (2020). Latest advances in MXene biosensors. *J. Phys. Mater.* 3 (3): 031001.

95 Huang, H., Jiang, R., Feng, Y. et al. (2020). Recent development and prospects of surface modification and biomedical applications of MXenes. *Nanoscale* 12 (3): 1325–1338.

96 Sajid, M. (2021). MXenes: are they emerging materials for analytical chemistry applications?—a review. *Anal. Chim. Acta* 1143: 267–280.

97 Mathew, M. and Rout, C.S. (2021). Electrochemical biosensors based on $Ti_3C_2T_x$ MXene: future perspectives for on-site analysis. *Curr. Opin. Electrochem.* 30: 100782.

98 Riazi, H., Taghizadeh, G., and Soroush, M. (2021). MXene-based nanocomposite sensors. *ACS Omega* 6 (17): 11103–11112.

99 Ganesh, P.-S. and Kim, S.-Y. (2022). Electrochemical sensing interfaces based on novel 2D-MXenes for monitoring environmental hazardous toxic compounds: a concise review. *J. Ind. Eng. Chem.* 109: 52–67.

100 Rizwan, K., Rahdar, A., Bilal, M., and Iqbal, H.M.N. (2022). MXene-based electrochemical and biosensing platforms to detect toxic elements and pesticides pollutants from environmental matrices. *Chemosphere* 291: 132820.

101 Zhang, C.J., Pinilla, S., McEvoy, N. et al. (2017). Oxidation stability of colloidal two-dimensional titanium carbides (MXenes). *Chem. Mater.* 29 (11): 4848–4856.

102 Huang, H., Dong, C., Feng, W. et al. (2022). Biomedical engineering of two-dimensional MXenes. *Adv. Drug Deliv. Rev.* 184: 114178.

103 Siwal, S.S., Sheoran, K., Mishra, K. et al. (2022). Novel synthesis methods and applications of MXene-based nanomaterials (MBNs) for hazardous pollutants degradation: future perspectives. *Chemosphere* 293: 133542.

104 Hong Ng, V.M., Huang, H., Zhou, K. et al. (2017). Recent progress in layered transition metal carbides and/or nitrides (MXenes) and their composites: synthesis and applications. *J. Mater. Chem. A* 5 (7): 3039–3068.

105 Rhouati, A., Berkani, M., Vasseghian, Y., and Golzadeh, N. (2022). MXene-based electrochemical sensors for detection of environmental pollutants: a comprehensive review. *Chemosphere* 291: 132921.

11

Advances in Ti_3C_2 MXene and Its Composites for the Adsorption Process and Photocatalytic Applications

Channe Gowda Sushma[1], Bangalore Nanjundappa Nagalaxmi[1], Raghavachari Kavitha[2], and Shivashankar Girish Kumar[3]

[1] *Department of Chemistry, Maharani Lakshmi Ammanni College for Women, Affiliated to Bengaluru City University, Bengaluru, Karnataka, India*
[2] *Department of Chemistry, Post-Graduation Studies, Vijaya College, Affiliated to Bengaluru City University, Bengaluru, Karnataka, India*
[3] *Department of Chemistry, RV College of Engineering, Affiliated to Visvesvaraya Technological University, Bengaluru, Karnataka, India*

11.1 Introduction

The removal and the treatment of various contaminants such as heavy metals, organic compounds and bacteria can be harmful to many living organisms and continues to be a foremost challenge in the wastewater purification. The adsorption and complete mineralization of toxic pollutants through exploring a variety of functional nanomaterials are at the forefront and tremendous progress are witnessed in this field over the years. The adsorption process expedites the separation of the pollutants from the aqueous suspension and further processing using various techniques renders them to reuse in the washing purposes. In parallel, semiconductor photocatalysis as one of the advanced oxidation processes has garnered major attention for the degradation of wide variety of organic pollutants due to their ability to completely mineralize the contaminants under the moderate operating conditions [1]. The diverse nanomaterials are explored as a stable adsorbent as well as for the photocatalytic treatment of wastewater. However, these nanomaterials mostly are obtained in the powder form with irregular morphology and are difficult to recover and reuse after the processing steps. In this context, MXenes aroused significant interest for their application both as adsorbent and co-catalysts in the photocatalytic reactions due to their larger specific surface area that complements with their prominent chemical reactivity arising from the reduced dimensionality and quantum confinement effects [2]. MXenes are synthesized by selectively removing the 'A' layer (main group IIIA/IVA elements) from the MAX phases ($M_{n+1}AX_n$, $n = 1, 2, 3\ldots$). The 'M' is ascribed to the presence of transition metals and the 'X' specifies the carbon and/or nitrogen elements [2–4]. In general, MXenes are represented by $M_{n+1}X_nT_x$, with 'T_x' relates for the surface terminating groups that are formed during the etching procedure with fluoride ions (HF, KF, $NaBF_4$). This process introduces polar terminating groups which eloquently alters the surface chemistry, textural properties

MXenes: Next-Generation 2D Materials: Fundamentals and Applications, First Edition.
Edited by Jay Singh, Kshitij RB Singh, Ravindra Pratap Singh, and Charles Oluwaseun Adetunji.

and the electronic band structures making them as spectacular substrate for the growth of plentiful nanomaterials [5, 6]. On the other hand, its anisotropic mobility, electronegative surface, and exceptional metallic conductivity increases the lifetime of excited charge carriers [7]. MXenes are intensively explored as adsorbent for the removal of pollutants through chemical and electrostatic interactions. Additionally, they can serve as cocatalysts to promote various reactions such as pollutants degradation, H_2 generation, CO_2 reduction and decomposition of NO_x under the light illuminating conditions [3]. Among the various MXenes, Ti_3C_2 has been in the limelight due to their admirable features like remarkable hydrophilicity, tailored surface acid-base properties, easy functionalization, biocompatibility with adsorptive and reductive characteristics and high negative zeta potential values. In this context, we emphasize the updated information on Ti_3C_2 based composites as an exemplar material with relevance to the treatment of wastewater. The synthesis methods and influence of various reaction parameters over the adsorption process as well as the photocatalytic reaction pathways are discussed at the relevant stages. Furthermore, the gap analysis of this exciting field is presented with an intention to design the efficient heterostructures in near future.

11.2 Ti_3C_2 as Adsorbent for the Metal Ions

The application of traditional adsorbents such as activated carbon for the adsorption of metal ions finds limited scope because of their meagre capacity and inability to be reused multiple times. In this context, multifaceted properties of Ti_3C_2 complemented with high removal efficiency forecasts an important role as adsorbent. The presence of surface functional groups on Ti_3C_2 facilitates a direct ion exchange process, which accelerates the metal ions adsorption from the aqueous medium. Pandey et al. [8] reported the high reduction capacity (321.8 mg BrO_3^-/g $Ti_3C_2T_x$) of delaminated $Ti_3C_2T_x$ nanosheets (NSs) for BrO_3^- ions at acidic pH conditions. This was attributed to the electrostatic force of attraction operating between the negatively charged bromate ions and positively charged $Ti_3C_2T_x$. In contrast, electrostatic repulsion forces dominate in the alkaline medium to reduce the adsorption process [9]. The reaction rate spiked at the starting stages (initial 15 min) and suddenly suppressed thereafter. This material exhibited significant selectivity (~92%) even with the coexistence of many impurities such as SO_4^{2-}, NO_3^- and PO_4^{3-} ions (5 mg L^{-1}). Unfortunately, the regeneration of $Ti_3C_2T_x$ was far from satisfaction due to the oxidation of MXene, which led to the destruction of its sheet-like morphology. Also, formation of TiO_2 and amorphous carbon during the oxidation of $Ti_3C_2T_x$ layer masks the surface-active sites and certainly hinders the adsorption process [8].

The leakage and the inappropriate disposal of radioactive wastes into the aqueous streams is a serious concern for the survival of aquatic species. Among the radionuclides, Cs(I)/Sr(II) ions are the most harmful by the virtue of its persisted half-life period and high degree of solubility in water. The strontium and caesium have chemical resemblance with that of calcium and potassium which are susceptible for accumulation in the human and animal bones leading to the blood and bone cancers. The $Ti_3C_2T_x$ functionalized with $-NH_2$ groups through treating with amino-propyl-isobutyl polyhedral oligomeric silsesquioxane (POSS-NH_2) triggered the instantaneous adsorption of Cs(I)/Sr(II) ions

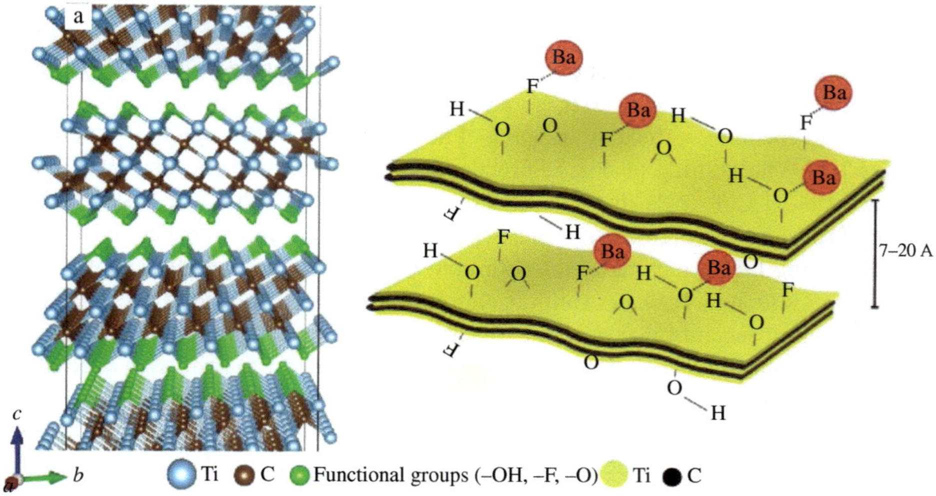

Figure 11.1 Mechanism of barium ions adsorption on MXene layers. Source: Reproduced with the permission from Fard et al. [11]. Copyright 2017, Elsevier.

and the corresponding adsorption equilibrium attained very rapidly (<30 min) [10]. The functionalized material also exhibited the notable selectivity in eliminating the Cs(I)/Sr(II) ions even with the dominance of other cations and retained their adsorption capacity for the three cycles. The synergistic effects arising from the increased inter-lamellar distances as well as the complex formation tendency between the functional groups and the cations resulted in the high adsorption capacity. Similarly, $Ti_3C_2T_x$ exhibited notable adsorption capacity towards Ba(II) ions compared to the activated carbon and carbon nanotubes (CNTs) [11]. The Dubinin–Radushkevich (DR)-model calculation implied that the mean adsorption energy for Ba(II) ions over the $Ti_3C_2T_x$ surface was significant (~50 kJ/mol) which is an indicative of chemisorption mechanism (Figure 11.1).

The charge density of the sorbate combined with their affinity towards the metal ions as well as the chemical entity of metal ions in the aqueous phase drives the overall adsorption process. The negative zeta potential of $Ti_3C_2T_x$ at neutral pH endorses the electrostatic force of attraction with Ba(II) ions, while the formation of $Ba(OH)_2$ suppressed the concentration of free Ba(II) ions for the adsorption process at the basic pH regime [12]. As the barium exists in Ba(II) state in all the forms such as $Ba(OH)_2$, $Ba(NO_3)_2$ and $BaCl_2$, they can be potentially separated through the chemical adsorption process. The complete removal of Ba(II) ions was achieved in less than an hour, whereas the same process was extended over the perlite surface [13]. The surface adsorption process over $Ti_3C_2T_x$ was found to be straightforward and well-described by the Freundlich isotherm, indicating a multi-layer adsorption mechanism. Mu et al. [14] reported threefold greater activity of alkalinized-$Ti_3C_2T_x$ (~46.46 mg g^{-1}) than the bare $Ti_3C_2T_x$ for the Ba(II) adsorption. This enhancement is attributed to the presence of more active sites available for the adsorption process, which resulted from the alkaline (NaOH) treatment.

The leakage of lead from industries like oil refineries, lead-acid batteries and paints pose adverse effects on the human health. The adsorption efficiency of amine functionalized

$Ti_3C_2T_x$ reached the highest of ~384.6 mg/g for Pb(II) ions via the Langmuir adsorption isotherm. This amine-functionalized $Ti_3C_2T_x$ (alk-$Ti_3C_2T_x$-NH_2) were synthesized by treating the $Ti_3C_2T_x$ with NaOH (5 wt%) at moderate temperature (~293 K, 4 h) followed by grafting with (3-aminopropyl) triethoxysilane (APTES). The unpaired electrons on the –NH_2 group encourages the easy trapping of heavy metal ions through the complexation process [15]. The grafting of organo-functional groups improved the interlayer spacing between the $Ti_3C_2T_x$ NSs and consequent enhancement within the distinct surface region of the adsorbent [16]. In another report, biosurfactant-functionalized $Ti_3C_2T_x$ was prepared using LiF as etching agent with non-ionic enzymatic hydrolysis lignin (EHL), anionic lignosulfonate (LS) and cationic chitosan (CS). The low toxicity and eco-friendly nature of biosurfactants makes them to serve as good adsorbents [17, 18]. The EHL showed admirable uptake of Pb(II) ions which is ascribed to the introduction of additional reaction sites over the Ti_2CT_x compared to LS and CS counterparts. The CS exhibited a competition with Pb(II) ions for the adsorption process over the negatively charged surface, while the electrostatic repulsion between LS and anionic $Ti_3C_2T_x$ inhibited the good interactions between them [18]. The self-polymerization of $Ti_3C_2T_x$ by an amino acid levodopa (tri-(hydroxymethyl)-aminethane) strengthened the adsorption for Cu(II) ions owing to the modification of the surface with carboxyl and amine groups [19]. The driving force for the adsorption process emanates from the ionic attraction between the positively charged Cu (II) ions and negatively charged $Ti_3C_2T_x$. At the acidic medium, competitive adsorption between H^+ or H_3O^+ and the target metal ion on the $Ti_3C_2T_x$ decreases the efficiency. Thus, the sorbent efficacy was elevated with shift in the pH values to the basic medium (pH 11) and then decrement was observed at a very high value (pH 13). This suggests that there exists an optimum pH to attain maximum adsorption beyond which the precipitation of metal hydroxides dominates over the adsorption process [20]. It should be noted that the precipitation of metal hydroxides depends on the nature of the metal ion/base, solution temperature and concentration of metal ions/base. For instance, high removal efficiency is normally observed at a low dosage of metal ions owing to the availability of surplus surface-active sites. In contrast, higher concentration of metal ions smoothens the adsorption at the initial stages, but the formation of metal hydroxides dominates after the attainment of adsorption equilibrium. The intercalation of guest ions or molecules enhances the inter-layer spacing in MXene increasing the active sites for the adsorption, thus improving the sorption process [20]. Mu et al. obtained layered morphologies for $Ti_3C_2T_x$ when Ti_3AlC_2 was exfoliated with HF under mild temperature (25–45 °C) and book-like morphology was observed at 45 °C [21]. The high-resolution transmission electron microscopy (HRTEM) analysis revealed the changes in the interlayer spacing from 0.216 to 0.313 nm with increase in exfoliating temperature. The rise in surface area with higher exfoliating temperatures demonstrated the direct correlation of morphology with specific surface area. The thermodynamic parameters suggested that the sorption of Pd(II) ions was indeed spontaneous and endothermic in nature. The excellent stability even after five cycles indicate that the morphology of $Ti_3C_2T_x$ can also influence the adsorption process [21].

The oxygen-functionalized Ti_3C_2 ($Ti_3C_2O_x$) exhibited superior adsorption kinetics, high selectivity and pH independent removal of Hg(II) ions [22]. This was the unique study to report the maximum removal capacity value of Hg (II) ions over the functionalized MXenes by fitting the isotherm data with the Langmuir model. The density functional theory (DFT) studies unveiled a significant interaction between HgClOH and the titanium facets, with the former undergoing radical scission to generate hydroxyl and mercurous chloride. The latter dimerizes to form water insoluble Hg_2Cl_2 and possesses relatively lower toxicity than Hg(II) ions [23].

$$2Ti^* + 2HgClOH \rightarrow Hg_2Cl_2 + 2\,Ti^*OH \tag{11.1}$$

Shahzad et al. [24] fabricated spherical MX-$SA_{4:20}$ adsorbent involving $Ti_3C_2T_x$ and sodium alginate with NH_4F as etching agent. The adsorption of Hg(II) ions process involved multiple mechanisms such as complex-formation, ion-exchange process, and electrostatic interactions. The inner-surface complexation between [Ti-O]-H^+ and Hg(II) along with the ion exchange reaction between Ca(II) and Hg(II) ions collectively promoted the adsorption of Hg(II) ions. The presence of varied functional groups such as hydroxyl (−OH), alkanes (−CH), and carboxyl (−COO) in the biopolymer alginate influences the adsorption process. Furthermore, the introduction of $CaCl_2$ during the synthesis step drives the formation of a [Ti-O]-Ca^{2+} complex species, which played a crucial role in accelerating the binding capacity of Hg(II) ions [24].

Thorium is used as the fuel for generating nuclear energy and its toxicity imposes serious effects towards physiological functions of the human body even when present at the minimal quantity. Li et al. suggested that the crucial factor for the adsorption of Th(IV) ions was the multi-layered Ti_2CT_x-hydrated material derived from the LiF-HCl etching process [25]. The intercalated water molecules and Li(I) ions collectively enhanced the removal of Th(IV) ions (Figure 11.2). The sorbent dosage is of prime importance for optimization as it determines the cost effectiveness of the adsorption process. As the sorbent dosage increases, the surface-active site also enhances proportionally leading to the higher removal efficiency through Freundlich isotherm model [25].

Uranium is an extremely toxic metal ion which pose negative effect on the human health leading to severe liver and the bone cancer. The sorption of uranyl ions to the surface oxygen sites occurs either through a monodentate or bidentate mode, involving

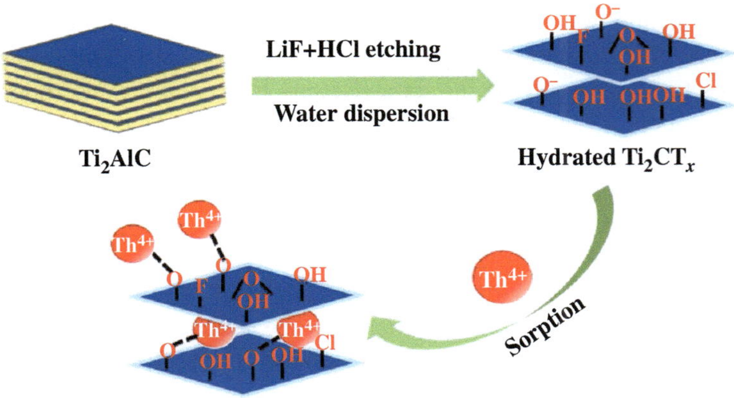

Figure 11.2 Adsorption mechanism of Thorium onto multi-layered Ti_2CT_x-hydrated originating from LiF-HCl etching process. Source: Reproduced with the permission from Li et al. [25]. Copyright 2019, Elsevier.

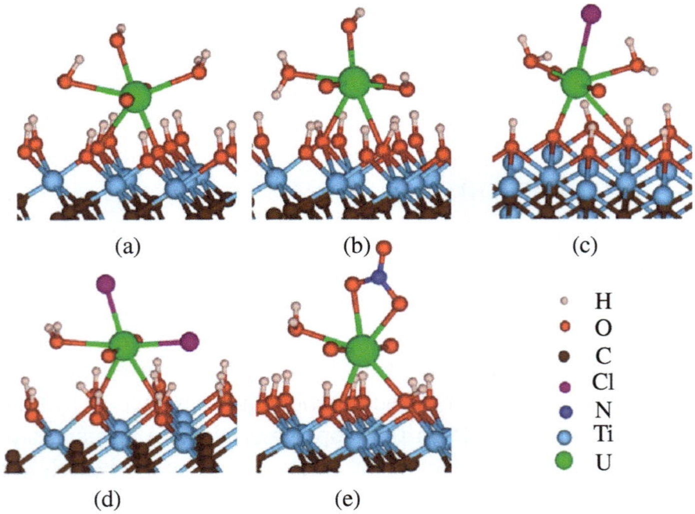

Figure 11.3 Uranyl adsorption configurations of (a) TiO_2-$UO_2(H_2O)_2(OH)$, (b) TiO_2-$UO_2(H_2O)(OH)_2$, (c) TiO_2-$UO_2(H_2O)_2Cl$, (d) TiO_2-$UO_2(H_2O)Cl_2$ and (e) TiO_2-$UO_2(H_2O)(NO_3)$. Source: Reproduced with the permission from Zhang et al. [26]. Copyright 2016, Elsevier.

the displacement of one or two water ligands from the primary coordination shell. The removal of two water molecules from penta-coordinated uranyl species $[UO_2(H_2O)_5]^{2+}$ generates TiO_2-$UO_2(H_2O)_3$ via binding with two surface deprotonated O_s moiety during the adsorption process. Under basic conditions, the deprotonation of surface –OH groups facilitate the formation of a bidentate adsorption configuration of TiO_2-$UO_2(H_2O)_3$. Conversely, the TiO-$UO_2(H_2O)_4$ monodentate adsorption configuration dominates under acidic pH owing to the protonation of complete surface –OH groups (Figure 11.3) [26].

Zhang et al. reported the limiting grafting ratio of carboxyl terminated diazonium salt to MXene as 5:1 for the adsorption of U(VI) and Eu(III) ions [27]. The enrichment of radionuclides onto the adsorbent surface was attributed to the high affinity of the carboxyl groups, which facilitates the formation of inner-sphere coordination complexes with U(VI) and Eu(III) ions. The increase of diazonium salt content resulted in the polymeric layer which distorts the spatial structure and masks the surface-active sites to prevent the adsorption [28]. The reported adsorption capacity of ~344.8 mg/g for U(VI) ions and ~97.1 mg/g for Eu(III) ions was higher compared to the other sorbents [29, 30]. The adsorption process exhibited different trend in the pH range of 6–10 for U(VI) and Eu(III) ions depending on the formation of hydroxides or anionic complexes. Interestingly, the percentage removal of U(VI) ions remains unaffected even with the presence of different types of cations [27].

The overall studies indicate that the further modifications of $Ti_3C_2T_x$ are essential to proficiently promote the adsorption process. The change in the structure-morphological features and toxicity of the material after the adsorption process require detailed investigation to carefully understand the respective associated mechanisms and biocompatibility. The surface modifications involving the introduction of organic functional groups may not be always appreciated as it involves the tedious preparation steps under insensitive conditions. Also, they can influence the hydrophilicity of MXene to affect their dispersion in the aqueous medium. Therefore, strategies concerning the design and development of Ti_3C_2 that can be operative at a wider pH range should be a focus of our further research.

11.3 Photocatalytic Degradation Mechanism of Organic Pollutants via Ti_3C_2 MXene and Its Derivatives

Photocatalysis is a promising technology for the wastewater purification under the light illuminated conditions. Depending on the band gap absorption properties and the band edge potentials, various semiconductors are investigated for the wastewater treatment [1]. The generation of excited charge carriers and drifting them apart to react with the surface adsorbed species to produce oxygenated free radicals comprises the basic principle and plays a crucial role in achieving the complete decomposition of organic pollutants. As an alternative to graphene, Ti_3C_2 are explored as cocatalysts for various photocatalytic reactions because of their layered structure that can aid for the nucleation of several catalytic materials, thus providing good intimate contact to form the stable heterostructures. Also, the layered structures are more susceptible for the adsorption of gaseous compounds and the dangling bonds over its surface possess substantial reactivity [31]. More importantly, obtaining $Ti_3C_2T_x$ from the Ti_3AlC_2 through conventional fluoride ion etching is facile as it is operative under the moderate reaction conditions. The surface Ti-sites are susceptible for robust redox reactions promoting the multiple electron reduction pathways to produce more oxygenated free radicals and consequently accelerates the pollutant degradation reactions. The prominent metallic conductivity of Ti_3C_2 makes it an ideal candidate to serve as electron sink similar to the graphene and other noble metals [32]. The Ti_3C_2 can promote the generation of oxygen vacancies, which in turn capture excited electrons, leading to a significant enhancement in the performance [33]. Also, charge carrier mobility of holes exceeds that of electrons in the MXene monolayers, crafting it as versatile material for suppressing the reunification of charge carriers through the hole trapping dynamics [34]. The interlayer pores in the Ti_3C_2 offers superior pathways for dissipating excited holes during the photocatalytic process [35]. Due to these multifaceted features, Ti_3C_2 is coupled with various semiconductors and explored in the wastewater treatment.

11.3.1 Heterostructuring the Ti_3C_2 with Metal Oxides

The TiO_2 is regarded as benchmark photocatalysts owing to their venerable features such as suitable band edge potentials to generate free radicals, photostability and exceptional resistance to corrosion across a broader pH range [1]. The anatase polymorph of TiO_2 is extensively studied for photocatalysis due to their high tendency for the oxygen adsorption [36]. Thus, design and fabrication of Ti_3C_2/TiO_2 has gained significant interests due to their facile preparation without using any external titanium source and the formation of stable contacts between them [37]. The Ti_3C_2/TiO_2 possess various advantages such as: (i) enhanced absorption of visible light in comparison to the pure TiO_2; (ii) exposure of more surface-active sites; (iii) excellent conductivity of the composite; (iv) utilization of TiO_2 valence band (VB) holes for the degradation reactions; (v) formation of stable core@shell structures [38]. The DFT computational studies unveiled that the Fermi level of Ti_3C_2 was positioned above (~0.18 eV vs. NHE) the conduction band (CB) of TiO_2 and hence these materials incline to form Schottky-type heterojunction. The electrons transfer from TiO_2 to Ti_3C_2 continues until the Fermi level equilibration takes place. The space charge layer generated on the side of semiconductor bends at the interface region thus inducing the bending of energy bands in an upward direction, leading to the formation of Schottky barrier. As a result, flow of electrons back into the TiO_2 CB layer is prohibited leading to the separation of the charge carriers [39, 40]. The high electrical conductivity and metallic property of Ti_3C_2 initiate the capturing of electrons from the CB of TiO_2 leaving positive charge accumulation on the TiO_2. The electrons resided at the Ti_3C_2 triggers the dioxygen reduction, while holes in the VB of TiO_2 prompts the oxidation of hydroxyl anions to form hydroxyl radicals. Therefore, this composite is extensively studied for the degradation of diverse organic dyes and phenolic compounds [4, 41–44]. As stated above, the degradation pathways are primarily mediated by superoxide and hydroxyl radicals compared to the VB holes [4, 42, 43]. The design of this heterostructure broadly classified into two types: (i) annealing Ti_3C_2 at moderate temperature to initiate the partial phase conversion to TiO_2; (ii) heterogeneous nucleation of TiO_2 on the preformed Ti_3C_2 through the wet chemical approaches [4, 41, 42]. In the latter case, either the external titanium precursor is added or the Ti_3C_2 itself is exposed to oxidation process in the liquid phase to facilitate the TiO_2 formation. Gao et al. synthesized TiO_2 on the surface of Ti_3C_2 utilizing the $TiOSO_4$ [41], while Hieu et al. [4] and Quyen et al. [42] obtained TiO_2 from the Ti_3C_2 itself via hydrothermal oxidation under alkaline and acidic medium with H_2O_2 and $NaBH_4$ respectively [4, 42]. The former resulted in irregular morphology [41] and the latter assisted the formation of flower-like morphology for TiO_2 [42]. It should be noted that the $H_2Ti_3O_7$ intermediate formation was observed which later transformed to TiO_2 upon the annealing step. The decrease of carbon content accompanied with increment in the oxygen attests the formation of Ti_3C_2/TiO_2 [42]. The use of reducing agent and the acidic medium favoured the nucleation of rutile TiO_2 [4], which is in accordance with the literature reports [45].

The solvothermal oxidation of $Ti_3C_2T_x$ using $NaBF_4$ as the co-surfactant resulted in the formation of (101) and (001) active facets of TiO_2 which were embedded in the layers of $Ti_3C_2T_x$ [43]. The nanoparticle formation was observed in the absence of $NaBF_4$ suggesting that the fluoride ions play a vital role in tailoring the active facets of TiO_2. This heterojunction promoted both the adsorption and decomposition of methyl orange (MO) and methylene blue (MB). More importantly, the composite exhibited stable performance up to five cycles [43]. The optical response of the composite was red shifted to an extent of ~0.9–2.4 eV through varying the reaction conditions in the preparation method. The narrow gap improves the photon absorption capacity and generates the surplus charge carriers during the bandgap excitation process under visible light [43, 44]. The optimum content of $NaBF_4$ (1.5 mmol) and solvothermal temperature (473 K) was crucial to modulate the more exposure of active facets [43, 44]. The {001}-$TiO_2/Ti_3C_2T_x$ was also active for the degradation of carbamazepine under the acidic conditions (pH 3) and the rate constant noticed under UV light illumination was appreciably superior compared to the natural solar light [46]. The degradation proceeded through the formation of several intermediates such as acridine, 2-aminobenzoic acid, 2-hydroxybenzoic acid and formaldehyde-acridine (Figure 11.4).

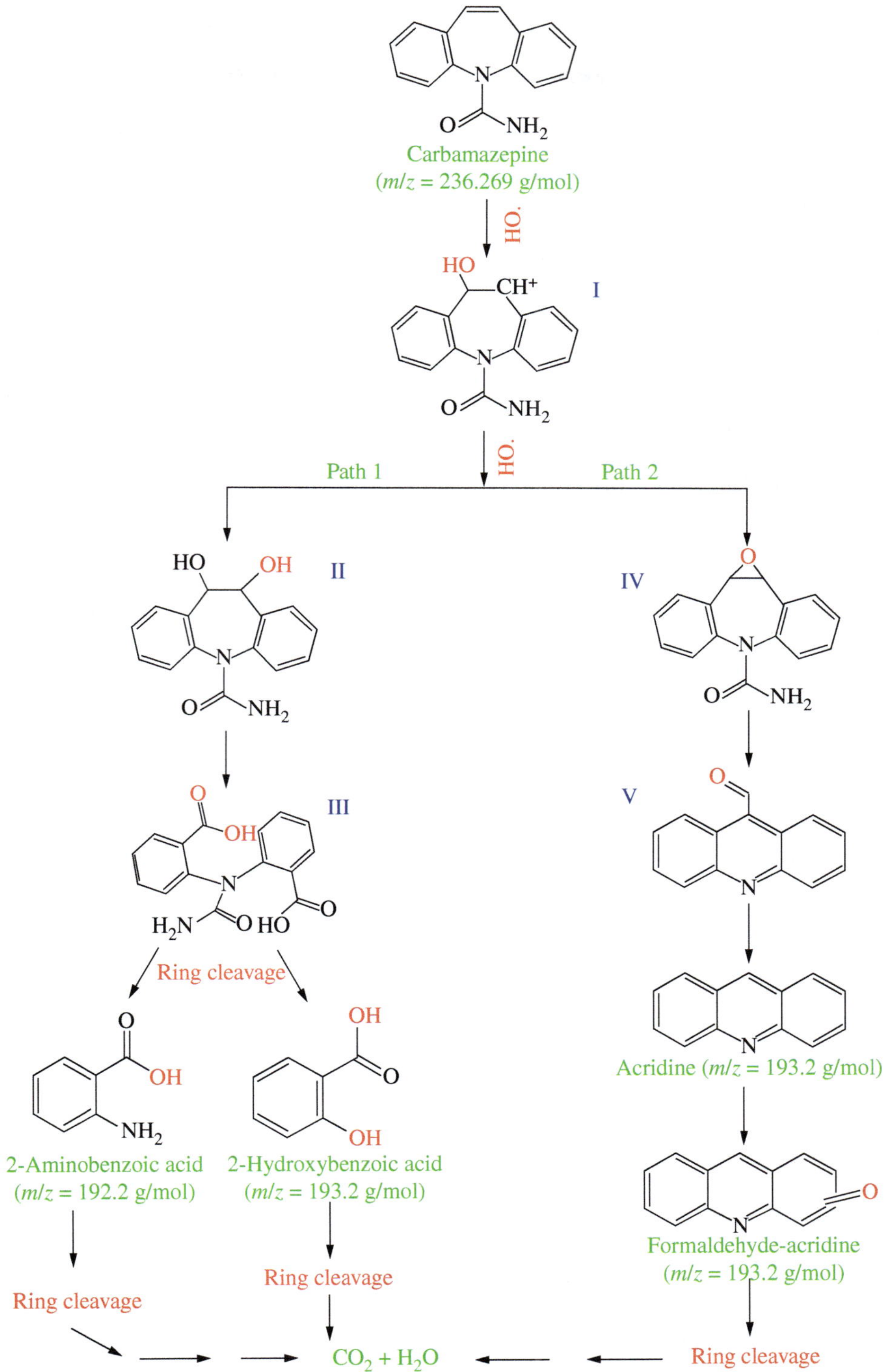

Figure 11.4 Degradation mechanisms of carbamazepine using {001}-$TiO_2/Ti_3C_2T_x$. Source: Reproduced with the permission from Shahzad et al. [46]. Copyright 2018, Elsevier.

At the interface of Ti_3C_2/TiO_2, Ti_3C_2 and TiO_2 can integrate at the atomic level and a new kind of Ti—O—Ti covalent bonds are likely to be formed. This chemical entity intensifies the orbital interactions between the '*d*' and '*p*' orbitals of 'Ti' and 'O' respectively [35]. The participation of hydroxyl radicals dominates the degradation pathways and is beneficial from the prospect of complete mineralization of contaminants. Due to the low work function of Ti_3C_2, holes traverse from TiO_2 to Ti_3C_2 and the Schottky barrier prevents the back flow of holes to the TiO_2. The high density of electrons accumulated over the {001}-TiO_2 commences the reduction of dissolved oxygen and generates superoxide and hydroxyl radicals thereby promoting the degradation kinetics [35]. The presence of oxygen vacancies in the TiO_2 promotes the charge carrier separation and initiates the formation of superoxide and hydroxyl radicals to fasten the decomposition of phenolic compounds [39]. The oxygen vacancies also decreased the Schottky barrier height to enhance the charge transfer from TiO_2 to Ti_3C_2 [39]. In addition, oxygen vacancies promote the adsorption of dioxygen as well as the electron rich pollutants which remains as an added merit to enhance the reaction rates. The density of the oxygen vacancies was modulated by annealing the composite under nitrogen atmosphere at different times [39]. The oxygen atoms gain thermal energy and escapes from the lattice sites resulting in the formation of oxygen vacancies. According to the X-ray photoelectron spectroscopy (XPS) analysis, fraction of oxygen vacancies varied from 1.2 to 11.6% complementing with increase in the annealing time from 0.5 to 4.0 h at 450 °C and the optimized content of 9.9% (2 h) resulted in good performance. Alternatively, high concentration of oxygen vacancies assists the charge carrier recombination process. The degradation of bisphenol A (BPA) was inhibited when the reaction solution was bubbled with nitrogen, while purging with oxygen gas promoted the same. This aspect highlights the crucial role of dissolved oxygen in generating superoxide radicals. Additionally, self-doping of TiO_2 via Ti(III) incorporation can induce the creation of oxygen vacancies which is driven by charge compensation mechanism. Such doping process not only promotes the degradation of CH_3CHO but also induces a visible light absorption edge for the composite [47, 48]. The Ti(III) ion formation was achieved by H_2O_2 treatment of Ti_3C_2 rather than using harsh conditions like annealing under hydrogen atmosphere and plasma treatment [48]. The presence of Ti(III) ions induces continuous vacancy band of electronic states within the forbidden states of TiO_2 and narrowed the band gap to an extent of ~2.0 eV [47]. The interfacial microstructure between the Ti_3C_2 and TiO_2 can be tailored by modulating the reaction conditions of hydrothermal treatment. The doping of Ti(III) ion was also achieved by treating {111}-TiO_2/Ti_3C_2 with hydrazine and subsequent hydrothermal treatment [47], wherein {111} facet was obtained using NH_4F. The content of fluoride ions regulated the morphology from rod-like to octahedron structure and proportion of {111} facet reached beyond 90%. It was also noted that the high surface area of the heterostructure with low proportion of {111} facets displayed lower activity. This viewpoint further outlines that the exposure of reactive facets dominate over the specific surface area to govern the activity [47]. The degradation pathways under visible light primarily involved hydroxyl radicals, followed by holes and superoxide radicals. The active facet of rutile TiO_2 possessed a work function of ~4.2 eV [49] and -OH terminated has ultralow work function of ~1.8 eV [50]. The {111} faceted self-doped rutile TiO_2 over the Ti_3C_2 was active for the degradation of MB and MO dyes under visible light [43].

Similarly, work function of {001}-anatase TiO_2 facet was reported to be ~4.9 eV [46]. In any case, the CB position of TiO_2 lies above the Fermi level of Ti_3C_2 and terminated (F or O) Ti_3C_2 to channelize the smooth flow of excited electrons [40]. Chen et al. reported the formation of {001} facet through the external addition of $Ti(OBu)_4$ in the HF medium over the Ti_3C_2 surface [40]. The *ex situ* formation of TiO_2 occurs not only on the surface but also within the layers of Ti_3C_2, leading to an enhanced proportion of Ti_3C_2/TiO_2 junctions. The photocatalytic degradation of Rhodamine B (RhB) dye sustained smoothly for lower fraction of Ti_3C_2 and declined at its higher content. The enhanced generation of superoxide and hydroxyl radicals were evidenced by Electron Spin Resonance (ESR) technique, while the photoluminescence (PL) spectral analysis confirmed the effective charge carrier separation process. Thus, modification of TiO_2 could be beneficial in modulating the Schottky barrier height, optical response, and photocatalytic properties. It should be noted that the work function of TiO_2 varies with respect to the exposed facets and therefore, the direction of charge carriers flow may also be altered accordingly. The future research must focus on exploring the growth kinetics and photocatalytic properties of TiO_2(B), brookite and mixed phases of titania over the Ti_3C_2 surface.

11.3.2 Heterostructuring the $Ti_3C_2/Ti_3C_2T_x$ with Metal Sulphides

The metal sulfides are well known cocatalysts for many redox reactions due to their narrow gap and high negative potential values for the CB edge. As an illustration, In_2S_3 nanoflakes/Ti_3C_2 initiated the reduction of Cr(VI) to Cr(III) ions under the visible light [51]. The adsorption of In(III) ions prevented the self-stacking of Ti_3C_2 sheets and that two-dimensional (2D)/2D heterostructure shortens the interface distance to alleviate the charge carrier flow from In_2S_3 to Ti_3C_2. It was interesting to note that all the Cr(III) ions were adsorbed on the surface of the heterojunction [51]. As the VB potential of In_2S_3 is not high enough to reoxidize the adsorbed Cr(III) ions, this hierarchical heterojunction would be promising for the effective reduction pathways of diverse metal ions. An optimized content of $CaIn_2S_4/Ti_3C_2T_x$ were valuable for the Cr(VI) ion reduction and tetracycline degradation [5]. In this study, $CaIn_2S_4$ hierarchical microspheres self-assembled by numerous interweaving sheets maximized the light harvesting capacity because of multiple scattering effects. The elevation in the band gap absorption towards longer wavelength region was evidenced and was proportional to the content of $CaIn_2S_4$. Similarly, decoration of 1D-CdS functionalized with APTES over Ti_3C_2 accelerated the reduction of U(VI) ions without any sacrificial agents and efficiency was retained over a broader pH range of 4–10 [52]. As the CB position of CdS is more negative than the reduction potentials of UO_2^{2+}/U^{4+} ions (0.267 V) and UO_2^{2+}/UO_2 (0.411 V), the reduction pathways of U(VI) species were driven by thermodynamic feasibility [51, 52]. Moreover, the interaction of Cd(II) ions with Ti_3C_2 inhibited the photo-induced corrosion of CdS. The functionalization of CdS with APTES resulted in the surface positive charge density to trigger the spontaneous self-assembly process with negatively charged Ti_3C_2. This might enable high degree of intimate contact among the integrated components which not only improve the charge transfer process but also promote the structural stability. Thus, increasing the content of Ti_3C_2 promotes the negative charge density to favour the adsorption of U(VI) species. In addition, the leaching of Cd(II) ions from the composite declined from 27 to 2 ppm with increase in solution pH from mild acidic to the alkaline medium [52]. The removal of U(VI) ions was still maintained at appreciable levels even with the presence of many co-existing anions, indicating the anti-interference ability of the 1D/2D composite systems. However, care must be taken to avoid the interference from molecular oxygen which competes for the trapping of excited electrons with the reduction pathways of these metal ions [52]. The stable interface achieved between alkalized Ti_3C_2 and CdS was active for the RhB degradation which was mediated by both holes and superoxide radicals. [53]. The –OH functionality associated with Ti_3C_2 was induced by dispersing them in the alkaline medium (NaOH) and CdS were heterogeneously grown under the hydrothermal conditions [53]. The transport and the diffusion of ions and the charge carriers were smoothened by the enlargement of interlayer spacing upon the alkali treatment [53]. However, Ti_3C_2 treated with high concentration of NaOH destroyed the surface-active centres and deteriorates their activity [53]. The participation of superoxide radicals, holes and hydroxyl radicals were evidenced in the degradation of rantidine using Ti_3C_2/MoS_2 and the composite was stable for efficient reuse [54]. The work function of MoS_2 and OH-terminated Ti_3C_2 are ~4.4 and ~1.8 eV respectively, which facilitates the holes transfer from the VB of MoS_2 to the Ti_3C_2. The electron triggers the reduction of dioxygen to generate superoxide radicals and reacts with protons forming H_2O_2. The as formed H_2O_2 is susceptible for the decomposition to produce hydroxyl radicals [54]. The major setback with metal sulphide stems from the low VB potential which lacks capability to generate hydroxyl radicals. Also, oxidation of sulphide ion to form sulphur precipitate leads to structural instability under the light illuminated conditions. Further research should prioritize in enhancing the photostability of metal sulfides in this kind of heterostructures.

11.3.3 Heterostructuring the $Ti_3C_2/Ti_3C_2T_x$ with Ag/Bi-based Semiconductors and Layered Double Hydroxides

In recent times, Ag and Bi-based semiconductors are explored as photocatalytic material for the wastewater treatment due to their visible light absorption sensitivity and easy preparation combined with their feasibility to form hierarchical nanostructures. The larger electronegativity of the terminal groups associated with $Ti_3C_2T_x$ provides a suitable platform for the adsorption of Ag(I) ions via electrostatically driven self-assembly process to favour the heterogeneous nucleation and growth of Ag-based semiconductors without any severe aggregation [55, 56]. The $Ti_3C_2T_x$ served as a template for the growth of Ag_3PO_4/AgBr and improved its quantum efficiency towards the Cr(VI) reduction and dye degradation [55]. Furthermore, $Ti_3C_2T_x$ increased the carrier concentration in the composite and extended the carrier lifetime [55]. It was proposed that the Z-scheme junction was established between AgBr and Ag_3PO_4 and the holes generated from these materials gets accumulated over the $Ti_3C_2T_x$ to mediate the degradation reaction. It should be noted that metallic silver can also be deposited over Ag_3PO_4/Ti_3C_2 surface via photo-deposition method to promote the photocatalytic efficiency. The Ag deposits shifts the optical response to the visible region due to surface plasmon resonance (SPR) effects and can capture excited electrons, hindering the reunification of charge carriers. [57].

Furthermore, the presence of lower valent titanium with strong reductive ability within the Ti_3C_2 promotes the formation of metallic silver [58]. The XPS results revealed that this lower valence titanium tends to disappear after compounding with Ag_3PO_4 suggesting an intense redox reaction [32]. The Ag_3PO_4 can form Schottky contact with Ti_3C_2 to steer the unidirectional electron flow from Ag_3PO_4 to Ti_3C_2. This process not only enhances the effective charge transfer pathways but also impede the photocorrosion of Ag-based semiconductors [32, 56]. In fact, performance of Ag_3PO_4/Ti_3C_2 towards the MO degradation is higher than Ag_3PO_4/RGO counterpart, which hints the potential coupling of Ti_3C_2 with photosensitized materials to improve their catalytic activity and stability. Similarly, Ag_2WO_4/Ti_3C_2 and AgI/Ti_3C_2 was active for the degradation of organic compounds [32, 56]. The VB holes and the superoxide radicals plays dominant role in the degradation mechanism using Ti_3C_2/Ag-based semiconductors, while the hydroxyl radicals generated from the multi-electron reduction pathways driven by lower valence titanium participates to a lesser extent [32]. Recently, Zhong et al. reported hollow sphere architecture constituting Ti_3C_2/Ag_2ZnGeO_4 Schottky type heterojunction that was effective for the mineralization of Acid Red-1 [59]. The preparation of Ag_2ZnGeO_4 hollow spheres involved the heating of respective precursors with ammonia under the moderate conditions (60 °C, 3 h). The self-assembly process of the formed nanoparticles resulted in the construction of hollow spheres. The DFT calculations revealed that the charge redistribution across the interface driven from the electron depletion and accumulation at Ag_2ZnGeO_4 and Ti_3C_2 renders them with positive and negative values respectively [59]. The incomplete product formation was observed at higher content of Ti_3C_2 suggesting the optimized fraction of individual component is essential to facilitate the epitaxial growth of semiconductors over the MXene. The leaching of silver ions and structure-morphological destabilization of the Ag-based semiconductors are prevalent due to the photocorrosion, making this composite unsuitable for the long-term operations [59]. The partial decomposition sometimes results in the formation of metallic silver on the composite surface, which may not always benefit the photocatalytic reactions.

The narrow gap and ease of synthesis with diverse morphologies, non-toxicity and good chemical stability associated with Bi-based semiconductors such as Bi_2WO_6 and BiOX (X=Cl, Br, I) have spotlighted them for coupling with MXenes towards the photocatalytic applications. The BiOX has a unique layered structure consisting of $[Bi_2O_2]^+$ and halogen atom slabs which creates in-built electric field to drive the separation of charge carriers and augments the photoconductivity [60–62]. The degradation of gaseous CH_3COCH_3 and HCHO was mediated with Ti_3C_2/Bi_2WO_6, wherein the composite was obtained by the adsorption of Ti_3C_2 onto the Bi_2WO_6 nanoplates [63]. The simulation results revealed that the electrons from CH_3COCH_3 and HCHO are extensively delocalized over the Ti_3C_2 to favour their chemical adsorption, while the Bi_2WO_6 showed weak physical adsorption for these organic compounds. Furthermore, ESR spectroscopy confirmed the pivotal role of superoxide and hydroxyl radicals in the degradation pathways. It was proposed that the electrons from the CB of Bi_2WO_6 transferred to the Ti_3C_2 triggered the effective charge carrier separation pathways [63]. In another study, Bi_2WO_6 NSs was in-situ grown on the Ti_3C_2 through hydrothermal method could detoxify the organic contaminants under the visible light [64]. The enhanced specific surface area of the composite provides numerous reaction sites and facilitates the adsorption-desorption dynamics of pollutants alongside intensifying the photon harvesting capacity. The utmost degradation of RhB was encountered in the pH range of 5–8 owing to the strong electrostatic force of attraction between the negatively charged composite surface and the cationic dye. In contrast, deprotonation of nitrogen-containing group and the dissociation of carboxylic group initiate the electrostatic repulsion with the catalyst surface thus disfavouring the adsorption and photocatalytic processes at higher pH values [64]. Similarly, tetracycline degradation rate was maximized at mild acidic conditions (pH 5.5), while the extreme acidic and alkaline medium was detrimental with $Bi_2O_2CO_3/Ti_3C_2$ [65]. The degradation pathways were influenced by the presence of dissolved oxygen, along with the involvement of superoxide radicals and holes, which comprised a series of reactions like ring opening, demethylation, deamination, dehydration, and so on, eventually leading to the formation of smaller non-toxic molecules (Figure 11.5).

The dimensional anisotropy of the semiconductors associated with the Ti_3C_2 is very important in promoting the required performance. As an illustration, Zhang et al. reported that the integration of 3D porous BiOI with Ti_3C_2 through solvothermal method was beneficial towards the dye degradation [66]. The NSs-like morphology of pure BiOI changed to nanoparticle form during the heterogeneous nucleation process, which outlines that the Ti_3C_2 can also influence morphological evolution of the semiconductors [66]. The nanoparticles were distributed over the lamellar layers to form uniform mesoporous coating with a pore size distribution of 2.5–20 nm, which resulted in the larger surface area of the composite and the porous nature facilitated the seamless movement of charge carriers. The theoretical calculation of frontier electron densities revealed that the reactive iodine species from BiOI played a unique role in the photocatalytic reactions [33]. The formation of metallic Bi and controlled oxygen vacancies during the growth of BiOI was achieved through the introduction of different content of Ti_3C_2 [33]. The presence of SPR active metallic Bi further extends the optical sensitivity towards the longer wavelength region [6, 33]. Although VB of BiOI is not conducive for the hydroxyl radical generation, the production of superoxide radicals and holes in the Ti_3C_2/BiOI facilitates the degradation reactions. The antipyrine degraded much faster than the BPA which emerged from the 'interlayer anion exchange' equilibrium effect of BiOI with the pollutant molecules. Furthermore, pyrazole ring in the antipyrine was more susceptible for the attack of superoxide radicals, thus leading to its faster decomposition. The formation of in-situ Bi alongside the BiOX was found to be feasible in the ethylene glycol medium under the hydro/solvothermal conditions [6, 33]. Similarly, Bi@BiOCl/Ti_3C_2 was also active for the mineralization of ciprofloxacin under visible light [6]. It was proposed that the SPR effect of Bi metal triggers its excitation and then initiates the unidirectional electron transfer from the Bi to BiOCl to Ti_3C_2. Due to the wide gap nature of BiOCl, they only serve as carrier sink without altering the optical properties of the composite. The increased photocurrent generation, reduced intensity in PL spectra and smaller arc radius from electrochemical impedance spectroscopic data collectively suggests that the charge carriers are effectively generated and possess longer carrier lifetime alongside the immediate transfer to the adsorbed species. The absence of hydroxyl radicals in the degradation mechanism can also be advantageous for the synthesis of organic compounds under the mild reaction conditions which should be the focus of the future study.

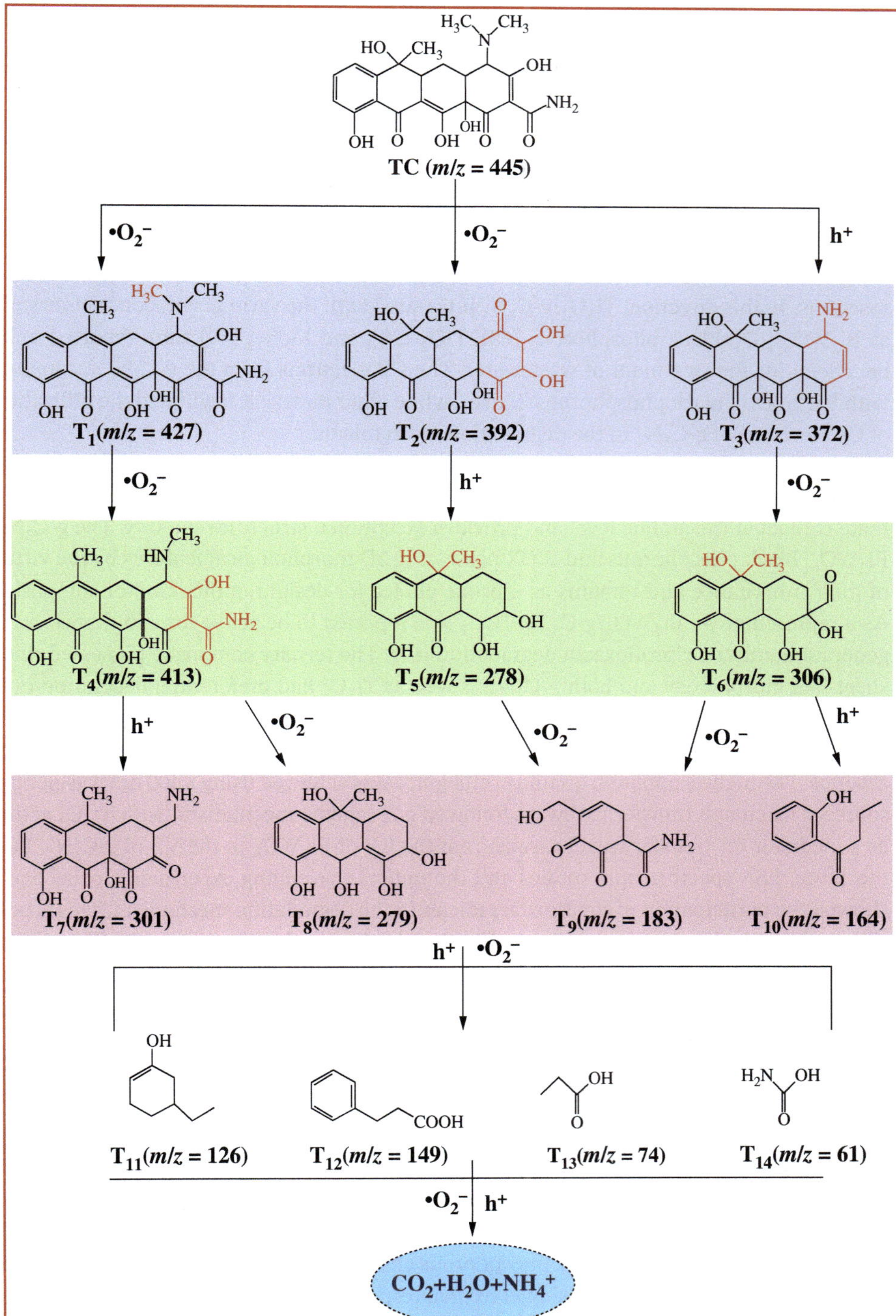

Figure 11.5 Degradation mechanisms of tetracycline using $Bi_2O_2CO_3/Ti_3C_2T_x$. Source: Reproduced with the permission from Tan et al. [65]. Copyright 2021, Elsevier.

11.4 Ternary Heterostructures based on the Ti_3C_2

The coupling of Ti_3C_2 with two different semiconductors possessing distinct structure-optical-morphological properties can merit the utilization of complete fraction of solar light and the corresponding photocatalytic reactions. In this context, 2D g-C_3N_4 is widely explored owing to their narrow gap and high CB edge potential that can easily generate superoxide radicals in the solution. Thus, integrating the g-C_3N_4 and Ti_3C_2 with another material to form Z-scheme heterojunctions can complement the effectual electron-hole separation and utilization of energetic charge carriers to catalyse the desired redox reactions. In this direction, Ti_3C_2/g-C_3N_4 integrated with the various semiconductors such as Bi_2WO_6 [67], black phosphorous [68], TiO_2 [38], and $MoSe_2$ [69] were reported to be beneficial for the treatment of wastewater. The CB electrons from the g-C_3N_4 recombine with VB holes of black phosphorous/$MoSe_2$, while other materials facilitated the utilization of CB electrons of g-C_3N_4 in the degradation mechanisms.

The advanced configuration in the ternary structures was realized by synthesizing the ternary composites possessing 2D structure. Such arrangement allows high degree of intimate contact at the atomic level and provides exceptional structural stability. The g-C_3N_4, Bi_2WO_6, black phosphorous and BiOX possessing 2D morphological features by the virtue of their inheritance and remains as a prime choice for designing this kind of composite. As a representative, Bi_2WO_6/g-C_3N_4/Ti_3C_2 was reported to be active for simultaneous H_2 generation and the ciprofloxacin degradation [67]. The ternary composite possessed thick sheet-like morphology and both g-C_3N_4 as well as Ti_3C_2 had preferential decoration over the {131} crystal plane of Bi_2WO_6. In the presence of the probe pollutant, the average H_2 evolution reached approximately ~37.8 μmol/h/g, which increased to ~55.4 μmol/h/g in its absence. The highest apparent quantum efficiency was achieved using a 400 nm wavelength source. The charge transfer pathways followed a Z-scheme mechanism, with Ti_3C_2 acting as a mediator for transferring electrons from the CB of Bi_2WO_6 to the VB of g-C_3N_4. Furthermore, ESR spectroscopic studies and the radical scavenging experiments collectively showed the participation of superoxide radicals in the degradation mechanism. This is possible only with the electrons generated in the g-C_3N_4, as the CB edge of Bi_2WO_6 is less negative for the same. The presence of $CO_3{}^{2-}$, $SO_4{}^{2-}$ and Ca(II) ions displayed detrimental effects on the degradation rate, while the Mg(II) ions did not influence the same. The Ca(II) ions form a stable metal complex with ciprofloxacin and shows some resistance to react with the free radicals. However, the carbonate and sulphate anion quench the hydroxyl radicals which are essential for the degradation reactions [67]. Similarly, black phosphorous/g-C_3N_4/Ti_3C_2 complimented with excellent electrical conductivity and unique electronic band structure resulted in the faster ciprofloxacin degradation under the visible light. The participation of singlet oxygen alongside the superoxide radicals suggests that the different component in the ternary system can distinctly alter the charge carrier pathways [68]. The electrons resided on the black phosphorous and holes on the g-C_3N_4 enabled the generation of superoxide radicals and singlet oxygen respectively. The composites were obtained by the simple mixing and annealing steps under the influence of nitrogen atmosphere. The existence of P—N/P=N bonds (P-bridging effect) between the black phosphorous and g-C_3N_4 assisted the good intimate contact between them and facilitated the generation of free radicals and ciprofloxacin degradation [68]. The BiOBr/faceted TiO_2/Ti_3C_2 was effective for the degradation of RhB dye under visible light [70]. The reactive facets of titania had lower band gap and therefore electron-hole pairs were generated on both BiOBr and TiO_2. Thus, holes traversed from BiOBr to the TiO_2 and electrons were channelized on the Ti_3C_2 surface. However, holes dominated the degradation pathways despite the possibility of hydroxyl radical generation from both TiO_2 and BiOBr. This result suggests that the structural properties of the composite, conditions adopted for the degradation reactions and the susceptibility of the pollutant to react with the free radicals govern the fate of the degradation mechanism [70].

The coupling of wide gap and narrow gap semiconductors prompt to explore full scale solar spectrum for the photocatalytic reactions. In this direction, $Ti_3C_2/ZnO/Bi_2WO_6$ was fabricated using two-step electrostatic assembly route that spanned the significant UV-visible region and strenuous holes and electrons mediated the degradation reactions of ciprofloxacin [71]. The functionalization of ZnO sheets with APTES induces negative charge which triggers ionic attraction with positively charged Bi_2WO_6. The subsequent addition of varying amounts of Ti_3C_2 led to the formation of ternary structures. The degradation increased with increment in the amount of Ti_3C_2 and later declined at its higher content. The one-way transfer of electrons from ZnO to Bi_2WO_6 to Ti_3C_2 and reverse hole transfer from Bi_2WO_6 to ZnO facilitated the charge carrier separation process. The presence of Ti_3C_2 quantum dots (QDs) resulted in the Z-scheme charge carrier transfer pathways upon integrating with 2D-In_2S_3 and 1D-WO_3. The 2D-In_2S_3 can increase the specific surface area, while 1D-WO_3 can avoid the aggregation of sheets and maintain the structural integrity [72]. The 0D Ti_3C_2 can be easily dispersed at the interface of 2D/1D junctions, which can further mobilize the charge carriers to initiate the BPA degradation and Cr(VI) ion reduction. The latter reaction proceeded effectively at acidic pH and BPA degradation remained same over the pH range of 4–10. The carbonate ions quenched the reaction rate of BPA and the presence of chloride ions exhibited the minimal effects. On the other hand, presence of organic acids suppressed the reduction of Cr(VI) species owing to the decline in the adsorption of $Cr_2O_7^{2-}$ over the catalyst surface (Figure 11.6).

The decline in the reaction rates with $AgNO_3$ and oxygen free environment suggested that the CB electrons driven from In_2S_3 participates extensively in the degradation pathways, which was possible only with the formation of Z-scheme junctions. The reduction of Cr(VI) ion was not affected in the BPA solution, while the degradation kinetics of BPA itself was inhibited to a certain extent. As the larger portion of electrons was utilized for the Cr(VI) ions reduction, the production of superoxide radicals will be declined which consequently suppresses the degradation of BPA. In addition, reduced Cr(III) species were adsorbed over the surface of the composite, which exerts competition for the adsorption of BPA. It should be noted that both these semiconductors possess narrow gap and therefore, utmost utilization of visible light photons promotes the high concentration of free radicals [72]. The Ti_3C_2 QDs not only inherit the good conductivity of 2D counterparts, but also features other properties such as high dispersion, better tenability in the physicochemical properties and good light stability [73]. The WO_3 rods were grown initially and then In_2S_3 sheets covered the WO_3-rods, while Ti_3C_2 QDs were also uniformly distributed over the rod surface [72].

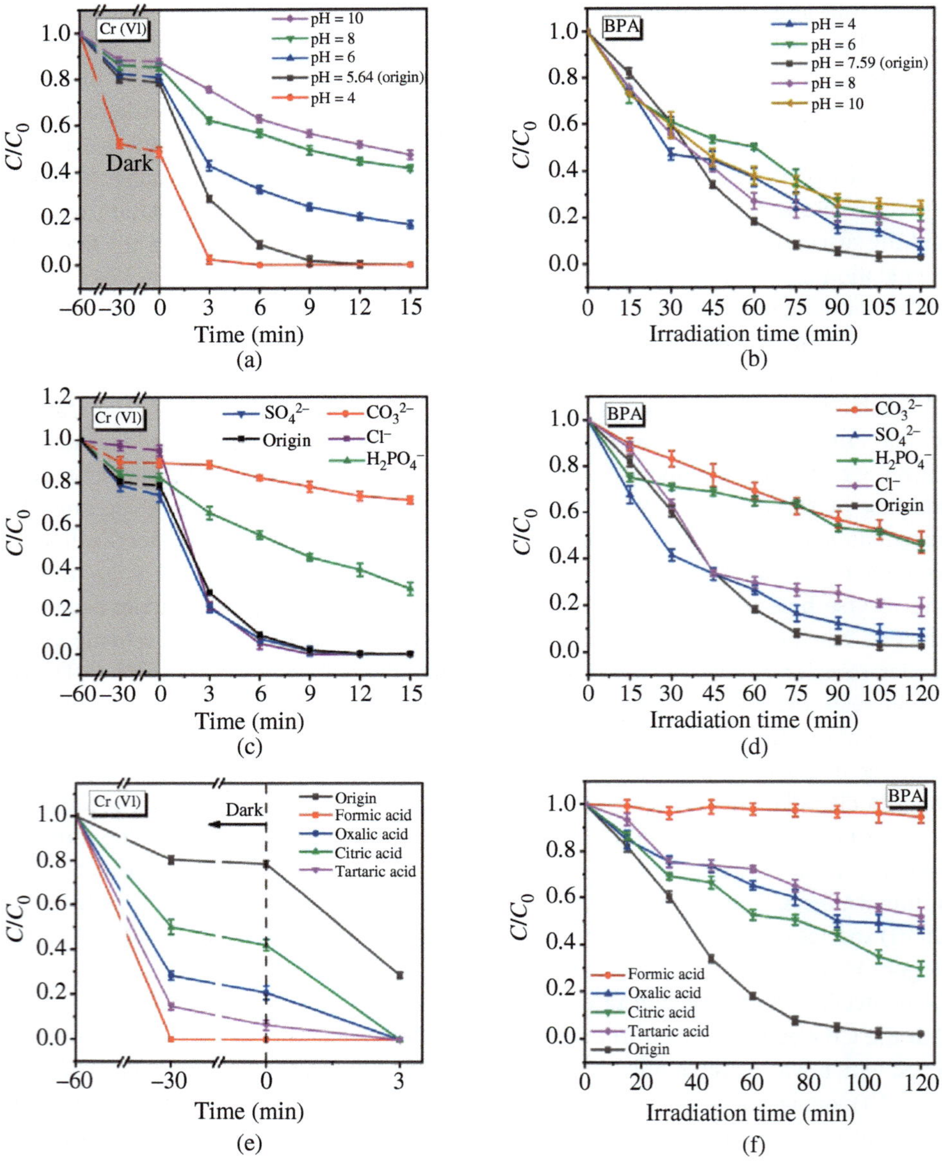

Figure 11.6 Influence of (a, b) solution pH, (c, d) inorganic electrolytes and (e, f) organic acids on the Cr(VI) reduction and degradation of BPA using WO_3/Ti_3C_2 quantum dots/In_2S_3 (1D/0D/2D) ternary system. Source: Reproduced with the permission from Yuan et al. [72]. Copyright 2021, Elsevier.

The intimate contact and the staggered configuration between pyromellitic diimide (PDI)@g-C_3N_4 and TiO_2@Ti_3C_2 in the quaternary composite boosted the charge carrier separation via S-scheme charge transfer and promoted the atrazine degradation with peroxy monosulfate as oxidising agent. The singlet oxygen and H_2O_2 generated was helpful to mediate the degradation of the pollutants [38]. This composite was obtained by calcining the mixture of g-C_3N_4, pyromellitic dianhydride and Ti_3C_2 at the moderate temperature (~325 °C) in static air. During the annealing step, polycondensation of pyromellitic dianhydride results in pyromellitic diimide which was succeeded by the dehydration reaction with g-C_3N_4 alongside the subsequent transformation of Ti_3C_2 to TiO_2. The integration of PDI shifted the VB and CB positions of g-C_3N_4 slightly upwards leading to the band gap reduction, which was attributed to the extensive π–π interactions between them. The exceptional photocatalytic performance with different water matrices and various inorganic electrolytes (chlorides, sulphates and nitrates) in a broader pH range (~3.03 to ~9.09) is an indicative of improved structural stability and boosted charge carrier transfer pathways [38]. The generated superoxide radical undergoes disproportion reaction to form H_2O_2, which was later confirmed by the colorimetric method. Therefore, singlet oxygen governed the degradation mechanism followed by holes, hydroxyl radicals and sulfate radicals [38]. The other fluoroquinolones-based pollutants such as moxifloxacin, ofloxacin, levofloxacin, norfloxacin, ciprofloxacin and gatifloxacin were also degraded using g-C_3N_4/Ti_3C_2/$MoSe_2$ [69]. The degradation rate displayed the sequence: levofloxacin > norfloxacin > ciprofloxacin > gatifloxacin. The degradation mechanism was primarily driven by superoxide radicals, clearly indicating the formation of Z-scheme heterojunction between the integrated components.

To expand the scope of the ternary composite, the widely explored materials such as ZnO, WO_3, CdS, In_2S_3, reduced graphene oxide (RGO) and ferrites are coupled with Ti_3C_2/TiO_2 system [73–78]. The combination of semiconductors having distinct band gap absorption spanning the entire UV-visible region would be an ideal strategy to form advanced ternary composites. The growth of other semiconductors over the Ti_3C_2/TiO_2 is convenient and the problems associated with the formation of other impurities and intimate contact may be avoided. In this direction, CdS [73], In_2S_3 [7], RGO [75] and BiOBr [70] are reported to be convenient to activate under the broader spectrum of solar light. The photocatalytic degradation of sulfachloropyridazine, phenol, RhB and MB were achieved using the CdS@Ti_3C_2@TiO_2 under visible light. It was proposed that the negative charge density at the surface and edges of Ti_3C_2 facilitates the adsorption of Cd(II) ions and promotes the growth of CdS. The Cd(II):S^{2-} ratio influenced the performance of the composite and followed the order: 2:1 > 3:1 > 4:1 > 1:1. The activation energy for the degradation of sulfachloropyridazine using this composite was found to be ~25 kJ/mol. The CB electrons in TiO_2 recombine with VB holes of CdS via Ti_3C_2 as electron mediator which avoids the oxidation of CdS [74]. The simultaneous formation of In_2S_3 and TiO_2 was evidenced under the hydrothermal conditions using Ti_3C_2, $In(NO_3)_3.xH_2O$ and CH_3CSNH_2 [7]. Such instantaneous nucleation avoids the tedious steps involved in the fabrication process and enables the formation of quasi-core@shell structures with well-defined double heterojunction structure (type-II and Schottky). The optical response spanned the range of 578–800 nm with increment in the Ti_3C_2 dosage in the preparation method, which indicates the favourable electronic interactions of the heterostructure. This composite promoted the dye degradation, and the dye sensitization mechanism was ruled out due to the poor adsorption of the dye [7]. The XPS data revealed that '*p*' orbital of oxygen atom interacts with the '*d*' orbitals of indium to form the In-O bond. Such chemical coupling increased the electron density on the oxygen atoms which is accompanied with decreased electron concentration on the indium atoms. Thus, interactions between the TiO_2 and In_2S_3 gets promoted and leads to good charge carrier transfer pathways [7]. Since the VB holes produced in In_2S_3 have a lower ability to produce hydroxyl radicals, they are not involved in the degradation mechanism. The successive electron transfer from In_2S_3 to TiO_2 to Ti_3C_2 attributes for the enhanced charge carrier separation and the surplus generation of superoxide radicals. The $BiOCl/Ti_3C_2/TiO_2$ promoted the activity towards RhB degradation in the pH range of 2–6 under the simulated solar light [76]. The Ti_3C_2 captures the electrons from both TiO_2 and BiOCl, leading to the generation of superoxide radicals. Concurrently, holes traverse from BiOCl to TiO_2, effectively preventing the charge carrier recombination pathways [76].

The hybridization of Ti_3C_2@TiO_2-nanowire with RGO was helpful to reduce the Cr(VI) ions from the wastewater, wherein RGO trap the electrons from Ti_3C_2 owing to their good conductivity [75]. The use of RGO is advantageous due to their chemical inertness, adjustable layered features, and 2D sheet like structure which collectively promoted the reaction rates. The maximum reduction was observed at acidic pH as the negatively charged chromium species ($Cr_2O_7^{2-}$) can adsorb on the positively charged composite surface. In contrast, formation of negatively charged surface density over the catalyst surface and the precipitation of $Cr(OH)_3$ under neutral and alkaline pH respectively refrain the adsorption and subsequent reduction process [75]. Interestingly, the Cr(VI) ion reduction proceeded faster in the presence of RhB as the VB holes directly react with the dye molecules which encourages the participation of more electrons in the reduction pathways. The coupling of Ti_3C_2 with distinct metal sulfides and metal oxides [71, 76–78] also benefits the degradation of pollutants. As an illustration to the former, Ti_3C_2 was integrated with In_2S_3 and CdS via hydrothermal method. Initially, Ti_3C_2 was alkalized with hydroxyl groups to facilitate the adsorption of metal cations and immediate sulfidation process elicited the formation of metal sulfides. An appreciable specific surface area (~68 m^2/g) was evidenced which encouraged the adsorption and degradation of the dyes containing a mixture of RhB and MO. It was interesting to note that the RhB degraded much faster than MO, while the trend was reversed with GO cocatalysts instead of Ti_3C_2. This finding underscores that the distinct materials present in the composite modify the degradation kinetics under certain reaction conditions. The Ti_3C_2 captured the electrons from both CdS and In_2S_3 leading to the production of superoxide radicals, while the holes accumulated at the VB edges of the sulphide semiconductors drives the direct oxidation of the pollutants [77]. The integration of the magnetic composite α-$Fe_2O_3/ZnFe_2O_4$ with Ti_3C_2 enabled the easy separation and promoted the structural stability during the recycling steps for RhB degradation and Cr(VI) ions reduction [78]. This composite was obtained by ultrasonicating the mixture of Ti_3C_2 and α-$Fe_2O_3/ZnFe_2O_4$ in the presence of nafion solution as the stabilizing agent. The longer wavelength response (~400–700 nm) was an added advantage for this composite [78].

The ternary composites encompass the benefits of distinct structure-optical properties of each semiconductor towards improving the photocatalytic reaction rates. However, it would be meaningful if one component contributes for light absorption properties and the other facilitates the adsorption of pollutants to make the ternary structure more relevant for the wastewater treatment. The preparation steps are always tedious and multiple heat treatment involved results in the aggregation of nanoparticles.

11.5 Gap Analysis

The major setback is the etching process involving the corrosive HF to obtain Ti_3C_2 from Ti_3AlC_2. Although many alternative fluoride precursors such as KF or $NaBF_4$ are also employed at lower concentration, it is essential to develop the etching processes that are free from fluoride ions. On the other hand, ionic liquids containing fluoride ions as well as organo fluorine containing compounds (for instance CF_3COOH) should be used in the etching process to realize their effects in altering the structural properties of Ti_3C_2. The direct correlation between the oxygen vacancies and Schottky barrier height is yet to be understood from the perspective of band bending. The extent to which lower valent titanium can influence the performance is not clear and there is no thumb rule for their proportion to maximize the activity. The surface terminating groups may alter the Fermi-levels and the work function of MXenes, whose exact impact in the heterojunction formation needs to be recognised [54, 79]. The formation of TiO_2 from Ti_3C_2 during the annealing step is inevitable and preparation methods should be tailored to avoid this phase transition pathways.

The degradation pathways clearly reveal the least influence of hydroxyl radicals in some of the composite structure and, therefore its efficiency leading to complete mineralization of the contaminants remain questionable. Thus, development of Ti_3C_2 based composites utilizing the hydroxyl radicals in the degradation should be given a major attention. As the superoxide radicals are less effective and mobility of holes can be varied, the mineralization of the pollutants may not be always favourable. Also, the distribution of intermediates during the degradation reaction mediated by the superoxide radicals is not clear and no idea if it differs from those driven by hydroxyl radicals. Such analysis is essential to realize the toxicity of by-products formed under these conditions. The application of Ti_3C_2 based composite in real wastewater containing electrolytes, phenolic compounds, pharmaceuticals and so on needs to be considered to attest their merit in the field of photocatalysis. The intimate contact between the integrated components is crucial, and even the slightest change in the preparation technique might change the characteristics of junction interface.

11.6 Conclusion

The development of composites is pursued with the aim of effective utilization of solar energy complimented with extended carrier lifetime and the structural stability. In this context, the construction of heterojunctions based on Ti_3C_2 offers significant opportunity for boosting the catalytic performance in the desired redox reactions (Figure 11.7). The diverse applications of Ti_3C_2 composites in the wastewater treatment strategy are predicted by the preparation techniques, distinctive electron-hole transfer pathways, and the participation of free radicals. We positively hope that our book chapter furnishes adequate information to bestow the possibility of developing Ti_3C_2 based composite photocatalysts for widespread environmental applications.

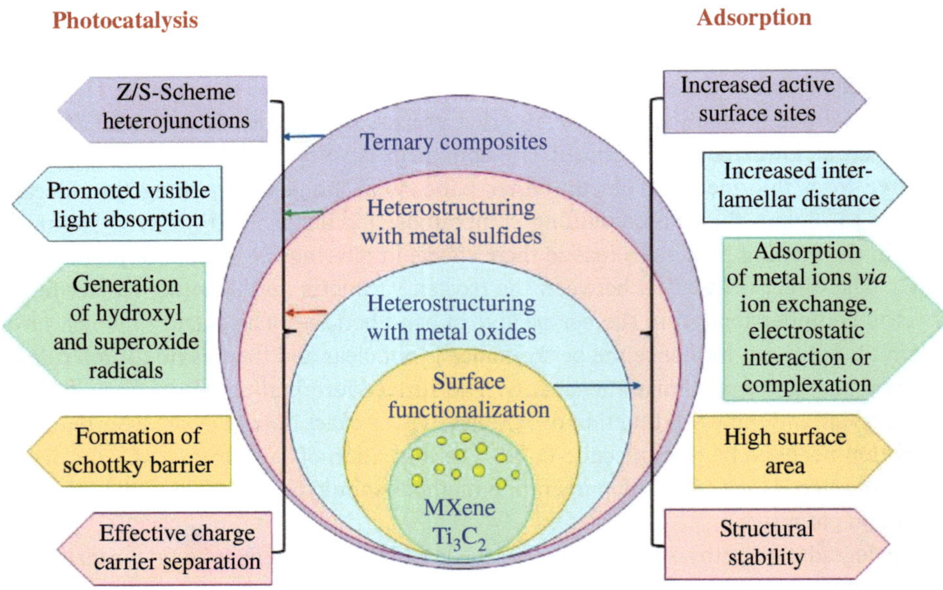

Figure 11.7 Ti_3C_2 MXene based composites for adsorption process and photocatalytic applications.

Acknowledgements

The author G. K. acknowledges the support from Prof. K.S.R. Koteswara Rao, Department of Physics, Indian Institute of Science, Bengaluru, for his valuable guidelines.

References

1 Kumar, S.G. and Rao, K.S.R.K. (2017). Comparison of modification strategies towards enhanced charge carrier separation and photocatalytic degradation activity of metal oxide semiconductors (TiO_2, WO_3 and ZnO). *Appl. Surf. Sci.* 391: 124.

2 Naquib, M., Kurtoglu, M., Presser, V. et al. (2011). Two-Dimensional Nanocrystals Produced by Exfoliation of Ti_3AlC_2. *Adv. Mater.* 23: 4248.

3 Yaqub, A., Shafiq, Q., Khan, A.R. et al. (2021). Recent advances in the adsorptive remediation of wastewater using two-dimensional transition metal carbides (MXenes): a review. *New J. Chem.* 45: 9721.

4 Hieu, V.Q., Phung, T.K., Nguyen, T.Q. et al. (2021). Photocatalytic degradation of methyl orange dye by Ti_3C_2-TiO_2 heterojunction under solar light. *Chemosphere* 276: 130154.

5 Zhuge, Z., Liu, X., Chen, T. et al. (2021). Highly efficient photocatalytic degradation of different hazardous contaminants by $CaIn_2S_4$-$Ti_3C_2T_x$ Schottky heterojunction: An experimental and mechanism study. *Chem. Eng. J.* 421: 127838.

6 Wu, S., Su, Y., Zhu, Y. et al. (2020). In-situ growing Bi/BiOCl microspheres on Ti_3C_2 nanosheets for upgrading visible-light-driven photocatalytic activity. *Appl. Surf. Sci.* 520: 146339.

7 Wang, H., Wu, Y., Xiao, T. et al. (2018). Formation of quasi-core-shell In_2S_3/anatase TiO_2@metallic $Ti_3C_2T_x$ hybrids with favorable charge transfer channels for excellent visible-lightphotocatalytic performance. *Appl. Catal. B* 233: 213.

8 Pandey, R.P., Rasool, K., Rasheed, P.A., and Mahmoud, K.A. (2018). Reductive sequestration of toxic bromate from drinking water using lamellar 2D $Ti_3C_2T_X$ (MXene). *ACS Sustain. Chem. Eng.* 6: 7910.

9 Wang, Q., Snyder, K.J., and Choi, H. (2009). Aqueous ethanol modified nanoscale zerovalent iron in bromate reduction: synthesis, characterization, and reactivity. *Environ. Sci. Technol.* 43: 3292.

10 Rethinasabapathy, M., Hwang, S.K., Kang, S. et al. (2021). Amino-functionalized POSS nanocage-intercalated titanium carbide ($Ti_3C_2T_X$) MXene stacks for efficient cesium and strontium radionuclide sequestration. *J. Hazard. Mater.* 418: 126315.

11 Fard, A.K., Mckay, G., Chamoun, R. et al. (2017). Barium removal from synthetic natural and produced water using Mxene as two dimensional (2-D) nanosheet adsorbent. *Chem. Eng. J.* 317: 331.

12 Jastrze bska, A.M., Karcz, J., Letmanowski, R. et al. (2016). Synthesis of RGO/TiO_2 nanocomposite flakes and characterization of their unique electrostatic properties using zeta potential measurements. *J. Alloys Compounds* 679: 470.

13 Torab-Mostaedi, M., Ghaemi, A., Ghassabzadeh, H., and Ghannadi-Maragheh, M. (2011). Removal of strontium and barium from aqueous solutions by adsorption onto expanded perlite. *Can. J. Chem. Eng.* 89: 1247.

14 Mu, W., Du, S., Yu, Q. et al. (2018). Improving barium ions adsorption on two-dimensional titanium carbide by surface modification. *Dalton Trans.* 47: 8375.

15 Liu, Y., Xu, J., Cao, Z. et al. (2020). Adsorption behaviour and mechanism of Pb(II) and complex Cu(II) species by biowaste-derived char with amino functionalization. *J. Colloid Interface Sci.* 559: 215.

16 Zhang, G., Wang, T., Xu, Z. et al. (2020). Synthesis of amino-functionalized $Ti_3C_2T_x$ MXene by alkalization-grafting modification for efficient lead adsorption. *Chem. Commun.* 56: 11283.

17 Peng, W., Chang, L., Li, P. et al. (2019). An overview on the surfactants used in ion flotation. *J. Mol. Liq.* 286: 110955.

18 Wang, S., Liu, Y., Lu, Q.F., and Zhuang, H. (2019). Facile preparation of biosurfactant-functionalized Ti_2CT_X MXene nanosheets with an enhanced adsorption performance for Pb(II) ions. *J. Mol. Liq.* 297: 111810.

19 Gan, D., Huang, Q., Dou, J. et al. (2020). Bioinspired functionalization of MXenes ($Ti_3C_2T_X$) with amino acids for efficient removal of heavy metal ions. *Appl. Surf. Sci.* 504: 144603.

20 Shahzad, A., Rasool, K., Miran, W. et al. (2017). Two-dimensional $Ti_3C_2T_X$ MXene nanosheets for efficient copper removal from water. *ACS Sustain. Chem. Eng.* 5: 11481.

21 Mu, W., Du, S., Li, X. et al. (2019). Removal of radioactive palladium based on novel 2D titanium carbides. *Chem. Eng. J.* 358: 283.

22 Fu, K., Liu, X., Yu, D. et al. (2020). Highly efficient and selective Hg(II) removal from water using multilayered $Ti_3C_2O_x$ MXene *via* adsorption coupled with catalytic reduction mechanism. *Environ. Sci. Technol.* 54: 16212.

23 Berkovic, A.M., Bertolotti, S.G., Villata, L.S. et al. (2012). Photoinduced reduction of divalent mercury by quinones in the presence of formic acid under anaerobic conditions. *Chemosphere* 89: 1189.

24 Shahzad, A., Nawaz, M., Moztahida, M. et al. (2019). $Ti_3C_2T_x$ MXene core-shell spheres for ultrahigh removal of mercuric ions. *Chem. Eng. J.* 368: 400.

25 Li, S., Wang, L., Peng, J. et al. (2019). Efficient thorium(iv) removal by two-dimensional Ti_2CT_x MXene from aqueous solution. *Chem. Eng. J.* 366: 192.

26 Zhang, Y., Lan, J., Wang, L. et al. (2016). Adsorption of uranyl species on hydroxylated titanium carbide nanosheet: A first-principles study. *J. Hazard. Mater.* 308: 402.

27 Zhang, P., Wang, L., Du, K. et al. (2020). Effective removal of U(VI) and Eu(III) by carboxyl functionalized MXene nanosheets. *J. Hazard. Mater.* 396: 122731.

28 Sun, Z., Li, Z., Gao, L. et al. (2019). Grafting benzenediazonium tetrafluoroborate onto $LiNi_xCo_yMn_zO_2$ materials achieves subzero-temperature high capacity lithium-ion storage *via* a diazonium soft-chemistry method. *Adv. Energy Mater.* 9: 1802946.

29 Wang, L., Tao, W., Yuan, L. et al. (2017). Rational control of the interlayer space inside two-dimensional titanium carbides for highly efficient uranium removal and imprisonment. *Chem. Commun.* 53: 12084.

30 Huan, S., Lin, W., Hong-Qing, W., and Wei-Qunn, S. (2020). Adsorption of Eu(III) on alkalized $Ti_3C_2T_x$ MXene studied by batch experiment and its mechanism investigation. *J. Inorg. Mater.* 35: 65.

31 Zhang, Y., Zhou, Z., Lan, J., and Zhang, P. (2019). Prediction of $Ti_3C_2O_2$ MXene as an effective capturer of formaldehyde. *Appl. Surf. Sci.* 469: 770.

32 Cai, T., Wang, L., Liu, Y. et al. (2018). Ag_3PO_4/Ti_3C_2 MXene interface materials as a Schottky catalyst with enhanced photocatalytic activities and anti-photocorrosion performance. *Appl. Catal. B* 239: 545.

33 Gao, P., Zhang, Z., Feng, L. et al. (2021). Novel Ti_3C_2/Bi@BiOI nanosheets with gradient oxygen vacancies for the enhancement of spatial charge separation and photocatalytic performance: The roles of reactive oxygen and iodine species. *Chem. Eng. J.* 426: 130764.

34 Zhang, X., Zhao, X., Wu, D. et al. (2015). High and anisotropic carrier mobility in experimentally possible Ti_2CO_2 (MXene) monolayer and nanoribbons. *Nanoscale* 7: 16020.

35 Peng, C., Yang, X., Li, Y. et al. (2016). Hybrids of two-dimensional Ti_3C_2 and TiO_2 exposing {001} facets toward enhanced photocatalytic activity. *ACS Appl. Mater. Interfaces* 8: 6051.

36 Kumar, S.G. and Devi, L.G. (2011). Review on modified TiO_2 photocatalysis under UV/visible light: selected results and related mechanisms on interfacial charge carrier transfer dynamics. *J. Phys. Chem. A* 115: 13211.

37 Low, J., Zhang, L., Tong, T. et al. (2018). TiO_2/MXene Ti_3C_2 composite with excellent photocatalytic CO_2 reduction activity. *J. Catal.* 361: 255.

38 Tang, R., Gong, D., Deng, Y. et al. (2022). π-π Stacked step-scheme PDI/g-C_3N_4/TiO_2@Ti_3C_2 photocatalyst with enhanced visible photocatalytic degradation towards atrazine via peroxymonosulfate activation. *Chem. Eng. J.* 427: 131809.

39 Miao, Z., Wang, G., Zhang, X., and Dong, X. (2020). Oxygen vacancies modified TiO_2/Ti_3C_2 derived from MXenes for enhanced photocatalytic degradation of organic pollutants: The crucial role of oxygen vacancy to schottky junction. *Appl. Surf. Sci.* 528: 146929.

40 Chen, L., Ye, X., Chen, S. et al. (2020). Ti_3C_2 MXene nanosheet/TiO_2 composites for efficient visible light photocatalytic activity. *Ceram. Int.* 46: 25895.

41 Gao, Y., Wang, L., Zhou, A. et al. (2015). Hydrothermal synthesis of TiO_2/Ti_3C_2 nanocomposites with enhanced photocatalytic activity. *Mater. Lett.* 150: 62.

42 Quyen, V.T., Ha, L.T.T., Thanh, D.M. et al. (2021). Advanced synthesis of MXene-derived nanoflower-shaped TiO_2@Ti_3C_2 heterojunction to enhance photocatalytic degradation of Rhodamine B. *Environ. Technol. Innovation* 21: 101286.

43 Zheng, H., Meng, X., Chen, J. et al. (2021). In situ phase evolution of TiO_2/$Ti_3C_2T_x$ heterojunction for enhancing adsorption and photocatalytic degradation. *Appl. Surf. Sci.* 545: 149031.

44 Chen, J., Zheng, H., Zhao, Y. et al. (2020). Preparation of facet exposed TiO_2/$Ti_3C_2T_x$ composites with enhanced photocatalytic activity. *J. Phys. Chem. Solids* 145: 109565.

45 Kumar, S.G. and Rao, K.S.R.K. (2014). Polymorphic phase transition among the titania crystal structures using a solution-based approach: from precursor chemistry to nucleation process. *Nanoscale* 6: 11574.

46 Shahzad, A., Rasool, K., Nawaz, M. et al. (2018). Heterostructural TiO_2/$Ti_3C_2T_x$ (MXene) for photocatalytic degradation of antiepileptic drug carbamazepine. *Chem. Eng. J.* 349: 748.

47 Peng, C., Wang, H., Yu, H., and Peng, H. (2017). (111) TiO_{2-x}/Ti_3C_2: Synergy of active facets, interfacial charge transfer and Ti^{3+} doping for enhance photocatalytic activity. *Mater. Res. Bull.* 89: 16.

48 Wang, X., Yang, Y., Lu, G. et al. (2020). In-situ preparation of Ti_3C_2/Ti^{3+}-TiO_2 composites with mosaic structures for the adsorption and Photo-degradation of flowing acetaldehyde under visible light. *Appl. Surf. Sci.* 531: 147101.

49 Imanishi, A., Tsuji, E., and Nakato, Y. (2007). Dependence of the work function of TiO_2 (Rutile) on crystal faces, studied by a scanning auger microprobe. *J. Phys. Chem. C* 111: 2128.

50 Khazaei, M., Arai, M., Sasaki, T. et al. (2015). OH-terminated two-dimensional transition metal carbides and nitrides as ultralow work function materials. *Phys. Rev. B.* 92: 075411.

51 Huang, H., Jiang, X., Li, N. et al. (2021). Noble-metal-free ultrathin MXene coupled with In_2S_3 nanoflakes for ultrafast photocatalytic reduction of hexavalent chromium. *Appl. Catal., B* 284: 119754.

52 Liang, P., Yuan, L., Du, K. et al. (2021). Photocatalytic reduction of uranium(VI) under visible light with 2D/1D Ti_3C_2/CdS. *Chem. Eng. J.* 420: 129831.

53 Fang, H., Pan, Y., Yin, M., and Pan, C. (2019). Enhanced visible light photocatalytic activity of CdS with alkalized Ti_3C_2 nano-sheets as co-catalyst for degradation of rhodamine B. *J. Mater. Sci.* 30: 14954.

54 Zou, X., Zhao, X., Zhang, J. et al. (2021). Photocatalytic degradation of ranitidine and reduction of nitrosamine dimethylamine formation potential over MXene-Ti_3C_2/MoS_2 under visible light irradiation. *J. Hazard. Mater.* 413: 125424.

55 Wang, S., Wang, Y., Wan, J. et al. (2020). Dual channel carrier transfer based on $Ti_3C_2T_x$ improves carrier utilization of Z-scheme Ag_3PO_4/AgBr heterojunction photocatalyst. *Sep. Purif. Technol.* 253: 117486.

56 Fang, Y., Cao, Y., and Chen, Q. (2019). Synthesis of an Ag_2WO_4/Ti_3C_2 Schottky composite by electrostatic traction and its photocatalytic activity. *Ceram. Int.* 45: 22298.

57 Sun, B., Tao, F., Huang, Z. et al. (2021). Ti_3C_2 MXene-bridged Ag/Ag_3PO_4 hybrids toward enhanced visible-lightdriven photocatalytic activity. *Appl. Surf. Sci.* 535: 147354.

58 Zou, G.D., Zhang, Z.W., Guo, J.X. et al. (2016). Synthesis of MXene/Ag composites for extraordinary long cycle lifetime lithium storage at high rates. *ACS Appl. Mater. Interfaces* 8: 22280.

59 Zhong, Q., Liu, J., Wang, Z. et al. (2022). Ti_3C_2 MXene/Ag_2ZnGeO_4 Schottky heterojunctions with enhanced photocatalytic performances: Efficient charge separation and mechanism studies. *Sep. Purif. Technol.* 278: 119560.

60 Sridharan, K., Shenoy, S., Girish Kumar, S. et al. (2021). Advanced Two-Dimensional Heterojunction Photocatalysts of Stoichiometric and Non-Stoichiometric Bismuth Oxyhalides with Graphitic Carbon Nitride for Sustainable Energy and Environmental Applications. *Catalysts* 11: 426.

61 Bacha, A.U.R., Nabi, I., Fu, Z. et al. (2019). A comparative study of bismuth-based photocatalysts with titanium dioxide for perfluorooctanoic acid degradation. *Chin. Chem. Lett.* 30: 2225.

62 Zou, Z.W., Xu, H.M., Li, D.Y. et al. (2019). Facile preparation and photocatalytic activity of oxygen vacancy rich BiOCl with 001 exposed reactive facets. *Appl. Surf. Sci.* 463: 1011.

63 Huang, G., Li, S., Liu, L. et al. (2020). Ti_3C_2 MXene-modified Bi_2WO_6 nanoplates for efficient photodegradation of volatile organic compounds. *Appl. Surf. Sci.* 503: 144183.

64 Zhao, D. and Cai, C. (2019). Layered Ti_3C_2 MXene modified two-dimensional Bi_2WO_6 composites with enhanced visible light photocatalytic performance. *Mater. Chem. Front.* 3: 2521.

65 Tan, B., Fang, Y., Chen, Q. et al. (2021). Construction of $Bi_2O_2CO_3/Ti_3C_2$ heterojunctions for enhancing the visible-light photocatalytic activity of tetracycline degradation. *J. Colloid Interface Sci.* 601: 581.

66 Zhang, H., Li, M., Wang, W. et al. (2021). Designing 3D porous BiOI/Ti_3C_2 nanocomposite as a superior coating photocatalyst for photodegradation RhB and photoreduction Cr (VI). *Sep. Purif. Technol.* 272: 118911.

67 Wu, K., Song, S., Wu, H. et al. (2020). Facile synthesis of $Bi_2WO_6/C_3N_4/Ti_3C_2$ composite as Z-scheme photocatalyst for efficient ciprofloxacin degradation and H_2 production. *Appl. Catal. A* 608: 117869.

68 Zhou, Y., Yu, M., Liang, H. et al. (2021). Novel dual-effective Z-scheme heterojunction with g-C_3N_4, Ti_3C_2 MXene and black phosphorus for improving visible light-induced degradation of ciprofloxacin. *Appl. Catal. B* 291: 120105.

69 Zhou, Y., Yu, M., Zhan, R. et al. (2021). Ti_3C_2 MXene-induced interface electron separation in g-C_3N_4/Ti_3C_2 MXene/$MoSe_2$ Z-scheme heterojunction for enhancing visible light-irradiated enoxacin degradation. *Sep. Purif. Technol.* 275: 119194.

70 Yang, L., Chen, J., Que, M. et al. (2021). Construction of 2D/2D/2D BiOBr/(001)-$TiO_2/Ti_3C_2T_x$ heterojunction for enhancing photocatalytic degradation of dye. *J. Alloys Compd.* 893: 162289.

71 Sharma, V., Kumar, A., Kumar, A., and Krishnan, V. (2022). Enhanced photocatalytic activity of two dimensional ternary nanocomposites of ZnO-Bi_2WO_6-Ti_3C_2 MXene under natural sunlight irradiation. *Chemosphere* 287: 132119.

72 Yuan, Z., Huang, H., Li, N. et al. (2021). All-solid-state WO_3/TQDs/In_2S_3 Z-scheme heterojunctions bridged by Ti_3C_2 quantum dots for efficient removal of hexavalent chromium and bisphenol A. *J. Hazard. Mater.* 409: 125027.

73 Chen, X., Li, J., Pan, G. et al. (2019). Ti_3C_2 MXene quantum dots/TiO_2 inverse opal heterojunction electrode platform for superior photoelectrochemical biosensing. *Sens. Actuators B* 289: 131.

74 Liu, Q., Tan, X., Wang, S. et al. (2019). MXene as non-metal charge mediator in 2D layered CdS@Ti_3C_2@TiO_2 composites with superior Z-scheme visible light-driven photocatalytic activity. *Environ. Sci. Nano* 6: 3158.

75 Xu, C., Yang, F., Deng, B. et al. (2021). RGO-wrapped Ti_3C_2/TiO_2 nanowires as a highly efficient photocatalyst for simultaneous reduction of Cr(VI) and degradation of RhB under visible light irradiation. *J. Alloys Compd.* 874: 159865.

76 Liu, H., Yang, C., Jin, X. et al. (2020). One-pot hydrothermal synthesis of MXene Ti_3C_2/TiO_2/BiOCl ternary heterojunctions with improved separation of photoactivated carries and photocatalytic behavior toward elimination of contaminants. *Colloids Surf A* 603: 125239.

77 Fang, H., Pan, Y., Yin, M. et al. (2019). Facile synthesis of ternary Ti_3C_2-OH/ln_2S_3/CdS composite with efficient adsorption and photocatalytic performance towards organic dyes. *J. Solid State Chem.* 280: 120981.

78 Zhang, H., Li, M., Zhu, C. et al. (2020). Preparation of magnetic α-Fe_2O_3/$ZnFe_2O_4$@Ti_3C_2 MXene with excellent photocatalytic performance. *Ceram. Int.* 46: 81.

79 Ye, M.H., Wang, X., Liu, E.Z. et al. (2018). Boosting the photocatalytic activity of P25 for carbon dioxide reduction by using a surface-alkalinized titanium carbide MXene as cocatalyst. *ChemSusChem* 11: 1606.

12

MXenes and its Hybrid Nanocomposites for Gas Sensing Applications in Breath Analysis

Sampada Koirala, Mahek Sadiq, and Danling Wang

Biomedical Engineering Program, Department of Electrical and Computer Engineering, North Dakota State University, Fargo, ND, USA

12.1 Introduction

Major advances in cancer control can be greatly aided by early diagnosis and effective treatment in its preinvasive state before metastasis. Lung cancer (small cell and non-small cell) is the second most common cancer in men and women and the first leading cause of cancer deaths in the United States [1]. The survival rate of lung cancer is very low because most patients (~75%) have been diagnosed at the late stage of disease (stage III or IV) [1]. Breath analysis has become popular in the area of respiratory and chronic diseases as a diagnosis tool due to its simple, noninvasive, easy sample access, and point-of-care process for monitoring diseases in patients [2]. Exhaled human breath consists of over 1000 Volatile organic compounds (VOCs) and microscopic aerosol particles, also called respiratory droplets [3]. Some VOCs such as isoprene, ammonia, and acetone are pointed out as biomarkers for specific diseases; for example, elevated acetone for diabetes, pentane for heart disease, ammonia for kidney disorders, nitrogen monoxide for asthma, and toluene for lung cancer [4]. Therefore, analyzing breath is very attractive and offers great promise for cancer early detection and diagnosis. Also, breath analysis can provide real-time monitoring, in-time guidance and then optimize cancer disease treatment [5]. In particular, breath analysis is noninvasive, and breath supply is nearly infinite (inexpensive). So, it will provide an entirely new way to access health conditions for early disease detection, precision medicine, real-time treatment, and disease prevention [6–10].

There is a list of biomarkers directly related to lung cancer based on their presence and concentration. Within these biomarkers, four chemicals have been selected for lung cancer-related diagnosis and screening. These are propanol, toluene, *o*-xylene, and cyclohexane [11]. Lung cancer patients show higher concentrations of these compounds than healthy people have in their breath. However, the detection of breath biomarkers is quite challenge since the concentrations of biomarkers are relatively quite low, ranging from 10 to 1000 parts-per-billions (ppb). Table 12.1 summarizes different concentrations in VOCs that are recorded in lung cancer patients and healthy individuals [11].

MXenes: Next-Generation 2D Materials: Fundamentals and Applications, First Edition.
Edited by Jay Singh, Kshitij RB Singh, Ravindra Pratap Singh, and Charles Oluwaseun Adetunji.

Table 12.1 Concentrations of propanol, toluene, *o*-xylene, and cyclohexane chemicals present in the breath of people in good health and lung cancer patients [11].

Biomarker	Concentrations for healthy people (ppb)	Concentrations for lung cancer patients (ppb)
Propanol	—	4–450
Toluene	5-20	20–30
O-Xylene	15	25
Cyclohexane	25	200

Source: Adapted from [11] Geoffrey Gregis, et al., 2022.

Exhaled gases can be analyzed accurately by gas chromatography (GC), mass spectrometry (MS), ion mobility spectrometry (IMS), or flame ionization detectors (FIDs). The diagnosis process is time-consuming, expensive, and bulky, making it challenging to achieve portable and real-time diagnosis [12]. The development of nanostructured metal–oxide semiconductors (NMOX), such as nanostructured SnO_2, WO_3, and MnO_3, enables rapid detection of a wide variety of gases with high sensitivity benefitting from their large surface area and porous structures [13–22]. The chemiresistive gas sensors based on NMOX are simple and easy to miniaturize for affordable devices. Although many strategies exist, including functionalizing the surface layer of the materials with noble metals [23], doping various impurities into metal oxides [24–27], and the use of thermal assistance with UV-illumination or heterostructures [28], the applications of NMOX-based chemiresistive sensors in breath analysis still face limitations and challenges in the final stage of device integration and simplicity because of their elevated operating temperature, limited selectivity, and strong humidity interference. The lack of an ideal material with excellent performance in terms of breath analysis has motivated us to search for new functionalized sensing materials and sensing techniques.

Two-dimensional (2D) MXenes are nanomaterials recently discovered that have multilayered structures sharing the structural formula of $M_{1+n}\,X_n$, where n is from 1 to 3 and M stands for early transition metal layer bonded to the X, the layer of carbon or nitrogen [29–31]. MXenes have similar morphology but superior electrical properties compared to other 2D nanomaterials, such as graphene [30]. This could lead to the alternative use of 2D materials in the fields of gas sensing [32], biosensing [33–35], device storage for energy, smart wear technology, and other high-performance sensing applications [30].

The synthesis of MXene is based on selectively etching the atomic A (an element such as aluminum or silicon) layer from a precursor ceramic MAX phase that is layered carbides and nitrides [36]. This phase has a general formula $M_{1+n}\,AX_n$, where X represents an element from the boron and carbon groups of the periodic table [37]. The starting MAX phase thus determines MXene's elemental composition. The most studied MXene, Ti_3C_2, is usually fabricated from the Ti_3AlC_2 MAX material. There are different etching methods that have been used to synthesize Ti_3C_2 MXenes, such as hydrofluoric acid (HF) or *in situ* HF [36]. These etching steps are essential as the chemical bonds between the aluminum and titanium elements in the MAX phase are very strong [38]. MXene has different functional groups on the surface layers such as –O, –F, or –OH, which act for T_x in the $Ti_3C_2T_x$ MXene chemical formula [36].

Numerous research studies have been conducted recently to try and improve the 2D Ti_3C_2 nanostructures synthesis for semiconductor material research. Nevertheless, MXenes still face difficulties in the synthesis process, and future research exploring synthesis procedures for chemical and structural diversity will allow for a wider variety of possibilities [39]. In our group, we first developed a "hot etching" method to synthesize Ti_3C_2 MXenes [40]. There are many advantages to synthesizing 2D MXene Ti_3C_2 at a high temperature in order to yield a high-quality MXene material efficiently. Creating an effective and universal etching method for MXene preparation is one of the advantages [41], as it will make it easier to comprehend the impact of etching on the purity, microstructure, and general morphology of as-synthesized MXenes [36], and finally, to study about the Ti-based MXene, whether it exhibits relatively higher temperature stability or not [42]. This chapter summarizes the most recent Ti_3C_2 MXene synthesis methods to explain how the synthesis conditions affect the material's properties [43].

There are two routes of etching methods to synthesize Ti_3C_2 from the Ti_3AlC_2 phase: HF and *in situ* HF (also called *in situ* hydrothermal etching) [37]. The HF route uses three concentrations: 5, 10, and 30% wt, along with three etching durations such as 24, 18, and 5 hours; the other method is the *in situ* HF method. In comparison to the direct use of hazardous HF, the *in situ* HF method is generated by using minimally intensive layer delamination (MILD) with etchants such as lithium fluoride/hydrochloric acid (LiF/HCl) [36, 44]. In the HF etching method, a distinct color difference (gray/black) from Ti_3AlC_2 powder (gray color) can be observed. The black layer indicates a denser, layered morphology [36]. However, multilayered Ti_3C_2 powders look different in morphology depending on the etchant concentration. For instance, in the MILD method, the flakes of Ti_3C_2 have more significant sizes, are more distinctive, and have straight edges [39].

There are several important parameters for MXene synthesis to determine their properties. Certain parameters such as etching concentration, etching temperature, and the reaction duration of the etchants play an important role in affecting MXene morphology and functionalities. Therefore, the investigation of these parameters, for instance, the use of a higher temperature for synthesizing Ti_3C_2, can reveal the change in morphology, the purity of the final products, and the changes in the product's crystal structure. Similarly, the concentration of HF affects the morphology of MXene, for instance, an increase in the HF concentration from 5 to 10% wt etched at 100 °C yields a pure MXene sample. However, if the concentration is increased further, then MXene samples may decompose or disappear. So, choosing an optimal concentration and temperature to synthesize an MXene sample to obtain the highest purity is critical. As for the etching reaction duration, the longer the reaction duration, the more complete the etching will be, which means that a pure MXene can be yielded. Here, since we focus on the identification of possible VOCs as biomarkers for early diagnosis of real-time disease and monitoring, synthesizing high-quality, tunable functional MXene as the sensing materials will be very important but quite challenging. An overview of the impact of MXenes' functionalities on the sensing performance of VOC detection will be discussed in this chapter focusing on the factors that can affect the relationship of material properties and structures in MXenes, such as HF concentration and temperature.

12.2 Discussion

Studies have shown that different biofluids such as skin and breath are the important sources that generate VOCs, which are about 54%, as shown in Figure 12.1 [45]. Developing wearable or portable sensors to detect VOCs from skin or breath will be feasible methods for a noninvasive, less risk of infection, and much more convenient technique for early disease detection and diagnosis [33].

Changes in an individual's lifestyle and diet can affect the VOC concentration in human breath. These are called exposomes. The smell of exposomes can vary from one individual to another. For instance, higher concentrations of acetonitrile and furans are due to smoking cigarettes [46]. Similarly, a higher isoprene level can cause exertion and affect pulmonary alveoli [47]. Also, increased levels of ethane and pentane in human breath result from chronic lung diseases [46]. The most challenging thing is the presence of exogenous compounds that can alter the measurement when detecting specific VOC from breath. To address these challenges, an effective sensing technique should be developed which can provide accurate results and is cost-effective.

As mentioned earlier, there are different sensing techniques for VOC detection, such as GC-MS. Among these techniques, nanomaterials-based chemiresistive sensors have been intensively developed for breath analysis due to their advantages, such as easy fabrication, simple structure, facile integration into small devices, and wide adoption of sensitive materials [48, 49]. Since this type of sensing technique relies closely on sensing materials' properties, the development of nanotechnology and nanomaterials offers great opportunities for changing nanomaterials' shape, size, and crystal structure during the synthesis process and forming hybrid structures using compositions to boost the sensors' function [50]. Furthermore, nanomaterials have a high ratio of surface area by volume, which can significantly improve sensitivity and stabilize the operation of sensing devices [29]. This chapter focuses on the chemiresistive gas sensors using MXenes for VOC biomarkers detection, their sensing mechanisms, limitations, and prospects.

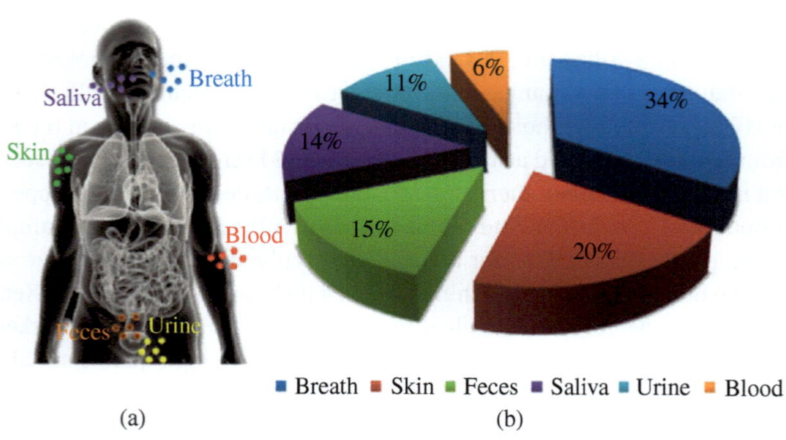

Figure 12.1 (a) Human body showing a variety of biofluids, (b) the percentage distribution of VOCs of healthy humans in different biofluids [45]. Source: Reproduce with permission from Ahmed H. Jalal, Fahmida Alam, Sohini Roychoudhury, et al. Copyright 2018, American Chemical Society.

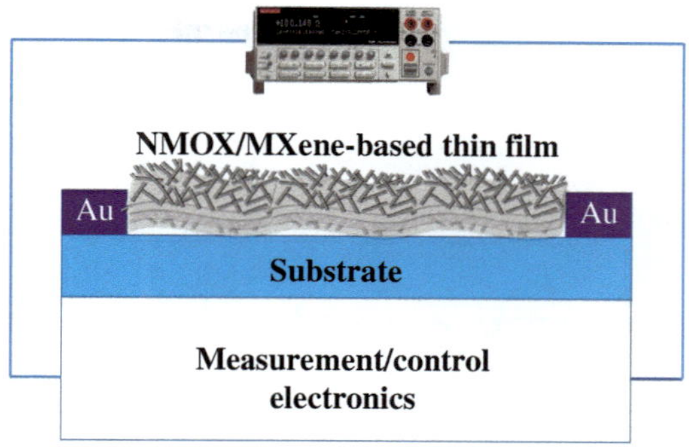

Figure 12.2 Illustration of chemiresistive gas sensor based on NMOX/MXene nanocomposites.

Chemiresistive gas sensors are used to analyze gas molecules, including exhaled VOCs present in breath due to diseases such as lung cancer [51]. This gas sensor structure includes a sensing layer, substrate, and electrodes, as shown in Figure 12.2. Two major parts must be contemplated to develop an efficient gas-sensing device. First, the sensor requires a porous sensing layer that allows gases to diffuse throughout their entirety and allows for gas-molecule interactions at the surface of individual pores, layer's borders, and interface between pores and electrodes [52]. Therefore, a *more porous layer* for a suitable breath VOC sensor is more acceptable [53]. Second, the *thickness of the sensing* layer is another critical factor affecting the sensor's performance [54, 55]. To achieve an optimal response, the thickness of the sensing layer region should be as close to that of the electron-depleted region as possible. Current chemiresistive gas sensors based NMOX, usually need to operate at high temperatures with limited stability and selectivity. In order to address challenges, the NMOX can be combined with other polymers or 2D nanomaterials to form nanocomposites, which can increase active surface sites for molecular absorption and interaction and offer higher signal-to-noise ratio due to the better conductivity (such as NMOX/MXene composites) to make sensors operating at room temperature with improved selectivity and sensitivity [52].

Recently, our research group has focused on the study of 2D MXene nanomaterials to form nanocomposites with NMOX in applications of VOCs detection. We have used field-emission scanning electron microscopy (FESEM) to understand the impacts of etching conditions on the morphology of Ti_3C_2 MXene, as shown in Figure 12.3 [56]. The as-synthesized Ti_3C_2 was etched at temperatures of 100, 120, and 150 °C using HF concentrations of 5, 10, and 15% wt, respectively. The results indicate that small increase in HF concentration (such as 5% wt to 10% wt) shows a negligible impact on the morphological change of Ti_3C_2 when etched at the temperature up to 150 °C [57, 58]. However, the morphology of MXene will significantly change if the HF etching concentration is 15% wt even if the etching temperature is 120 °C, if. We observed MXene has been successfully synthesized at 150 °C with HF concentrations of 5 and 10% wt. However, there was no yield if the etchant concentration was 15% wt. This means that etchant concentration and temperature both play important roles to cause the change of morphology/structure in as-synthesized MXenes. Figure 12.3 also shows that the increase in temperature leads to a more exfoliated morphology. The samples were etched at 150 °C, and 5% wt shows a new accordion-like structure with a larger interlayer. However, more impurities were involved when the samples were etched at 150 °C with a higher HF concentration of 10% wt and

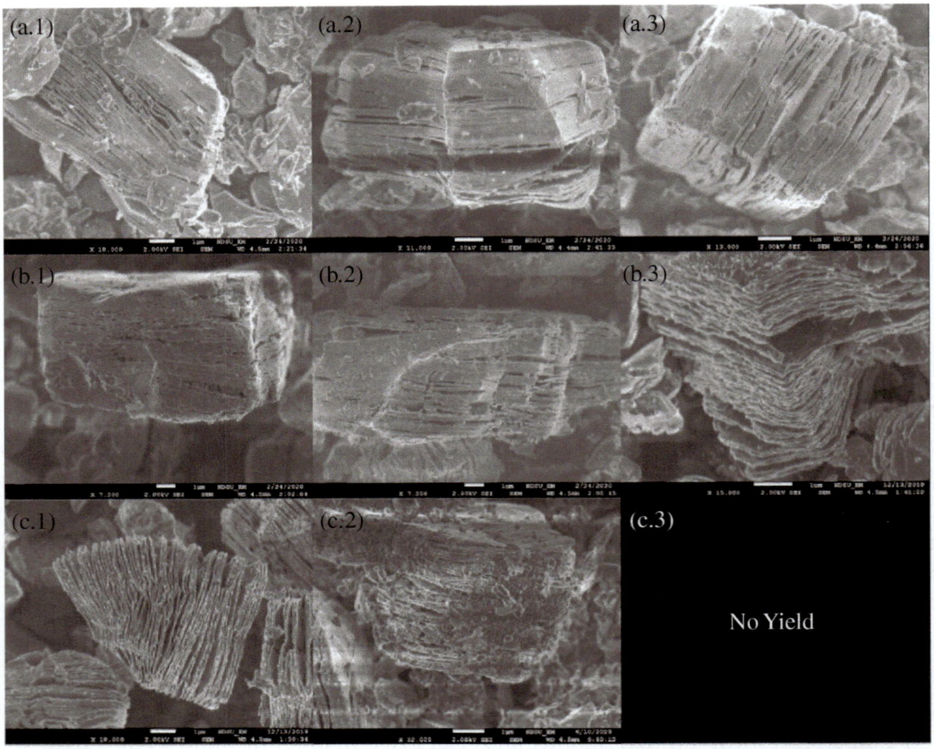

Figure 12.3 SEM images of Ti_3C_2 samples etched at (a) 100 °C, (b) 120 °C and (c) 150 °C, where (1), (2) and (3) correspond to 5, 10, and 15% wt HF concentration, respectively.

completely destroyed when HF concentration was 15% wt. All these findings indicate that the optimal etching condition (temperature and concentration) is critical to synthesize high quality of MXene.

In addition, the results reveal that tuning the temperature and HF concentration conditions could allow us to control interlayer spacing in MXene.

X-ray diffraction (XRD) was used to study the Ti_3C_2 MXene's crystal structure and purity [56]. The relative intensities of the characteristic MXene peaks (at ~8.9°) are used to identify the purity: the relative higher peak means higher purity. The peaks at 36° and 42° indicated the impurity caused by TiC, which is originally from the as-synthesized MAX phase. Figure 12.4 presents the XRD of all samples etched at 100 °C. It can be seen that the HF concentration of 10% wt yields the highest purity of MXene and results in a complete etching.

As shown in Figure 12.5, the samples etched at 120 °C show the purity of all samples and showed little variance when HF concentration was 5% wt and 10% wt, respectively. While the etching concentration increased to 15% wt, the MXene phase began to decompose. It became even worse when the etching temperature was at 150 °C, and the concentration was 15% wt as shown in Figure 12.6. The XRD presents no MXene material and only TiC peaks. Combining the FESEM images and XRD data, as shown previously, we can tell

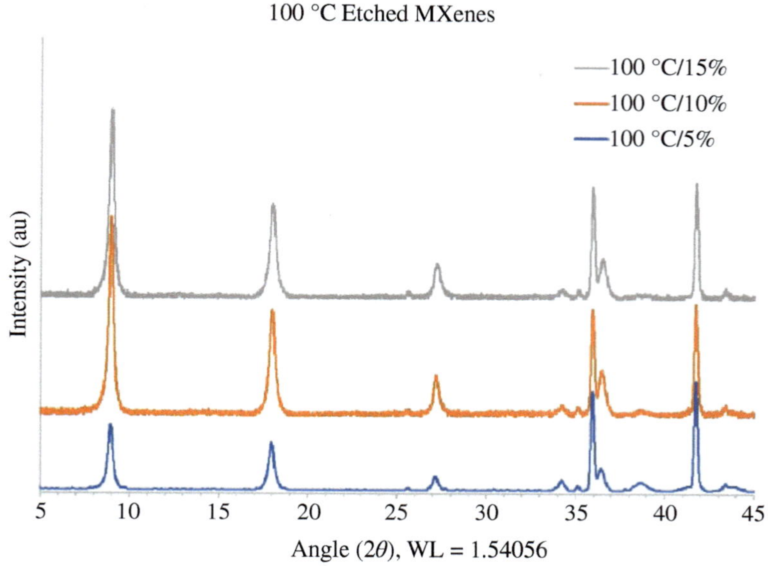

Figure 12.4 XRD of MXenes etched at 100 °C using 5, 10, and 15% wt HF concentrations.

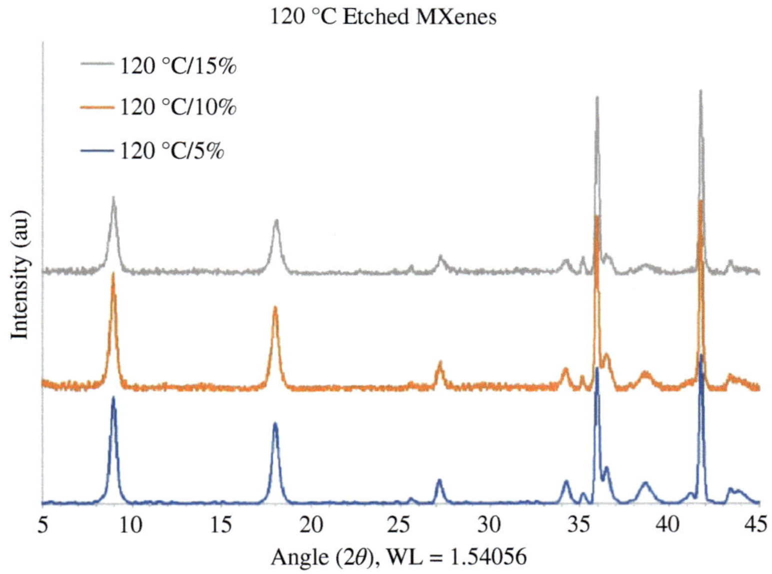

Figure 12.5 XRD of samples etched at 120 °C using 5, 10, and 15% wt HF concentration.

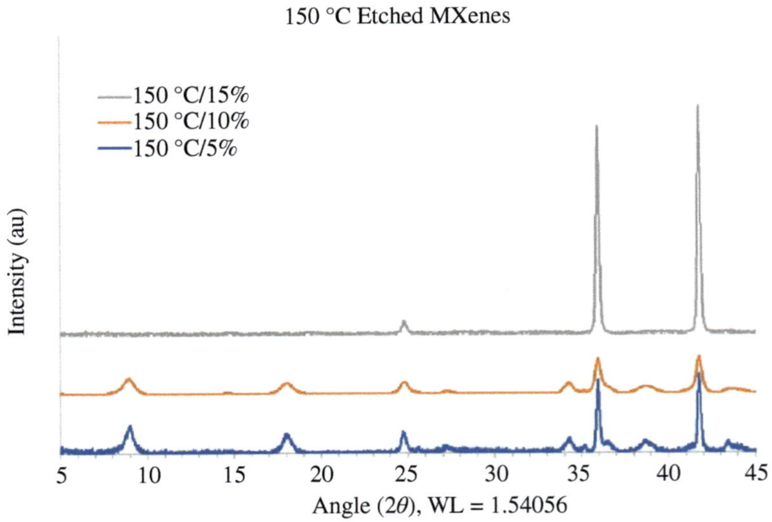

Figure 12.6 XRD of samples etched at 150 °C using 5, 10, and 15% wt HF concentration.

the optimal etching conditions: 10% wt etchant concentration at 120 °C can synthesize the highest quality of MXenes.

Based on the above study, the greatest advantage of using HF etching methods at higher temperatures is that it helps in developing an orderly and general etching route to prepare the good quality of MXenes with high purity, clear morphology, and favorite interlayer spacing [41]. By studying about the factors that affect MXene structure and characterization (e.g. HF concentration and temperature), it will allow us to think about developing high-performance sensing devices without affecting any alteration in the sensing performance. The pure MXene and its stability lay a foundation of sensing materials to fabricate high-performance efficient sensing devices [41].

As mentioned previously, because of the complexity of human breath, containing over 1000 compounds with trace concentrations (1 ppb level) and extremely variable in humidity, the detection of disease-related VOCs with high sensitivity, excellent selectivity, and satisfactory reliability is extremely challenging. Additionally, most gas sensors are influenced by the changes in the environmental conditions and exogenous compounds [59–61]. To design an idea sensing system for breath VOC detection, the sensors should satisfy the following requirements [62]:

(a) Strong in accuracy and reliability: Gas sensors must be resistant to greater humidity and other gases present in exhaled breath in order for breath analyses to be accurate and reliable.
(b) High sensitivity and low limit of detection (LOD): Sensors should be able to recognize trace-level concentrations of biomarkers (~ ppt_v to ppm_v).
(c) High stability: Gas sensing systems should exhibit consistent results to the same gas sample during the operational period.

(d) Rapid detection of specific VOCs: Gas sensors must be capable of responding fast to biomarkers and having the capability of providing in-time health information and guidance.
(e) Simple and user-friendly devices: An intuitive human/device interface is necessary for gas sensing systems. The process should be straightforward, the experimental parameters should be configurable, and the test findings should be obvious and visible.
(f) Smart devices act as a family doctor to provide in-time health guidance and continuous monitoring: The combination of a sensor array (GSA) and AI algorithms should offer intelligent pattern recognition and sensing signal classification for noninvasive, complete analysis of various VOCs in the breath.

The as-synthesized 2D Ti_3C_2 MXene and its composites are promising candidates of sensing materials in breath analysis with satisfying above requirements. This is because MXene and its nanocomposites have distinctive structures and tunable material features: outstanding conductivity to realize room-temperature operation with much better signal to noise ratio, huge surface and interface area to achieve excellent sensitivity, rich and controllable surface chemistry and interlayer spacing with improved selectivity, and excellent mechanical flexibility for wearable sensing capability [44, 63-65]. By integrating the Internet of Things (IoT) into the sensors, the IoT enabled sensors can realize real-time monitoring, remote sensing, and continuous monitoring of health status without causing discomfort and invasiveness [63].

12.3 Conclusion

Detecting breath VOC biomarkers can provide an easy, straight, non-invasive, and robust monitoring of some diseases like lung cancer [45]. Nevertheless, many challenges of sensing techniques for VOC detection still need to be addressed, such as selectively detecting disease-related biomarkers from complex human breath, sensitive response to trace concentrations of VOC gases, and reduced humidity interference. In this chapter, we have described a 2D layered nanomaterial, Ti_3C_2 MXene, for the potential applications in breath analysis. Due to MXene's unique properties, 2D MXene-based sensors have offered great solutions to address the challenges in breath analysis and detect low-concentration breath VOC biomarkers selectively and sensitively for early disease detection. Also, MXene-based sensors have the flexible capability of IoT integration to realize real-time sensing and monitoring. We also have discussed that the material properties of MXenes can be adjusted through synthesis. These offer an effective and flexible solution for designing high-performance breath sensors by controlling sensing material, 2D MXene's structure.

References

1 Sadiq, M., Pang, L., Johnson, M. et al. (2020). 2D nanomaterial, Ti_3C2 MXene-based sensor to guide lung cancer therapy and management. *Multidiscip. Digit. Pub. Inst. Proc.* 60 (1): 29.

2 Li, W., Liu, H., Xie, D. et al. (2017). Lung cancer screening based on type-different sensor arrays. *Sci. Rep.* 7 (1): 1–2.

3 Mutlu, G.M., Garey, K.W., Robbins, R.A. et al. (2001). Collection and analysis of exhaled breath condensate in humans. *Am. J. Respir. Crit. Care Med.* 164 (5): 731–737.

4 Alizadeh, N., Jamalabadi, H., and Tavoli, F. (2019). Breath acetone sensors as non-invasive health monitoring systems: a review. *IEEE Sensors J.* 20 (1): 5–31.

5 Guntner, A.T., Abegg, S., Konigstein, K. et al. (2019). Breath sensors for health monitoring. *ACS Sens.* 4 (2): 268–280.

6 Weitz, Z.W., Birnbaum, A.J., Skosey, J.L. et al. (1991). High breath pentane concentrations during acute myocardial infarction. *Lancet* 337 (8747): 933–935.

7 Timmer, B., Olthuis, W., and Van Den Berg, A. (2005). Ammonia sensors and their applications—a review. *Sensors Actuators B Chem.* 107 (2): 666–677.

8 Peng, G., Tisch, U., Adams, O. et al. (2009). Diagnosing lung cancer in exhaled breath using gold nanoparticles. *Nat. Nanotechnol.* 4 (10): 669–673.

9 Güntner, A.T., Kompalla, J.F., Landis, H. et al. (2018). Guiding ketogenic diet with breath acetone sensors. *Sensors* 18 (11): 3655.

10 Righettoni, M., Amann, A., and Pratsinis, S.E. (2015). Breath analysis by nanostructured metal oxides as chemo-resistive gas sensors. *Mater. Today* 18 (3): 163–171.

11 Gregis, G., Sanchez, J.B., Bezverkhyy, I. et al. (2018). Detection and quantification of lung cancer biomarkers by a micro-analytical device using a single metal oxide-based gas sensor. *Sens. Actuators B Chem.* 255: 391–400.

12 Tai, H., Wang, S., Duan, Z., and Jiang, Y. (2020). Evolution of breath analysis based on humidity and gas sensors: potential and challenges. *Sens. Actuators B Chem.* 318: 128104.

13 Konvalina, G. and Haick, H. (2014). Sensors for breath testing: from nanomaterials to comprehensive disease detection. Accounts of chemical research. *Acc. Chem. Res.* 47 (1): 66–76.

14 Wang, C., Yin, L., Zhang, L. et al. (2010). Metal oxide gas sensors: sensitivity and influencing factors. *Sensors* 10 (3): 2088–2106.

15 Chen, H., Qi, X., Ma, J. et al. (2020). Breath-borne VOC biomarkers for COVID-19. *MedRxiv*.

16 Risby, T.H. and Solga, S.F. (2006). Current status of clinical breath analysis. *Appl. Phys. B Lasers Opt.* 85 (2): 421–426.

17 Bai, H. and Shi, G. (2007). Gas sensors based on conducting polymers. *Sensors* 7 (3): 267–307.

18 Anderson, J.C., Lamm, W.J., and Hlastala, M.P. (2006). Measuring airway exchange of endogenous acetone using a single-exhalation breathing maneuver. *J. Appl. Physiol.* 100 (3): 880–889.

19 Tassopoulos, C.N., Barnett, D., and Fraser, T.R. (1969). Breath-acetone and blood-sugar measurements in diabetes. *Lancet* 293 (7609): 1282–1286.

20 Minh, T.D., Blake, D.R., and Galassetti, P.R. (2012). The clinical potential of exhaled breath analysis for diabetes mellitus. *Diabetes Res. Clin. Pract.* 97 (2): 195–205.

21 Sun, Y.F., Liu, S.B., Meng, F.L. et al. (2012). Metal oxide nanostructures and their gas sensing properties: a review. *Sensors* 12 (3): 2610–2631.

22 Tricoli, A. and Pratsinis, S.E. (2010). Disperse nanoelectrode devices. *Nat. Nanotechnol.* 5: 54.

23 Joshi, N., Hayasaka, T., Liu, Y. et al. (2018). A review on chemiresistive room temperature gas sensors based on metal oxide nanostructures, graphene and 2D transition metal dichalcogenides. *Microchim. Acta* 185 (4): 1–6.

24 Moon, H.G., Jung, Y., Han, S.D. et al. (2016). Chemiresistive electronic nose toward detection of biomarkers in exhaled breath. *ACS Appl. Mater. Interfaces* 8 (32): 20969–20976.

25 Qi, Q., Zhang, T., Liu, L. et al. (2008). Selective acetone sensor based on dumbbell-like ZnO with rapid response and recovery. *Sens. Actuators B Chem.* 134 (1): 166–170.

26 Chen, J., Pan, X., Boussaid, F. et al. (2017). Breath level acetone discrimination through temperature modulation of a hierarchical ZnO gas sensor. *IEEE Sens. Lett.* 1 (5): 1–4.

27 Li, W.Q., Ma, S.Y., Luo, J. et al. (2014). Synthesis of hollow SnO2 nanobelts and their application in acetone sensor. *Mater. Lett.* 132: 338–341.

28 Deng, L.L., Zhao, C.X., Ma, Y. et al. (2013). Low cost acetone sensors with selectivity over water vapor based on screen printed TiO_2 nanoparticles. *Anal. Methods* 5 (15): 3709–3713.

29 Amiri, V., Roshan, H., Mirzaei, A. et al. (2020). Nanostructured metal oxide-based acetone gas sensors: a review. *Sensors* 20 (11): 3096.

30 Michael, J., Qifeng, Z., and Danling, W. (2019). Titanium carbide MXene: synthesis, electrical and optical properties and their applications in sensors and energy storage devices. *Nanomater. Nanotechnol.* 9: 1847980418824470.

31 Lee, J., Lee, K., and Lee, J.H. (2021). Nonlinear absorption property investigation into MAX phase Ti2 AlC at 1.9 μm. *Opt. Mater. Exp.* 11 (10): 3556–3566.

32 Buszewski, B., Kęsy, M., Ligor, T., and Amann, A. (2007). Human exhaled air analytics: biomarkers of diseases. *Biomed. Chromatogr.* 21 (6): 553–566.

33 Zhou, L., Zhang, X., Ma, L. et al. (2017). Acetylcholinesterase/chitosan-transition metal carbides nanocomposites-based biosensor for the organophosphate pesticides detection. *Biochem. Eng. J.* 128: 243–249.

34 Liu, H., Duan, C., Yang, C. et al. (2015). A novel nitrite biosensor based on the direct electrochemistry of hemoglobin immobilized on MXene-Ti3C2. *Sens. Actuators B Chem.* 218: 60–66.

35 Wu, L., Lu, X., Wu, Z.S. et al. (2018). 2D transition metal carbide MXene as a robust biosensing platform for enzyme immobilization and ultrasensitive detection of phenol. *Biosens. Bioelectron.* 107: 69–75.

36 Alhabeb, M., Maleski, K., Anasori, B. et al. (2017). Guidelines for synthesis and processing of two-dimensional titanium carbide (Ti3C2T x MXene). *Chem. Mater.* 29 (18): 7633–7644.

37 Aguedo, J., Lorencova, L., Barath, M. et al. (2020). Electrochemical impedance spectroscopy on 2D nanomaterial MXene modified interfaces: application as a characterization and transducing tool. *Chem. Aust.* 8 (4): 127.

38 Parajuli, D. and Samatha, K. (2019). MXene as topological insulator. *J. Emerg. Technol. Innov. Res.* 6 (4): 689–706.

39 Liu, P., Yue, M., Xue, T. et al. (2019). Preparation conditions and electrical properties of Ti3C2Tx nanosheets. *Mater. Sci. Semicond. Process.* 104: 104683.

40 Michael, J., Wang, D.L., and Zhang, Q.F. Synthesis of high yield, pure Ti_3C_2 MXene using high temperature etching. Invited to submit to. *Nanomaterials.*

41 Peng, C., Wei, P., Chen, X. et al. (2018). A hydrothermal etching route to synthesis of 2D MXene (Ti3C2, Nb2C): enhanced exfoliation and improved adsorption performance. *Ceram. Int.* 44 (15): 18886–18893.

42 Amann, A., de Lacy Costello, B., Miekisch, W. et al. (2014). The human volatilome: volatile organic compounds (VOCs) in exhaled breath, skin emanations, urine, feces and saliva. *J. Breath Res.* 8 (3): 034001.

43 Chen, H., Ma, H., and Li, C. (2021). Host–guest intercalation chemistry in MXenes and its implications for practical applications. *ACS Nano* 15 (10): 15502–15537.

44 Gao, L., Li, C., Huang, W. et al. (2020). MXene/polymer membranes: synthesis, properties, and emerging applications. *Chem. Mater.* 32 (5): 1703–1747.

45 Jalal, A.H., Alam, F., Roychoudhury, S. et al. (2018). Prospects and challenges of volatile organic compound sensors in human healthcare. *ACS Sens.* 3 (7): 1246–1263.

46 Jalal AH. 2018 Multivariate Analysis for the Quantification of Transdermal Volatile Organic Compounds in Humans by Proton Exchange Membrane Fuel Cell System.

47 Broza, Y.Y. and Haick, H. (2013). Nanomaterial-based sensors for detection of disease by volatile organic compounds. *Nanomedicine* 8 (5): 785–806.

48 Kim, N.H., Choi, S.J., Yang, D.J. et al. (2014). Highly sensitive and selective hydrogen sulfide and toluene sensors using Pd functionalized WO3 nanofibers for potential diagnosis of halitosis and lung cancer. *Sensors Actuators B Chem.* 193: 574–581.

49 Han, L., Shi, X., Wu, W. et al. (2005). Nanoparticle-structured sensing array materials and pattern recognition for VOC detection. *Sensors Actuators B Chem.* 106 (1): 431–441.

50 Mercante, L.A., Andre, R.S., Mattoso, L.H., and Correa, D.S. (2019). Electrospun ceramic nanofibers and hybrid-nanofiber composites for gas sensing. *ACS Appl. Nano Mater.* 2 (7): 4026–4042.

51 Miao, J. *Improving the Sensing Performance of Semiconductor Metal Oxide Gas Sensors Through Composition and Nanostructure Design* (Doctoral Dissertation. Arizona State University.

52 Johnson, M., Koirala, S., Rudie, A. et al. Nanomaterial-based sensing technology for the application in breath analyzer as for early disease detection and prevention. *Recent Trends Biotechnol.*

53 Sakai, G., Baik, N.S., Miura, N., and Yamazoe, N. (2001). Gas sensing properties of tin oxide thin films fabricated from hydrothermally treated nanoparticles: dependence of CO and H_2 response on film thickness. *Sensors Actuators B Chem.* 77 (1-2): 116–121.

54 Johnson, M., Koirala, S., Zhang, R.A., and Wang, Q. (2021). Nanomaterial-based sensing technology for the application in breath analyzer as for early disease detection and prevention. *Recent Trends Biotechnol. MedDocs Publ.* 2: 1–8.

55 Dent, A.G., Sutedja, T.G., and Zimmerman, P.V. (2013). Exhaled breath analysis for lung cancer. *J. Thorac. Dis.* 5 (Suppl 5): S540.

56 Hong J. 2018 Two-dimensional materials: synthesis and application in lithium/sodium ion batteries.

57 Seredych, M., Shuck, C.E., Pinto, D. et al. (2019). High-temperature behavior and surface chemistry of carbide MXenes studied by thermal analysis. *Chem. Mater.* 31 (9): 3324–3332.

58 Johnson, M., Zhang, Q., and Wang, D. (2019). Room-temperature ferroelectric K2W7O22 (KWO) nanorods as a sensor material for the detection of acetone. *Med. Dev. Sensors* 2 (3-4): e10044.

59 Kim, J., Jeerapan, I., Imani, S. et al. (2016). Noninvasive alcohol monitoring using a wearable tattoo-based iontophoretic-biosensing system. *ACS Sens.* 1 (8): 1011–1019.

60 Kim, S.S., Young, C., Vidakovic, B. et al. (2009). Potential and challenges for mid-infrared sensors in breath diagnostics. *IEEE Sensors J.* 10 (1): 145–158.

61 Vaddiraju, S. and Gleason, K.K. (2010). Selective sensing of volatile organic compounds using novel conducting polymer–metal nanoparticle hybrids. *Nanotechnology* 21 (12): 125503.

62 Zhou, X., Xue, Z., Chen, X. et al. (2020). Nanomaterial-based gas sensors used for breath diagnosis. *J. Mater. Chem. B* 8 (16): 3231–3248.

63 Hermawan, A., Amrillah, T., Riapanitra, A. et al. (2021). Prospects and challenges of MXenes as emerging sensing materials for flexible and wearable breath-based biomarker diagnosis. *Adv. Healthc. Mater.* 10 (20): 2100970.

64 Koh, H.J., Kim, S.J., Maleski, K. et al. (2019). Enhanced selectivity of MXene gas sensors through metal ion intercalation: in situ X-ray diffraction study. *ACS Sens.* 4 (5): 1365–1372.

65 Potyrailo, R.A. (2011). Bio-inspired device offers new model for vapor sensing. *SPIE Newsroom* 10 (2.1201103): 003568.

13

MXenes for Catalysis and Electrocatalysis

Abel Inobeme[1], John Tsado Mathew[2], Alexander Ikechukwu Ajai[3], Charles Oluwaseun Adetunji[4], Abdullahi Mann[3], John Olusanya Jacob[3], Elijah Yanda Shaba[3], Stanley Osita Okonkwo[3], Gregory Edema[5], Efosa John Osarenren[6], Esther Bernard[7], Evbuomwan L.[8], and Tedunjaiye O.H.[9]

[1]*Department of Chemistry, Edo State University Uzairue, Iyamho, Nigeria*
[2]*Department of Chemistry, Ibrahim Badamasi Babangida University, Lapai, Nigeria*
[3]*Department of Chemistry, Federal University of Technology Minna, Minna, Nigeria*
[4]*Applied Microbiology, Biotechnology and Nanotechnology Laboratory, Department of Microbiology, Edo University, Iyamho, Auchi, Edo State, Nigeria*
[5]*Department of Physical Science Laboratory Technology, Federal Polytechnic Auchi, Edo State, Nigeria*
[6]*Department of Biochemistry, Federal Polytechnic Auchi, Auchi, Nigeria*
[7]*Department of Chemical Engineering, Federal University of Technology Minna, Minna, Nigeria*
[8]*Department of Microbiology, Edo State University Uzairue, Nigeria*
[9]*Edo State University Uzairue, Nigeria*

13.1 Introduction

Electrocatalysis has been recognized as one of the most reliable approaches for overcoming the global crisis of energy and related environmental issues. Efficiency of an electrocatalytic system significantly depends on the design of highly efficient catalysts. MXenes are highly promising in this regard. Electrocatalysis takes advantage of the kinetic and thermodynamic tuning of various electrochemical processes taking place at the surface of the electrodes as a function of the external voltage applied [1]. The catalyst alters the rate of the reaction in comparison to inert electrode through the provision of more favorable sites of reaction, with high alteration of the energies of adsorption or giving rise to a reaction route that is more favorable. Thus, it becomes possible for achieving a faster rate of reaction at relatively lower overpotentials. The selectivity and activity of the electrocatalysts are improved through surface functionalization and exposed facets modification. On the basis of the reaction targeted, MXenes can be used efficiently as catalysts as a result of their high conductivity, and optimum strength of adsorption for major reaction intermediates [2]. There are several other applications that have been put forward based on experimental studies on the uniqueness of MXenes in terms of their magnetic, electronic, electrochemical, and optical properties (Table 13.1).

After the preparation of MXenes was first achieved in 2004 through the method of mechanical exfoliation, they have gained remarkable attention on the basis of their chemical and physical properties. The materials are made up of thin sheets of atoms

MXenes: Next-Generation 2D Materials: Fundamentals and Applications, First Edition.
Edited by Jay Singh, Kshitij RB Singh, Ravindra Pratap Singh, and Charles Oluwaseun Adetunji.

Table 13.1 A summary on specific applications of MXenes in various processes.

Catalyst (materials)	Application	Performance (mV)	Synthetic route	Refs.
MXenes activated using quantum dots $Ti_3C_2O_2$@GQDs	OER	530	Lattice oxygen oxidation	[3]
$Ti_3C_2T_x$	TMO/MXene OER	266	Hydrothermal	[4]
Hybrid MXene with Co_3O_4	OER	300	Solvothermal method	[5]
2D Ti_3C_2MXene	OER	83.7	Facile hydrothermal method	[6]
NiFe	Reduced graphene oxide, OER	235	Solvothermal	[7]
MXene with FeNi-LDH/$Ti_3C_2T_x$	OER		Co-precipitation	[8]

having large surface areas and remarkable conductivity hence have been employed in various areas such as biotherapy, energy storage, catalysis, electrocatalysis, membrane separation, and photocatalysis (Table 13.2) [9]. The MAX phases consist of a class of special ternary multilayered nitrides and carbides. Their basic uniqueness emerged from their unique integration of various properties, as they are a combination of both ceramics and metals properties. For example, like ceramics, they possess a very high stiffness, creep resistance, high corrosion, and oxidation. Like metals, they have high thermal and electrical conductivity and are highly malleable. Aside MXenes, there are related materials such as boron nitrides, silicones, germanene, antimonenes, and layered double hydroxides. More recently some other materials such as carbonitrides, nitrides, and metal carbides have been added to the family of MXenes [10]. The unique features of MXenes make them suitable for several potential applications such as electrochemical energy storage, transparent and flexible electrodes, thermal heaters, sensors, antibacterial films, and diodes. Generally, MXenes are more commonly represented using the formula $M_{n+1}X_nT_x$, where n can be simple whole number such as 1, 2, 3, M stands for an early transition element, X stands for carbon or nitrogen, or both, and T_x represents the functional groups on the surface such as fluorine, hydroxyl, or oxygen. MXenes representative structures include M_2XT_x, $M_3X_2T_x$, and $M_4X_3T_x$ in which n layers of the X elements are enclosed by $n+1$ layers of M. There are more than 19 different types of MXenes that have been synthesized since 2011 [11]. More recently, the existence and remarkable stability of numerous MXenes have been documented together with the hexagonal nitrides of boron, silicene, phosphorene, thin oxide layers, and dichalcogenides of transition metals [12]. This chapter presents the applications of MXenes in various catalytic and electrocatalytic reactions. Their methods of synthesis and strategies for improving their efficiencies are also discussed. It also highlights some of the future trends in this regard.

Table 13.2 A summary of synthetic approaches of MXenes and their limitations.

Method of synthesis	Principles	Advantage	Limitations
Fluoride etching	Etching of MAX phase in presence of Conc. HF with mechanical agitation	Several types of multilayer MXenes can be prepared using this method	HF is highly toxic and difficult to handle. Also the high reactivity of Al toward HF is an issue of concern
Alkali etching	Takes advantage of the suitable reactivity of alkali toward amphoteric aluminum	Nanoscale tablets with single-layer MXenes can be efficiently synthesized	The kinetics of this process is not that favorable. The yield is also of concern
Electrochemical etching	On the basis of the differences in the reactivity existing between M—Al and M—C bonds	Does not require the use of HF hence not of safety concern	There is overheating of the MAX phase into the carbons in the carbines
Water-free etching	The MAX phase of is achieved through substitution process, followed by transformation into MXenes through gradual increase of the combining ratios of MAX and zinc chloride	Varieties of MXenes are synthesized using this technique	Very high temperature is required to achieve this process and this restricts its commercialization

13.2 Application of MXene for Catalytic Processes

MXene has been employed in various processes (Table 13.1). One of the prominent areas is in catalysis (Figure 13.1). This is an area that has gained much attention because CO_2 is electrocatalytically converted into products of higher value through reduction. CO_2RR has been identified as a starting material with the formation of various intermediates from which a multistep process gives rise to different products. Finding the most excellent catalyst for this process also requires the need for channeling the catalyst to ensure the selectivity of the reaction toward a specific targeted product. The extent of complexity of the carbon dioxide reduction reaction has led to the development of more advanced methods beyond the computational hydrogen electrode [12]. For the exploitation of MXene in the area of catalysis, there is a need for an in-depth investigation focusing on the evolution of the surface state on the basis of the required intermediates. In heterogeneous catalytic processes involving MXenes, the bifunctional catalyst usually outperforms the one-component electrocatalysts. The activity is highly regulated by the sizes, morphology, and mode of distribution of the particles. The electrochemical versatility in their intrinsic constituents (X and M sites), potential of introducing a fourth metal through the process

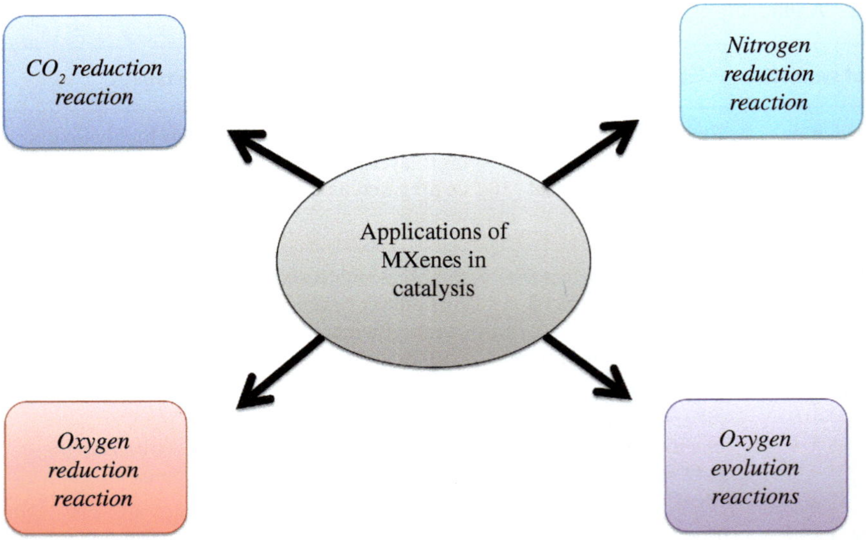

Figure 13.1 Catalytic applications of MXenes.

of alloying, and the surface terminations give rise to the unique potential of MXenes for the catalytic reduction of carbon dioxide [12].

13.2.1 CO_2 Reduction Reaction

Electrocatalysis is a reliable approach for the production of various renewable energy and chemical raw materials using carbon (IV) oxide-based reduction reactions (CO_2RR). It has however been observed that conventional electrocatalysts that are based on the utilization of transition metals are not effective enough due to their remarkable overpotential as well as relatively lower turnover. More recently, the two-dimensional MXenes have been shown to be highly promising for the catalysis of CO_2RR; however, there is a need for systematic exploration into their catalytic mechanisms, most especially on MXenes that are hydroxyl terminated within aqueous environment [13]. In some studies, first principal simulations were adopted for the systematic screening and exploration of various physicochemical properties of MXenes for the catalysis of CO_2RR into CH_4 from both kinetics and thermodynamic perspectives. results showed that $Sc_2C(OH)_2$ was relatively more promising having the lowest negative limiting value of −0.53 V. This was attained through a different reaction route, where the species that were adsorbed underwent stabilization through the capturing of hydrogen atoms from the hydroxyl termination of the MXene. New scaling processes that are based on the sharing of hydrogen interactions between MXenes and reactive intermediates have also been established [14].

13.2.2 Nitrogen Reduction Reaction

Nitrogen gas has been identified as one of the most abundant elements within the atmosphere of the Earth. It therefore serves as a vital natural raw material for the production of various intermediate chemical materials most especially ammonia and hydrazenes. The production of ammonia is vital because it is used in the production of fertilizers. The production from MXenes electrocatalysis is being considered as a more feasible route for the conversion of nitrogen to these desired products because the traditional Haber Bosch processes required a high amount of energy. Various recent investigations have documented the potentials of MXenes and their derivative compounds in the absorption, activation, and conversion of atmospheric nitrogen into ammonia through photocatalytic and electrocatalytic pathways involving nitrogen reduction reactions (NRRs) [15]. The nitrogen reduction process is seen as an ecofriendly approach for the production of ammonia through the utilization of nitrogen gas and molecules of water under conditions that are mild to attain. Unlike the present-day Haber technology which directly utilizes fossil with high energy demand, the overall yield as well as efficiency of the NRR involving the use of MXenes catalytic routes is very low hence not capable of meeting the practical needs. It is paramount to comprehend the detailed mechanisms of the NRR reactions as well as their primary challenges. Various studies have provided detailed summaries of the promising features of MXenes in this regard. The protocols of synthesis for the obtaining of turnable morphological features have also been explored [16].

13.2.3 Oxygen Reduction Reaction

Reactions involving oxygen reduction (ORR) are vital, but very slow in various usages, including metal–oxygen batteries and fuel cells. At present, the standard catalysts for oxygen-based reduction processes are platinum-based catalytic materials. Such catalysts are scarce and also highly expensive. It becomes paramount for sourcing catalysis that is cheap and more readily available. The utilization of MXene has been found to be highly promising in this direction. They have recently contributed to the advancement of the overall efficiencies of sensors, batteries, and capacitors [17]. As a result of their unique electrochemical properties, they are expected to highly thrive in reactions involving ORR and photocatalytic processes using MXene as support. MXenes have been shown to have remarkable stability and high efficiency under different pH conditions making them suitable for applications in metal–air batteries. With regard to electrochemical energy conversion involving the use of fuel cells, one of the most desirable pathways involving the use of MXene is the four-electron route, which involves more electron transfer. Most studies have used platinum-based catalysts. These catalysts however are exorbitant in the cost and also have undesireable long-term features such as dissolution in various solvents and poisoning in the presence of carbon monoxide. This therefore paved way for MXene as a better replacement in this regard. The oxygen reduction process on the cathode of the cell is the primary basis of the electrochemical reduction process hence the performance of the ORR electrocatalysts determines the overall efficiencies of various related components [18].

13.2.4 Oxygen Evolution Reactions

Oxygen evolution processes form the basis for the production of various forms of renewable energies. However, there is a pressing need for cheap and highly efficient electrocatalysts for the purpose of overcoming the remarkably high overpotential and extremely slow process kinetics. MXenes have been widely employed primarily in the production of electrodes in supercapacitors and battery components. More recently their unique physicochemical properties have made them highly promising for their potential utilization in components for catalytic layers during reactions that involve the oxygen evolution reaction (OER). Various studies have reported that the hybrid catalysts involved in this process are made of transition metal oxides and MXenes. Also, the hydrophilic feature of MXenes makes it possible for their usage in the full covering of hydroxyl ions within the surfaces of the catalysts from the electrolytic material hence aiding in the formation of oxygen. Also, their high tendencies of being tuned also make them significantly useful for the improvement of their potential of releasing oxygen. Various experimental and computational procedures have been used to establish the relevance of MXenes in this regard [19]. Some of the available studies have documented a novel kind of non-precious metallic material for OER through the synergistical coupling of double hydroxide with MXenes [8].

13.3 Strategies for Optimization of Catalytic Potential of MXenes

Two major methods have been put forward for enhancing the catalytic properties of MXenes. First, the various functional groups such as fluorine, hydroxyl group, and oxygen present on the surface as well as the terminal sites exposed can function as the site for the catalytic activities as proven from theoretical basis and experimentations. It is therefore clear that moderating the surface chemistry of such molecules is a reliable approach for enhancing the electrocatalytic properties. Some of the major strategies for the improvement of the catalytic abilities of MXenes are highlighted.

13.3.1 Termination Modification

This is important in aiding the performance of MXene catalytic efficiency and involves improvement of surface termination and conductivity. This also helps in ensuring other parameters such as electronic structure are optimized. There are studies on termination modification of various synthesized MXenes with theoretical and experimental findings. The two-dimensional MXenes having hydroxyl and oxygen terminal points showed remarkable metallic properties hence aiding efficient charge transfer. Through termination modification, there is improvement of exchange current and rise in the evolution of hydrogen gas. The presence of atoms of oxygen on the surface brings about an improvement in the interaction between MXene and hydrogen. The surface terminated using oxygen tends to be more effective than the surface terminated using fluorine and hydroxyl groups. On experimental basis, the absence of moderation of surface termination hinders the production of ideal MXenes [17].

13.3.2 Nanostructuring

This is another reliable approach for enhancing the catalytic application of various MXenes such as nanodots and nanoribbons. MXenes that have nanodimension have rapid electrochemical feedback mechanisms, unique electronic structure, and numerous active sites which make them useful as catalyst. Nanostructuring makes it possible to increase the area of applications of MXenes by organizing the active sites. There are various strategies for the preparation of MXenes in the form of nanodots and nanoribbons some of which include ball mining, shaking treatment, and hydrothermal treatment [8].

13.3.3 Hybridization

The combination of MXenes with several other active substances to form hybrid has been recognized as another approach in improving their application as electro-catalyst. Some of the hybrid substances include phosphides, layered hydroxides, and chalcogenides. The special tunning can be achieved through integration with several two-dimensional materials like phosphorene, graphene, MXenes, their derivative compounds, and dichacolgenides. There are also numerous emerging studies focusing on theoretical and computational models. There is need to further investigate such theoretical findings at experimental and commercial scale. There are limited studies on the catalytic mechanisms that should be elucidated [20].

13.3.4 Metal Atom Doping

Various atoms have been doped into the two-dimensional MXenes some of which include graph diyne, MoS_2, and graphene. The primary target is to enhance the performance of the MXene as an electrocatalyst through modification of the electronic structure, altering the elemental constituents, and moderating the surface chemistry. In the material lattice of MXenes, and the coordination environment, the stabilization of the unit atom catalyst can be achieved through strong covalent bonds, which can act as active sites. Single metal atoms can therefore aid the improvement of the catalytic applications of MXene [21]. The impacts of the modification metallic atoms such as zinc, iron, amongst other on the catalytic activity have been documented using models such as Ti_3CNO_2 [22] fabricated CoP-3D MXenes hybrids and investigated their catalytic application. The nanohybrid synthesized showed a very low overpotential of 168 mV with a current density of 10 mA/cm^2. The remarkable electrocatalytic property was attributed to the three-dimensional architecture materials, which lead to an enhancement of their surface area thereby promoting a better charge transfer, and increased their diffusion speed. Modulation of the surfaces of the surface morphology of electrocatalysts for the purpose of improving the catalytic property has gained much attention in recent times. This can be achieved using defect engineering as a reliable strategy for optimization of the topical electronic structure of the catalyst [8].

13.4 Conclusion and Future Trend

The prospect of MXenes as electrocatalysts and catalysts for various processes cannot be overemphasized which is due to some of their unique properties such as remarkable hydrophilic potential, high conductivity, and exceptionally large surface area. This chapter has highlighted the applications of MXene in the area of catalysis. It discussed their synthetic routes, usages, and limitations. There is need to choose the most appropriate synthetic technique since different etching conditions give rise to varying products for different applications. Although several studies on synthesis and fabrication of MXene for catalytic applications have been carried out, most of the studies have been on at pilot scale not on a large scale. This aspect should be looked into by researchers through collaborations with industries. There is also a need for further investigations focusing on increasing the yield MXene since most of the findings from some available studies have shown a low yield.

References

1 Biswas, S. and Alegaonkar, P. (2022). MXene: evolutions in chemical synthesis and recent advances in applications. *Surfaces* 5 (1): 1–34. https://doi.org/10.3390/surfaces5010001.

2 Iqbal, A., Hong, J., Ko, T., and Koo, C. (2021). Improving oxidation stability of 2D MXenes: synthesis, storage media, and conditions. *Nano Converg.* https://doi.org/10.1186/s40580-021-00259-6.

3 Ma, Y., An, Y., Xu, Z. et al. (2022). Activating lattice oxygen of two-dimensional $M_nX_{n-1}O_2$MXenes via zero-dimensional graphene quantum dots for water oxidation. *Sci. China Mater.* 10.1007.

4 Shinde, P.V., Mane, P., Chakraborty, B., and Rout, C.S. (2021). Spinel $NiFe_2O_4$ nanoparticles decorated 2D Ti_3C_2MXene sheets for efficient water splitting: experiments and theories. *J. Colloid Interface Sci.* 602: 232–241.

5 Zhu, Z., Xu, C., Wang, Y. et al. (2022). The high performance nife layered double hydroxides@ $Ti_3C_2T_x$/reduced graphene oxide hybrid catalyst for oxygen evolution reaction. *J. Alloys Compd.* 894.

6 Liu, Y., Bai, L., Liu, H. et al. (2022). MXene-supported NiMn-LDHs as efficient electrocatalysts towards enhanced oxygen evolution reactions. *Mater. Adv.* 3 (10).

7 Cui, B., Hu, B., Liu, J. et al. (2018). He Solution-plasma-assisted bimetallic oxide alloy nanoparticles of pt and pd embedded within two-dimensional $Ti_3C_2T_x$ nanosheets as highly active electrocatalysts for overall water splitting. *ACS Appl. Mater. Interfaces* 10: 23858–23873.

8 Yu, M., Zhou, S., Wang, Z. et al. (2018). Boosting electrocatalytic oxygen evolution by synergistically coupling layered double hydroxide with MXene. *Nano Energy* 44: 181–190.

9 Etim, U., Zhang, C., and Zhong, Z. (2021). Impacts of the catalyst structures on CO_2 activation on catalyst surfaces. *Nanomaterials (Basel)* 11 (12): 3265. https://doi.org/10.3390/nano11123265.

10 Xie, C., Dafeng, Y., Chen, W., and Wang, S. (2019). Insight into the design of defect electrocatalysts: from electronic structure to adsorption energy. *Mater. Today* 31: 22. https://doi.org/10.1016/j.mattod.2019.05.021.

11 Champagne, A. and Charlier, J. (2020). Physical properties of 2D MXenes: from a theoretical perspective. *J. Phys. Mater.* 3: 3J.

12 Peng J., Chen X., Ong W., Zhao X., and Li N. Surface and Heterointerface Engineering of 2D MXenes and Their Nanocomposites: Insights into Electro- and Photocatalysis. 2018 https://doi.org/10.1016/j.chempr.2018.08.037.

13 Shen, J., Wu, Z., and Li, C. (2021). Emerging applications of MXene materials in CO_2 photocatalysis. *FlatChem* 28 (2): 100252. https://doi.org/10.1016/j.flatc.2021.100252.

14 Nguyen, T., Tusan, D., Tran, D., and Le o. (2020). MXenes: applications in electrocatalytic, photocatalytic hydrogen evolution reaction and CO_2 reduction. *Mol. Catal.* 486: 11085.

15 Lv, Z., Kou, L., and Frauenheim, T. (2021). Hydroxyl-boosted nitrogen reduction reaction: the essential role of surface hydrogen in functionalized MXenes. *ACS Appl. Mater. Interfaces* 13 (12): 14283–14290. https://doi.org/10.1021/acsami.1c00871.

16 Amrillah T., Hermawan A. and Yin S. MXenes and Their Derivatives as Nitrogen Reduction Reaction Catalysts: Recent Progress and Perspectives. 2021, 100864 https://doi.org/10.1016/j.mtener.2021.100864.

17 Huang, X., Song, M., Zhang, J. et al. (2022). Investigation of MXenes as oxygen reduction electrocatalyst for selective H_2O_2 generation. *Nano Res.* 15: 3927–3932. https://doi.org/10.1007/s12274-021-4057-9.

18 Yoo, R., Pranda, E., Johnson, D. et al. (2022). Review—the oxygen reduction reaction on MXene-based catalysts: progress and prospects. *J. Electrochem. Soc.* 169 (6): 063513.

19 Yang, S. et al. (2018). Fluoride-free synthesis of two-dimensional titanium carbide (MXene) using a binary aqueous system. *Angew. Chem. Int. Ed.* 57: 15491–15495.

20 Sing, M., Kumar, A., and Sing, H. (2017). Performance improvement of photovoltaic: utilization of two-dimensional $Ti_3C_2T_x$ MXene. *Surf. Interfaces* 27, 12: 100169. https://doi.org/10.1016/j.surfin.2021.101566.

21 Sinha S., Kim H., and Robertson A. Preparation and application of 0D-2D nanomaterial hybrid heterostructures for energy applications. 2022, https://doi.org/10.1016/j.mtadv.2021.100169.

22 Xiu, L.Y., Wang, Z.Y., and Yu, M.Z. Aggregation-resistant 3D MXene-based architecture as efficient bifunctional electrocatalyst for overall water splitting. *ACS Nano* 12 (8): 8017–8028.

14

MXene and Its Hybrid Materials for Photothermal Therapy

Chansi, Karan Hadwani, and Tinku Basu

Amity Centre for Nanomedicine, Amity University Uttar Pradesh, Noida, India

14.1 Introduction

Photothermal and photodynamic therapy (PTT) emerged as minimally invasive therapeutic methods and sought to be a new research center in cancer diagnosis and prevention [1]. It has exceptional real-world value due to its ease of function, short treatment time, and rapid recovery [2]. PTT stimulates cancer cell death by generating thermal radiation with near-infrared (NIR) light. Various NIR absorbers or photothermal agents (PTAs) are used to accelerate efficient heat production [3]. Frequently, photoacoustic therapies (PTAs) within the NIR biowindows, specifically NIR-I (650–1000 nm) and NIR-II (1000–1700 nm), have gained popularity for tumor treatment due to their ability to cause minimal harm to healthy cells, exhibit low self-absorption, and offer deep tissue penetration. When exposed to light, PTAs produce photons and light stimulation, leading to the creation of mobile charge carriers due to the electric field generated by the light [4]. The energy acquired by these carriers is subsequently transformed into heat energy. Based on their chemical arrangement, PTAs can be classified as organic and inorganic nanomaterials [5]. Organic materials often used for PTT are small-molecule dyes and polymer-based nanoparticles (NPs). Several semiconducting polymers, including polydopamine (PDA) and polypyrrole (PPy), which also have NIR absorption properties, have also been investigated for PTT [6]. Among inorganic nanomaterials such as carbon-based nanomaterials, gold (Au) NPs, transition metal nanomaterials, black phosphorus, and transition metal carbides (MXene), have been extensively studied due to their exceptional properties like photothermal conversion efficiency (PCE) and excellent photothermal stability [7]. However, when contrasted with zero-dimensional (0D), one-dimensional (1D), and bulk substances, two-dimensional (2D) nanomaterials employed in photothermal applications [8, 9] have several benefits, including ultrathin thickness, preparation of large-area photothermal devices with minimal concentration, ultimately reducing overall cost, fast response, light, and tunable bandgaps by adjusting the number of layers of the laminar stack, it enables a wider absorption of the solar spectrum. In addition, the excellent in-plane electron mobility enriched the PCE of 2D PTAs for converting light to thermal energy [10].

MXenes: Next-Generation 2D Materials: Fundamentals and Applications, First Edition.
Edited by Jay Singh, Kshitij RB Singh, Ravindra Pratap Singh, and Charles Oluwaseun Adetunji.

MXenes are multifunctional 2D nanomaterials consisting of transition metal carbides and carbonitride with many attractive properties [11]. They are denoted by the general chemical formula $M_{n+1}X_n$, where M signifies an initial transition metal and X stands for either carbon or nitrogen. The value of n can be 1, 2, or 3. The M atoms are uniformly distributed in a densely packed arrangement, while the X atoms are situated within octahedral interstitial sites. (Here, M represents a transition metal and X refers to carbon and/or nitrogen [12]. MXenes are typically hexagonal in the crystal structure, and they are produced by etching a particular atomic layer from MAX phases, which are ternary compounds of transition metals, carbon or nitrogen, and group A elements (mostly group IIIA and IVA) [13]. More than 70 MAX phases have been recognized, and a total of 20 MXenes have been produced using elements such as Ti, V, Nb, Mo, Ta, and Zr [14, 15]. MXenes are a versatile class of 2D materials with a wide range of potential applications. They offer greater structural variability than other 2D materials, such as graphene, and this allows them to be tailored to specific applications. For example, MXenes can be made with different metal compositions and surface terminations, which can change their properties such as electrical conductivity, surface area, and bandgap. This makes MXenes a very promising material for a variety of applications, including electronics, energy storage, catalysis, and biomedicine [16–18].

Recent studies have demonstrated the unique and attractive photothermal conversion behavior of MXenes [18]. They have excellent absorption in the NIR region and have an increased PCE, indicating broad photothermal applications. For example, MXenes have the ability to increase the temperature at the tumor site by irradiating light, which can eliminate cancer cells with minimal effects on the surrounding environment in a controlled [19] manner. Ti_3C_2 nanosheets were reported as one of the first MXenes found to be highly effective for photothermal therapy (PTT) [20]. A further increase in photothermal absorption was achieved by organic base-controlled intercalation and delamination of Ti_3C_2 sheets. However, until now, these materials have mainly been investigated for cancer treatment, and a precise understanding is still in its infancy.

With this background in mind, this chapter summarizes the basic mechanism, namely localized surface plasmon resonance (LSPR) effect, electron–hole effect, and hyperconjugation for photothermal conversion with special emphasis on different materials under investigation. Furthermore, an overview of the unique optical and thermal properties of MXenes is discussed, highlighting their role in their photothermal properties. A broad overview of the various mechanisms proposed for their photothermal conversion is discussed, and the origin of MXenes has been highlighted. Supported by the exceptional optical and thermal properties of MXenes, they have been explored for various biomedical purposes. Finally, the applications of MXene in PTT, as well as its synergistic therapy together with chemotherapy and immunotherapy, are discussed in detail. Finally, a critical analysis of current challenges, technological barriers, and future opportunities in this field is discussed.

14.2 Photothermal Conversion

Photothermal conversion refers to the utilization of solar energy into thermal energy through absorption and collection systems [21]. The infrared band of the solar spectrum is known to produce heat, while the other spectrums do not. Therefore, it has become mandatory to use light-absorbing materials to utilize the phenomenon of photothermal conversion. As the light-absorbing material is irradiated with light, it leads to the generation of photon energy by interacting with the crystal lattice, amplifying the vibrations, and ultimately increasing the temperature. Ideally, the photothermal material exhibits efficient absorption over a wide range of the solar spectrum (contained in the 0.3–3 μm range) along with a high ability to convert sunlight into heat. Recently, many photothermal materials have become very popular and have been extensively researched for their unique energy conversion efficiency. Among them are plasmonic metal NPs, semiconductors, nanocarbon materials, MXenes, as well as several organic polymers. When irradiated with sunlight, most of these materials assimilate photons upon irradiation, leading to photoexcitation and ultimately generating mobile charge carriers through a light-induced electric field that definitively converts solar energy into heat [22, 23]. A few methods have been adopted to increase the efficiency of light-to-heat conversion by tuning the structural configuration, hybridizing light-absorbing materials, and optimizing the absorber/reactant configuration. Material selection for photothermal applications is highly dependent on their optical properties, such as absorption coefficient, adaptation of material absorption to incident light, heat transfer, and losses. In addition, the inherent properties of the materials, such as thermal stability and mechanical integrity, also play an integral role. Due to the factors discussed above, the conversion mechanism for different photothermal-absorbing materials is different and varies depending on their response to electromagnetic radiation associated with different intrinsic bandgaps and electronic configurations. In general, it can be classified as (i) the phenomenon of LSPR, (ii) the creation of electron vacancies, and (iii) the impact of conjugation or hyperconjugation. This segment provides a brief overview of these mechanisms.

14.2.1 Localized Surface Plasmon Resonance Effect (LSPR)

Plasmonic metal nanostructures exhibit exceptional optical behavior, the so-called LSPR effect [24]. It is explained that when the energy of the incoming light aligns with the resonance frequency of the unbound electrons located at the surface of the NPs situated at the interface of the dielectric and the metal, the photoinduced oscillation of the charges [25] occurs, as shown in Figure 14.1.

A surface plasmon is a collective oscillation of the free electrons in a metal that is caused by the interaction of the electrons with electromagnetic fields. It is a quasiparticle, which means that it is not a real particle, but it behaves like a particle in some respects [26]. Surface plasmons can decay in two ways: radiative decay, where the energy of the plasmon is emitted as light, and non-radiative decay, where the energy of the plasmon is converted into heat. Non-radiative decay is the most common way for surface plasmons to decay, and it is this process that is responsible for the photothermal effect. When a surface plasmon undergoes non-radiative decay, it generates energetic electrons referred to as "hot electrons." These electrons subsequently transfer their energy to the nearby environment as thermal energy. The photothermal effect can be used to generate heat in a variety of materials, including metals, semiconductors, and graphene. This heat can then be used for a variety of applications, such as cancer therapy, water purification, and food processing. Noble metals (gold and silver NPs), doped semiconductors, and graphene are the most common plasmonic

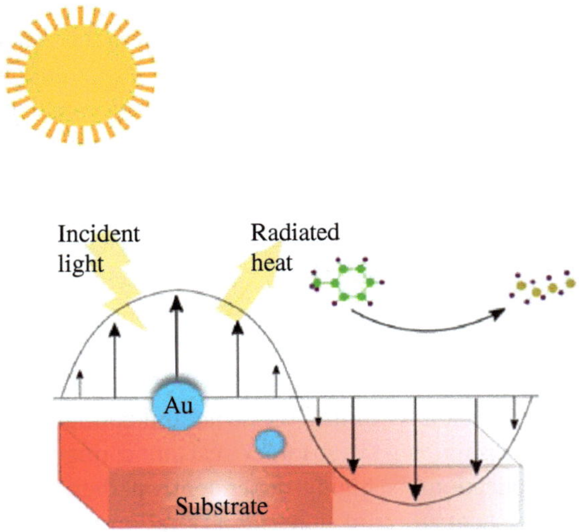

Figure 14.1 Schematic representation of LSPR Effect in plasmonic nanostructures.

materials used for photothermal applications. This is because they have strong surface plasmon resonances that can be excited by light in the NIR region. The NIR region is ideal for photothermal applications because it can penetrate deep into tissue, making it possible to target cancer cells with heat [27]. Gold NPs can heat up to 75 °C when their plasmon resonance wavelength matches that of the illumination laser. There are a few methods for enhancing the LSPR effect, such as decreasing the scattering cross-section and increasing the PCE. The size (volume) of a NP affects both its absorption and scattering cross-sections, but to different degrees. The scattering cross-section (σ_{sca} ($\propto V$^2)) is more sensitive to changes in V than the absorption cross-section (σ_{abs} ($\propto V$)). As the size of the NPs increases, the scattering cross-section eventually becomes equal to the extinction cross-section. Gold NPs with smaller dimensions have better absorption capabilities across the entire visible light spectrum (400–800 nm) and therefore have the advantage of better light-to-energy conversion. A reliance on the dimension of a nanomaterial of different geometries (Au nanorods, nanoshells) has been observed for displaying different absorption and their scattering cross-section. Synthesis of nonspherical NPs with precisely controlled aspect ratio has attracted attention recently [28]. For AuNR, an increase in volume is seen, and absorption does not have strong relevance, further confirming that the required PCE can be achieved through smaller-size nanostructures. In addition to gold, other metal nanomaterials (Al, Ag, Ge, etc.) and some bimetals, such as Au–Ag, have also been investigated as light-absorbing materials for solar thermal evaporation. Among 2D materials, the transition metal carbide $Ti_3C_2T_x$ MXene [29] exhibits a LSPR effect on optical radiation and electron ejection into the conduction band, which increases photogenerated charge carriers, and consequently, PEC performance is better. Recent advances in simulation, theoretical modeling, and synthetic methodologies have led to the development of novel photothermal agents that are designed specifically for unique applications. These new photothermal agents often incorporate characteristics distinctive to plasmonic NPs, like sensitivity to polarization, catalytic reactivity, and autonomous adjustment of thermal distortion, enabling the creation of more intricate and utilitarian substances. This results in improved photothermal attributes.

Plasmonic coupling stands out as a highly promising strategy for expanding the absorption range of plasmonic NPs to align with the energy distribution found in the solar spectrum. This can be achieved by combining different plasmonic NPs with different compositions, sizes, and morphologies. It is important to create procedures with a suitable configuration to improve the desired coupling effects. This can be done by controlling the size, shape, and spacing of the plasmonic NPs. Some additional benefits of using plasmonic NPs for photothermal applications are as follows:

- They have a high absorption efficiency, which means that they can convert more light into heat.
- They can be tuned to different wavelengths, which makes them suitable for a variety of applications.
- They are relatively easy to synthesize.

Overall, plasmonic NPs are a promising new class of photothermal agents with a wide range of potential applications. However, the challenge [30] of using plasmonic nanomaterials is their cytotoxicity toward healthy cells. Cytotoxicity is the ability of a substance to kill cells. Plasmonic nanomaterials can be cytotoxic because they can generate heat when exposed to light. This heat can damage healthy cells as well as cancer cells. The development of stable and low-toxic plasmonic nanomaterials is therefore extremely important. Researchers are working to develop these materials by using different materials and surface coatings. Another approach to reducing the cytotoxicity of plasmonic nanomaterials is to incorporate multiple elements into a single NP. This can be done by using core–shell NPs or by using NPs with different shapes. Multifunctional NPs have the potential to create a cohesive platform for diagnosis and treatment. For example, a NP could be used to deliver drugs to cancer cells and to image the tumor at the same time. Overall, plasmonic nanomaterials have the potential to be used for a variety of biomedical applications. However, more research is needed to develop stable, low-toxic, and multifunctional plasmonic nanomaterials.

14.2.2 Electron–Hole Generation

In a typical semiconductor material, electron–hole pair generation and relaxation can occur [31]. When a semiconductor is hit with light that has a wavelength equal to or shorter than its bandgap, it can absorb the light and generate electron–hole pairs. The generation of electron–hole pairs occurs when a photon is absorbed, causing an electron from the valence band to transition to the conduction band, as shown in Figure 14.2. This leaves a hole in the valence band, which can be filled by an electron from another valence band state. The electron and hole can then recombine, either radiatively or non-radiatively. Radiative recombination occurs when the electron and hole emit a photon as they recombine. Non-radiative recombination occurs when the electron and hole lose their energy to the crystal lattice in the form of heat. The non-radiative recombination of electron–hole pairs is what causes the temperature of the semiconductor to increase. This is because the energy lost by the electron and hole is transferred to the crystal lattice, which causes the lattice to vibrate more. The temperature propagation in a semiconductor can be studied by measuring the change in optical absorption and non-radiative bulk/surface

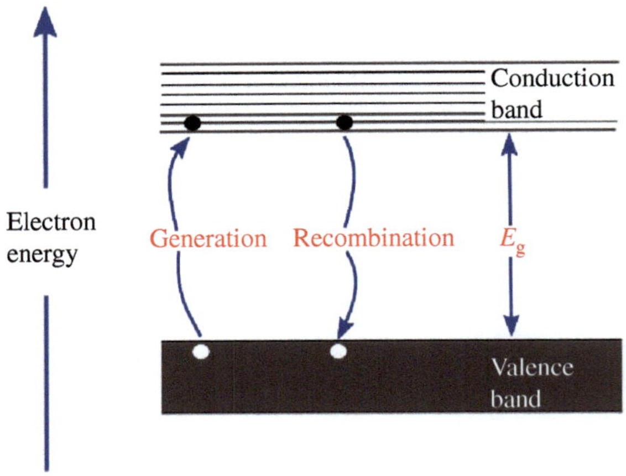

Figure 14.2 Electron–hole generation in semiconducting materials.

recombination. Low-bandgap semiconductors such as CuS, Ti_2O_3, and MoS_2 show very efficient photoconversion efficiency [32]. In wide-gap semiconductors, recombination of electron–hole pairs occurs simply because almost all electrons accept energy and re-emit as photons, reducing the efficiency of photothermal conversion. Within metal oxide semiconductors, the occurrence of electron–hole pairs can be observed [33] when light possessing energy similar to the bandgap energy is taken by the substance.

Low-bandgap semiconductors can absorb more energy from sunlight than wide-bandgap semiconductors because the energy of most photons from incident sunlight is greater than the bandgap energy of low-bandgap semiconductors. When an electron–hole pair is generated in a low-bandgap semiconductor, the electron and hole can recombine non-radiatively, releasing their energy into the crystal lattice in the form of heat. This makes low-bandgap semiconductors well-suited for applications where heat is needed, such as solar evaporation. Narrow bandgap semiconductors such as hydrogenated black titanium, Ti_2O_3 NPs, oxygen-deficient MoO quantum dots, and Fe_3O_4 have been studied for their photothermal properties in recent years. Bimetallic oxides such as $MnFe_2O_4$, $CoFe_2O_4$, and $ZnFe_2O_4$ have also shown promise as photothermal materials. Wang and her team recently developed a 3D light absorber design that can achieve both high light absorption and thermal efficiency. This design could be used to improve the performance of solar evaporation and other photothermal applications.

14.2.3 Hyperconjugation Effect

In a conjugated system, the π electrons in adjacent atoms are delocalized, or spread out over the entire system. This delocalization of electrons is what gives conjugated systems their unique properties. (Figure 14.3). The interaction between the σ-bond electrons and the neighboring empty or partially filled p-orbital can lead to the delocalization of electrons, which is known as the hyperconjugation effect. This effect, along with the conjugation effect, can allow conjugated systems to absorb light in the NIR region. When the electrons

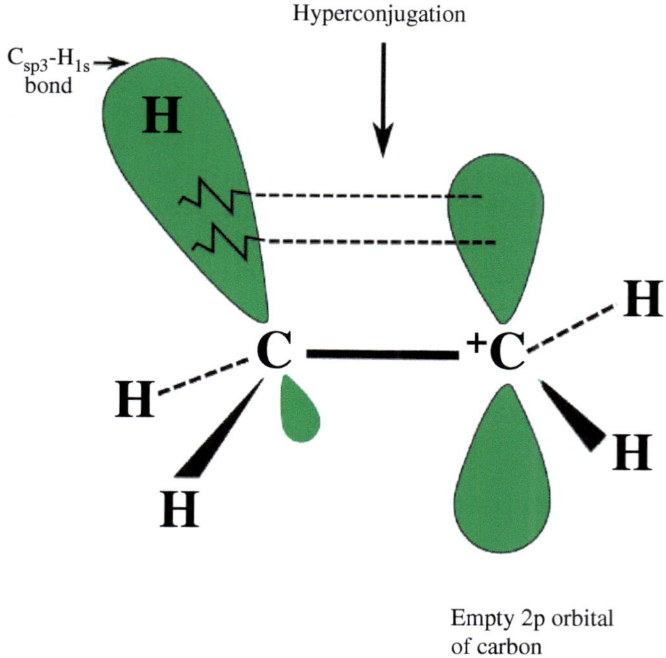

Figure 14.3 Hyperconjugation effect.

in the π orbital are excited by light, they can jump to the π* orbital. This releases heat as the electrons return to the ground state. The hyperconjugation effect and the conjugation effect also accelerate electron mobility in conjugated systems. This is because the delocalization of electrons makes it easier for electrons to move from one atom to the next. These two phenomena play a role in the photothermal conversion observed in numerous conjugated carbon nanomaterials, including carbon nanotubes (CNTs) and graphene. When these materials are exposed to light, they can absorb the light and convert it into heat. This heat can then be used for a variety of applications, such as cancer therapy and water purification.

CNTs consist of a large number of sp^2 carbons and are considered a standard 1D nanomaterial [34, 35]. CNTs are classified as SWCNTs or MWCNTs. They are potentially toxic and hydrophobic but can be chemically modified to be more soluble in water. Du et al. decorated the surface of SWCNTs with amino groups using HNO_3 oxidation and amination methods to increase their dispersibility [36]. CNTs can be functionalized with molecules to improve their photothermal properties. One way to do this is to non-covalently functionalize them by simply adsorbing molecules onto the surface of the CNTs. Wang et al. modified SWCNTs with a polyethylene glycol (PEG)-grafted amphiphilic polymer to improve their dispersibility and solar heat conversion efficiency. This greatly increased the efficiency of the solar heat conversion in sunlight. In other cases, CNTs have been non-covalently modified through straightforward adsorption; these materials can be functionalized with a wide range of molecules, including polymers, biomolecules like carbohydrates/polysaccharides, proteins, enzymes, and DNA. Wang et al. conducted the modification of single-walled carbon nanotubes (SWCNTs) using an amphiphilic polymer grafted with PEG [37]. PEG-coated SWCNTs showed strong absorbance in the NIR region.

Graphene is a 2D material made up of carbon atoms. It is hydrophobic but can be made hydrophilic by functionalizing it with nitrogen atoms [38] or incorporation of hydrophilic functional groups (e.g. hydroxyl and carboxyl) [39]. Robinson et al. developed a method for producing biocompatible reduced graphene oxide (RGO) sheets with improved NIR absorbance. This could lead to the development of more efficient RGO-based photothermal agents. [40]. Arginine modification can improve the stability of RGO in aqueous solutions and make it more photothermally active. This makes it a promising material for a variety of applications [41]. Graphene sheets can be assembled into a 3D film with vertically oriented graphene sheets. This material exhibits excellent light absorption throughout the solar spectrum (250–2500 nm), which makes it a promising candidate for photothermal applications. Photothermal applications use light to generate heat, which can be used for a variety of purposes, such as cancer therapy and water purification. The 3D graphene-based material has a high photothermal transduction efficiency of 86.5%, which means that it can efficiently convert light into heat. This makes it a highly promising material for photothermal applications [42]. Organic materials, especially polymeric ones, are promising photothermal materials because of their biocompatibility, flexibility, low cost, excellent light-harvesting ability, and PCE. Conjugated polymers are large molecules distinguished by π-conjugated frameworks and widely recognized as advanced materials for NIR absorption in photothermal applications. Among the well-known biocompatible photothermal materials based on conjugated polymers are polyaniline [42], poly (3,4ethylenedioxythiophene): poly(4-styrene sulfonate) (PEDOT: PSS) [43, 44], PPy [45, 46], PDA [47], and donor–acceptor (D–A) structured polymer molecules [48]. COFs are a class of crystalline porous organic polymers with light elements, high surface area, and thermal/mechanical stability. They are easy to functionalize, customizable, and can be predeterminable [49]. Consequently, COFs have found extensive application as effective tools for the controlled and predictable incorporation or containment of functional NPs. Tan et al. synthesized core–shell COF–Fe_3O_4 microspheres using an amorphous-to-crystalline process by encapsulating Fe_3O_4 nanoclusters in an imine-linked COF network [50]. COF–Fe_3O_4 microspheres are made up of a core of Fe_3O_4 NPs surrounded by a shell of covalent organic frameworks (COFs). The COF shell gives the microspheres their NIR absorbance and high molar extinction coefficient. The PCE of COF–Fe_3O_4 microspheres is the percentage of light energy that is converted into heat energy. The high PCE of these microspheres makes them promising for applications in PTT, which uses heat to destroy cancer cells. Hu et al. have developed a method for fabricating COF–CuSe nanocomposites with NIR photothermal conversion properties. This method is simple and can be performed under ambient conditions, which makes it more scalable and cost-effective than other methods for fabricating COF–CuSe nanocomposites [51]. When COF–CuSe nanocomposites were irradiated with a laser, they exhibited a photothermal effect with a conversion efficiency of 26.34%.

14.3 Optical and Thermal Properties of Mxenes

The essentiality of the optical properties of an object relates to its electronic and structural compositions. Advantageous properties like metallic conductivity, surface hydrophilicity, and excellent mechanics are assets of MXene family. Other than that, MXene is the type of material that allows meticulous tuning of optoelectronic properties when changing intercalation products between MXene. The bandgap ranges from 0.92 to 1.75 eV. Few Mxenes are reported to have absorption in the NIR-II region (1000–1350 nm) [52]. MXenes offer high metallic conductivity and a large surface area allowing their use for enhancement of signals for Raman scattering and bioimaging-guided therapeutic applications.

Hypothetically, there are two components one, is a real component (Re) and the other is an imaginary component (Im) of a material that has a dielectric constant $\omega(\varepsilon)$ with a ruling frequency (ε). The presence of a terminal group in MXene can have a sharp effect on the MXene structure ($Ti_3C_2T_x$) and electronic properties, and therefore it regulates the optical properties. Pointedly, the imaginary component of the dielectric constant is affected by inter/intraband electronic shifts. This process is heavily accustomed to fundamental structural differences between the hydroxylated/fluorinated terminations accepts only one electron. These terminations have a comparable absorption within the visible range. Oxygen, on the other hand, requires two electrons to be steadied for its termination. Recently, Berdiyorov has attempted to evaluate the optical properties of MXenes that are tagged with functional units. This study reveals that the functionalization of the structure has led to a considerable reduction by two times the static dielectric function. When the absorption of the fluorinated and oxidized samples was compared, the oxidized samples exhibited higher absorption than the fluorinated samples. When oxygen is introduced for the termination of MXenes, there is a change in density of states (DOS) near their Fermi level, which results in a change in their optical properties. A minor adsorption and reflectivity are observed in the transparent hydroxylated/fluorinated structures. This indicates the discovery of $Ti_3C_2T_x$ surface in a partially oxidized state when it is subjected to a wet atmosphere for a long period of time.

MXenes that are tagged with oxygen are capable of having varied applicability and relatively higher thermodynamic stability as compared with –F and –OH functionalized types. As the absorption of MXenes varies from 760 to 268 nm, their intensity can be tuned further by the terminal groups present in the structure. When substitution of a carbon atom is replaced with nitrogen, it results in anomalous absorption, which is observed at 670 nm compared to their built-in 780 nm peak. The difference in distribution of electrons in nitrogen can lead to a change in conductivity. For example, in Ti_3CNT_x a less conductivity is observed in comparison to $Ti_3C_2T_x$, which indicates a lower energy band far from the Fermi level.

Jhon et al. have correctly estimated the optical absorption and the extinction coefficient of MXene, and it ranges to Terahertz, which is around 1 trillion hertz. They have gone through several calculations by employing density functional theory. In their work, HF_2CO_2 has been reported to circumstantially possess a moderate bandgap and deliver a thermal conductivity that exceeded that of MoS_2 and phosphorene. The rise in thermal conductivity happens generally by increasing the atomic number of M of M_2CO_2 (M=Ti, Zr, Hf) MXenes, and all three MXenes offer high carrier mobilities as well. On the contrary, Ti_2CO_2 is found to possess much diminished thermal conductivity but higher hole mobility than that of Zr_2CO_2 and Hf_2CO_2. Exempting the surface modification factor, surface imperfections could potentially arise during the synthesis procedure or due to the incorporation of external elements, which is believed to impact the overall material conductivity. When number of surface defects is increased, it affects the free movement of electrons because it gets lowered, as well as causing a loss of balance in electronic density within the metallic layer. There are other attributes that MXenes possess, such as the ability to produce electrons and vacancies under specific stimuli, such as interaction with light or heat. Lastly, MXene expresses an excellent optical property from UV to THz ranges; this points toward the possibility that various other beneficial applications could be explored by manipulating the optical properties when intrinsic structural and immediate environment modifications are done. In future, scientists might meddle with a well-defined singular termination and finely tune the metal ratio, which will help to determine the optical properties of MXenes.

14.4 Photothermal Conversion Mechanism of MXenes

Photothermal conversion is one of the important aspects of MXenes. The mechanism of photothermal conversion in MXenes is not fully understood, as research on the photothermal properties of MXenes is still in its early stages. Figure 14.4 illustrates the general pathway for the absorption of light energy by MXene nanosheets and their composites. First, the MXene nanosheets absorb electromagnetic waves. The energy of the photons is then transferred to the lattice motion inside the structure. As more photons are absorbed, the temperature of the MXene increases. A series of studies have been conducted to gain a deeper understanding of the photothermal conversion mechanism in MXenes. These studies have shown that the PCE of MXenes is influenced by a number of factors, including the type of MXene, the size and morphology of the MXene nanosheets, and the presence of defects in the MXene structure. The research on the photothermal properties of MXenes is ongoing, and it is expected that a better understanding of the photothermal conversion mechanism in MXenes will lead to the development of new and improved applications for MXenes in the fields of PTT, biosensing, and solar energy harvesting.

Shahzad et al. have shown that MXenes have electromagnetic interference shielding properties. This means that MXenes can absorb electromagnetic waves without reflecting them. The un-reflected electromagnetic waves travel freely throughout the lattice structure of MXenes and go through several internal reflections amongst their layers. These waves are gradually absorbed in the lattice structure of MXenes with unnoticeable emission. This property of MXenes makes them promising for applications in solar energy harvesting. MXenes can absorb light energy over a wide range of the solar spectrum, which makes them more efficient than traditional solar absorbers. The droplet laser heating method at the wavelengths of 473 or 785 nm has revealed 100% light-to-heat conversion for Ti_3C_2 MXene. MAX ceramics can impact the properties of the exfoliated MXene nanosheets

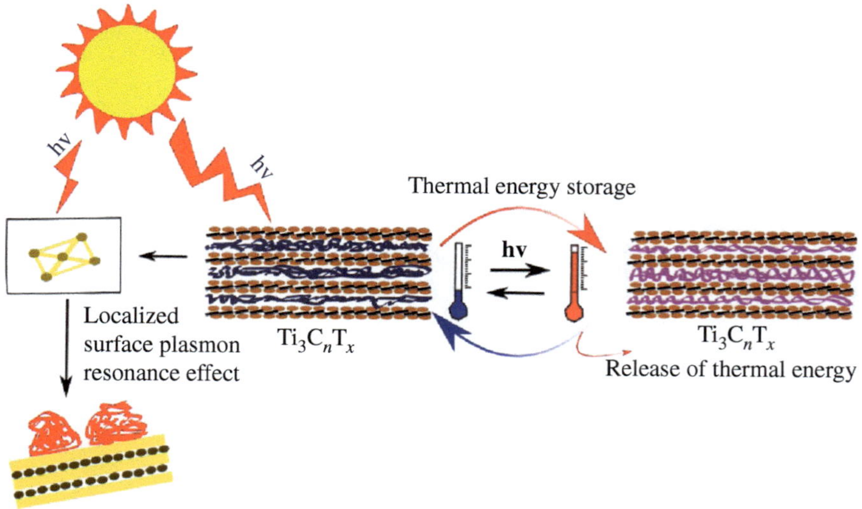

Figure 14.4 Photothermal Conversion in MXene.

giving them a metal-like property that works like a "semimetal" nature with an extremely sharp metallic conductivity. The presence of carbon atoms within the MXenes combines both the ceramic and metallic properties into their structure.

Fan et al. [53] have clearly stated that the absorption bands of PEG/$Ti_3C_2T_x$ nanocomposites in visible and NIR spectral ranges are amplified by the LSPR phenomenon exhibited by $Ti_3C_2T_x$. This leads to the rapid harvesting of solar energy for photothermal conversion and storage. When compared to other 2D materials such as graphene oxide nanosheets, the broad absorption bands of PEG/$Ti_3C_2T_x$ nanocomposites in the UV to NIR range make MXene a superior light absorber.

Some other research groups have also reported that the photothermal behavior of MXenes arises due to their inherited LSPR effect. High extinction coefficient (α) (as discussed before) possessed by the MXene is eminently responsible for their light-harvesting ability.

The absorption coefficient of Ti_3C_2 nanosheets at 808 nm is 25.2 Lg^{-1} cm^{-1}, which is higher than that of GO nanosheets (3.6 L g^{-1} cm^{-1}). This means that Ti_3C_2 nanosheets have a stronger absorption of NIR light. When Ti_3C_2 nanosheets are dispersed in water and irradiated with 808 nm laser light, the temperature of the suspension increases to 57 °C within 6 minutes. The temperature of pure water does not change under the same conditions. This shows that Ti_3C_2 nanosheets have excellent photothermal performance. Chaudhuri et al. [54] confirmed that arrays of nanodisks made of $Ti_3C_2T_x$ exhibit very high PCE. Because we know that MXene films absorb sunlight to reach temperatures of up to 70 °C, this facilitates the use of this property in anticancer research, which has significantly advanced biomedical applications.

Considering all the above prospects, the photothermal property of MXenes was fundamentally advantageous for PTTs due to their inherited photoconversion efficiency and wide range of applicability, as shown in Figure 14.5. Recently published studies have shown that biocompatible Ti_3C_2 nanosheets have great potential for photothermal

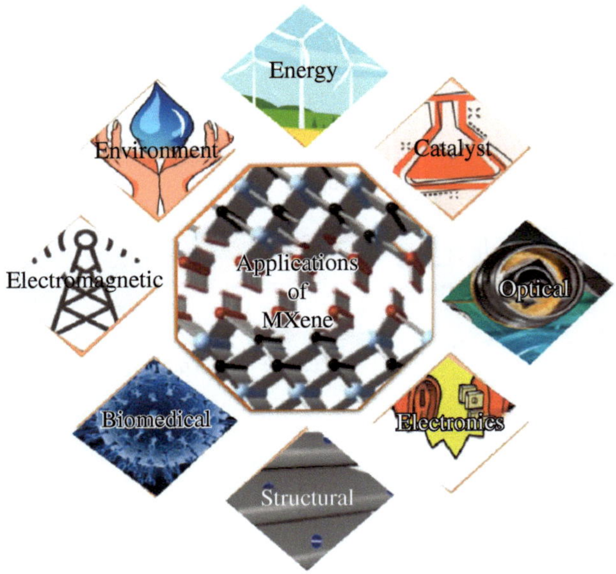

Figure 14.5 Applications of MXene.

elimination of tumors. This is achieved by applying local heat to the pathological site under the influence of laser irradiation. The photothermal applications of MXenes are summarized in Table 14.1.

14.5 Applications of MXenes in Photothermal Therapy

14.5.1 Photothermal Therapy

A process of utilizing light to be converted into heat that helps in curing a range of medical ailments like tumors is known as PTT. Laser-induced thermal ablation was traditionally adopted as a process that is undependable and has inadequate competence because it uses abnormally high power for its laser initiation. This can result in the devastation of the healthy tissue as they are vulnerable to the heightened magnitude of the laser, where the PTT comes to play a highly crucial role. This can aid in reducing the unnecessarily high intensity of the laser and converging it to a specific site that is affected by utilizing the materials, which can be triggered by light. Another way to save healthy tissue is by reducing the wavelength of the laser being used to obtain a low extinction coefficient for live tissues.

Photothermal agents and external light are the two deciding factors that can help or deteriorate the performance of PTT. The photothermal agents are known for their exceptional demonstration to absorb light, ideally based on the coefficient of extinction (ε) along with the capability to produce heat on external light irradiation, also termed as PCE (η). Usually, these agents accumulate throughout the tumor vicinity, generating localized heat when exposed to light and, as a result causing a hyperthermic impact that kills cancer cells.

Table 14.1 Photothermal applications of MXene.

S. No.	Material	Formula	Dimension	Peak values	Application	Classification of synthesis	Ref.
1.	$Nb_4C_3T_x$	M_4X_3	2-D	0.0046 Ω m (Resistivity)	Electrodes in batteries and supercapacitors	Etching and delamination	[55]
2.	Ti_3AlC_2 MXene clay	$M_{n+1}X_nT_x$	2-D	900 farads/cm^3 (capacitance)	Supercapacitor electrodes	Etching and delamination	[56]
3.	Ti_3AlC_2/PtNPs	Pt/e-TAC	3-D	0.180 A mg_{Pt}^{-1} (mass activity)	Catalyst support	Etching and Delamination	[57]
4.	Ti_3AlC_2	$Ti_3C_2T_x$	2-D	Areal capacitance = 220 mF cm^{-2}; Volumetric capacitance = 439 F cm^{-3}	Supercapacitor	Etching and delamination	[58]
5.	M_2C, where M represents Sc, Ti, V, Cr, Zr, Nb, Mo, Hf, Ta, etc.	M_2C	2-D	Gravimetric capacity >400 mAh/g (Sc, Ti, V and Cr)	Rechargeable batteries	Etching and delamination	[59]
6.	Intercalation of $Ti_3C_2 + Ti_3CN$ and TiNbC	f-MXene	2-D	Capacity = 410 mA h g^{-1}	Energy storage	Etching and delamination	[60]
7.	V_2CT_x and Ti_3CNT_x	Multilayered MXene	2-D		Capacitance	Etching and delamination	[61]
8.	Ti_3AlC_2-stacked sheets	$M_{n+1}AX_n$	2-D	Mass extinction coefficient = 29.1 Lg^{-1} cm^{-1}	Photothermal agent against tumor cells	Etching and Delamination	[62]
9.	Mo_2CT_x	$M_{n+1}AX_n$	2-D	Capacitance = 700 F cm^{-3}	Supercapacitors	Etching and Delamination	[63]

(Continued)

S. No.	Material	Formula	Dimension	Peak values	Application	Classification of synthesis	Ref.
10.	Ti_2N	M_2X	2-D		SERS application	Etching and Delamination	[64]
11.	Ti_3AlC_2 flakes	$M_{n+1}X_nT_x$	2-D	Field electron mobility = $2.6 \pm 0.7\,cm^2V^{-1}\,s^{-1}$ Resistivity = $2.31 \pm 0.57\,\mu\Omega\,m$	Sensing application	Etching and delamination	[65]
12.	V_2C		2-D	Absorption in NIR range = 48% under 808 laser irradiation	*In vivo* ablation of tumors	Etching and delamination	[33]
13.	$Ti_3C_2T_x$	$M_{n+1}AX_n$	2-D	Electrochemical capacitance value is $520\,F\,cm^{-3}$ at $2\,mV\,s^{-1}$	Energy storage electrodes	Surface modification	[66]
14.	Ti_3C_2	$Ti_3C_2T_x$	2-D	Shielding effectiveness	EM shielding and absorption materials with high performance	Surface modification	[67]
15.	Ti3C2-DOX	$Ti_3C_2T_x$	2-D	Mass extinction coefficient = $28.6\,Lg^{-1}\,cm^{-2}$ at 808 nm	Stimuli-responsive drug release behavior and cancer cell killing	Surface modification	[68]
16.	Ti_3C_2/CuS	$Ti_3C_2T_x$	2-D	Specific capacity of the battery is $169.5\,C\,g^{-1}$	High performance electrochemical energy storage devices	Hybridization of nanocomposites	[69]
17.	AgNWs + MXene		1-D + 2-D	Sheet resistance = $0.8 \pm 0.8\,\Omega\,sq.^{-1}$	Intelligent garments, sensors, actuators, energy generators, and EMI shielding	Hybridization of nanocomposites	[70]

18.	V_2C@PDMAEMA	2-D	Transmittance at 25 °C and 45 °C = 15% and 75%: Net increment of conductivity = 76 μS cm^{-1}	Smart-responsive MXene systems	Hybridization of nanocomposites	[71]
19.	$Ti_3C_2/BiWO_6$	2-D/2-D		Cocatalyst in photocatalytic reactions	Hybridization of nanocomposites	[72]
20.	rGO/$Ti_3C_2T_x$	2-D/2-D	Volumetric and gravimetric capacitance = 370 F cm^{-3} and 405 F g^{-1}	Supercapacitors	Hybridization of nanocomposites	[73]
21.	$Ti_3C_2T_x$/CNT	1-D/2-D	Volumetric capacitance = 390 F cm^{-3}	Electrochemical energy storage and generation devices	Hybridization of nanocomposites	[74]
22.	PEG/$Ti_3C_2T_x$	2-D	Photothermal storage efficiency = 94.5%	Photothermal storage	Hybridization of nanocomposites	[53]

PTT has shown potential and promise as the prime entity to exterminate tumors with minimal complications.

The efficacy of the photothermal conversion can be elevated by various photo-thermal materials served by materials like carbon, plasmon-metal-type nanostructures, other versatile materials, and materials containing organic components, and inorganic materials, all of which are currently explored. Although, they have several advantages and achievements, they feature a rather low PCE and an inconspicuous coefficient of extinction, hence hindering their wide application for more beneficial PTT applications. In recent times, MXenes have emerged as a new material with quite noticeable PCE when compared to the traditionally studied materials because they exert considerable absorption within the NIR spectrum, large absorption of biological opening segments, comparatively simplified post adjustments, as well as their good stability in biological environments required for a long-term circulation and low cell damage guarantee human health is safeguarded. These unique features imply their use as efficient photothermal agents. Figure 14.6 illustrates a few MXene-based materials applied for PTT.

Lin et al. have described that the employment of etching by using HF in bulk MAX Ti_3AlC_2 through outer area of soybean (SP) alteration results in 2D biologically compatible Ti_3C_2 nanosheets [75]. The synthesized Ti_3C_2-SP nanosheets are excessively thin and are singularly or few layers thick, have a strikingly high coefficient of extinction, i.e. $\varepsilon = 25.1\,g^{-1}\,cm^{-1}$, that is dominating graphene oxide 2-D sheets in nanodimensions or gold nanorods, which is 3.6 and 13.9 $g^{-1}\,cm^{-1}$. A very extensive light absorption efficiency ranging from 400 to 900 nm is associated with the abovementioned nanosheets indicating toward a favorable NIR laser-absorption characteristic. Even when the laser radiation was inflicted upon the Ti_2C_3-SP nanosheets for merely 6 minutes, the temperature of the solution reached 55 °C, which demonstrates that the nanosheets are highly capable

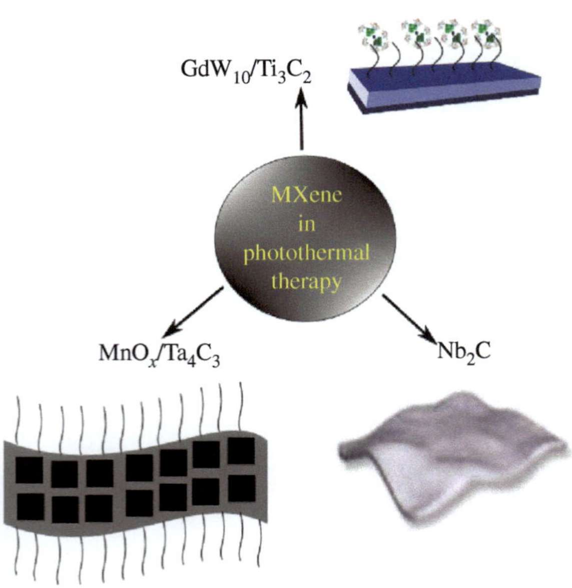

Figure 14.6 Various MXenes used for Photothermal therapy.

of converting the NIR light to heat. Moving on, the live-dead cell straining carried with the Ti_2C_3-SP nanosheet, where the green fluorescence exhibits the living cells and the red shows the dead cells, shows that a vast number of 4T1 cells have been eroded via hyperthermia after Ti_2C_3-SP nanosheet was used for treatment. The same observation has been noted after the 20 mg kg^{-1} Ti_2C_3-SP nanosheets were used on a mouse model containing tumor xenografts, as they amassed adjacent to the tumor's presence. The curative effect is further confirmed after reporting the transformation in tumor volume, and the results indicated that nanosheets exhibit impactful tumor inhibition abilities.

The predominant MXene nanosheet studies are fixated at the very initial NIR window, which is in the line between 750–1000 nm applied in PTT. On the other hand, the subsequent infrared region, which falls between 1000–1350 nm, never even came into consideration. As opposed to the first NIR window, the NIR-II window exerts a comparatively deeper tissue penetration; this, in turn, leads to the lowering of tissue spreading and claims to have the least side effects because the biological environments have self-heating abilities. They are traditionally known and used for graphene, MoS_2, and phosphorus (black), which are known to demonstrate insignificant absorbance in the second NIR window, which damages their chance of being applied to the NIR-II bio-window.

In recent times, Lin et al. [76] and team communicated that niobium carbide sheets can work within NIR I & II windows altogether through the exceedingly unusual efficiency, as shown in Figure 14.7. The nanosheets feature PCE of almost 47% at 1065 nm laser irradiation and 36.4% upon 808 nm laser irradiation. Ideally, Nb_2C-PVP nanosheets perform can provide absorption in each instance of first and second windows, which is further applied as a phototherapeutic agent. After careful heightening, the intensity of the laser power that is inflicted on the affected site demonstrates that the viability of the 4T1 cells lowers with each window. Also, the laser irradiation provided in 808 or 1064 nm was compared with respect to their various tissue penetration ability. The results show that NIR-II laser irradiation exerts a better conversion rate when it reaches a particular point *in vivo* tissue because of the advantages provided by lowered accumulation rates in the tissues. All these results, toward the combination that testifies the potential of *In vivo* photothermal elimination of tumors in NIR windows. The multifunctional agent as a theranostic application has been developed using titanium carbide type of MXene (MnO_x/Ti_3C_2) [77] with high efficiency for photoacoustic dual modality imaging and is highly efficient for monitoring tumor growth pattern and ablation. A heterostructure [78] designed over titanium carbide–cobalt nanowires (Ti_3C_2–CoNWs), can serve as a synergistic anticancer agent with features such as good magnetic properties and stimuli-responsive drug release. The nanocarrier offers excellent photothermal properties with laser irradiation (808 nm) and better drug-storing capacity (225%). Release behavior for Adriamycin is altered at different pH(4–6). Another recent study demonstrates Ta_4C_3 [79] MXene sheets are designed by a rational choice in composition of functional counterparts and MXene. MnO_x/Ta_4C_3 functioned as exceptionally well contrast substances in contrast-enhanced CT scans, meanwhile the incorporated MnO_x segment acted as contrast agents responsive to the tumor microenvironment in T1-weighted magnetic resonance imaging. It has wide applications in developing functionalization modes for the synthesis of material with theranostic applications.

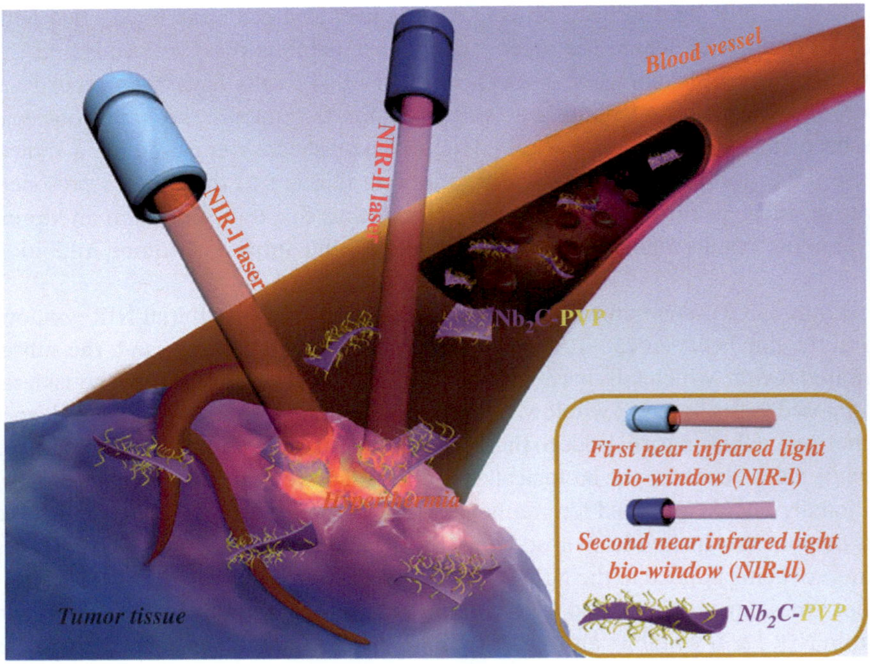

Figure 14.7 Schematic representation of 2D biodegradable Nb_2C modified with a polymeric covering for in vivo photothermal tumor ablation in both NIR windows. Source: Reproduced with permission [76] Copyright © 2017, American Chemical Society.

14.5.2 PTT-Coupled Chemotherapy

MXenes have emerged as an important material in nanomedicine exerting superior efficiency and effectiveness for synergistic therapy, especially during cancer therapy. Improved medical results were achieved by showcasing the effective merging of the Ti_3C_2 MXene's structural aspect (2D planar structure for drug delivery) and its demonstrated physicochemical characteristic (photothermal conversion for hyperthermia), which can achieve simultaneous chemotherapy and photothermal removal of cancer.

Han et al. [80] have reported that Ti_3C_2 MXene can act as a nanosized system bringing an outstanding drug-storing capability and stimulus-driven drug-offloading efficiency when engaged from within or without. The produced 2D Ti_3C_2 nanosheets have successfully achieved comprehensive and cooperative elimination of tumors in living organisms by incorporating the PTT-based treatment, as illustrated in Figure 14.8. A two-step process was used to synthesize nanosheets. Ti_3AlC_2 MAX-phase ceramics were subjected to elevated sintering temperatures. Subsequently, the central aluminum (Al) layer was removed by means of hydrofluoric acid etching. This process exfoliated the MAX-phase ceramics, but the resulting nanosheets were still aggregated. Further, a second exfoliation step was performed by using a tetra propylammonium hydroxide (TPAOH) to achieve an interposition reaction, finally reducing the size of particles. The soybean phospholipid (SP) modification of Ta_4C_3-SP nanosheets improves their stability, dispersibility, and ability to reach the bloodstream. Doxorubicin (Dox) can also be loaded onto the SP-modified Ta_4C_3-SP

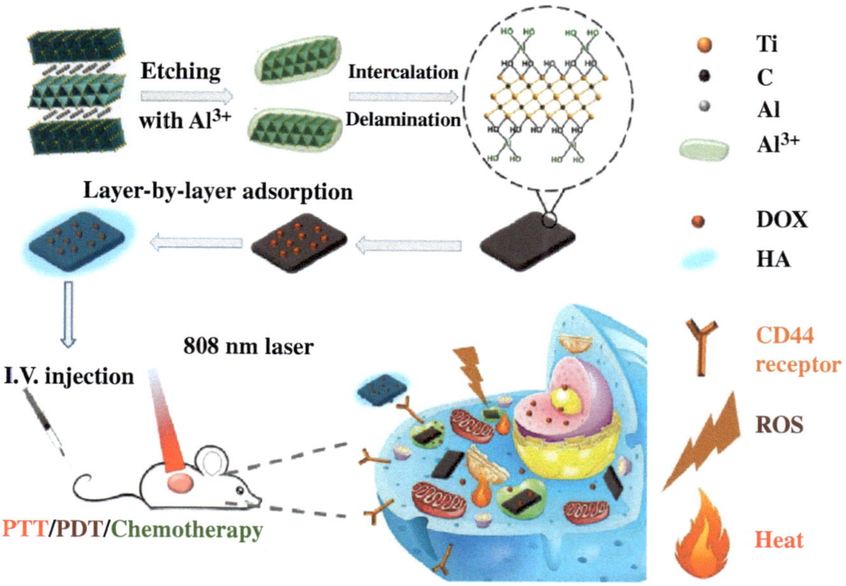

Figure 14.8 A schematic representation of Ti_3C_2-based nanoplatform with synergistic effects. Source: Reproduced with permission [68] Copyright © 2017, American Chemical Society.

nanosheets. When irradiated with a laser at 808 nm for 5 minutes, the temperature quickly reaches about 55 °C. The Ta_4C_3-SP nanosheets retain their thermal stability after repeated heating and cooling cycles. When laser irradiation is not applied, investigations into cell cytotoxicity revealed that Ta_4C_3-SP nanosheets exhibit minimal impact on the viability of 4T1 cells, even at elevated concentrations. Ta_4C_3-SP nanosheets can be administered intravenously or intratumorally in mice, and the temperature of tumor is raised to 60-to-68-°C after 6 minutes of irradiation with an 808-nm laser. As a result, Ta_4C_3-SP nanosheets can be used to effectively eradicate tumors after irradiation with an 808 nm laser.

Xu et al. [81] have developed a new type of nanomaterial called Ti_3C_2-PVP@DOXjade that can be used for chemo-PTT. This nanomaterial is made up of two layers: a layer of Ti_3C_2 nanosheets and a layer of a drug called doxorubicin (DOX). The Ti_3C_2 nanosheets are able to absorb light and convert it into heat, which can kill cancer cells. The DOX drug is able to kill cancer cells by disrupting their DNA. When Ti_3C_2-PVP@DOXjade is exposed to light, it releases both heat and DOX, which can kill cancer cells in a synergistic way. Dong et al. have reported a synergistic therapy based on Ti_3C_2 nanosheets that combines PTT, photodynamic therapy (PDT), and chemotherapy. The therapy uses a multilayer synthesis method to load the chemotherapy drug DOX into the nanosheet [68]. The load efficiency is high, at 84.2%. The Ti_3C_2-DOX nanosheets are effective at absorbing NIR light and converting it into heat. This heat can kill cancer cells in an acidic environment. The release of DOX is also triggered by heat, which can kill cancer cells by disrupting their DNA. The generation of reactive oxygen species (ROS) has been measured using the fluorescence intensity of the ROS probe dichloro-dihydro-fluorescein diacetate (DCFH-DA) on NIR laser irradiation. In vivo, studies have shown that the Ti_3C_2-DOX nanosheets can ablate tumors when administered at a dose of 2 mg/kg. The synergistic effect of PTT, PDT, and chemotherapy has been

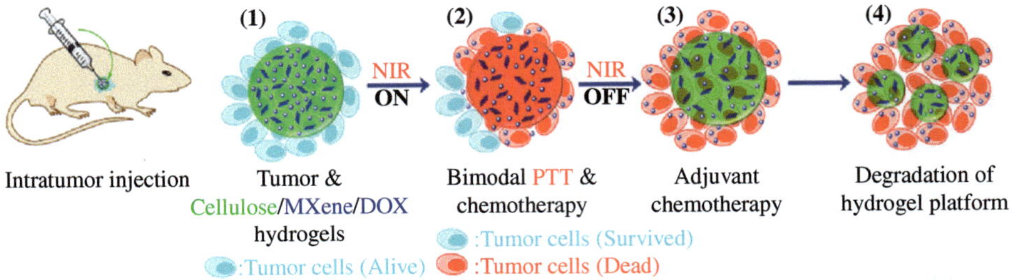

Figure 14.9 A graphical representation of PTT/chemotherapy therapy by injecting Cellulose/MXene/DOX hydrogels within tumors of mice. The therapy consists of four steps: (1) Injection of cellulose/MXene/DOX hydrogels within tumor site. (2) Near-infrared (NIR) irradiation activating the PTT and chemotherapy components of the hydrogels (3) Turning off NIR irradiation but Chemotherapy continues to work (4) Cellulose hydrogel platform is degraded, ultimately releasing the chemotherapy drugs. Source: Reproduced with permission [82] Copyright © 2018, American Chemical Society.

shown to be effective in killing cancer cells. Xing et al. [82], as shown in Figure 14.9, have reported that a 2D MXene (Ti_3C_2)-integrated cellulose hydrogel nanoplatform has been developed that exhibits light-induced swelling and bimodal photothermal/chemotherapy anticancer activity. The nanoplatform is a smart, three-dimensional network that can be used to deliver chemotherapy drugs to tumors. When the nanoplatform is exposed to light, it swells and releases chemotherapy drugs. The drugs then kill cancer cells by disrupting their DNA. The nanoplatform is effective in destroying tumors and preventing tumor relapse. This suggests that the nanoplatform has potential for application in cancer therapy.

14.5.3 PTT Coupled Immunotherapy

For the considerable actions toward the region affected by pathogenic activity, there have been a captious requirement for the accumulation of therapeutic agents that are capable of executing the PTT effects on a noticeable scope. Generally, the passive targeting method is preferred for the transportation of the abovementioned agents to the affected section. The active-type targeting mode is highly favorable if the goal is to deliver a high dosage of the required site as compared to the former mode of targeting. Specifically targeting tumor-affected sites by modifying the surface properties. For instance, when exterior part of Ti_3C_2 is altered by attaching hyaluronic acid, tumor recognition in this system becomes better because of the interaction between HA and $CD4^+$, the major component that is overly expressed near a tumor microenvironment. The setup can be further loaded with an anti-cancer drug making the unit complete and effective against tumors all two of the *in vivo* and *in vitro* experiments, providing a solid base for PTT [68].

Ti_3C_2@mMSN-RGD, a nanocomposite structure that consists of a two-dimensional MXene-Ti_3C_2 nanosheets within nanodimensions and a porous topography that is highly suitable for drug loading and is further functionalized with RGD for targeting tumor reported in recent times. This nanosized material has shown an excellent capabilities for being a suitable photothermal agent [83]. Another type of material whose composition is Ti_3C_2@mMSNs, which is further modified with RGD, is known to possess a cell suppression effect more competitively when correlated against the materials that do not contain RGD conjugation on their surface giving it an upper hand in making the material highly specific. This exceptionally curated active-targeting capability is because of the strong binding between RGD peptides and the receptors overexpressed on the surfaces of hepatocellular carcinoma (HCC) cells, which can be used to deliver therapeutic agents to these cells through endocytosis. This is a promising approach for the treatment of HCC, as it allows for the targeted delivery of drugs to the cells that need them most. The overall effect of the composite Ti_3C_2@mMSN-RGD was observed in the in vivo settings where this material demonstrated outstanding targeting of the tumor, where RGD is responsible for this action, and on the other hand, the rest of the material successfully ablated tumor cells and gave an irreversible and permanent outcome. Out of the whole PTT application, MXenes are also strongly equipped and suitable to regulate and note the advancement of healing. Other applications also include three-dimensional imaging, which can reach deep into the affected tissues and is an exceptional tool known for its highly known spatial recognition. All these features and functionalities enable MXene for guided theranostic applications.

14.6 Conclusion

An incomparable insight into the various mechanisms dominating the photothermal conversion along with a significant emphasis on the PT modification for 2D MXene and MXene-type nanocomposites has been communicated. The MAX phases are etched to develop and synthesize a series of MXenes. A diverse range of etchants modifies the surface chemistry of MXene to improve physicochemical properties, finally modulating the surface as per the targeted applications. The fundamental photoconversion efficiency and mechanism have been highlighted. Few research groups attempted to isolate the exact mechanism stating LSPR is responsible for light-to-heat conversion; however, the exact mechanism is still not clear. As a matter of fact, within a short span of years, there is a constant rise in the research where photothermal applications have contrived advancements for solar steam production and various biomedical operations. There is still an array of problems that are supposed to be addressed, and there are wide opportunities for researchers to explore. The entire studies that have been performed have exclusively calculated the stellar absorbing capacity along with the light-to-heat alteration capability by the medium of a laboratory arrangement, which calls for bridging the gap between the laboratory and real-life practical applicability of the MXenes. Further photothermal coupling with other nanostructured materials will enhance the photothermal performance. Though the studies ascribed for the photothermal conversion of MXene due to LSPR, owing to its metallic character, not all of them have a metallic behavior and display semiconducting behavior. The versatility in chemical composition of MXenes demands detailed discovery of a general understanding of photothermal working, needs to be addressed sequentially and intelligently. Even though MXenes are capable of converting heat from light inexplicably high, there is always a certain amount of energy dissipated in the form of heat toward the environment, which is almost inevitable. If a favorable architectural adjustment could be made it might be extremely helpful for heightening solar retention. The majority of prevailing research demonstrates that single MXene flakes could be turned into singular sheets during the process of delamination, and the way to store this material will always have a major effect on its shelf life for long-term applicability, which is mainly for photothermal applications. Hence, special attention is required to study the oxidation behavior and reproducibility before considering their commercial applicability. Exploring theoretical modeling and computational studies for rational design and synthesis of MXene with a greater light-to-heat conversion efficiency. As soon as a favorable material is designed, through the critical inputs of interdisciplinarian intellect, the MXene will become highly dominant and one of the most outstanding materials expanding its roots for several commercial applicability for the betterment of the human race.

Acknowledgment

The authors are thankful to Amity University, Noida, for providing valuable support and facilities.

Conflict of interest

The authors state that they have no known competing interests between them.

References

1 Nomura, S., Morimoto, Y., Tsujimoto, H. et al. (2020). Highly reliable, targeted photothermal cancer therapy combined with thermal dosimetry using a near-infrared absorbent. *Sci. Rep.* 10 (1): 9765. https://doi.org/10.1038/s41598-020-66646-x.

2 Gao, G., Sun, X., and Liang, G. (2021). Nanoagent-promoted mild-temperature photothermal therapy for cancer treatment. *Adv. Funct. Mater.* 31 (25): 2100e738. https://doi.org/10.1002/adfm.202100738.

3 Zou, L., Wang, H., He, B. et al. (2016). Current approaches of photothermal therapy in treating cancer metastasis with nanotherapeutics. *Theranostics* 6 (6): 762–772. https://pubmed.ncbi.nlm.nih.gov/27162548.

4 Lv, S., Miao, Y., Liu, D., and Song, F. (2020). Recent development of photothermal agents (PTAs) based on small organic molecular dyes. *Chem. Bio. Chem.* 21 (15): 2098–2110. https://doi.org/10.1002/cbic.202000089.

5 Chen, J., Ning, C., Zhou, Z. et al. (2019). Nanomaterials as photothermal therapeutic agents. *Prog. Mater Sci.* 99: 1–26. https://www.sciencedirect.com/science/article/pii/S0079642518300732.

6 Pu, K. (2020). Polymers for photothermal applications in biology. *ACS Appl. Polym. Mater.* 2 (10): 4160–4161. https://doi.org/10.1021/acsapm.0c00994.

7 Wang, J., Wu, X., Shen, P. et al. (2020). Applications of inorganic nanomaterials in photothermal therapy based on combinational cancer treatment. *Int. J. Nanomed.* 15: 1903–1914. https://pubmed.ncbi.nlm.nih.gov/32256067.

8 Ma, H. and Xue, M. (2021). Recent advances in the photothermal applications of two-dimensional nanomaterials: photothermal therapy and beyond. *J. Mater. Chem. A* 9 (33): 17569–17591. https://doi.org/10.1039/D1TA04134G.

9 Liu, S., Pan, X., and Liu, H. (2020). Two-dimensional nanomaterials for photothermal therapy. *Angew. Chem. Int. Ed.* 59 (15): 5890–5900. https://doi.org/10.1002/anie.201911477.

10 Savchuk, O.A., Carvajal, J.J., Massons, J. et al. (2016). Determination of photothermal conversion efficiency of graphene and graphene oxide through an integrating sphere method. *Carbon N Y* 103: 134–141. https://www.sciencedirect.com/science/article/pii/S0008622316301683.

11 Naguib, M., Kurtoglu, M., Presser, V. et al. (2011). Two-dimensional nanocrystals produced by exfoliation of Ti_3AlC_2. *Adv. Mater.* 23 (37): 4248–4253. https://doi.org/10.1002/adma.201102306.

12 Naguib, M., Mochalin, V.N., Barsoum, M.W., and Gogotsi, Y. (2014). 25th Anniversary article: MXenes: a new family of two-dimensional materials. *Adv. Mater.* 26 (7): 992–1005. https://doi.org/10.1002/adma.201304138.

13 Peng, J., Chen, X., Ong, W.-J. et al. (2019). Surface and heterointerface engineering of 2D MXenes and their nanocomposites: insights into electro- and photocatalysis. *Chem* 5 (1): 18–50. https://www.sciencedirect.com/science/article/pii/S2451929418303905.

14 Alhabeb, M., Maleski, K., Anasori, B. et al. (2017). Guidelines for synthesis and processing of two-dimensional titanium carbide (Ti3C2Tx MXene). *Chem. Mater.* 29 (18): 7633–7644. https://doi.org/10.1021/acs.chemmater.7b02847.

15 Anasori, B., Lukatskaya, M.R., and Gogotsi, Y. (2017). 2D metal carbides and nitrides (MXenes) for energy storage. *Nat. Rev. Mater.* 2 (2): 16098. https://doi.org/10.1038/natrevmats.2016.98.

16 Xiong, D., Li, X., Bai, Z., and Lu, S. (2018). Recent advances in layered Ti (3) C(2) T(x) MXene for electrochemical energy storage. *Small* 14 (17): e1703419.

17 Li, R., Zhang, L., Shi, L., and Wang, P. (2017). MXene Ti3C2: An effective 2D light-to-heat conversion material. *ACS Nano* 11 (4): 3752–3759. https://doi.org/10.1021/acsnano.6b08415.

18 Lin, H., Wang, X., Yu, L. et al. (2017). Two-dimensional ultrathin MXene ceramic nanosheets for photothermal conversion. *Nano Lett.* 17 (1): 384–391. https://doi.org/10.1021/acs.nanolett.6b04339.

19 Jiao, E., Wu, K., Liu, Y. et al. (2021). Robust bioinspired MXene-based flexible films with excellent thermal conductivity and photothermal properties. *Compos. Part A Appl. Sci. Manuf.* 143: 106290. https://www.sciencedirect.com/science/article/pii/S1359835X21000208.

20 Gazzi, A., Fusco, L., Khan, A. et al. (2019). Photodynamic therapy based on graphene and MXene in cancer theranostics. *Front. Bioeng. Biotechnol.* 7: 295. https://pubmed.ncbi.nlm.nih.gov/31709252.

21 He, Y., Cao, Y., and Wang, Y. (2018). Progress on photothermal conversion in the second NIR window based on conjugated polymers. *Asian J. Org. Chem.* 7 (11): 2201–2212. https://doi.org/10.1002/ajoc.201800450.

22 Cao, S., Jiang, Q., Wu, X. et al. (2019). Advances in solar evaporator materials for freshwater generation. *J. Mater. Chem. A* 7 (42): 24092–24123. https://doi.org/10.1039/C9TA06034K.

23 Bogaerts, W.F. and Lampert, C.M. (1983). Materials for photothermal solar energy conversion. *J. Mater. Sci.* 18 (10): 2847–2875. https://doi.org/10.1007/BF00700767.

24 Kim, M., Lee, J.-H., and Nam, J.-M. (2019). Plasmonic photothermal nanoparticles for biomedical applications. *Adv. Sci.* 6 (17): 1900471. https://doi.org/10.1002/advs.201900471.

25 Han, B., Chen, L., Jin, S. et al. (2021). Modulating mechanism of the LSPR and SERS in ag/ITO film: carrier density effect. *J. Phys. Chem. Lett.* 12 (31): 7612–7618. https://doi.org/10.1021/acs.jpclett.1c01727.

26 Zhang, J., Zhang, L., and Xu, W. (2012). Surface plasmon polaritons: physics and applications. *J. Phys. D: Appl. Phys.* 45 (11): 113001. https://doi.org/10.1088/0022-3727/45/11/113001.

27 Jain, P.K., Huang, X., El-Sayed, I.H., and El-Sayed, M.A. (2008). Noble metals on the nanoscale: optical and photothermal properties and some applications in imaging, sensing, biology, and medicine. *Acc. Chem. Res.* 41 (12): 1578–1586. https://doi.org/10.1021/ar7002804.

28 Singh, M., Harris-Birtill, D.C.C., Zhou, Y. et al. (2016). Application of gold nanorods for photothermal therapy in ex vivo human oesophagogastric adenocarcinoma. *J. Biomed. Nanotechnol.* 12 (3): 481–490.

29 Hussein, E.A., Zagho, M.M., Rizeq, B.R. et al. (2019). Plasmonic MXene-based nanocomposites exhibiting photothermal therapeutic effects with lower acute toxicity than pure MXene. *Int. J. Nanomed.* 14: 4529–4539. https://pubmed.ncbi.nlm.nih.gov/31417256.

30 Borghei, Y.-S., Hosseinkhani, S., and Ganjali, M.R. (2022). "Plasmonic nanomaterials": an emerging avenue in biomedical and biomedical engineering opportunities. *J. Adv. Res.* 39: 61–71. https://www.sciencedirect.com/science/article/pii/S2090123221002162.

31 Chen, G. and Borca-Tasciuc, T. (1998). Applicability of photothermal radiometry for temperature measurement of semiconductors. *Int. J. Heat Mass Transfer* 41 (15): 2279–2285. https://www.sciencedirect.com/science/article/pii/S0017931097003591.

32 Xiao, J., Zhang, Y., Chen, H. et al. (2018). Enhanced performance of a monolayer MoS(2)/WSe(2) heterojunction as a photoelectrochemical cathode. *Nano-micro Lett.* 2018/07/03 10 (4): 60. https://pubmed.ncbi.nlm.nih.gov/30393708.

33 Ouyang, W., Teng, F., He, J.-H., and Fang, X. (2019). Enhancing the photoelectric performance of photodetectors based on metal oxide semiconductors by charge-carrier engineering. *Adv. Funct. Mater.* 29 (9): 1807672. https://doi.org/10.1002/adfm.201807672.

34 Saifuddin, N., Raziah, A.Z., and Junizah, A.R. (2013). Carbon nanotubes: a review on structure and their interaction with proteins. Urriolabeitia EP, editor. *J. Chem.* 2013: 676815. https://doi.org/10.1155/2013/676815.

35 Rao, R., Pint, C.L., Islam, A.E. et al. (2018). Carbon nanotubes and related nanomaterials: critical advances and challenges for synthesis toward mainstream commercial applications. *ACS Nano* 12 (12): 11756–11784. https://doi.org/10.1021/acsnano.8b06511.

36 Du, X., Xu, J., Deng, S. et al. (2019). Amino-functionalized single-walled carbon nanotubes-integrated polyurethane phase change composites with superior photothermal conversion efficiency and thermal conductivity. *ACS Sustainable Chem. Eng.* 7 (21): 17682–17690. https://doi.org/10.1021/acssuschemeng.9b03853.

37 Wang, C., Xu, L., Liang, C. et al. (2014). Immunological responses triggered by photothermal therapy with carbon nanotubes in combination with anti-CTLA-4 therapy to inhibit cancer metastasis. *Adv. Mater.* 26 (48): 8154–8162. https://doi.org/10.1002/adma.201402996.

38 Ito, Y., Tanabe, Y., Han, J. et al. (2015). Multifunctional porous graphene for high-efficiency steam generation by heat localization. *Adv. Mater.* 27 (29): 4302–4307. https://doi.org/10.1002/adma.201501832.

39 Yang, J., Pang, Y., Huang, W. et al. (2017). Functionalized graphene enables highly efficient solar thermal steam generation. *ACS Nano* 11 (6): 5510–5518. https://doi.org/10.1021/acsnano.7b00367.

40 Robinson, J.T., Tabakman, S.M., Liang, Y. et al. (2011). Ultrasmall reduced graphene oxide with high near-infrared absorbance for photothermal therapy. *JACS* 133 (17): 6825–6831. https://doi.org/10.1021/ja2010175.

41 Hashemi, M., Omidi, M., Muralidharan, B. et al. (2017). Evaluation of the photothermal properties of a reduced graphene oxide/arginine nanostructure for near-infrared absorption. *ACS Appl. Mater. Interfaces* 9 (38): 32607–32620. https://doi.org/10.1021/acsami.7b11291.

42 Zhang, P., Li, J., Lv, L. et al. (2017). Vertically aligned graphene sheets membrane for highly efficient solar thermal generation of clean water. *ACS Nano* 11 (5): 5087–5093. https://doi.org/10.1021/acsnano.7b01965.

43 Yang, J., Choi, J., Bang, D. et al. (2011). Convertible organic nanoparticles for near-infrared photothermal ablation of cancer cells. *Angew Chemie Int Ed* 50 (2): 441–444. https://doi.org/10.1002/anie.201005075.

44 Zhou, J., Lu, Z., Zhu, X. et al. (2013). NIR photothermal therapy using polyaniline nanoparticles. *Biomaterials* 34 (37): 9584–9592. https://www.sciencedirect.com/science/article/pii/S0142961213010533.

45 Yang, K., Xu, H., Cheng, L. et al. (2012). In vitro and in vivo near-infrared photothermal therapy of cancer using polypyrrole organic nanoparticles. *Adv. Mater.* 24 (41): 5586–5592. https://doi.org/10.1002/adma.201202625.

46 Ke, K., Lin, L., Liang, H. et al. (2015). Polypyrrole nanoprobes with low non-specific protein adsorption for intracellular mRNA detection and photothermal therapy. *Chem. Commun.* 51 (31): 6800–6803. https://doi.org/10.1039/C5CC01129A.

47 Liu, Y., Ai, K., Liu, J. et al. (2013). Dopamine-melanin colloidal nanospheres: an efficient near-infrared photothermal therapeutic agent for in vivo cancer therapy. *Adv. Mater.* 25 (9): 1353–1359. https://doi.org/10.1002/adma.201204683.

48 Guo, B., Sheng, Z., Hu, D. et al. (2017). Molecular engineering of conjugated polymers for biocompatible organic nanoparticles with highly efficient photoacoustic and photothermal performance in cancer theranostics. *ACS Nano* 11 (10): 10124–10134. https://doi.org/10.1021/acsnano.7b04685.

49 Bessinger, D., Ascherl, L., Auras, F., and Bein, T. (2017). Spectrally switchable photodetection with near-infrared-absorbing covalent organic frameworks. *JACS* 139 (34): 12035–12042. https://doi.org/10.1021/jacs.7b06599.

50 Tan, J., Namuangruk, S., Kong, W. et al. (2016). Manipulation of amorphous-to-crystalline transformation: towards the construction of covalent organic framework hybrid microspheres with NIR photothermal conversion ability. *Angew. Che. Int. Ed.* 55 (45): 13979–13984. https://doi.org/10.1002/anie.201606155.

51 Hu, C., Zhang, Z., Liu, S. et al. (2019). Monodispersed CuSe sensitized covalent organic framework photosensitizer with an enhanced photodynamic and photothermal effect for cancer therapy. *ACS Appl. Mater. Interfaces* 11 (26): 23072–23082. https://doi.org/10.1021/acsami.9b08394.

52 Huang, Z., Cui, X., Li, S. et al. (2020). Two-dimensional MXene-based materials for photothermal therapy. *Nanophotonics* 9 (8): 2233–2249.

53 Fan, X., Liu, L., Jin, X. et al. (2019). MXene Ti3C2Tx for phase change composite with superior photothermal storage capability. *J. Mater. Chem. A* 7 (23): 14319–14327. https://doi.org/10.1039/C9TA03962G.

54 Chaudhuri, K., Wang, Z., Alhabeb, M. et al. (2019). *Optical Properties of MXenes BT - 2D Metal Carbides and Nitrides (MXenes): Structure, Properties and Applications* (ed. B. Anasori and Y. Gogotsi), 327–346. Cham: Springer International Publishing https://doi.org/10.1007/978-3-030-19026-2_17.

55 Tan, Y., Zhu, Z., Zhang, X. et al. (2021). $Nb_4C_3T_x$ (MXene) as a new stable catalyst for the hydrogen evolution reaction. *Int. J. Hydrogen Energy* 46 (2): 1955–1966. https://www.sciencedirect.com/science/article/pii/S036031992033812X.

56 Ghidiu, M., Lukatskaya, M.R., Zhao, M.-Q. et al. (2014). Conductive two-dimensional titanium carbide "clay" with high volumetric capacitance. *Nature* 516 (7529): 78–81.

57 Xue, Y., Zheng, Y., Wang, E. et al. (2022). Ti(3)C(2)T(*x*) (MXene)/Pt nanoparticle electrode for the accurate detection of DA coexisting with AA and UA. *Dalton Trans.* 51 (11): 4549–4559.

58 Shuck, C.E., Han, M., Maleski, K. et al. (2019). Effect of Ti_3AlC_2 MAX phase on structure and properties of resultant $Ti_3C_2T_x$ MXene. *ACS Appl. Nano Mater.* 2 (6): 3368–3376. https://doi.org/10.1021/acsanm.9b00286.

59 Ma, S., Fan, X., An, Y. et al. (2019). Exploring the catalytic activity of MXenes $M_{n+1}CnO_2$ for hydrogen evolution. *J. Mater. Sci.* 54 (17): 11378–11389. https://doi.org/10.1007/s10853-019-03712-4.

60 Mashtalir, O., Naguib, M., Mochalin, V.N. et al. (2013). Intercalation and delamination of layered carbides and carbonitrides. *Nat. Commun.* 4: 1–18.

61 Fagerli, F.H., Wang, Z., Grande, T. et al. (2022). Removing fluoride-terminations from multilayered V_2CT_x MXene by gas hydrolyzation. *ACS Omega* 7 (27): 23790–23799. https://doi.org/10.1021/acsomega.2c02441.

62 Dai, B., Zhao, B., Xie, X. et al. (2018). Novel two-dimensional $Ti_3C_2T_x$ MXenes/nano-carbon sphere hybrids for high-performance microwave absorption. *J. Mater. Chem. C* 6 (21): 5690–5697. https://doi.org/10.1039/C8TC01404C.

63 Guo, Y., Jin, S., Wang, L. et al. (2020). Synthesis of two-dimensional carbide Mo_2CT_x MXene by hydrothermal etching with fluorides and its thermal stability. *Ceram.*

Int. 46 (11, Part B): 19550–19556. https://www.sciencedirect.com/science/article/pii/S0272884220312931.

64 Soundiraraju, B. and George, B.K. (2017). Two-dimensional titanium nitride (Ti2N) MXene: synthesis, characterization, and potential application as surface-enhanced Raman scattering substrate. *ACS Nano* 11 (9): 8892–8900. https://doi.org/10.1021/acsnano.7b03129.

65 von Treifeldt, J.E., Firestein, K.L., Fernando, J.F.S. et al. (2021). The effect of Ti_3AlC_2 MAX phase synthetic history on the structure and electrochemical properties of resultant Ti_3C_2 MXenes. *Mater. Des.* 199: 109403. https://www.sciencedirect.com/science/article/pii/S0264127520309394.

66 Li, X., Bai, Y., Shi, X. et al. (2021). Applications of MXene ($Ti_3C_2T_x$) in photocatalysis: a review. *Mater. Adv.* 2 (5): 1570–1594. https://doi.org/10.1039/D0MA00938E.

67 Chen, X., Shi, Z., Tian, Y. et al. (2021). Two-dimensional Ti_3C_2 MXene-based nanostructures for emerging optoelectronic applications. *Mater. Horiz.* 8 (11): 2929–2963. https://doi.org/10.1039/D1MH00986A.

68 Liu, G., Zou, J., Tang, Q. et al. (2017). Surface modified Ti_3C_2 MXene nanosheets for tumor targeting photothermal/photodynamic/chemo synergistic therapy. *ACS Appl. Mater. Interfaces* 9 (46): 40077–40086. https://doi.org/10.1021/acsami.7b13421.

69 Pan, Z., Cao, F., Hu, X., and Ji, X. (2019). A facile method for synthesizing CuS decorated Ti_3C_2 MXene with enhanced performance for asymmetric supercapacitors. *J. Mater. Chem. A* 7 (15): 8984–8992. https://doi.org/10.1039/C9TA00085B.

70 Bi, L., Yang, Z., Chen, L. et al. (2020). Compressible AgNWs/$Ti_3C_2T_x$ MXene aerogel-based highly sensitive piezoresistive pressure sensor as versatile electronic skins. *J. Mater. Chem. A* 8 (38): 20030–20036. https://doi.org/10.1039/D0TA07044K.

71 Chen, T. (2014). CO_2 and Temperature dual stimuli-responsive "smart". *Mxene Phases.*

72 Zhao, D. and Cai, C. (2019). Layered Ti_3C_2 MXene modified two-dimensional Bi_2WO_6 composites with enhanced visible light photocatalytic performance. *Mater. Chem. Front.* 3 (11): 2521–2528. https://doi.org/10.1039/C9QM00570F.

73 Zhou, X., Qin, Y., He, X. et al. (2020). $Ti_3C_2T_x$ Nanosheets/$Ti_3C_2T_x$ quantum dots/RGO (reduced graphene oxide) fibers for an all-solid-state asymmetric supercapacitor with high volume energy density and good flexibility. *ACS Appl. Mater. Interfaces* 12 (10): 11833–11842. https://doi.org/10.1021/acsami.9b21874.

74 Cai, Y., Shen, J., Ge, G. et al. (2018). Stretchable $Ti_3C_2T_x$ MXene/carbon nanotube composite based strain sensor with ultrahigh sensitivity and tunable sensing range. *ACS Nano* 12 (1): 56–62. https://doi.org/10.1021/acsnano.7b06251.

75 Hope, M.A., Forse, A.C., Griffith, K.J. et al. (2016). NMR reveals the surface functionalisation of Ti_3C_2 MXene. *Phys. Chem. Chem. Phys.* 18 (7): 5099–5102. https://doi.org/10.1039/C6CP00330C.

76 Lin, H., Gao, S., Dai, C. et al. (2017). A two-dimensional biodegradable niobium carbide (MXene) for photothermal tumor eradication in NIR-I and NIR-II biowindows. *JACS* 139 (45): 16235–16247. https://doi.org/10.1021/jacs.7b07818.

77 Dai, C., Lin, H., Xu, G. et al. (2017). Biocompatible 2D titanium carbide (MXenes) composite nanosheets for pH-responsive MRI-guided tumor hyperthermia. *Chem. Mater.* 29 (20): 8637–8652. https://doi.org/10.1021/acs.chemmater.7b02441.

78 Liu, Y., Han, Q., Yang, W. et al. (2020). Two-dimensional MXene/cobalt nanowire heterojunction for controlled drug delivery and chemo-photothermal therapy. *Mater. Sci. Eng., C* 116 (April): 111212. https://doi.org/10.1016/j.msec.2020.111212.

79 Dai, C., Chen, Y., Jing, X. et al. (2017). Two-dimensional tantalum carbide (MXenes) composite nanosheets for multiple imaging-guided photothermal tumor ablation. *ACS Nano* 11 (12): 12696–12712. https://doi.org/10.1021/acsnano.7b07241.

80 Han, X., Huang, J., Lin, H. et al. (2018). 2D Ultrathin MXene-based drug-delivery nanoplatform for synergistic photothermal ablation and chemotherapy of cancer. *Adv. Healthcare Mater.* 7 (9): e1701394.

81 Xu, Y., Wang, Y., An, J. et al. (2022). 2D-ultrathin MXene/DOXjade platform for iron chelation chemo-photothermal therapy. *Bioact. Mater.* 14: 76–85. https://www.sciencedirect.com/science/article/pii/S2452199X21005831.

82 Xing, C., Chen, S., Liang, X. et al. (2018). Two-dimensional MXene (Ti_3C_2)-integrated cellulose hydrogels: toward smart three-dimensional network nanoplatforms exhibiting light-induced swelling and bimodal photothermal/chemotherapy anticancer activity. *ACS Appl. Mater. Interfaces* 10 (33): 27631–27643. https://doi.org/10.1021/acsami.8b08314.

83 Li, Z., Zhang, H., Han, J. et al. (2018). Surface nanopore engineering of 2D MXenes for targeted and synergistic multitherapies of hepatocellular carcinoma. *Adv. Mater.* 30 (25): 1706981. https://doi.org/10.1002/adma.201706981.

15

MXenes and Its Composites for Biomedical Applications

Aykut Arif Topçu[1], *Seçkin Kılıç*[2], *and Adil Denizli*[2]

[1]*Medical Laboratory Program, Vocational School of Health Service, Aksaray University, Aksaray, Turkey*
[2]*Department of Chemistry, Hacettepe University, Ankara, Turkey*

15.1 Introduction

MXenes are 2D materials (Figure 15.1 that have taken the attention of scientists since 2011, and the first MXene ($Ti_3C_2T_x$) was synthesized via the selective etching of the "A" layer of the MAX phase, which are carbides, nitrides, and carbonitrides [1–6].

Typically, interactions of M–A atoms are weak due to the metal bonds [2]. In the MAX phase, the interactions of M–X atoms are strong due to the covalent and ionic bonds. So, the chemically reactive A layer from MAX phase using strong acid or other methods such as hot-press sintering and hydrothermal method [2, 7] results in the fabrication of MXene (Figure 15.2).

These emerging 2D materials MXenes with the formula of $M_{n+1}X_nT_x$ (M; represents more than 12 early transition metals, X; represents C or N atoms, T; the functional –OH, –F, –O groups) are fabricated by top–down approaches, including selective etching (wet acid, acid-free etching, molten salt etching), *in situ* chemical conversion, and bottom-up methods, including chemical vapor deposition and salt template method [2]. Among these, the wet etching approach is the earliest approach for the preparation of MXene via selective etching of "A" layer in the MAX phase by using hydrofluoric acid (HF) [2, 3]. However, optimizing reaction conditions (the etching time, temperature) and the need for further treatments, e.g., intercalation and sonication, play crucial roles in developing high-quality MXenes [8, 9].

In recent studies, MXenes with different terminal groups, e.g., F, O, OH, CI, NH, Br, have been reported [3, 10–13]. Consequently, modifying these MXene terminal groups with various molecules combined with the superior features of MXenes, such as good biocompatibility, excellent electric conductivity, and hydrophilicity, make them favorable materials used for biomedical purposes [5, 14]. So, in this section, we introduce MXenes and their composites for biomedical purposes, including biosensor platforms, theranostic studies, drug delivery applications, and antibacterial properties, with recent and relevant articles.

MXenes: Next-Generation 2D Materials: Fundamentals and Applications, First Edition.
Edited by Jay Singh, Kshitij RB Singh, Ravindra Pratap Singh, and Charles Oluwaseun Adetunji.

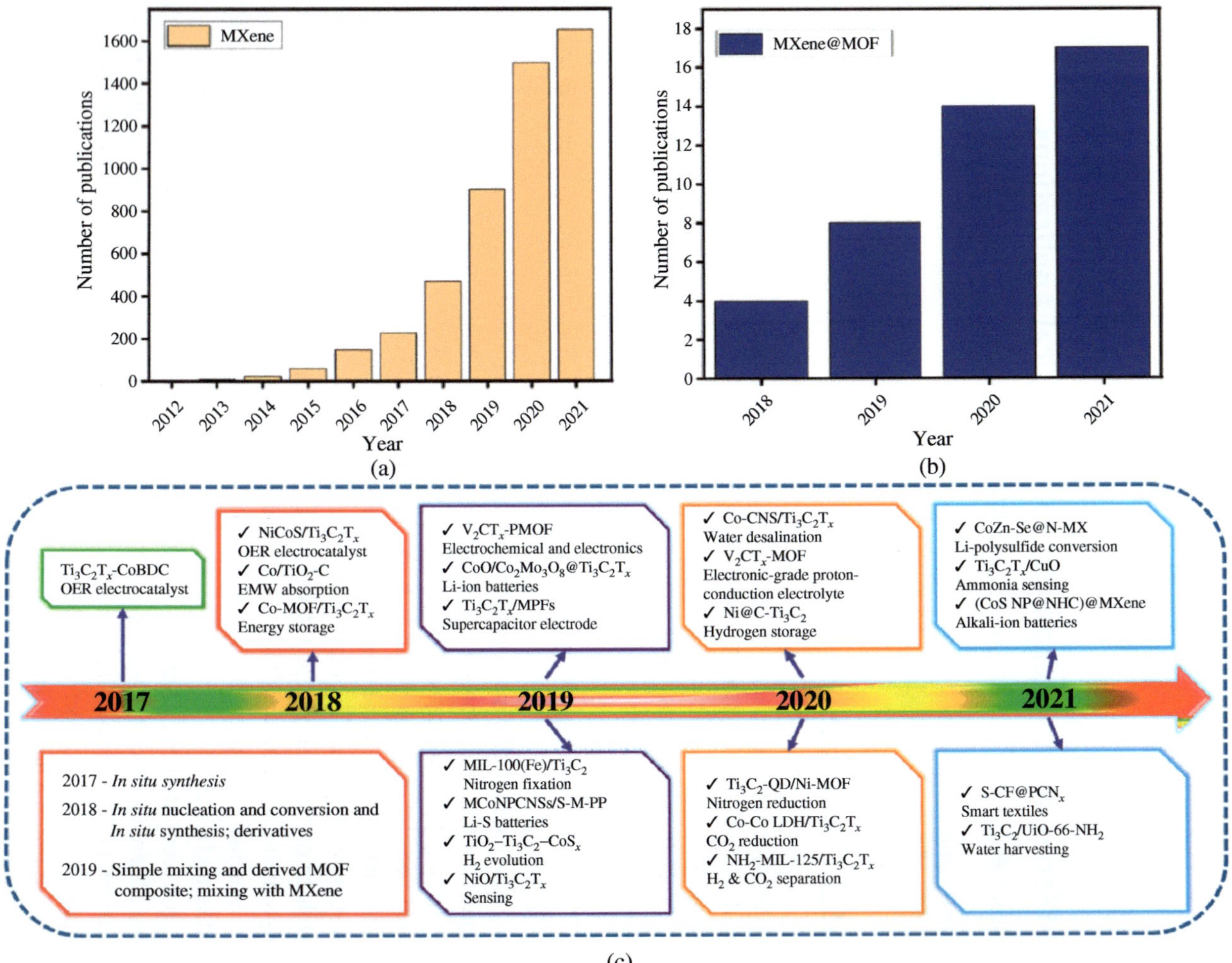

Figure 15.1 (a, b) Show the reported studies of MXene and MXene@MOF, and (c) shows the preparation methods of MXene@MOF materials. Source: Reproduced from Saini et al. [1] with the permission of the American Chemical Society, 2022.

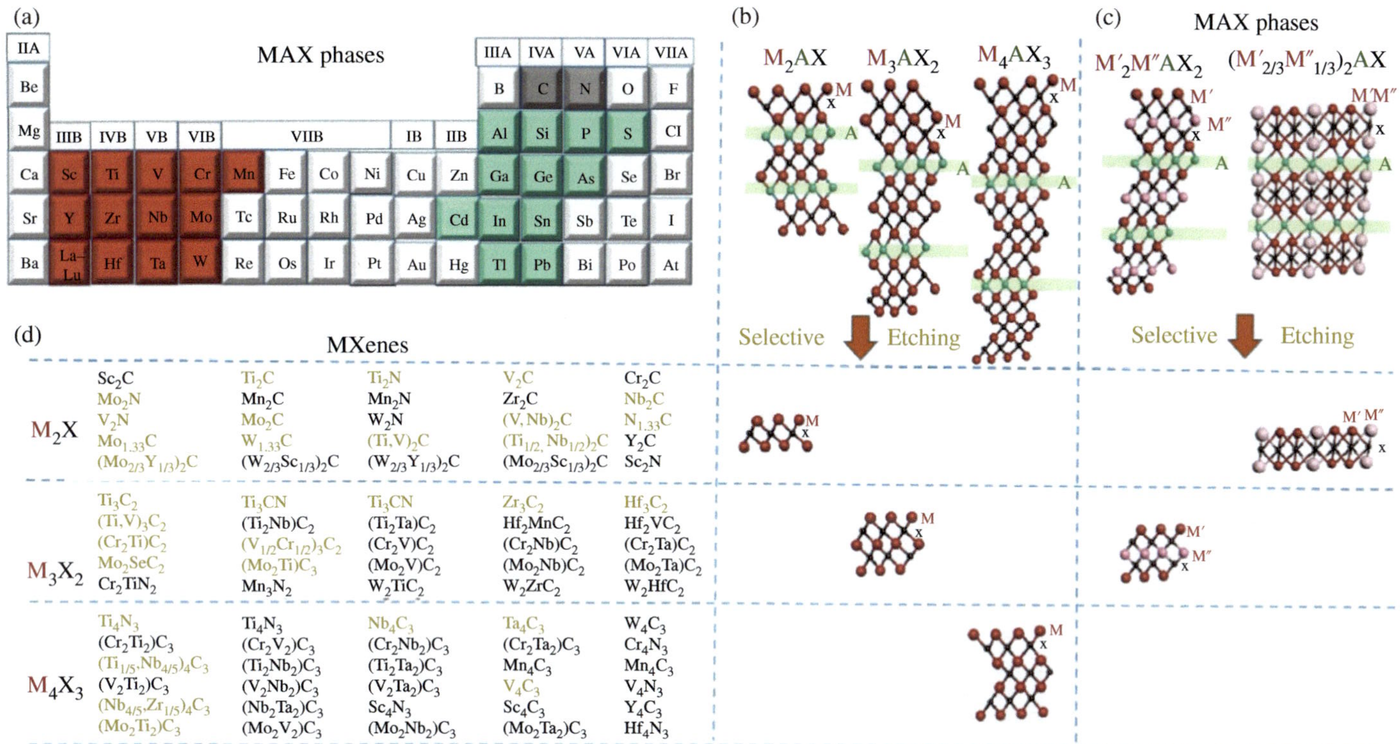

Figure 15.2 Element composition of MAX phase and MXenes. (a) Represents the 12 transition metals with red color except for La, 12 groups of "A" elements with dark green color, and 2 "X" elements with grey color. (b) Represents close to 70 MAX phase structures of M_2AX, M_3AX_2, or M_4AX_3. (c) Crystal structure of MAX phases, and (d) shows the more than 70 MXenes, which have been experimentally fabricated. Source: Reproduced from Qin et al. [2], with the permission of Elsevier, 2022.

15.2 Various Biomedical Applications of MXenes

15.2.1 Biosensor Applications

MXenes have excellent mechanical, optical, and electrical properties, ease of surface functionalization, negative surface charge, high hydrophilicity, large surface area, and good biocompatibility. These make them emerging materials for sensing and biosensing applications [15, 16].

Wang et al. [17] made an electrochemiluminescence (ECL) sensor by utilizing $Ti_3C_2T_x$ to detect the polynucleotide kinase (PNK) enzyme that is responsible for phosphorylation-related DNA repair. Before developing the ECL sensor, AuNPs were anchored on $Ti_3C_2T_x$ nanosheets to fabricate $Ti_3C_2T_x$-AuNPs, and then the luminophore, Ru $(bpy)_3{}^{2+}$, was immobilized on MXene via the electrostatic interactions between $Ti_3C_2T_x$ and the luminophore. After that, the probe Rue$Ti_3C_2T_x$-AuNPs was functionalized with phosphorylated DNA, and the sensing performance of the ECL sensor toward the PNK enzyme was carried out. According to the experimental results, the fabricated ECL sensor could sense PNK with a low detection limit (LOD) of 0.0002 U mL^{-1} and good selectivity. Furthermore, the fabricated sensing platform was feasible for sensing PNK enzyme in cell lysate.

Additionally, Goveia et al. [18] investigated the suitability of Ti_2CO_2 MXene as a nucleobase sensor with the aid of computational approaches. For that purpose, density functional theory (DFT) was used to analyze the configurations of the nucleobases (Figure 15.3) and calculate the adsorption energies.

Experimental findings showed that Ti_2CO_2 MXene exothermically adsorbs the nucleobases, and the nucleobase sits away around 2.5 Å from the MXene without structural deformation. Furthermore, the adsorption energy values of nucleobases are negative, and the adsorption energy results are in accordance with the Van der Walls volumes of the molecules. Recently, MXene (Ti_3C_2) nanosheets were used to construct an electrochemical DNA biosensor for sensing one of the most toxic mycotoxins called gliotoxins [19]. Here, MXene nanosheets' titanium moiety served as a robust electrode modification agent to coordinate with the phosphate groups of DNA. The electron transfer was facilitated owing to the high electron transfer capability of MXenes. Using MXenes as an immobilization matrix led to a better orientation of aptamer on the electrode surface and enhanced the measured electrochemical signals with the low LOD of 5 pm.

Wearable sensing platforms are personalized smart monitoring systems worn close to or onto the skin to obtain real-time information about the status of an individual's activities and health, for instance, heart rate, body temperature, and electrocardiograms. In particular, wearable electrochemical biosensors fabricated for perspiration analysis show an attractive strategy for noninvasive biomarker monitoring in sweat. Nah et al. [20] developed a wearable impedimetric immunosensor (Figure 15.4) to sense cortisol in human sweat. Cortisol, a stress hormone, is mainly secreted due to psychological and emotional stress, and its concentration level in body fluids varies between 0.004 and 70 μM [21]. The high concentration level of cortisol causes high blood pressure and muscle tissue damage; furthermore, chronic stress keeps the cortisol level continuously high, affecting the working of body

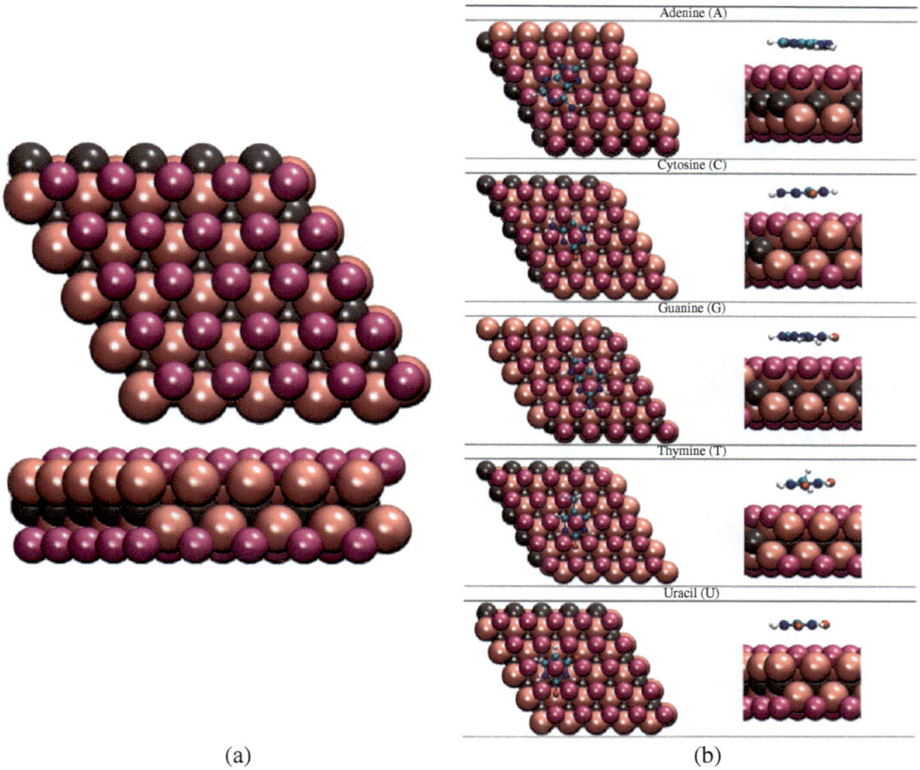

Figure 15.3 (a) The top and bottom images of the Ti_2CO_2 and (b) the stable configurations of the five nucleobases on Ti_2CO_2. Source: Reproduced from Gouveia et al. [18], with the permission of Elsevier, 2022.

systems and causing chronic diseases [20]. So, monitoring cortisol levels is of significant importance to prevent public health and unwanted situations.

The wearable sensor patch was fabricated by combining polydimethylsiloxane (PMDS). The sensing performance of the sensor was enhanced with $Ti_3C_2T_x$ due to its excellent electrical properties. The laser-burned graphene (LBG) electrode was transferred to the PDMS substrate, and MXene was coated on the working electrode during the fabrication process. After that, the cortisol antibody was immobilized onto the working electrode via "click chemistry." Finally, the underside of the patch was mounted onto the skin's surface, and the cortisol level was monitored in human sweat. The excellent sensing performance, the lower detection limit (3.88 pM), and the higher selectivity made this fabricated wearable immunosensor a promising point-of-care platform for cortisol detection.

Another enzymatic wearable, an electrochemical biosensor, was designed for sweat analysis, combining an MXene ($Ti_3C_2T_x$) and Prussian blue (PB) in a composite structure to sense glucose and lactate in sweat as well as the pH profile during physical exercise [22]. Due to its 2D morphology and metallic conductivity, $Ti_3C_2T_x$/PB-based biosensors could measure sweat lactate, glucose, and pH simultaneously in an ultrasensitive manner.

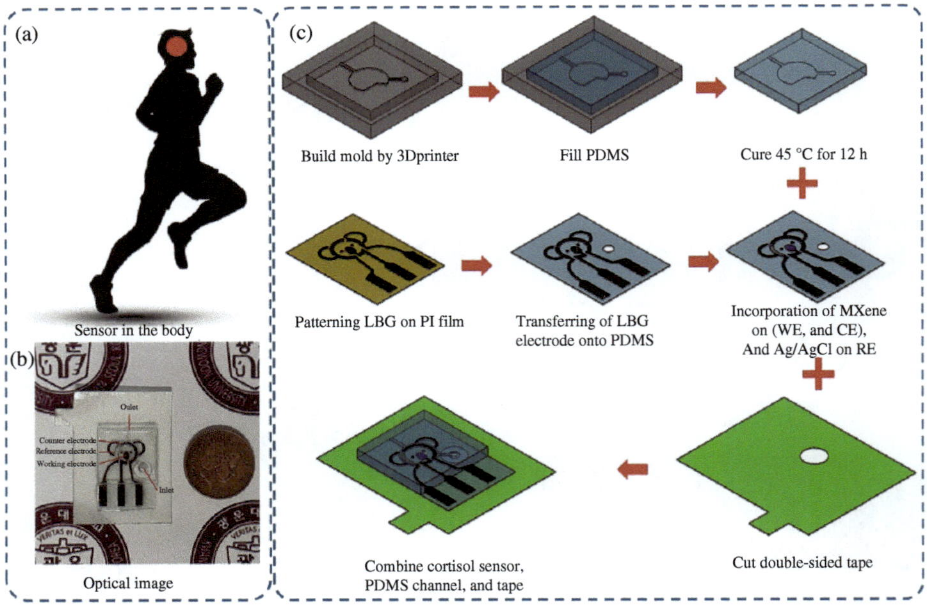

Figure 15.4 The integration of patch sensor (a), the optical image of the wearable patch (b), and the integration of the wearable sensor with the microfluid system (c) Source: Reproduced from Nah et al. [20], with the permission of Elsevier, 2022.

In the following study, a point-of-care-test (POCT) system was constructed by incorporating $Ti_3C_2T_x$-based screen-printed electrode and microfluid chip to determine some clinical analytes in human blood [21]. $Ti_3C_2T_x$ was fabricated with the etching of Ti_3AlC_2 and was integrated into the bottom layer of the chip for continuous online monitoring of the biomarkers in human blood.

An interesting study was reported by Zaho et al. [23], who designed durable and stable electronic skin as a flexible pressure sensor by combining polyvinyl alcohol (PVA) and $Ti_3C_2T_x$ MXene. For that purpose, the hybrid PVA/MXene films were fabricated with the removal of the Al layer from $Ti_3C_2T_x$. After that, the flexible sensor was constructed via sandwiching of hybrid films on the polyethylene terephthalate (PET) surface with the Au electrode. Finally, the proposed sensing system and the exposed Ag wire were encapsulated with PDMS. The flexible sensor was integrated into the BALB/c mouse's heart and gastric wall to examine the *in vivo* experiments (Figure 15.5).

The results showed that the fabricated sensor could be potentially used in biomedical studies due to its good biocompatibility. Moreover, the hybrid film with high-pressure sensitivity showed more stable, more robust character features than MXene film alone and could protect its stability in harsh environments. The other MXene-based biosensor platforms and their purposes are shown in Table 15.1.

15.2.2 Cancer Treatment

Photothermal therapy (PTT) is an alternative treatment method for tumor eradication. During PTT (Figure 15.6), the photothermal agent accumulates with the tumor cell region

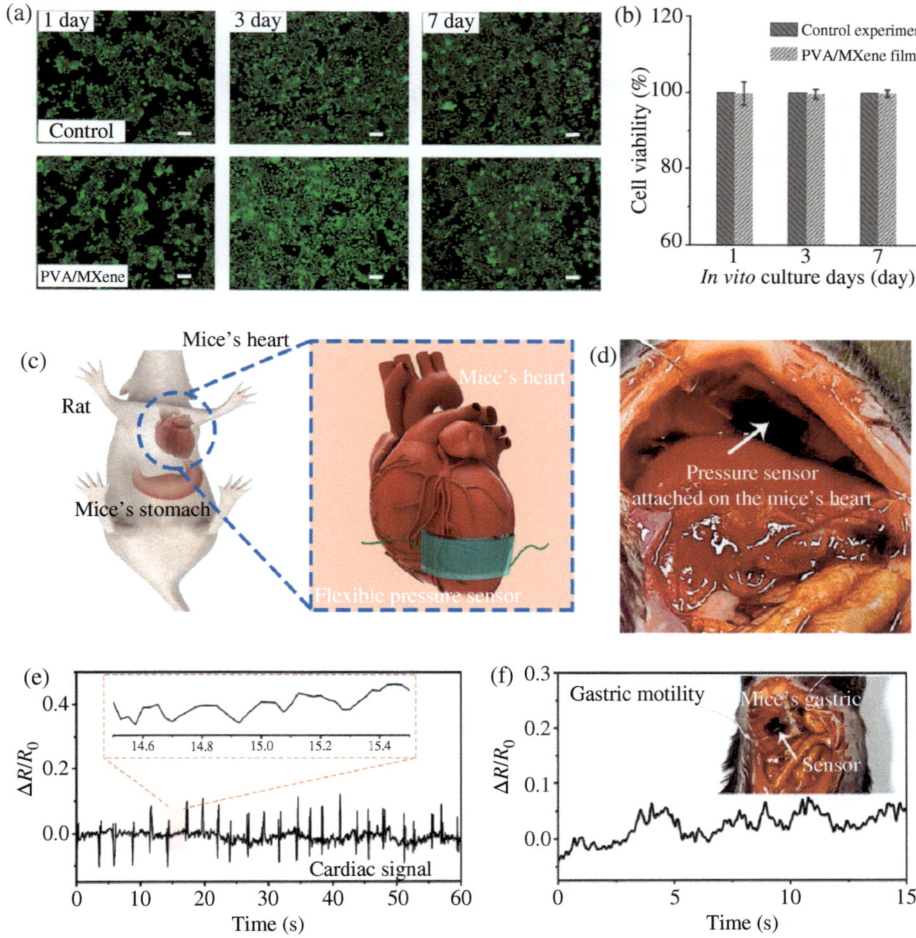

Figure 15.5 (a, b) Show the fluorescent image and cell viability of HUVECs cultured on PVA/MXene hybrid films, respectively. (c, d) Represent the schematic images and photographs of the pressure sensor attached to a mouse's heart, respectively. (e, f) Show the measured electronic signal and gastric dynamics study of the pressure sensor attached to a balb/c mice's heart and gastric wall, respectively. Source: Reproduced from Zhao et al. [23], with the permission of Elsevier, 2022.

and kills the cancer cells under light irradiation [35–37]; however, the performance of PTT is dependent on the light absorption capacity and the PCE of the photothermal agents [36].

MXenes with high PCE, owing to the strong absorption in the near-infrared (NIR) range, are emerging photothermal agents for tumor ablation. For instance, Ti_3C_2 MXene nanosheets were fabricated with a two-step exfoliation strategy of Ti_3AlC_2 [38], and the photothermal tumor ablation performance was examined. The use of the two-exfoliation strategy reduces the lateral sizes of MXenes and makes the developed MXene a photothermal agent for biomedical studies.

Additionally, the MXene-based theranostic composite agent (MnO_x/Ti_3C_2) was produced for multi-purposes and cancer treatment [39]. For that purpose, the MXene composite was prepared by seeding MnO_x on the Ti_3C_2 surface to act as a pH-responsive contrast agent

Table 15.1 MXene-based biosensor platforms and their sensing performances toward the various analytes.

Sensing platform	Analyte	Method	Detection range	Limit of detection	Refs.
PI/$Ti_3C_2T_x$/AuNP	CEA	SAW	0.1–100 ng/mL	0.001 ng/mL	[24]
MCQMs	Humidity sensing	Electrochemical	1–98%	No data	[25]
Ti_3C_2-MXene/AuNPs composite	CEA	SPR	2×10^{-16}–2×10^{-8}	0.07 fM	[26]
$Ti_3C_2T_x$/AuNPs	OncomiRs	Electrochemical	138–204 aM	500 aM–50 nM	[27]
$Ti_3C_2T_x$/Chit/CHO_x composite	Cholesterol	Electrochemical	0.3–4.5 nM	0.11 nM	[28]
Ti_3C_2 MXene/DNA/Pd/Pt composite	Dopamine	Electrochemical	0.2–1000 nM	30 nM	[29]
Ti_3C_2MQDs composite	Uric acid	Colorimetric	1.2–75 μM	125 nM	[30]
$Ti_3C_2T_x$/ZIF-8 composite	HIV-1	ECL	1 fM–1 nM	0.3 fM	[31]
MXene-TiO_2/$BiVO_4$	CD-44	Photochemical	2.2×10^{-4} – 3.2 ng/mL	1.4×10^{-2} pg/mL	[32]
MXene/CRISPR-Cas-12a	Siglec-5	ECL	20 fM–100 pM	20.22 fM	[33]
Ti_3C_2 MXene/antigen composite	PSA	Electrochemical	1–50 ng/mL	0.031 ng/mL	[34]

Abbreviations: PI, polyimide; AuNP, gold nanoparticles; CEA, carcinoembryonic antigen; SAW, surface acoustic wave; MCQMs, $Ti_3C_2T_x$/chitosan-quercetin; SPR, surface plasmon resonance; $Ti_3C_2T_x$/Chit/CHO_x, $Ti_3C_2T_x$/chitosan/cholesterol oxidase; Ti_3C_2MQDs, Ti_3C_2MXene Quantum Dots; ZIF-8, zeolitic imidazolate framework-8; HIV-1, immunodeficiency virus protein 1; ECL, electrochemical luminescence; PSA, prostate-specific antigen.

for T1-weighted magnetic resonance (MR) imaging. After that, the composite was utilized to detect and cook tumor cells due to the photoacoustic (PA) efficiency and PCE, respectively. The experimental findings revealed that the biocompatible MXene-based theranostic composite agent could be potentially used in tumor imaging and tumor ablation.

In another study, the biocompatible titanium carbide MXene quantum dots (QDs) were developed as a PTT agent via fluorine-free method because of its safety and simplicity [40]. Furthermore, the fluorine-free method can increase the absorption of QDs in the NIR region due to the abundant Al oxoanions. According to the experimental results, PCE and the nano agent's extinction coefficient (808 nm) were found at 52.8 $L\,g^{-1}\,cm^{-1}$ and 52.2%, respectively. The biocompatible MXene QDs (2 nm thickness) could be used in photoacoustic (PA) imaging and showed a good PTT performance for tumor cells.

Moreover, Ti_3C_2 nanosheets (lateral sizes ~100 nm) were etched in HCl (containing LiF and Al atoms) and intercalated with tetrapropylammonium hydroxide (TMAHO) [41]. After that, the anticancer drug doxorubicin (DOX) was loaded on intercalated nanosheets,

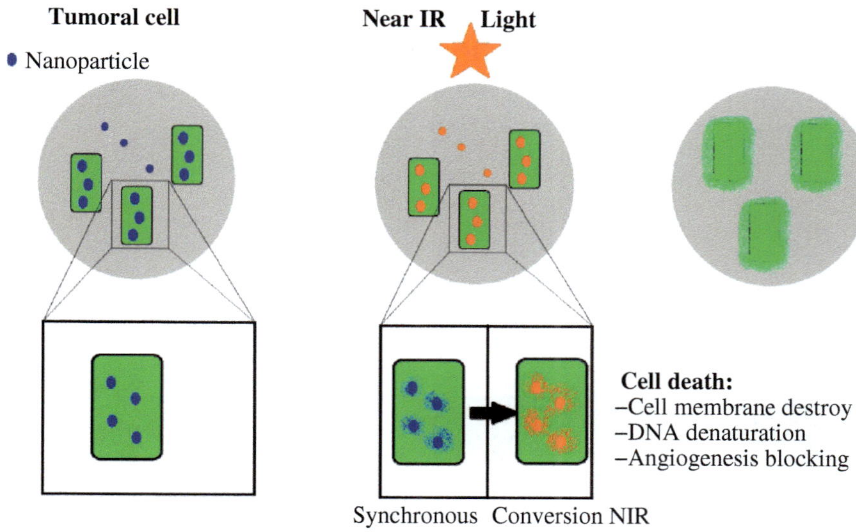

Figure 15.6 The schematic representation of PTT for cancer treatment Source: Reproduced from Sivasankarapillai et al. [35], with the permission of Elsevier, 2022.

and the surface of the drug carrier was modified with hyaluronic acid (HA) to enhance its biocompatibility. Finally, various capabilities of Ti_3C_2 nanosheets were investigated, such as their drug release performance, generation reactive oxygen species (ROS) ability, PTT performance, and PCE efficiency.

Meanwhile, Dai et al. [42] developed a composite Ta_4C_3 MXene material for multiple imaging and tumor ablation studies. During the preparation of the MXene composite, the Al layer was removed from Ta_4AlC_3 ceramics via HF. Subsequently, MnO_x NPs were integrated (*in situ* growth) onto the surface Ta_4C_3 nanosheets (MnO_x/Ta_4C_3). Finally, the MnO_x/Ta_4C_3 composite was modified with SP (MnO_x/Ta_4C_3-SP) to enhance the biocompatibility of the developed MXene for *in vivo* studies (Figure 15.7). The developed biocompatible MnO_x/Ta_4C_3-SP did not show only a potential for a high contrast imaging agent but also could be used as a PTT agent for cancer tumor ablation. The extinction coefficient and PCE of the hydrophilic MnO_x/Ta_4C_3 were 8.67 L g^{-1} cm^{-1} (808 nm) and 34.9%, respectively.

Additionally, the biocompatible superparamagnetic theranostic agent Ti_3C_2 MXene was prepared in three steps [43]. In the first step, Ti_3C_2 MXenes were fabricated with a two-step exfoliation strategy, and then, Ti_3C_2-iron nanoparticles (Ti_3C_2-IONPs) were synthesized using $FeSO_4{\cdot}7H_2O$ and ammonia solution. Finally, the surface of superparamagnetic materials was coated with SP to increase the stability in physiological environments and biocompatibility of Ti_3C_2-IONPs-SPs. After the fabrication process, the experimental studies (including MR imaging) and PTT performances of Ti_3C_2-IONPs-SPs were investigated.

In a subsequent research study, Ta_4C_3-based MXene was prepared with a two-step liquid exfoliation strategy using HF etching and probe sonication [44]. Later, its surface was modified with SB to enhance the stability and biocompatibility of the theranostic agent. After the modification, the dual-mode photoacoustic/computed tomography imaging

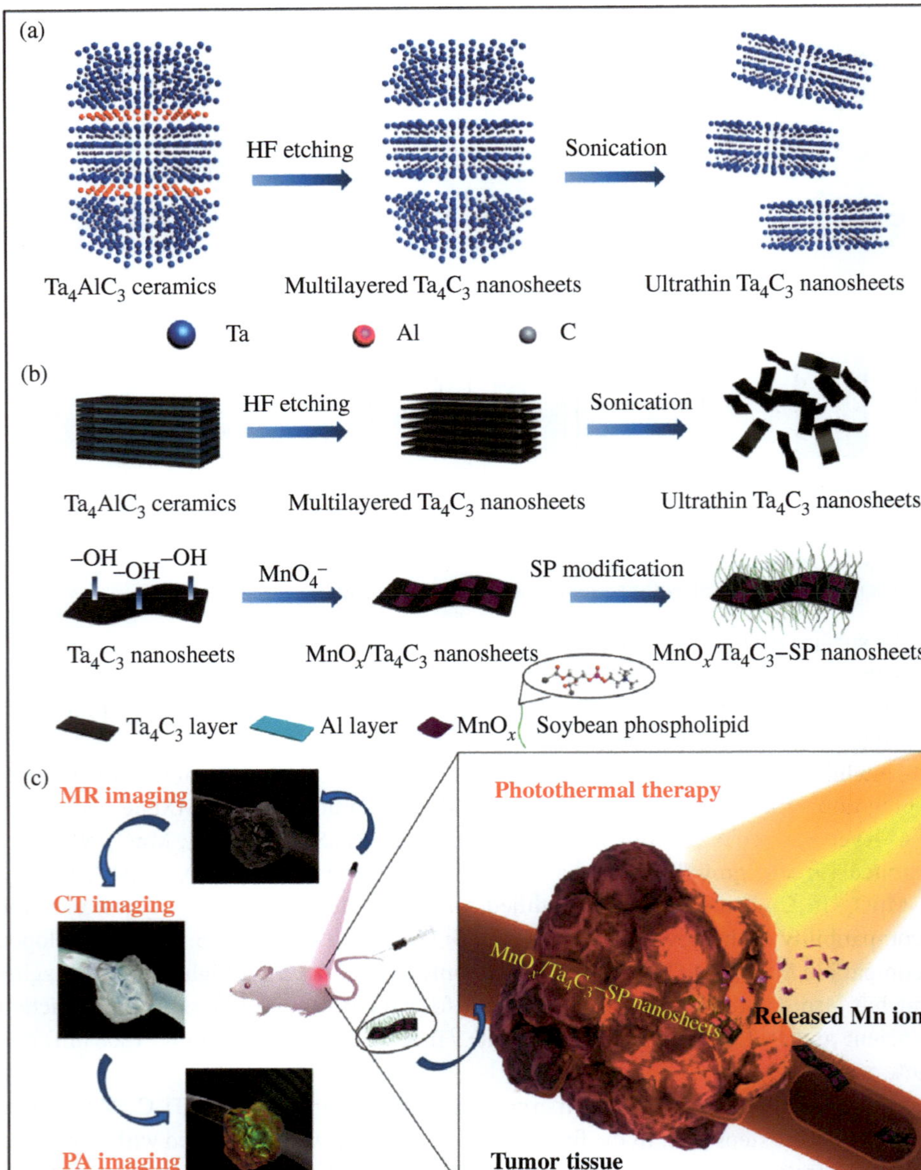

Figure 15.7 (a, b) Represent the preparation of the intercalated MXene, and (c) shows the imaging and PTT efficiency of the prepared MXene. Source: Reproduced from Dai et al. [42], with the permission of American Chemical Society, 2022.

(PA/CT) performances with PCE were examined. According to the experimental findings, the extinction coefficient and PCE of MXene were found to be 4.06 $L\,g^{-1}\,cm^{-1}$ and 44.7%, respectively. The prepared MXene protected its photothermal stability during five laser on/off cycles, and the surface modification did not affect the PTT performance. The boost of laser power density increased the number of dead cells incubated with MXene-SP nanosheets. Furthermore, the SP-modified MXene nanosheets showed significant toxicity during the *in vivo* and *in vitro* experimental studies and exhibited good performances during the imaging of tumors.

The researchers synthesized a 2D niobium carbide (Nb_2C) MXene and aimed to investigate the feasibility of thermodynamic therapy [45]. The Nb_2C MXene was fabricated with a two-step exfoliation approach using HF and tetrapropylammonium hydroxide (TPAOH), respectively. After that, its surface was covered with a silica layer (Si@ NB_2C@mSiO_2) and azobis (2-(2-imidazolin-2-yl) propane) dihydrochloride (AIPH) as a free-radical generating molecule to develop NB_2C@mSiO_2 MXene (AIPH@ NB_2C@mSiO_2). The extinction coefficient and the PCE of AIPN@ NB_2C@mSiO_2 NPs were measured at 21.11 $L\,g^{-1}\,cm^{-1}$ and 27.03%, respectively, at 1064 nm. The release of AIPH from AIPN@ NB_2C@mSiO_2 was triggered upon exposure to 1064 nm NIR-II laser producing free radicals, resulting in enhancing cancer cell apoptosis.

15.2.3 Antibacterial Properties

Antimicrobial resistance (AMR) is a global problem affecting human health worldwide. The misuse of antibiotics plays a significant role in increasing the emergence of antibiotic-resistant bacterial strains [46]. Hence, MXenes, with their antibacterial properties due to the PCE, could be a better choice for antibacterial studies and decrease the multi-drug resistance bacterial strains. For example, $TiVCT_x$ MXene was prepared via acid etching, and its antibacterial performance towards two bacterial strains was investigated [47]. The experimental results revealed that MXene nanosheets could disrupt the cell membrane of *Escherichia coli* (*E.coli*) thanks to its edge monolayer and are capable of killing *E.coli* in the absence of NIR laser irradiation. Moreover, its low dose showed a good antibacterial performance against *E.coli* (100%) and *Bacillus subtilis* (98%) by combining physical and photothermal sterilization.

More recently, COVID-19 is still a severe problem that causes millions of deaths worldwide, and scientific communities make great efforts to cope with this global pandemic [48]. So, the researchers (Figure 15.8) aimed to investigate the antiviral properties of four different forms of MXene with the aid of computational studies and proteomic studies against the four genotypes of the SARS-COV-2 virus [48].

The MXenes were fabricated by selectively etching Al layers from MAX phases. Their antiviral properties towards four SARS-COV-2 clades (GR, GH, S, and wild type) were examined in infected VERO E6 cells. The MXenes, $Ti_3C_2T_x$, and $Mo_2Ti_2C_3T_x$ with the different surface layers exhibited antiviral activities against SARS-COV-2/clade GR, and the researchers suggested that the surface chemistry of MXenes and the genotype and the mutations play vital roles in the antiviral property of MXenes.

In the following study, the combination of Au NPs and MXene composite ($Ti_3C_2T_x$-Au NPs) was prepared to sense bacteria sensing using surface-enhanced Raman spectroscopy (SERS) and photothermal sterilization of *E. coli* and *B. subtilis*, as well [49]. The MXene composite was synthesized via acid etching, and its surface was functionalized with (3-amino-propyl) triethoxysilane (APTES) solution to integrate Au NPs. Afterward, the sensing performance and the antibacterial activity of the composite were investigated. PCE of the composite material was measured as 43.40%, and the photothermal sterilization

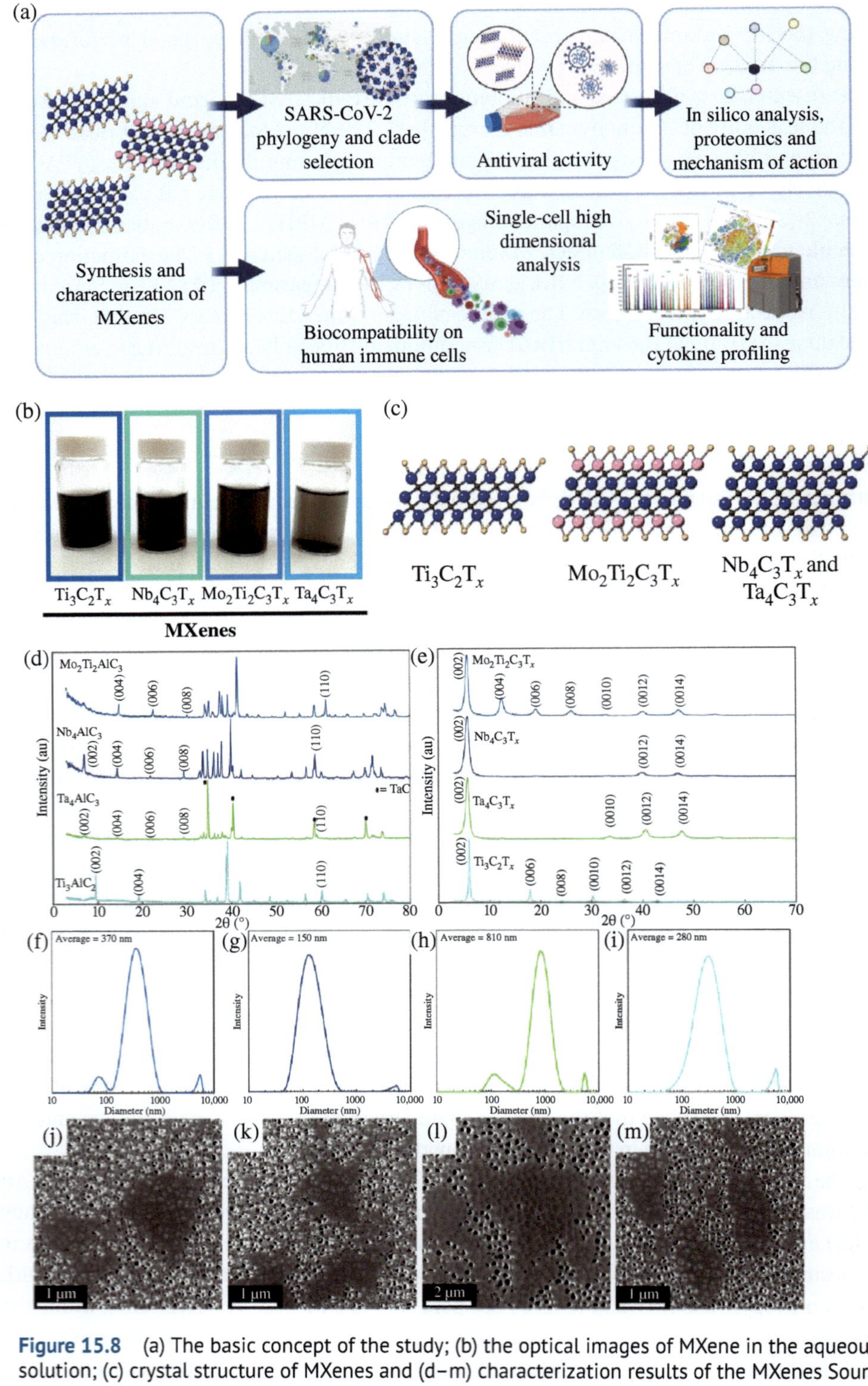

Figure 15.8 (a) The basic concept of the study; (b) the optical images of MXene in the aqueous solution; (c) crystal structure of MXenes and (d–m) characterization results of the MXenes Source: Reproduced from Unal et al. [48], with the permission of Elsevier, 2022.

rates of the composite were determined to be 99.25% for *E. coli* and 100% for *B. subtilis*. The sensing performance of $Ti_3C_2T_x$-Au NPs against the model microorganisms was highly satisfactory, and the increasing amount of composite led to a decline in the number of bacterial cells.

15.2.4 Drug Delivery

Using MXenes as an alternative approach to drug delivery systems (DDS), will possibly minimize conventional side effects, as well as increase the efficiency of a drug at the target site, and maintain the drug concentration at therapeutic levels [50]. DDS can be used as an alternative approach to minimize the side effects of a drug, increase the efficiency of a drug at the target site, and maintain the drug concentration at therapeutic levels [50]. Therefore, the large surface area of MXene allows the loading of chemotherapeutic drugs or biomolecules and makes it a promising nanocarrier for DDS. For example, Wu et al. [51] developed a skillful, customizable, green, remotely controlled drug delivery method using chitosan/hyaluronic acid-based hollow microcapsules equipped with MXene/gold nanorods. Synthesized Ti_3C_2 MXenes, AuNRs, [(hollow hydroxyapatite (HAP)/chitosan (CS)/hyaluronic acid (HA)] microcapsules were used to obtain HAP/CS/HA/MXene/AuNRs microcapsules. The microcapsules were loaded into DOX aqueous solution to prepare drug loading microcapsules. Then, their drug release behaviors were tested and calculated. Microcapsules were characterized via various spectroscopic approaches. The synergistically enhanced photothermal responsiveness behavior between AuNRs and MXene was observed during photothermal property analyses. Upon inducing NIR irradiation, the PTT responsiveness of MXene microcapsules showed outstanding results and stability. Even after the NIR laser was shut down, monitoring the descending temperature revealed that MXene showed a good photothermal conversion performance. The hollow structure of hybrid microcapsules indeed provided ample space for high drug loading efficiency, controlled via their charge. On the other side of the coin, decreased pH increased the cumulative release of the drug carrier, and NIR irradiation accelerated DOX diffusion, increasing its release. By controlling those factors, targeted drug delivery was achieved.

Further research showed that a 12 h-long NIR irradiation by itself was not affecting the structure, but with a lower pH of 4.5, it was possible to collapse the structure altogether. Furthermore, cytotoxicity evaluation of microcapsules with and without DOX showed that the microcapsules had good biocompatibility. Also, with the gradual release of DOX from the microcapsules, the cell inhibition rate was increased, indicating the excellent chemotherapy ability of the microcapsules. Moreover, adjusting the NIR laser and acid medium could control the release of DOX, which shows potential for remotely controlled drug delivery.

A recent study done by Dong et al. [52] introduced a new photothermally controlled drug delivery method with the help of MXene/agarose hydrogels, which could work with localized cancer treatment. The fabricated nanosheets were embedded into the temperature-responsive hydrogel matrix to achieve that ability. In order to prepare MXene nanosheets, MXene (Ti_3AlC_2) was etched with HCl and LiF. Acquired Ti_3C_2. and the MAX phase (Ti_3AlC_2), were analyzed with SEM, XRD, transition electron microscope

Table 15.2 MXene-based drug nanoplatforms.

Drug carrier	Drug	Refs.
Ti_3C_2-based Au nanorods	Doxorubicin	[53]
Ti_3C_2 composite	Doxorubicin	[54]
Ti_3C_2/PAM composite	Chloramphenicol	[55]
Ti_3C_2-CoNWs	Doxorubicin	[56]
Ti_3C_2	Doxorubicin	[57]
MOF-5/Ti_3C_2	Doxorubicin	[58]
Ti_3C_2 composite	Doxorubicin	[59]

Abbreviations: Ti_3C_2/PAM, Ti_3C_2/polyacrylamide; Ti_3C_2-CoNWs, titanium carbide-cobalt nanowires; MOF-5/Ti_3C_2, metal organic frame work 5/Ti_3C_2.

(TEM), and UV–vis spectrophotometer. DOX-loaded MXene@Hydrogels were obtained by introducing various concentrations of DOX into MXene nanosheets containing agarose solution as well as some MXene@hydrogels were prepared without DOX. Photothermal analyses of hydrogels and nanosheets followed this process via NIR laser under various power densities, which tested the photothermal properties of both MXene nanosheets and MXene@Hydrogels. Following that, DOX-loaded-MXene@Hydrogels' behavior under irradiation was tested while monitoring the concentration of DOX via UV–vis absorbance spectra, where MXene showed high PCE under 808 nm NIR radiation, which showed great promise for biomedical applications. While observing MXene@Hydrogel and MXene's behavior to rising temperature was almost the same and exhibited the same photothermal activity; however, adjusting concentration and temperature could melt the hydrogel without causing substantial harm to the skin. Subsequently, DOX-loaded MXene@Hydrogel with different agarose concentrations was exposed to NIR radiation. Correspondingly, slow and fast drug releases were experienced with high and low agarose concentrations, respectively.

Conversely, without NIR, there was almost no DOX release occurred. Hence, the NIR-responsiveness of MXene@Hydrogel could work as a switch and possibly prevent multiple drug injections. Moreover, witnessing DOX release from the hydrogel under NIR irradiation was promising for controlled drug delivery in localized cancer treatment. In Table 15.2, we summarized some drug delivery studies using MXene-based drug carriers.

15.3 Conclusion

MXenes, 2D emerging materials, were discovered in 2011 and have taken the attention of the scientific community owing to their superior features, e.g. hydrophilicity, optical, mechanical, large surface area, and photothermal properties. These new generation 2D materials are synthesized via top–down, bottom–up approaches to introduce the surface terminal groups, and the modification of these terminal groups with the various biomolecules, QDs, and nanoparticles make MXenes favorable candidates for biomedical purposes, e.g., biosensors, especially wearable sensing platforms, cancer treatment, and drug delivery systems. Interestingly, the functionalized MXenes could be potentially used in multipurposes such as imaging and ablation of tumors during medical treatment. Additionally, MXenes with antibacterial and antiviral properties can be potentially used in combating infectious diseases and can open the door to decreasing the drug resistance of microorganisms. These emerging 2D materials have paved the way for biomedical applications for these reasons.

In this context, we highlighted the preparation protocols of MXenes, their modification strategies, and their suitabilities for biomedical purposes by representing the recent and relevant articles. We hope these 2D materials will be expanded for biomedical purposes, and the developed new MXenes can open a new avenue soon.

References

1 Saini, H. et al. (2021). Emerging MXene@metalorganic framework hybrids: design strategies toward versatile applications. *ACS Nano* 15: 18742–18776.

2 Qin, R., Shan, G., Hu, W., and Huang, W. (2021). Two-dimensional transition metal carbides and/or nitrides (MXenes) and their applications in sensors. *Mater. Today Phys.* 21: 100527.

3 Naguib, M., Kurtoglu, M., Presser, V. et al. (2011). Two-dimensional nanocrystals produced by exfoliation of Ti_3AlC_2. *Adv. Mater.* 23: 4248–4253.

4 Barsoum, M.W. (2013). *MAX Phases: Properties of machinable ternary carbides and nitrides. Wiley-VCH Verlag GmbH & Co. KGaA.*

5 Zamhuri, A., Lim, G.M., Ma, N.L. et al. (2021). MXene in the lens of biomedical engineering: synthesis, applications and future Outlook. *BioMed. Eng.* 20: 33.

6 Kang, M.H., Lee, D., Sung, J. et al. (2019). Structure and chemistry of 2D materials. *Comprehens. Nanosci. Nanotechnol.* 1–5: 55–90.

7 Kannan, K., Sadasivuni, K.K., Abdullah, A.M., and Kumar, B. (2020). Current trends in MXene-based nanomaterials for energy storage and conversion system: a mini review. *Catalysts* 10: 1–28.

8 Hantanasirisakul, K. and Gogotesi, Y. (2018). Electronic and optical properties of 2D transition metal carbides and nitrides (MXenes). *Adv. Mater.* 30: 1–30.

9 Li, X., Wang, C., Cao, Y., and Wang, G. (2018). Functional MXene materials: progress of their applications. *Chem. Asian. J.* 13: 2742–2757.

10 Ghidiu, M., Lukatskaya, M.R., Zhao, M.Q. et al. (2015). Conductive two-dimensional titanium carbide 'clay' with high volumetric capacitance. *Nature* 516: 78–81.

11 Ghazaly, A.E., Ahmed, H., Rezk, A.R. et al. (2021). Ultrafast one-step, salt-solution-based acoustic synthesis of Ti_3C_2 MXene. *ACS Nano* 15: 4287–4293.

12 Sun, W., Shah, S.A., Chen, Y. et al. (2017). Electrochemical etching of Ti_2AlC to Ti_2CT_x (MXene) in low-concentration hydrochloric acid solution. *J. Mater. Chem. A.* 5: 21663–21668.

13 Kamybayev, V., Filatov, A.S., and Talapin, D.V. (2020). Covalent surface modifications and superconductivity of two-dimensional metal carbide MXenes. *Science* 369: 979–983.

14 Mei, J., Ayoko, G.A., Hu, C., and Sun, Z. (2020). Thermal reduction of sulfur-containing MAX phase MXene production. *Chem. Eng. J.* 395: 125111.

15 George, S.M. and Kandasubramanian, B. (2020). Advancements in MXene-polymer composites for various biomedical applications. *Ceram. Int.* 46: 8522–8535.

16 Sinha, A., Dhanjai, Zhao, H. et al. (2018). MXene: an emerging material for sensing and biosensing. *Trends. Analyt. Chem.* 105: 424–435.

17 Wang, L., Zhang, H., Zhuang, T. et al. (2022). Sensitive electrochemiluminescence biosensing of polynucleotide kinase using the versatility of two-dimensional $Ti_3C_2T_X$ MXene nanomaterials. *Anal. Chim. Acta.* 1191: 339346.

18 Gouveia, J.D., Leruth-Novell, G., Vines, F. et al. (2021). The Ti_2CO_2 MXene as a nucleobase 2D sensor: a first-principles study. *Appl. Surf. Sci.* 544: 148946.

19 Wang, H., Li, H., Huang, Y. et al. (2019). A label-free electrochemical biosensor for highly sensitive detection of gliotoxin based on DNA nanostructure/MXene nanocomplexes. *Biosens. Bioelectron.* 142: 111531.

20 Nah, J.S., Barman, S.C., Zahed, M.A. et al. (2021). A wearable microfluidics-integrated impedimetric immunosensor based on $Ti_3C_2T_x$ MXene incorporated laser-burned graphene for noninvasive sweat cortisol detection. *Sens. Actuators B Chem.* 329: 129206.

21 Liu, J., Jiang, X., Zhang, R. et al. (2019). MXene-enabled electrochemical microfluidic biosensor: applications toward multicomponent continuous monitoring in whole blood. *Adv. Funct. Mater.* 29: 1807326.

22 Lei, Y. et al. (2019). A MXene-based wearable biosensor system for high-performance in vitro perspiration analysis. *Small* 15: 1901190.

23 Zhao, L., Wang, L., Zhen, Y. et al. (2021). Highly-stable polymer-crosslinked 2D MXene-based flexible biocompatible electronic skins for in vivo biomonitoring. *Nano Energy* 84: 105921.

24 Jandas, P.J., Prabakaran, K., Luo, J. et al. (2021). $Ti_3C_2T_x$ MXene-Au nanoparticles doped polyimide thin film as a transducing bioreceptor for real-time acoustic detection of carcinoembryonic antigen. *Sens. Actuators B Chem.* 331: 112998.

25 Li, X., Lu, Y., Shi, Z. et al. (2021). Onion-inspired MXene/chitosan-quercetin multilayers: enhanced response to H_2O molecules for wearable human physiological monitoring. *Sens. Actuators B Chem.* 329: 129209.

26 Wu, Q. et al. (2019). A 2D transition metal carbide MXene-based SPR biosensor for ultrasensitive T carcinoembryonic antigen detection. *Biosens. Bioelectron.* 144: 111697.

27 Mohammadniaei, M., Koyappayil, A., Sun, Y. et al. (2020). Gold nanoparticle/MXene for multiple and sensitive detection of oncomiRs based on synergetic signal amplification. *Biosens. Bioelectron.* 159: 112208.

28 Xia, T., Liu, G., Wang, J. et al. (2021). MXene-based enzymatic sensor for highly sensitive and selective detection of cholesterol. *Biosens. Bioelectron.* 183: 113243.

29 Zheng, J., Wang, B., Ding, A. et al. (2018). Synthesis of MXene/DNA/Pd/Pt nanocomposite for sensitive detection of dopamine. *Biosens. Bioelectron.* 816: 189–194.

30 Liu, M., He, Y., Zhou, J. et al. (2020). A "naked-eye" colorimetric and ratiometric fluorescence probe for uric acid based on Ti_3C_2 MXene quantum dots. *Anal. Chim. Acta.* 1103: 134–142.

31 Wang, Y., Sun, W., Li, Y. et al. (2021). Imidazole metal-organic frameworks embedded in layered $Ti_3C_2T_x$ Mxene as a high-performance electrochemiluminescence biosensor for sensitive detection of HIV-1 protein. *Microchem. J.* 167: 106332.

32 Soomro, R.A., Jawaid, S., Kalawar, N.H. et al. (2020). In-situ engineered MXene-TiO_2/$BiVO_4$ hybrid as an efficient photoelectrochemical platform for sensitive detection of soluble CD44 proteins. *Biosens. Bioelectron.* 166: 112349.

33 Zhang, K., Fan, Z., Yao, B. et al. (2021). Exploring the trans-cleavage activity of CRISPR-Cas12a for the development of a Mxene based electrochemiluminescence biosensor for the detection of Siglec-5. *Biosens. Bioelectron.* 178: 113019.

34 Chen, J., Tong, P., Huang, L. et al. (2019). Ti_3C_2 MXene nanosheet-based capacitance immunoassay with tyramine-enzyme repeats to detect prostate-specific antigen on interdigitated micro-comb electrode. *Electrochim. Acta.* 319: 375–381.

35 Sivasankarapillai, V.S., Somakumar, A.K., Joseph, J. et al. (2020). Cancer theranostic applications of MXene nanomaterials: recent updates. *Nano-Struct. Nano-Objects.* 22: 100457.

36 Huang, Z., Cu, X., Li, S. et al. (2020). Two-dimensional MXene-based materials for photothermal therapy. *Nanophotonics* 9: 2233–2249.

37 Huang, P. et al. (2014). Triphase interface synthesis of plasmonic gold bellflowers as near-infrared light mediated acoustic and thermal theranostics. *J. Am. Chem. Soc.* 136: 8307–8313.

38 Lin, H., Wang, X., Yu, L. et al. (2017). Two-dimensional ultrathin MXene ceramic nanosheets for photothermal conversion. *Nano Lett.* 17: 384–391.

39 Dai, C., Lin, H., Xu, G. et al. (2017). Biocompatible 2D titanium carbide (MXenes) composite nanosheets for pH-responsive MRI-guided tumor hyperthermia. *Chem. Mater.* 29: 8637–8652.

40 Yu, X., Cai, X., Cui, H. et al. (2017). Fluorine-free preparation of titanium carbide MXene quantum dots with high near-infrared photothermal performances for cancer therapy. *Nanoscale* 9: 17859.

41 Liu, G., Zou, J., Tang, Q. et al. (2017). Surface modified Ti_3C_2 MXene nanosheets for tumor targeting photothermal/photodynamic/chemo synergistic therapy. *ACS Appl. Mater. Interfaces* 9: 40077–40086.

42 Dai, C., Chen, Y., Xiangxiang, J. et al. (2017). Two-dimensional tantalum carbide (MXenes) composite nanosheets for multiple imaging-guided photothermal tumor ablation. *ACS Nano* 11: 12696–12712.

43 Liu, Z., Zhao, M., Lin, H. et al. (2018). 2D magnetic titanium carbide MXene for cancer theranostics. *J. Mater. Chem. B* 6: 3541–3548.

44 Lin, H., Wang, Y., Gao, S. et al. (2018). Theranostic 2D tantalum carbide (MXene). *Adv. Mater.* 30: 1703284.

45 Xiang, H., Lin, H., Yu, L., and Chen, Y. (2019). Hypoxia-irrelevant photonic thermodynamic cancer nanomedicine. *ACS Nano* 13: 2223–2235.

46 Prestinaci, F., Pezzotti, P., and Pantosti, A. (2015). Antimicrobial resistance: a global multifaceted phenomenon. *Pathog. Glob. Health.* 109: 309–318.

47 He, Q., Hu, H., Han, J., and Zhao, Z. (2022). Double transition-metal $TiVCT_x$ with dual-functional antibacterial capability. *Mater. Lett.* 308: 131100.

48 Unal, M.A. et al. (2021). 2D MXenes with antiviral and immunomodulatory properties: a pilot study against SARS-CoV-2. *Nano Today* 38: 101136.

49 Yu, Z., Liu, R., Zhao, W. et al. (2021). Versatile self-assembled MXene-Au nanocomposites for SERS detection of bacteria, antibacterial and photothermal sterilization. *Chem. Eng. Sci.* 426: 131914.

50 Borandeh, S., Bochove, B.V., Teotia, A., and Seppala, J. (2021). Polymeric drug delivery systems by additive manufacturing. *Adv. Drug Deliv. Rev.* 173: 349–373.

51 Wu, Z., Shi, J., Song, P. et al. (2021). Chitosan/hyaluronic acid based hollow microcapsules equipped with MXene/gold nanorods for synergistically enhanced near infrared responsive drug delivery. *Int. J. Biol. Macromol.* 183: 870–879. (2021).

52 Dong, Y., Li, S., Li, X., and Wang, X. (2021). Smart MXene/agarose hydrogel with photothermal property for controlled drug release. *Int. J. Biol. Macromol.* 190: 693–699.

53 Zhu, B., Shi, J., Liu, C. et al. (2021). In-situ self-assembly of sandwich-like Ti3C2 MXene/gold nanorods nanosheets for synergistically enhanced near-infrared responsive drug delivery. *Ceram. Int.* 47: 24252–24261.

54 Xing, C. et al. (2018). Two-dimensional MXene (Ti_3C_2)-integrated cellulose hydrogels: toward smart three-dimensional network nanoplatforms exhibiting light-induced swelling and bimodal photothermal/chemotherapy anticancer activity. *ACS Appl. Mater. Interfaces* 10: 27631–27643.

55 Zhang, P., Yang, X.J., Li, P. et al. (2020). Fabrication of novel MXene (Ti_3C_2)/polyacrylamide nanocomposite hydrogels with enhanced mechanical and drug release properties. *Soft Matter* 16: 162.

56 Liu, Y., Han, Q., Yang, W. et al. (2020). Two-dimensional MXene/cobalt nanowire heterojunction for controlled drug delivery and chemo-photothermal therapy. *Mater. Sci. Eng. C.* 116: 111212.

57 Han, X., Hunag, J., Lin, H. et al. (2018). 2D ultrathin MXene-based drug delivery nanoplatform for synergistic photothermal ablation and chemotherapy of cancer. *Adv. Healthcare Mater.* 7: 1701394.

58 Rabiee, N., Bagherzadeh, M., Jouyandeh, M. et al. (2021). Natural polymers decorated MOF-MXene nanocarriers for co-delivery of doxorubicin/pCRISPR. *ACS Appl. Bio Mater.* 4: 5106–5121.

59 Li, Z., Zhang, H., Han, J. et al. (2018). Surface nanopore engineering of 2D MXenes for targeted and synergistic multitherapies of hepatocellular carcinoma. *Adv. Mater.* 30: 1706981.

16

MXenes for Point of Care Devices (POC)

Fulden Ulucan-Karnak[1], Sachin Mishra[2], Sukirti Tiwari[3], and Cansu İlke Kuru[4,5]

[1]*Department of Medical Biochemistry, Institute of Health Sciences, Ege University, Bornova, Izmir, Turkey*
[2]*Department of Electrical Electronic Engineering and Automation, Rovira i Virgili University, Tarragona, Spain*
[3]*Department of Nanotechnology, Amity University, Noida, Uttar Pradesh, India*
[4]*Department of Biotechnology, Graduate School of Natural and Applied Sciences, Ege University, İzmir, Turkey*
[5]*Buca Municipality Science and Art Center, Izmir, Turkey*

16.1 Introduction

A class of transition metal carbides, nitrides, and carbonitrides collectively referred to as MXenes consists of MAX-based element compositions (Figure 16.1) created via the chemical delamination of 3D quaternary compounds. They may also be created from layered compounds, which have the following general formula: [1].

$$M_nX_{n+1}T_x, (n = 1 - 3) \tag{16.1}$$

In the above equation, M denotes transition metals, X is carbon and/or nitrogen, and T_x shows the hydroxyl, oxygen, or fluorine terminations derived from the synthesis procedures [2]. Etching-based procedures, chemical vapor deposition (CVD), the template approach, and other methods are often used to create MXenes. For the fabrication of MXenes-based sensing platforms, the etching approach is preferred [3, 4].

MXenes have salient properties that allow their research to gain more importance over the next 15 years. These properties can be listed as:

(i) electrical conductivity,
(ii) mechanical properties of transition metal carbides/nitrides,
(iii) tunable electronic band gap,
(iv) stabilization of colloidal structures,
(v) plasmonic behavior,
(vi) efficient absorption of electromagnetic waves [5-7].

With the aid of these great properties, MXenes started to be applicable in biomedical applications in 2012 [8]. Currently, MXenes can show tremendous features in different platforms such as biosensors, theranostics, and cancer therapy. [6] Since MXene's discovery in 2011, it has increased. Now, more than 40 MXene compositions and hundreds of studies on using MXenes alone or as additives exist in the literature. However, detailed experimental studies of the mechanisms regarding increased performance are still needed. Modeling,

MXenes: Next-Generation 2D Materials: Fundamentals and Applications, First Edition.
Edited by Jay Singh, Kshitij RB Singh, Ravindra Pratap Singh, and Charles Oluwaseun Adetunji.

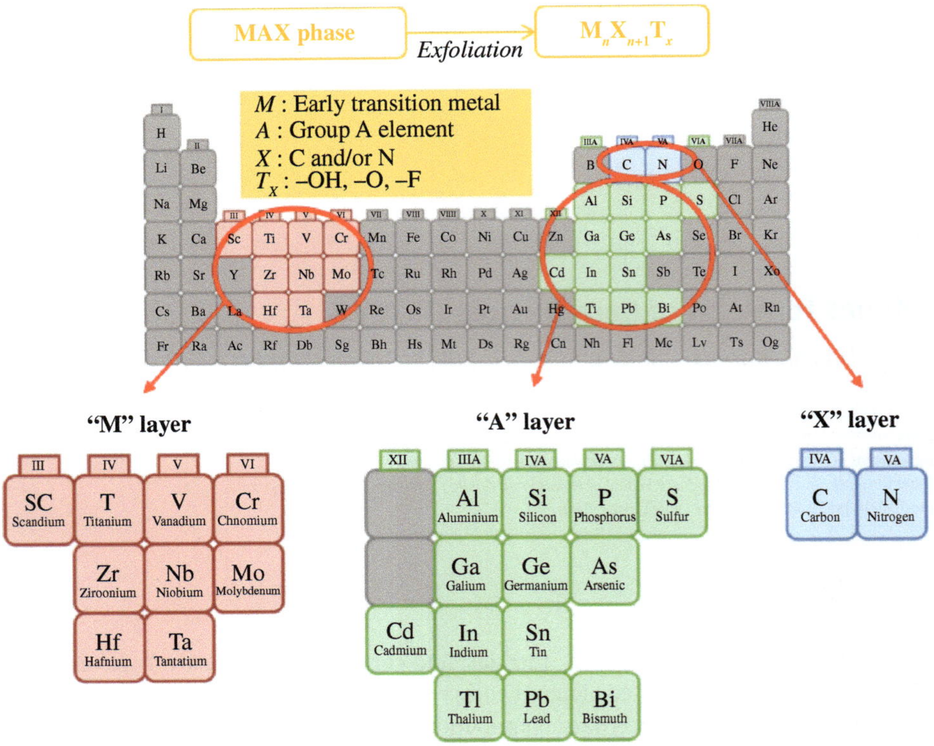

Figure 16.1 Element chart of MAX phase and MXene's compositions. Source: Zamhuri, A et al., 2021/Springer Nature/CC BY 4.0.

simulation, and machine learning processes are useful tools to predict their structural and chemical properties and performance in different applications [9].

16.2 Characteristics of MXenes on Biosensing

MXenes have a promising hope as potential advanced material on biosensing platforms due to their great functionalities in terms of biocompatibility, metallic conductivity, surface area, and ease of surface functionalization [10, 11]. There are three types of biosensing strategies in the literature that use MXenes based on interactions: (i) physical adsorption, (ii) anchoring antibodies, (iii) DNA sensors. Physical adsorption and antibody anchoring are easy to fabricate but not stable; also, DNA sensors are not sufficiently sensitive. They all needed to investigate in terms of *in vitro*/*in vivo* cytocompatibility, long-term toxicity, distribution, detoxification, and accumulation [10]. Many improvements in the design of MXene biosensors could provide the development of state-of-the-art detection techniques with electrochemistry, fluorescence, optical, luminescence, and colorimetry for intended selective and point-of-care detection of various biomarkers [12, 13].

16.2.1 Advantages of MXene and its Derivatives for Biosensing

MXenes have several advantages, especially in biosensing applications, as explained above. If explaining more detail is needed, their electronic properties allow sufficient flexibility for synthesis and modification processes. With the aid of these advantageous properties, MXenes are able to detect specific biomarkers and ions, and they can be effective in the field of drug deliveries. Their optical properties provide photoconversion for photoresponsive nanocomposites in the application of imaging-guided diagnosis. Also, their active multi-layer structures can be applied to enhance the sensitivity of the biosensing elements [14].

MXenes can be designed in very different forms and structures. 2D geometry and ultrathin thickness allow MXene production as thin films [15]. Also, their electrical and mechanical properties are tunable with cost-effective synthesis, which provides them with conductivity and hydrophilicity [16, 17]. It can be hydrophilic due to the addition of surface termination groups (like = O, —OH, and —F), and it is very useful for electrochemical biosensors and developing MXene-polymer nanocomposites [18]. The number of layers "n" in $M_{n+1}X_n$ also has a great influence on the mechanical properties (strength, bending, stiffness, etc.) of MXene due to elastic constant and Young's modulus [19].

16.2.2 Disadvantages of MXene and its Derivatives for Biosensing

The disadvantages of MXenes could be that the synthesis process can be more complicated along with low oxidation stability [20]. In terms of the synthesis process, control size, ingredients, and surface functionalization of the MXenes are the main challenges in developing their characteristic performances [21].

In biosensors, obtaining physical information due to the limiting positions in the body and mechanical durability, such as lack of comfort, low sensitivity, and low human health management of wearable sensors could be mentioned [22]. MXenes can also be used as the matrix of enzymes with the aid of their unique morphology and the capability to be split into multiple one-dimensional planes. Although they can provide a good matrix for immobilized enzymes or other proteins, that allow remaining sufficient catalytic activity, they can only be provided [23]. Environmental biosafety, vulnerability, and synthesis diversity must be carefully investigated to develop the stability of MXene biosensors [24]. MXene/polymer nanocomposites' mechanical and electrical properties continue to present difficulties, particularly for the development of wearable biosensors. These can be explained by resistivity, the optimal concentration of MXene in these kinds of nanocomposites, and a decrease in flexibility and stretchability with an increase in MXene concentration [25].

16.2.3 Sensing Mechanism of MXene Wearables

The MXene production technique, sensor design, and sensing mechanism that is chosen will have a significant impact on the degree of sensitivity that may be achieved for wearable measurements. Wearable sensing systems can be classified as physical or chemical sensors depending on the bodily parameter being monitored. The majority of MXene strain/pressure sensors operate on the piezoresistivity concept. The phrase

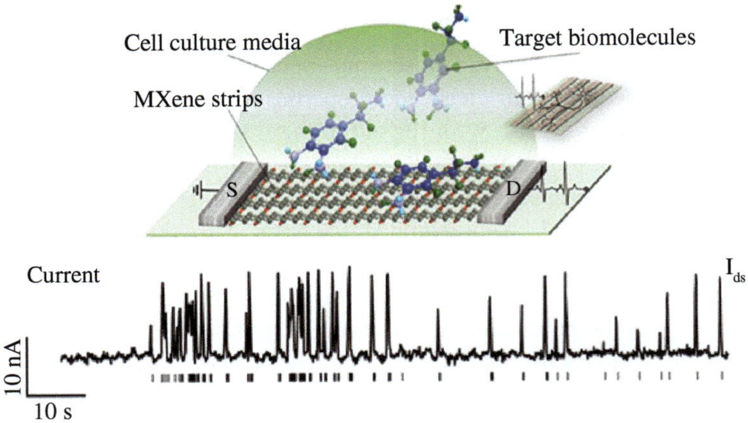

Figure 16.2 The MXene sensing mechanism. The electrocatalytic characteristics of the MXene-based devices are changed and produce signal peaks when they interact with the target biomolecules. Source: Bingzhe Xu et al., 2020/IOP Publishing/CC BY 4.0.

"piezoresistive effect" describes any mechanical distortion of conductive material that alters its electrical resistance. When the MXene layer is placed on porous substrates with a large internal surface area and elasticity, such as a sponge or paper, it exhibits similar strain-sensing characteristics. Electrochemical sensors use electrodes to transform a chemical reaction into an electrical output. The key sensing methods found in MXene-based gas sensors are Schottky barrier (SB) modulation and surface charge transfer from adsorbed gas molecules [26]. According to the concentration of the target signals, the MXene biosensor mechanisms typically make use of the special electrocatalytic capabilities of the MXene sheets. The electrical characteristics and current signal are changed when the biological targets attach to the MXene sheets, as classified in Figure 16.2 [27].

16.3 Point-of-Care Diagnosing COVID-19: Methods Used to Date

The COVID-19 pandemic has placed a tremendous strain caused by the SARS-CoV-2 (severe acute respiratory syndrome coronavirus 2) virus, which has infected millions around the globe. The World Health Organization (WHO) reported a pandemic stain that caused by a respiratory disease named "COVID 19"on11th March 2020. The SARS-CoV-2 virus is the seventh to infect humans and the first from the family Coronaviridae to produce a pandemic. Using metagenomic sequence analysis, the virus belongs to the β-coronavirus lineage, which exists in rodents and bats [26]. It is spread by aerosols by infected people who are presymptomatic, asymptomatic, or symptomatic. The condition can be life-threatening in many situations (often in individuals with accompanying comorbidities), ranging from asymptomatic to symptomatic. As of now, the symptoms may be divided into three groups: digestive, musculoskeletal, and respiratory issues. The viral structure has a positive sense RNA strand enveloped by a circular capsid with a regular

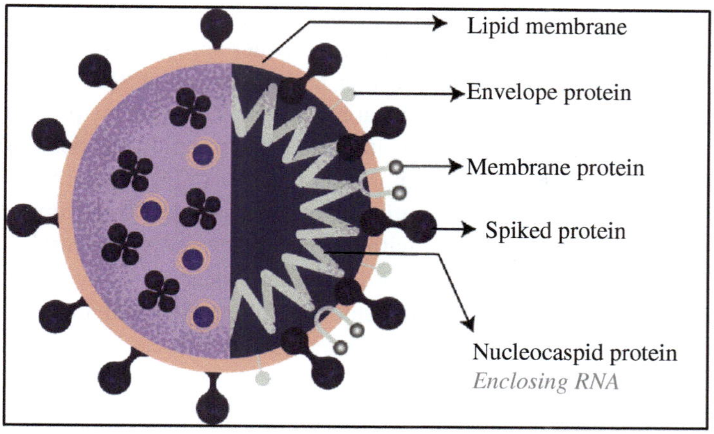

Figure 16.3 Illustration structural view of SARS-CoV-2.

arrangement of nucleocapsid proteins called spike (S) proteins. The S proteins are trimeric, with each monomeric unit comprising S1 and S2 subunits (Figure 16.3). The viral capsid contains a significant number of S glycoproteins, which bind to the angiotensin-converting enzyme 2 (ACE2) receptor on the host cell [27]. The virus has infected a large number of individuals due to its high transmissibility from infected individuals to healthy ones, as indicated by its transmission ratio (R0). Very few cases of infectivity have been reported in children below the age of 15, and most fatal cases have been reported in individuals aged 60 and above [28]. Further, viruses mutate, which leads to the formation of variants. The emergence of newer COVID-19 variants, such as Delta and Omicron, continues to be an added threat globally.

Limiting continuing infection and re-infection is critical to bringing the epidemic to a conclusion. To do this, it is critical to restrict viral shedding in order to confine the pandemic. Furthermore, the duration of COVID-19 infection varied between healthy individuals and those with concomitant comorbidities. Nucleic acid, protein, and antibody-based tests are the three types of diagnostic instruments. Despite a negative nasopharyngeal swab, the RT-PCR of the anal-fecal swab tests positive for weeks in individuals with comorbidities. RT-PCR is a diagnostic method for viral sequences. To generate a quantitative report, the total operation takes around two days [29]. However, the total procedure necessitates specialized knowledge, sensitive reagents, high-quality nucleic acid extraction, and costly equipment. Despite its sequence specificity, the test may provide false negatives when RNA sequence variants evolve. As a result, in addition to RT-PCR testing, quick, sensitive, portable diagnostics in radiology, immunology, and nucleic acid-based assays are available. In extreme instances of COVID-19, substantial respiratory and neurological impairment may ensue. These may be detected by MRI, chest X-ray (CXR), and CT-scan as supplementary tests [30, 31]. These may lead to conclusive diagnostics in asymptomatic individuals. Colorimetric assays have been developed using CRISPR-Cas12a to detect viral nucleic acid [32]. A group has developed an amplification-based plasmonic detection system. In this work, reverse transcriptase coupled with recombinase polymerase was employed, which led to abundant dsDNA production. The dsDNA then activates CRISPR, which degrades the AuNP-bound DNA strands. A change in SPR is then detected, giving visual and UV–Vis-based confirmation [32]. Other approaches have used electrochemical sensing tools, using AuNPs and polymer matrices to detect SARS-CoV-2. Overcoming the requirement of nucleic acid amplification, AuNP and graphene platform conjugated with complementary sequences give a voltage-dependent rapid detection [33]. Recently, Hussein et al. have developed a virus-imprinted polymer (VIP) matrix, which is independent of RNA extraction and amplification steps. Instead, the impedimetric biosensor detects all SARS-CoV-2 virus particles with high accuracy and specificity [34]. Sewage tests to track viral shedding may be another important aspect of containing the pandemic. Live viruses have been isolated from the stool test of infected individuals. In recent work, a group of researchers developed a rapid sensor that detects SARS-CoV-2 particles in waterbodies. The sensor used a lanthanide-based fluorescent technique, where the fluorescent patterns accurately discriminated between the virus and other microbes. The sensor doesn't require expert cleaning or handling and gives rapid real-time detection in infected reservoirs [35].

16.4 Applications of MXenes as PoCs

In recent times, MXene has achieved a remarkable position in terms of being a potent material for the construction of biosensing POCs. The underlying reasons for this are their unique and outstanding chemical, electrochemical, optical, and physical properties; the presence of active sites (such as —OH, CO, and —F groups) for immobilization of biomolecules; multilayered structure for providing larger surface area for electrocatalytic activities; anti passive and anti-fouling property; and outstanding biocompatibility [36–38]. Wang et al. employed HF etching to produce an organ-like structure of MXene-Ti_3C_2 nanomaterial to construct a biosensor that is mediator-free by hemoglobin immobilization on the surface. This sensor was prepared for the detection of H_2O_2. The prepared sensor exhibited 20 nM as the detection limit and a linear range of 0.1–260 μM [39]. Wang et al. fabricated a biosensor based on MXene-Au @Pt/5′-nucleotidase/xanthine oxidase/bovine serum albumin (BSA) for the inosine monophosphate and H_2O_2 detection [40]. Mohsen et al. fabrication was done by the amplification system to identify microRNAs in plasma. This sensor was composed of duplex-specific nuclease and AuNP-coated MXene-$Ti_3C_2T_x$. The biosensor exhibited a rapid processing time of 80 minutes, high specificity and sensitivity, an antifouling feature, a broad linear range of 500–50 nM, and a low detection limit [41].

Multiple researchers have also worked on MXene-based aptasensors owing to their high affinity, sensitivity, stability, and easier synthesis. Zhou et al. fabricated a multicomponent hybrid biosensor by embedding $Ti_3C_2T_x$ nanosheets and PMo_{12} nanoparticles hybrid nanostructure in a polypyrrole (PPy) matrix. This biosensor was fabricated for the detection of osteopontin (a biomarker of cancer). This sensor possessed outstanding sensitivity, stability, renewability, low detection limit (0.98 fg mL^{-1}), and a broad linear range [42]. Other MXene-based electrical sensors have also been fabricated to identify glucose, epinephrine, L-cysteine, nitrite, organophosphate pesticides, etc. [13]. Apart from electrical and optical biosensing PoCs, other principles, such as near-infrared (NIR) photothermal immunoassay and chemiluminescence, were also used to create MXene-based platforms.

Cai et al. fabricated a biosensor for the identification of prostate-specific antigens by employing Ti_3C_2 QDs-encapsulated liposomes. This sensor exhibited 0.4 ng mL^{-1} as a limit of detection and a linear range of detection, along with visual and thermal data with the help of NIR imaging cameras [43].

16.4.1 Cancer Diagnosis

In terms of cancer diagnosis, MXene provides multiple outstanding physiochemical attributes, such as ample electrical conductivity, magnetism, and unique luminescence, along with high surface area, biocompatibility, and low toxicity. Thus, owing to these properties, MXenes are best suited for biosensing and bioimaging of cancer disorders [44]. Dai et al. grew MnOx nanosheets on Ti_3C_2 surface, *in situ*, to fabricate a Ti_3C_2-based composite MXene biosensor to provide magnetic resonance (MR) imaging for cancer diagnosis [45].

MXene-based optical biosensors were based on the principles of electrochemiluminescence and surface plasmon resonance. Zhang et al. fabricated an aptameric biosensor associated with electrochemiluminescence signal amplification by employing Ti_3C_2 MXenes nanosheets for the detection of exosomes. This sensor exhibited 125 particles μL^{-1} as a detection limit [46]. Wu et al. fabricated a conjugate of Ti_3C_2-MXene (sensing platform), multi-walled carbon nanotube-polydopamine-silver nanoparticle (signal enhancer), and polyclonal anti-CEA antibody (MWPAg-Ab_2) for the detection of CEA (carcinoembryonic antigen), which is an early marker for the detection of cancer. This sensor has a detection limit of 0.07 fM and a linear detection range [47]. Zhang et al. created a biosensor for detecting exosomes based on Cy3-coated CD63 aptamer (Cy3-CD63 aptamer)/Ti_3C_2. The limit of detection for this sensor was 1.4103 particles/mL [48]. Because of the superparamagnetic feature of Ti_3C_2-MXenes-Fe_3O_4 nanocrystals, Liu et al. presented a biosensor based on them. This detector was designed to detect cancer [49]. Zong et al. constructed GdW10-based polyoxometalates associated with Ti_3C_2 MXene nanosheets to detect tumor hyperthermia [50]. Wang et al. designed a biosensor for the detection of long noncoding RNA of prostate cancer gene 3 based on entropy-driven RNA explorer (EDRE) coated on TAT peptide-functionalized titanium carbide MXenes (Ti_3C_2-TAT) [51]. Wang et al. constructed a biosensor for the detection of biomarkers from various cellular positions. The biosensor was based on a chimeric DNA functionalized nanoprobe associated with Ti_3C_2 MXene [52].

16.4.2 Diagnosis of Bacterial and Viral Diseases

MXenes have also become a prominent tool for creating diagnostic platforms for pathogenic diseases, such as bacterial and viral diseases. Peng et al. created a MXene-graphene-based FET for diagnosing influenza and SARS-CoV-2. In this diagnostic platform, MXene contributed extraordinary sensitivity, and graphene contributed a large surface area for detection. The amalgamation of graphene with MXene provided ultra-sensitive virus-detection materials. MXene-based biodevice is characterized by rapid response time (~50 ms; faster than RT-PCR) within the low detection limit, high sensitivity and high specificity [53].

Peng et al. created a biosensor to identify HPV on MXene nanosheet (Ti_3C_2 NS)-based fluorescence DNA. They utilized MXene's enhanced capability of quenching the fluorescent dyes and their differential affinities for single-stranded and double-stranded DNA. This device displays lower fluorescent emission to the ssDNA probe, but an enhanced fluorescent emission after the duplex formation of the probe with the target DNA (HPV-18). This sensor is also characterized by high specificity and a low detection level [53].

16.5 Current Challenges and Future Outlook

In the realm of nanotechnology, MXenes are capable of producing 2D functional materials with cutting-edge characteristics, including strong metallic conductivity, a sizable surface area, water dispersibility, simplicity of large-scale production, and surface modification. MXenes have recently become a preferred material for researching biosensors with desired performance because of their strong electro-conductivity, hydrophilicity, and tunable surface chemistry [54]. A range of enzymes, metalloproteinases, and nucleic acid structures may easily be immobilized on MXenes thanks to their distinctive shape, which promotes direct electron transfer and higher electrochemical activity [55]. Due to their excellent hydrophilicity, 2D structure, high mechanical strength, and distinctive optical characteristics, MXenes are a prospective family of electrode materials for creating diverse supercapacitor devices. Despite substantial development in MXenes for supercapacitor applications, many problems and unforeseen hurdles remain to be solved to improve MXenes' electrochemical performance and promote its practical implementation [6].

MXenes provide good fundamental and transdisciplinary research due to their complex physics and chemistry, as well as their deep structural and compositional diversity. Nonetheless, MXenes synthesis, including uniform and pure surface termination, has remained an experimental difficulty, and it is a roadblock to achieving the full potential of applications that have been proposed. As a result, it is important to look at MXene's mixed functionalization to see whether any applications are not affected by surface termination. Mixed functionalization can also result in numerous interfaces at the surface of MXenes, including sharp and diffusive interfaces. Isolation of virgin MXenes for comprehensive characterization and subsequently controlled functionalization should be attempted in the laboratory [56].

16.6 Conclusion

In recent years, there has been a remarkable rise in the number of study areas and publications in the MXenes field, as shown in Figure 16.4 [57].

By mixing various compounds, such as those from the large and understudied MXene family, fresh research may uncover novel routes in the biosensing area, which is still in its infancy [54]. We believe that MXene research is still in its infancy and that there are many fascinating discoveries yet to be made because of the enormous, undiscovered family of MXenes and their distinctive combination of properties that open the door to a wide variety of applications [57]. The most recent development in MXenes, known as Green-MXenes,

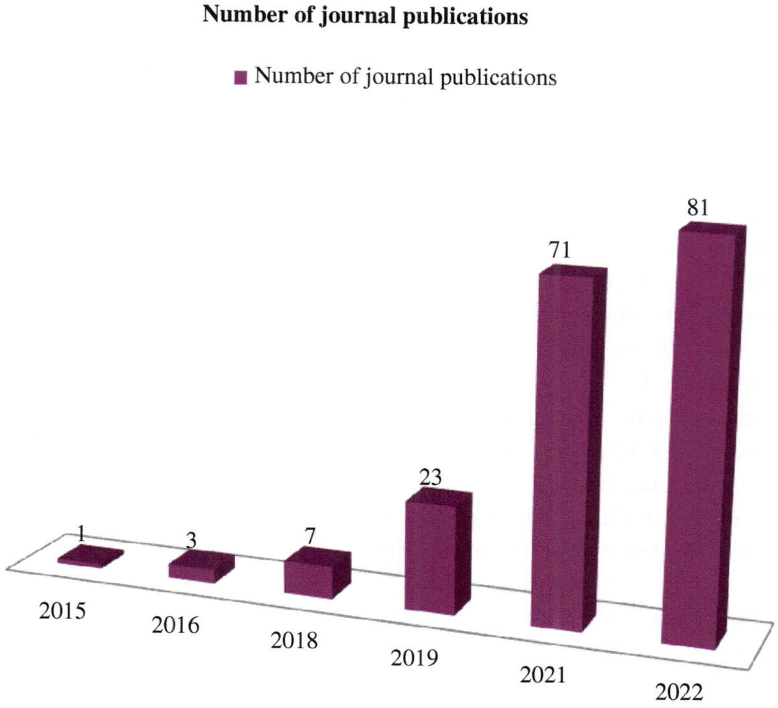

Figure 16.4 Number of journal publications on MXenes biosensors, searched with keywords "MXene biosensor."

may be characterized as low-cost, biocompatible, and functionalized 2D nanosystems that might be the next generation of nanomaterials for biosensors and point-of-care devices.

References

1 Ibrahim, Y., Mohamed, A., Abdelgawad, A.M. et al. (2020). The recent advances in the mechanical properties of self-standing two-dimensional MXene-based nanostructures: deep insights into the supercapacitor. *Nanomaterials* 10 (10): 1916. Available from: https://www.mdpi.com/2079-4991/10/10/1916.

2 Ronchi, R.M., Arantes, J.T., and Santos, S.F. (2019). Synthesis, structure, properties and applications of MXenes: current status and perspectives. *Ceram. Int.* 45 (15): 18167–18188. Available from: https://linkinghub.elsevier.com/retrieve/pii/S0272884219316153.

3 Xin, M., Li, J., Ma, Z. et al. (2020). MXenes and their applications in wearable sensors. *Front. Chem.* 8. Available from: https://www.frontiersin.org/article/10.3389/fchem.2020.00297/full.

4 Salim, O., Mahmoud, K.A., Pant, K.K., and Joshi, R.K. (2019). Introduction to MXenes: synthesis and characteristics. *Mater. Today Chem.* 14: 100191. Available from: https://linkinghub.elsevier.com/retrieve/pii/S2468519419302034.

5 Champagne, A. and Charlier, J.-C. (2021). Physical properties of 2D MXenes: from a theoretical perspective. *J. Phys.: Mater.* 3 (3): 032006. Available from: https://iopscience.iop.org/article/10.1088/2515-7639/ab97ee.

6 Gogotsi, Y. and Anasori, B. (2019). The rise of MXenes. *ACS Nano* 13 (8): 8491–8494. Available from: https://pubs.acs.org/doi/10.1021/acsnano.9b06394.

7 Papadopoulou, K.A., Chroneos, A., Parfitt, D., and Christopoulos, S.-R.G. (2020). A perspective on MXenes: their synthesis, properties, and recent applications. *J. Appl. Phys.* 128 (17): 170902. Available from: http://aip.scitation.org/doi/10.1063/5.0021485.

8 Zamhuri, A., Lim, G.P., Ma, N.L. et al. (2021). MXene in the lens of biomedical engineering: synthesis, applications and future outlook. *Biomed. Eng. Online* 20 (1): 33. Available from: https://biomedical-engineering-online.biomedcentral.com/articles/10.1186/s12938-021-00873-9.

9 Naguib, M., Barsoum, M.W., and Gogotsi, Y. (2021). Ten years of progress in the synthesis and development of MXenes. *Adv. Mater.* 33 (39): 2103393. Available from: https://onlinelibrary.wiley.com/doi/10.1002/adma.202103393.

10 Xu, B., Zhi, C., and Shi, P. (2020). Latest advances in MXene biosensors. *J. Phys.: Mater.* 3 (3): 031001. Available from: https://iopscience.iop.org/article/10.1088/2515-7639/ab8f78.

11 Khan, R. and Andreescu, S. (2020). MXenes-based bioanalytical sensors: design, characterization, and applications. *Sensors* 20 (18): 5434. Available from: https://www.mdpi.com/1424-8220/20/18/5434.

12 Sinha, A., Dhanjai, Z.H., Huang, Y. et al. (2018). MXene: an emerging material for sensing and biosensing. *TrAC Trends Anal. Chem.* 105: 424–435. Available from: https://linkinghub.elsevier.com/retrieve/pii/S0165993618301390.

13 Huang, J., Li, Z., Mao, Y., and Li, Z. (2021). Progress and biomedical applications of MXenes. *Nano Select* 2 (8): 1480–1508. Available from: https://onlinelibrary.wiley.com/doi/10.1002/nano.202000309.

14 Lin, X., Li, Z., Qiu, J. et al. (2021). Fascinating MXene nanomaterials: emerging opportunities in the biomedical field. *Biomater. Sci.* 9 (16): 5437–5471. Available from: http://xlink.rsc.org/?DOI=D1BM00526J.

15 Li, X., Ran, F., Yang, F. et al. (2021). Advances in MXene films: synthesis, assembly, and applications. *Trans. Tianjin Univ.* 27 (3): 217–247. Available from: https://link.springer.com/10.1007/s12209-021-00282-y.

16 Shukla, V. (2020). The tunable electric and magnetic properties of 2D MXenes and their potential applications. *Mater. Adv.* 1 (9): 3104–3121. Available from: http://xlink.rsc.org/?DOI=D0MA00548G.

17 Jeon, J., Yang, Y., Choi, H. et al. (2020). MXenes for future nanophotonic device applications. *Nanophotonics* 9 (7): 1831–1853.

18 Gui, J.-C., Han, L., and Cao, W. (2021). Lamellar MXene: a novel 2D nanomaterial for electrochemical sensors. *J. Appl. Electrochem.* 51 (11): 1509–1522. Available from: https://link.springer.com/10.1007/s10800-021-01593-7.

19 Biswas, S. and Alegaonkar, P.S. (2021). MXene: evolutions in chemical synthesis and recent advances in applications. *Surfaces* 5 (1): 1–36. Available from: https://www.mdpi.com/2571-9637/5/1/1.

20 Yoon, J., Shin, M., Lim, J. et al. (2020). Recent advances in MXene nanocomposite-based biosensors. *Biosensors* 10 (11): 185. Available from: https://www.mdpi.com/2079-6374/10/11/185.

21 Soleymaniha, M., Shahbazi, M.-A., Rafieerad, A.R. et al. (2019). Promoting role of MXene nanosheets in biomedical sciences: therapeutic and biosensing innovations. *Adv. Healthcare Mater.* 8 (1): 1801137. Available from: https://onlinelibrary.wiley.com/doi/10.1002/adhm.201801137.

22 Yuan, X., Zhang, M., and Wu, Y. (2020). MXene-based sensors for detecting human physiological information. *Sens. Mater.* 32 (12): 4047–4065.

23 Ramanavicius, S. and Ramanavicius, A. (2020). Progress and insights in the application of MXenes as new 2D nano-materials suitable for biosensors and biofuel cell design. *Int. J. Mol. Sci.* 21 (23): 9224. Available from: https://www.mdpi.com/1422-0067/21/23/9224.

24 Li, X., Lu, Y., and Liu, Q. (2021). Electrochemical and optical biosensors based on multifunctional MXene nanoplatforms: progress and prospects. *Talanta* 235: 122726. Available from: https://linkinghub.elsevier.com/retrieve/pii/S0039914021006470.

25 Riazi, H., Taghizadeh, G., and Soroush, M. (2021). MXene-based nanocomposite sensors. *ACS Omega* 6 (17): 11103–11112. Available from: https://pubs.acs.org/doi/10.1021/acsomega.0c05828.

26 Liu, Y.C., Kuo, R.L., and Shih, S.R. (2020). COVID-19: the first documented coronavirus pandemic in history. *Biom. J.* 43: 328–333. https://doi.org/10.1016/j.bj.2020.04.007.

27 Huang, Y., Yang, C., Xu, X.f. et al. (2020). Structural and functional properties of SARS-CoV-2 spike protein: potential antivirus drug development for COVID-19. *Acta Pharmacol. Sin.* 41: 1141–1149. https://doi.org/10.1038/s41401-020-0485-4.

28 Fang, Y., Nie, Y., and Penny, M. (2020). Transmission dynamics of the COVID-19 outbreak and effectiveness of government interventions: a data-driven analysis. *J. Med. Virol.* 92: 645–659. https://doi.org/10.1002/jmv.25750.

29 Xiao, A.T., Tong, Y.X., and Zhang, S. (2020). Profile of RT-PCR for SARS-CoV-2: a preliminary study from 56 COVID-19 patients. *Clin. Infect. Dis.* 71: 2249–2251. https://doi.org/10.1093/cid/ciaa460.

30 Hosseiny, M., Kooraki, S., Gholamrezanezhad, A. et al. (2020). Radiology perspective of coronavirus disease 2019 (COVID-19): lessons from severe acute respiratory syndrome and middle east respiratory syndrome. *Am. J. Roentgenol.* 214: 1078–1082. https://doi.org/10.2214/AJR.20.22969.

31 Zhou, Z., Guo, D., Li, C. et al. (2020). Coronavirus disease 2019: initial chest CT findings. *Eur. Radio.* 30: 4398–4406. https://doi.org/10.1007/s00330-020-06816-7.

32 Zhang, W.S., Pan, J., Li, F. et al. (2021). Reverse transcription recombinase polymerase amplification coupled with CRISPR-Cas12a for facile and highly sensitive colorimetric SARS-CoV-2 detection. *Anal. Chem.* 93: 4126–4133. https://doi.org/10.1021/acs.analchem.1c00013.

33 Alafeef, M., Dighe, K., Moitra, P., and Pan, D. (2020). Rapid, ultrasensitive, and quantitative detection of SARS-CoV-2 using antisense oligonucleotides directed electrochemical biosensor chip. *ACS Nano* 14: 17028–17045. https://doi.org/10.1021/acsnano.0c06392.

34 Hussein, H.A., Kandeil, A., Gomaa, M. et al. (2021). SARS-CoV-2-impedimetric biosensor: virus-imprinted chips for early and rapid diagnosis. *ACS Sens.* 6: 4098–4107. https://doi.org/10.1021/acssensors.1c01614.

35 Alafeef, M., Dighe, K., Moitra, P., and Pan, D. (2022). Monitoring the viral transmission of SARS-CoV-2 in still waterbodies using a lanthanide-doped carbon nanoparticle-based sensor array. *ACS Sustain. Chem. Eng.* 10: 245–258. https://doi.org/10.1021/acssuschemeng.1c06066.

36 Peng, X., Zhang, Y., Lu, D. et al. (2019). Ultrathin Ti_3C_2 nanosheets based "off-on" fluorescent nanoprobe for rapid and sensitive detection of HPV infection. *Sens. Actuators B* 286: 222–229.

37 Kalambate, P.K., Gadhari, N.S., Li, X. et al. (2019). Recent advances in MXene-based electrochemical sensors and biosensors. *Trends Anal. Chem.* 120: 115643.

38 Liu, J., Jiang, X., Zhang, R. et al. (2019). MXene-enabled electrochemical microfluidic biosensor: applications toward multicomponent continuous monitoring in whole blood. *Adv. Funct. Mater.* 29: 1807326.

39 Wang, F., Yang, C., Duan, C. et al. (2014). An organ-like titanium carbide material (MXene) with multilayer structure encapsulating hemoglobin for a mediator-free biosensor. *J. Electrochem. Soc.* 162 (1): B16.

40 Wang, G., Sun, J., Yao, Y. et al. (2020). Detection of inosine monophosphate (IMP) in meat using double-enzyme sensor. *Food Anal. Methods* 13 (2): 420–432.

41 Mohammadniaei, M., Koyappayil, A., Sun, Y. et al. (2020). Gold nanoparticle/MXene for multiple and sensitive detection of oncomiRs based on synergetic signal amplification. *Biosens. Bioelectron.* 159: 112208.

42 Zhou, S., Gu, C., Li, Z. et al. (2019). $Ti_3C_2T_x$ MXene and polyoxometalate nanohybrid embedded with polypyrrole: ultra-sensitive platform for the detection of osteopontin. *Appl. Surf. Sci.* 498: 143889.

43 Cai, G., Yu, Z., Tong, P., and Tang, D. (2019). Ti_3C_2 MXene quantum dot-encapsulated liposomes for photothermal immunoassays using a portable near-infrared imaging camera on a smartphone. *Nanoscale* 11 (33): 15659–15667.

44 Iravani, S. and Varma, R.S. (2021). MXenes for cancer therapy and diagnosis: recent advances and current challenges. *ACS Biomater. Sci. Eng.* 7 (6): 1900–1913.

45 Dai, C., Lin, H., Xu, G. et al. (2017). Biocompatible 2D titanium carbide (MXenes) composite nanosheets for pH responsive MRI-guided tumor hyperthermia. *Chem. Mater.* 29: 8637–8652.

46 Zhang, H., Wang, Z., Zhang, Q. et al. (2019). Ti_3C_2 MXenes nanosheets catalyzed highly efficient electrogenerated chemiluminescence biosensor for the detection of exosomes. *Biosens. Bioelectron.* 124: 184–190.

47 Wu, Q., Li, N., Wang, Y. et al. (2019). A 2D transition metal carbide MXene-based SPR biosensor for ultrasensitive carcinoembryonic antigen detection. *Biosens. Bioelectron.* 144: 111697.

48 Zhang, Q., Wang, F., Zhang, H. et al. (2018). Universal Ti_3C_2 MXenes based self-standard ratiometric fluorescence resonance energy transfer platform for highly sensitive detection of exosomes. *Anal. Chem.* 90: https://doi.org/10.1021/acs.analchem.8b03083.

49 Liu, Z., Zhao, M., Lin, H. et al. (2018). 2D magnetic titanium carbide MXene for cancer theranostics. *J. Mater. Chem. B* 6: 3541–3548.

50 Zong, L., Wu, H., Lin, H., and Chen, Y. (2018). A polyoxometalate-functionalized two-dimensional titanium carbide composite MXene for effective cancer theranostics. *Nano Res.* 11: https://doi.org/10.1007/s12274-018-2002-3.

51 Wang, S., Song, W., Wei, S. et al. (2019). Functional titanium carbide mxenes-loaded entropy-driven RNA explorer for long noncoding RNA PCA3 imaging in live cells. *Anal. Chem.* 91: 8622–8629.

52 Wang, S., Wei, S., Wang, S. et al. (2019). Chimeric DNA-functionalized titanium carbide mxenes for simultaneous mapping of dual cancer biomarkers in living cells. *Anal. Chem.* 91: 1651–1658.

53 Li, Y., Peng, Z., Holl, N.J. et al. (2021). MXene-graphene field-effect transistor sensing of influenza virus and SARS-CoV-2. *ACS Omega* 6 (10): 6643–6653. https://doi.org/10.1021/acsomega.0c05421. PMID: 33748577; PMCID: PMC7944395.

54 Khunger, A., Kaur, N., Mishra, Y.K. et al. (2021). Perspective and prospects of 2D MXenes for smart biosensing. *Mater. Lett.* 304: 130656.

55 Mohammadniaei, M., Nguyen, H.V., Tieu, M.V., and Lee, M.H. (2019). 2D materials in development of electrochemical point-of-care cancer screening devices. *Micromachines* 10 (10): 662.

56 Khazaei, M., Mishra, A., Venkataramanan, N.S. et al. (2019). Recent advances in MXenes: from fundamentals to applications. *Curr. Opin. Solid State Mater. Sci.* 23 (3): 164–178.

57 Hu, M., Zhang, H., Hu, T. et al. (2020). Emerging 2D MXenes for supercapacitors: status, challenges and prospects. *Chem. Soc. Rev.* 49 (18): 6666–6693.

17

MXenes and Their Hybrids for Electromagnetic Interference Shielding Applications

Kadavil Subhash Lakshmy[1], *Ananthu Prasad*[2], *Anandu M. Nair*[1], *Suji Mary Zachariah*[2,3], *Yves Grohens*[3], *and Sabu Thomas*[1,2,4]

[1] *School of Energy Materials, Mahatma Gandhi University, Kottayam, Kerala, India*
[2] *International and Inter University Centre for Nanoscience and Nanotechnology (IIUCNN), Mahatma Gandhi University, Kottayam, Kerala, India*
[3] *University of South Brittany (UBS), Centre de Recherche Christian Huygens, Rue de Saint-Maudé, Lorient, France*
[4] *Department of Chemical Sciences, University of Johannesburg, Doornfontein, Johannesburg, South Africa*

17.1 Introduction

Recently, a modern pollution problem known as electromagnetic interference (EMI) has emerged because of the miniaturization of numerous electronic gadgets and circuits. Prolonged exposure to EMI has resulted in various health issues, data loss from computers, malfunctioning devices, etc. Thus, its mitigation has become a crucial issue that is possible using shielding materials. Metals like Al, Cu, and Ni were widely used as shielding materials, but their weight, processability, corrosion susceptibility, etc., have limited their application. Various fillers like carbon nanotubes (CNT), reduced graphene oxide (RGO), ferrites, barium titanate, titanium dioxide, borophenes, and MXenes have been developed to surpass the disadvantages of metals. Among them, MXenes are very strong candidates due to their strong attenuation ability arising from their conductivity, mechanical attributes, and chemically active surface properties.

MXenes are a set of two-dimensional (2D) transition metal nitrides/carbides/carbonitrides prepared by selective etching from their parent MAX phase. Generally, they can be expressed as $M_{n+1}X_nT_x$. Here, M can be any primitive transition metal like Titanium, Vanadium, and Molybdenum. X represents carbon and/or nitrogen, $n = 1, 2$, or 3, and T_x can be any facet eliminations like hydroxide (OH), oxide (O), and/or fluoride (F) groups. The utilization of MXenes covers a wide area, like energy storage and conversion [1–4], water purification [5–7], and sensors [8–10] indicating its applicability in a variety of areas. In this chapter, MXene synthesis is introduced first. Next, various properties of MXenes are explained in detail. Finally, various hybrids for microwave absorption and EMI shielding have been discussed in detail.

MXenes: Next-Generation 2D Materials: Fundamentals and Applications, First Edition.
Edited by Jay Singh, Kshitij RB Singh, Ravindra Pratap Singh, and Charles Oluwaseun Adetunji.

17.2 Properties of MXenes

17.2.1 Stability

Because of the close connection between characteristics and terminations, it is vital to acknowledge the structures and stable states of terminated-MXenes. As a result, multiple research studies have been progressing to assess the configurations and influences of various functional groupings. On account of the six links between the M and X atoms generating M_2XT_2 molecules, computational simulations revealed the characteristic six-coordination attributes of transition metals. Besides, the most stable configuration was noticed when the termination "T" atoms were placed differently than the "M" and "X" atoms. [11–16]. Nevertheless, in several compounds, say specially arranged double-M elements, the "T" atoms are placed over the "X" atoms to enhance their electrical synergy. [17, 18]. The randomly distributed termination groups in $Ti_3C_2T_x$ and V2CT_x compounds [19–23] were determined experimentally, as predicted by density functional theory (DFT) [24], but a deeper experimental study centered on surface characterization and control is needed to entirely confirm the simulation predictions. [25, 26].

MXene structure has been explored using high-resolution transmission electron microscopy (HRTEM), scanning transmission electron microscopy (STEM), and electron energy loss spectroscopy (EELS) [27–33]. These strategies can disclose the atomic structures, innate flaws, and surface terminations of MXenes [34]. To evaluate their stoichiometric ratio AND chemical configuration, X-ray photoelectron spectroscopy (XPS) is also widely used. Lastly, a few investigations used nuclear magnetic resonance (NMR) spectroscopy to learn further about the surface functionalization of MXenes and how various synthesis strategies (such as HF and LiF-HCl) affect it.

Eventually, their storage and application stability must be highlighted, as studies have shown that MXenes deteriorate if exposed to moisture or extreme heat [35]. A comprehensive Nb_2CT_x MXene research, for example, revealed that during exposure, the Nb adatoms located at the hcp surface locations link to the available oxygen, unstabilizing and finally dismantling the compound complex. Furthermore, a $Ti_3C_2T_x$ FET exposed to air for 70 hours lost around 20% of its initial conductivity. To achieve better stability for the colloids, store them in an oxygen-less surrounding with limited exposure to light, in a vigorous polarity solvent (say, DMSO and nmethyl-2-pyrrolidinone [NMP]), or use solution filtration techniques [36].

17.2.2 Electrical Conductivity

One of the most interesting aspects of MXenes is the electric characteristics. MXenes pressed discs have electric conductivities similar to multilayered graphene and greater than CNT and reduced graphene oxide materials [37] in experiments. The number of layers and the presence of functional groups [10, 38] can also enhance resistivity values. Thus, the simulated conductivities are often higher than those measured experimentally [39, 40].

MXenes with relatively low HF concentration levels and etching intervals have lower defects and greater lateral diameters, causing an increase in the electric conductivity (e.g. large-sized flakes resulted in conductivities five times higher than MXenes with relatively lower sizes). Furthermore, a damp environment can affect their conductivities, indicating the need for relative humidity sensing material. Surface modification with thermal as well as alkaline treatments is an effective method for enhancing electrical properties. They show that conductivities can grow by two orders of magnitude. This elevation is caused by the addition and/or removal of intercalated molecules and functional groups, precisely fluoride groups [41, 42].

17.2.3 Magnetic Properties

In contrast to MAX phases, studies on MXenes' magnetic properties have been extended because of the possibility of magnetization. Numerous pristine components, such as Ti_4C_3, Ti_3CN, Fe_2C, Cr_2C, Ti_3N_2, Ti_2N, Zr_2C, and Zr_3C_2, are expected to have magnetic moments. However, after termination, every MXene and its functionalization group should be monitored individually. For example, Ti_3CNT_x and $Ti_4C_3T_x$ are nonmagnetic due to the functional groups, whereas Cr_2CT_x and Cr_2NT_x are made ferromagnetic by the OH and F groups at ambient temperature. Mn_2NT_x, independent of surface terminations, is ferromagnetic. The identified magnetic moments, on the other hand, are merely theoretical predictions that are to be validated experimentally. This is attributable to a shortage of surface chemistry control and restricted availability of MXene materials (especially pristine ones) [43, 44].

17.2.4 Dielectric Properties

MXenes contain interesting physiochemical features as well as good dielectric qualities. They are fillers that are highly conductive and are utilized with suitable flexible polymers to improve or adjust their dielectric properties while keeping their flexibility. Polymer composites based on flexible MXene, with a large dielectric constant (ε) and less dielectric loss ($\tan \delta$), have recently been developed in a number of ways. The prevalence of these properties is due to [1] MXenes' excellent metallic conductivity, which paves the way to enhanced interfacial polarisation due to the elevated electric conductivity imbalance between the MXenes and the polymer matrices, and [2] nanocapacitor-network development due to MXene's good dispersion in many polymer matrices [45].

17.3 Various MXene Hybrids For EMI-Hielding

17.3.1 Textile-based

With the rapid increase in EMI a genuine thought for human health and the dependency on precision equipment, wearable EMI shielding garments are extremely desirable. Due to the typically dissenting need for the thickness of the coating, controlling SE values and fabric flexibility continue to remain tricky. To address these issues numerous works have been carried out in this field by incorporating metal nanoparticles, carbonaceous materials, intrinsically conducting polymers, etc. The metal nanomaterials' high electric conductivity makes them desirable for EMI shielding fabrics; even so, their weight, expensiveness, and rigidity confine their applications. CNTs, on the other hand, have a high price as well as bad dispersion. At the same time, intrinsic conducting polymers' lower electrical conductivity limits their use. RGO has enormous potential due to its distinctive properties of tunable surface functional groups, high specific surface area, and strong interactions with substrate fibers. The remaining oxygen-functional groups of RGO, on the other hand, harmed the electrical performance. MXenes, on the other hand, have proven to be a better alternative in textile-based shielding materials owing to their remarkable dispersion, layered architecture, and significant electrical conductivity. Zheng et al. [46] designed a novel PEDOT/MXene coated cotton fabric via vapor phase polymerization and spray coating technique.

Cotton fabric benefits from the adsorption of functional materials, owing to its greater hydrophilicity compared to synthetic fabrics. The samples underwent different counts of spray coating up to five times (namely PMF-1, PMF-2, PMF-3, PMF-4, and PMF-5). The amount of PEDOT/MXene spray-coated is also reflected in the performance of the shielding fabrics over a frequency range of 8.2–12.4 GHz (X-band). A notable total shielding effectiveness value (SE_T) of 36.62 dB could be observed in PMF-5. Greater SE_T implies an elevated electrical conductivity and ultrahigh shielding effectiveness of MXene (nearly 92 dB). The PEDOT/MXene-covered fabric showed multiple applications, including strain sensing, joule heating, and electrochemical energy storage. The so-evolved PMF demonstrated excellent strain sensing and joule heating performance of 193.1 °C at a 12 V applied voltage and a large specific capacitance of 1000.2 mF/cm^2, with a current density of 0.5 mA/cm^2 [47].

Yin et al. [48] developed a wearable, light and durable fabric by a layer-by-layer (L-b-L) assembly technique. The strategy involved MXene sheets and polyaniline (PANI) alternately arranged over carbon fiber fabric via strong interfacial interactions. The so-fabricated PANI/MXene/CF fabric with 0.55 mm thickness (after five rounds of L-b-L assembly) flaunted high electrical conductivity (nearly 24.57 S/m) and a high shielding of about 26 dB across a frequency range of 8.2–12.4 GHz. Moreover, the fabric showed exceptional joule heating performances corresponding to the applied voltage. An applied voltage of 3 V generated a saturation temperature of around 66 °C. It was also found that PANI/MXene/CF fabric has the upper hand over similar EMI shielding fabrics in terms of flexibility, joule heating performance, and durability.

L-b-L assembly is a popular strategy in developing EMI shielding textiles. Yin and co-workers [49], in a different work, designed excellent wearable and multifunctional EMI shielding textiles by decorating a carbon fiber fabric (CFF) with a combination of 2D MXene-nanosheets and 1D PANI nanowires by the aforementioned technique. The L-b-L assembled sample subsequently underwent PDMS coating to attain substantial hydrophobicity and durability. On the other hand, charged PANI and MXene could form a robust bond, which, in due course, diminishes the peeling issue and thus aids in durability. An optimum result could be obtained for a sample that went through 50 cycles of L-b-L assembly (PDMF/PMCFF-50). The so-synthesized fabric revealed an outstanding effective shielding of nearly 35.3 dB in the X-band and high electrical conductivity of around

325 S/m at a thickness of 0.376 mm. The fabric befitted commercial purposes due to its notable air permeability of 346 mm/s. It also revealed a contact angle of 135.2°, indicating its hydrophobic properties. The PDMF/PMCFF-50 fabric also showed immense structural stability, incredible mechanical properties, and thermal management. A fine joule heating performance of 53° at 3 V applied voltage was also noted.

Cheng et al. [50] processed a flame-resistant, highly adaptable EMI shielding plating over a cotton fabric. The team developed the fabric using solution impregnation and dip coating techniques. The cotton fabric was coated with each layer of polyacetimidate (PEI), ammonium polyphosphate (APP), and MXene. The PEI/APP coating enhanced fire retardancy by boosting the production of phosphate and ammonia through the degradation of the PEI/APP coating, which eventually aided in the development of an efficient flame retardant. The improved flame retardancy was due to the increased limited oxygen index (LOI). The LOI value showed a tremendous increase of about 17.5% after the coating process. The increasing amount of MXene improved the EMI shielding effect; a significant level of electrical conductivity of 670.3 S/m could be observed when the fabric had a high content of MXene (5.2 mg/cm^2). The fabric also achieved a SE of 31.04 dB in the X-band, ranging from 8.2 to 12.4 GHz. An applied voltage of around 4 V resulted in the joule heating performance of around 64.3 °C. Altogether, their work brought up a practical idea for processing multifunctional EMI shielding fabrics.

A similar value of SE was obtained in a novel investigation by Zheng et al. [51] in which they fabricated a CNT/MXene coated fabric (CMF). The strategy involved roll-to-roll (R2R) L-b-L assembly and resulted in a bark-shaped CMF. The pristine fabric had a clean surface, and CNT-coated fabric with a relatively rough surface was named CMF-0.5. On rolling the CMF-0.5 fabric once with MXene, a CMF-1 sample could be obtained. Subsequently, after 9 R2R L-b-L cycles (CMF-9), the sample showed a bark-like appearance. CMF-9 displayed a shielding efficiency of 30 dB over the X-band region, adequate flexibility, reasonable electrical conductivity, sufficient air permeability, and ample joule heating of around 70.9 °C at 5 V. The bark-shaped CMF remarkably upgraded the reflectivity (above 90%) and SE. The fabric could be beneficial in monitoring human motions when arranged between a polydimethylsiloxane film and an interdigitated electrode and could be used in smart wearable electronics accordingly.

Spectacular values ranging from 57.8 dB to 90.2 dB in the X-band (across a range of 8.2–12.4 GHz) could be achieved for EMI SE in research by Jia et al. [52]. The group worked to develop a sustainable and multifunctional MXene coated wood-pulp fabric grid (FG) via vacuum filtration method, along with the hydrophobic methyl-trimethoxy-silane (MTMS) coating using sol–gel technique. The outstanding EMI SE values occurred due to the hydrophobic Si—O—Si bonds, along with the highly conductive MXene. An obtuse angle of 132°–138° substantiated its hydrophobicity. The fabric also endowed incredible mechanical stability, due to the strong fabric grid framework. The MTMS-M/FG (MXene-coated MTMS fabric grid) showed a fair joule heating performance, which is more suitable for low voltage-driven applications. Compared to other MXene-based fabrics, which showed an average joule heating performance of around 60 °C at 3 V, the MTMS-M/FG exhibited a reading around 75 °C at 3 V and even reached 95 °C at 4 V. Phonon excitations via scattering led to a rapid surge in joule heating. The fabric can turn into a fire retardant if incorporated with ammonium dihydrogen phosphate (ADP). The fabric is apt to be used even in extreme weather conditions including rain and snow.

17.3.2 Insulating Polymer-based

An effective EMI shielding is inevitable in most modern electronic devices to avert the radiated emissions, which would cause interference and affect the device performance. Moreover, EM waves have an adverse effect on human beings. It is not viable to put a stop to the EM radiations by limiting the usage of electronic equipment, and thereupon, we focus on the reduction of the penetration of EM waves released from electronic devices. To abate the penetration, a shield must be used. Thin sheets of elementary metals like copper, aluminum, silver, and gold have high electrical conductivity and are typically used to prevent EM pollution. However, their poor resistance to acids, alkalis, oxidation (corrosive nature), and high price pose a con and limit their application in EMI shielding [53].

Carbon-based materials like CNT and graphene have gained plenty of attraction as shielding materials owing to their low density, corrosion resistance, and high conductivity. However, a carbon filler-polymer matrix composite demands thickness in the millimeter range to attain adequate shielding efficiency. Hence, its applications cannot be fully beneficial in aerospace and microelectronic fields. So, to obtain a fair shielding efficiency at low density and thickness, MXenes have been utilized lately for their high specific capacitance, good aqueous dispersion, and excellent hydrophilicity. MXene, along with insulating polymers, was found to be a potential electromagnetic interference shielding material. For example, Shazad et al. [54] synthesized an MXene-sodium alginate film via vacuum-assisted filtration, which reported an excellent shielding of 57 dB. The composite had a height profile of 8 mm and 10 wt% of sodium alginate in it and exhibited low density, high conductivity, and good mechanical flexibility.

Since the shielding mechanism incorporated in metals and conductive polymers caused secondary pollution, and the introduction of carbon materials like CNT and graphene did not satisfy the norms of feasibility due to poor scalability and flexibility, MXenes have been an appropriate choice as a shielding material [55]. Since MXenes can easily integrate with any material, they, along with insulating polymers, can be excellent shielding materials. There is a good balance between the highly conductive MXene and insulating polymers. Insulating polymer-based EMI shielding materials can be in several forms, some of them being aerogels, hydrogels, thin films, electrospun sheets, MXene-based papers, and multilayered composites.

17.3.3 Aerogels, Hydrogels, and Foams

Compressible and weightless EMI shielding nanocomposites are vital to resolving the hiking EM pollution. MXenes, along with nonconducting polymers like polyvinyl alcohol, polypropylene, polylactic alcohol, and polystyrene, are ideal for fabricating efficient composites for outstanding EMI shielding effects. A pristine MXene sheet usually exhibits the SE around 92 dB. MXene polymer composites, however, do not exhibit such a superior value as SE, but their mechanical properties can be enhanced, and hence it is an appropriate strategy to integrate MXene with polymers to attain long-lasting shielding materials.

Guo et al. [56] utilized a PVA-MXene composite processed via directional freezing and freeze-drying to fabricate an aerogel using adequate electromagnetic shielding and mechanical properties. PVA is water-soluble, and the hydroxyls in PVA can interact with MXene to form strong bonds. The composite had a depth of 3 mm and revealed distinct results with varying compositions. As the MXene content was 70 wt%, the SE was discovered to be 40.6 dB, and a 90 wt% composite possessed an extremely high SE value of around 53.3 dB. The presence of PVA in the composite led to the polarization and interfacial reflection between the components, which benefitted the shielding ability.

Wang et al. [57] investigated an economical strategy to develop self-interlocked networks of MXene and PVA and subsequently joined those networks to make a composite membrane that is highly engineered. This prototype possessed extraordinary mechanical strength, nearly thrice better than the control PVA. Moreover, it exhibited excellent fire-retardant properties with a nearly 80% cutback in the peak heat liberation rate compared to pristine PVA. Its ability to shield the EM radiations was found to be around 41 dB. Intending to obtain an optimum result, PVA/MXene membranes with sandwich structures were used, and they were named "random PVA-MXene structure" (RPMS) and "oriented PVA-MXene structure" (OPMS). OPMS and RPMS could form 3D conductive structures owing to the presence of MXene nanolayers. However, their surface resistances were found to be low compared to the PVA-MXene blend composites (with 35 wt% of MXene). RPMS membranes had lower heat release rate (HRR) and total heat release (THR) values, indicating that they are a superior flame retardant than OPMS or PVA-MXene 35 membranes. RPMS also showed a larger absorption shielding mechanism (SE_A) than OPMS and PVA-MXene 35 blended composites.

Han and coworkers [58] reported a 3D porous aerogel made of MXene fabricated via bidirectional freezing of colloidal solutions of MXenes. The resultant products happened to be lightweight, compressible, and robust aerogels that excelled in EM wave shielding. Three different samples ($Ti_3C_2T_x$, Ti_2CT_x, and Ti_3CNT_x) were prepared with 1 mm thickness and a density equivalent to 11 mg/cm^3. SE of these samples reached 70.5 dB, 69.2 dB, and 54.1 dB. Since the morphology involved an aligned, lamellar architecture, the samples were found to be extremely effective in controlling the EM wave reflection and absorption. These aerogels sustained the shielding efficiency over the X-band. They are suitable to be used in flexible and portable devices for shielding purposes.

Lately, Zhou et al. [59] developed stretchable hydrogels that possessed remarkable mechanical strength and modulus, ultra-high electrical conductivity, and astounding shielding efficiency. They used a novel approach to synthesize aramid nanofiber (ANF) reinforced PVA hybrid hydrogels. The silver nanowire-PVA layer was electrospuned over ANF/PVA fiber and then filtered via vacuum assistance. The so-synthesized hybrid hydrogels also showed an improved crystal structure, strong hydrogen bonds, and an in-plane arrangement of the nanofiber structure. The SE was found to be 52 dB in the X-band range for the little content of silver nanowires. Apparently, the electrical conductivity was found to be very high, around 1.66×10^4 S/m, and the tensile strength was reported to be in the range of 3.3–5.5 MPa. High conductivity is an inference toward excellent joule heating performance. The composite showed an average equilibrium temperature from 40 °C to

55 °C when the applied voltage varied from 1 V to 1.5 V. Thus, the prototype is a potential material that can also be used in soft robots, wearable electronics, artificial skin, etc., besides EMI shielding.

Kim et al. developed an MXene/auxetic polyurethane (MX/APU) composite foam [60], which was beneficial for impact attenuation and multifunctional EMI shielding. The material exhibited an exceptional ability to shield the EM waves, around 76.2 dB. It also maintained an extremely low density, around 0.063 g/cm^3. MXene/APU composite also unveiled a negative Poisson's ratio, which can be utilized to safeguard electronic devices from EMI and impact attenuation, and applied in soft robotics and aerospace fields as well.

17.3.4 Polymer Thin Films

MXenes have been the focus of attention ever since their initial development. Owing to its amazing electrochemical properties, MXene composites have been largely used in EMI shielding, energy storage, field-effect transistors, etc. [47]. The cons of MXene include low flexibility, poor stability, and undemanding restacking, due to which pristine MXene is not suitable for many applications. Polymers are being widely utilized to fabricate MXene-based compounds due to their inexpensiveness, facile synthesis, and tunable features. Synthesized MXene/polymer composites usually exhibit outstanding conductivity, stability, and mechanical properties.

Lipton et al. [61] fabricated a freestanding and high-strength polymer nanocomposite, via a dip layer-by-layer assembly strategy, using MXene nanosheets, PVA, and clay nanoplatelets. The sample showed excellent tensile strength, around 200–250 MPa, with an average shielding capacity of 27 dB. SE was measured in the X-band, and the nanocomposite comprised montmorillonite (which is a clay nanoplatelet), MXene, and polyvinyl alcohol (PVA). Besides very high tensile strength, it exhibited a relatively lower electrical conductivity, ranging between 53 S/m and 125 S/m. The slightly insulating property was mainly because of the intrinsic resistance between the junctions of MXene and the fillers. The so-fabricated polymer nanocomposite is a potential shielding material that can be utilized in several applications. However, the effect of shielding in the composite is contributed by MXene alone, with other components being insulating.

Iqbal et al. [62] demonstrated that an MXene-polystyrene (PS) composite film enhanced EM wave absorption. The fabrication process included combining conductive, flakey MXene with insulating PS beads in an aqueous phase, subsequent vacuum filtration, and hot pressing. The shielding capability increased as the surface area of the multiple interfaces grew. The experiment took place in the X-band frequency range, and the sample was 40 micrometers thick on average. The arrangement of MXene sheets relied on the size of the PS beads; as the size of PS beads changed from 0.15 to 2 micrometers, the MXene sheet arrangement reduced slightly. The components were mixed in a 9:1 ratio by weight, as well as a 7:3 ratio. The electrical conductivities were found to be 2730 S/cm, 2840 S/cm, 2995 S/cm and 3263 S/cm for the former ratio for beads with 0.15, 0.5, 1, and 2 micrometers, respectively. The composite with the latter ratio displayed electrical conductivities of 671 S/cm, 786 S/cm, 842 S/cm, and 914 S/cm. In the 9:1 composites, an EM SE_T value of 74 dB was observed for the 0.15-micrometer bead containing the sample, and it was reduced with the increasing bead size; a 2-micrometer bead-composite possessed a SE_T value of 48 dB. The 7:3 composites exhibited smaller SE_T values relative to the 9:1 nanocomposites. The synthesized nanocomposites are appropriate for applications that demand lightweight, multifunctional shielding materials.

Wang et al. [63] prepared a layered double-percolated nanocomposite for EMI shielding. It came out to be one of the most efficient materials for shielding, owing to its memeager-iller contents. The composite contained polyvinylidene fluoride (PVDF), PS, and MXene in a co-continuous manner. An optimum result was attained when the MXene content was 12 wt%, and the thin film was 1.35 mm thick. A layered double-percolated sample showed an SE of 55 dB and high AC and DC conductivity. The frequency range used was the X-band. The composite contained numerous polar bonds that aided in the promotion of EM wave absorption via dipolar polarization.

Ma et al. [64] researched the absorption of the EM wave that caused the microwave heating nature in MXene. MXene nanosheets, when flocculation assembled onto polypropylene (PP) surface, induced selective heating that caused a compact and continuous network of segregated MXene/PP nanocomposite. The composite was utilized as an interfacial solder and possessed incredible EMI shielding effectiveness in the range of 75.12 dB to 78.85 dB over the X-band. An immensely less MXene quantity of nearly 1.138 vol% was used in the composite, and the synthesized sheet ascribed exceptional anti-dripping features. Thus, the synthesized composite is a potent compound for ultra-efficient shielding, for the properties possessed even at extremely low MXene content.

Recently, Wan and group [65] fabricated a high-performance shielding composite made of MXene, PEDOT (poly(3,4-ethylenedioxythiophene)), and PSS (polystyrene sulfonate). A total of 70 wt% of MXene in the composite with a thickness of around 6.6 micrometers was attributed to a shielding efficiency of 40.5 dB. The shielding capacity was measured in the frequency ranges of 8.2–12.4 GHz and 11.5—18 GHz. Apart from adequate shielding efficiency, the fabricated sample displayed excellent tensile strength ranging from 35.6 to 41.4 MPa. Their study paved a new strategy to process flexible and lightweight shielding materials. The increased tensile strength was due to the acid wash treatment (concentrated sulfuric acid) over the fabricated composite. However, the acid treatment had negligible effects on the SE of the sample. The obtained data imply that the shielding and mechanical properties are well balanced, and the prototype can become a promising material that can be utilized in electronic devices.

17.3.5 Electrospun Mats

The rising demand in the field of portable and wearable electronic equipment paves the way for multifunctional materials that are able to store energy efficiently and possess an enhanced electrical conductivity. MXene is one such material that can integrate easily with fibers and textiles. An electrospun nanofiber membrane behaves as a shielding material, due to the huge amount of reflecting surface between the MXene integrated fibers [66]. Therefore, MXene/nanofiber-based electrospun composites are potential shielding materials with high flexibility and performance.

Du et al. [67] designed a multiple-layered MXene/PLA compound membrane for shielding. The membrane was synthesized by assembling MXene over an electrospun PLA nanofiber material. They observed an efficient shielding larger than 55 dB in the X-band, as well as an excellent joule heating performance by the composite, with a heating response of 31 °C to 52.8 °C in a flash at an applied voltage of 1.5 V. When used in a triboelectric nanogenerator (TENG), the developed membrane demonstrated an impressive peak-power density of 457.1 mW/m^2. As a result, their work is a clever approach that can be used in wearable and portable devices, multifunctional machinery, and energy harvesters.

Wang et al. worked on wood-based materials, precisely MXene-cellulose composites, with the goal of developing shielding materials that are lighter in weight, sustainable, and have high mechanical properties [68]. Their strategy involved a scaffold assembly of MXene and cellulose, where the composites were 2 mm thick, displayed good mechanical strength of around 288 MPa, and a fair shielding capability of around 39 dB in the X-band. They thrived in fabricating a feasible and scalable green composite with a flexural modulus of 8.6 GPa and high mechanical robustness. The electrical conductivity was found to be 6333 S/m. The proposed strategy is a practical way to synthesize multifunctional composite materials, which can be used in electronic devices and structural materials.

Cui et al. [69] developed a hydrophilic, 2D, and ultrathin MXene nanosheet that exhibited high electrical conductivity. The 2D MXene sheets, along with electrospun cellulose nanofibers, showed excellent shielding features and were synthesized by a vacuum-assisted filtration process. The SE was observed to be around 42 dB for a 15-micrometer-thick composite at frequencies ranging from 2 to 18 GHz. They showed adequate mechanical properties and excellent conductivity of 46.3 S/cm. Hence, the synthesized material is a successful alternative for ultrathin, flexible EMI shielding materials.

Yang et al. [70] fabricated a sandwich-structured PVDF layer and a filler layer comprised of MXene and silver nanowires via electrospinning technique, simple vacuum-assisted filtration, and hot compression method. Its ability to shield was found to be at 47.8 dB in the X-band frequency range. PVDF/AgNW/MXene composite also showed an excellent joule heating effect; the equilibrium temperature was around 77 °C at an applied voltage of 2.5 V. This facile strategy is aimed at paving the way for more investigations on flexible multilayer EM shielding materials.

17.3.6 Paper-Based Composites

Polymers, when incorporated into MXene, can upgrade their properties. MXene/polymer composite papers have been one of the strategies to meet the challenges faced during EMI [71]. The studies on MXene/polymer-based papers have made enormous strides.

Hu et al. [72] researched a highly conductive MXene/cellulose nanocomposite paper through the facile dip-coating technique. The resultant material displayed an exceptional electrical conductivity of nearly 2750 S/m when the nanosheet loading was 1.89 vol%. Following that, coating the paper with a PDMS (polydimethylsiloxane) layer made 0.2 mm thick, and the SE was measured to be over 43 dB in the X-band and Ku-band frequencies. MXene loading was observed to be 1.07 vol%. The paper also exhibited an in-plane thermal conductivity of 3.89 W/mK. Hence, the fabricated paper has numerous applications in next-generation electronic devices, aerospace, flexible devices, and stealth weapon equipment.

Zhan et al. [73] studied the influence of 2,2,6,6-tetramethyl-1-piperidinyloxy (TEMPO) oxidized cellulose nanofibers (TOCNF) over an MXene film. A flexible MXene/TOCNF composite paper was fabricated via vacuum-assisted assembly. The synthesized paper displayed an exceptional Young's modulus (nearly 7 GPa) and mechanical strength (around 212 MPa). This was due to the strong interaction between the molecules and the aligned structure of the components. The processed MXene/TOCNF paper showed a large value for electrical conductivity, around 2830 S/m, and the efficiency to shield was found to be 39.6 dB for a 38-micrometer thick composite over the X-band frequency range. The MXene/TOCNF composite paper paves a new way for the progress of electronic equipment based on MXene.

Zhan et al. [74] developed highly efficient MXene/cellulose (MC) based shielding papers. MXene sheets were oriented on a single side of cellulose filter paper, which built up a continual layer that is conductive. The MC paper, which was 0.192 mm thick, ascribed high electrical conductivity of 240 S/m and a decent shielding effectiveness of 29.3 dB at the X band frequency range. Four such stacked sheets could even reach 40.5 dB. All the aforementioned properties could be achieved when the MXene content was merely 0.72 vol%; therefore, the fabricated MC paper can be advantageous in many electronic devices.

Xie and coworkers [73] successfully synthesized a mechanically strong, ultrathin MXene/aramid nanofiber paper (ANF) with superior shielding effects. The interfaces of MXene and ANF had good adhesion, owing to the hydrogen bonds between them. The MXene/ANF paper was processed via vacuum-assisted filtration. The conductivity of the composite paper was around 170 S/cm, and it showed an adequate shielding performance of around 28 dB when the thickness was around 17 micrometers. The composite paper had an extremely low density, around 1.25 g/cm^3. Apart from low density, thickness, and a reasonable SE, the sample also displayed excellent mechanical properties. It has potential applications as shielding materials in electronic devices.

Cao et al. [74] worked on ultrathin and flexible composite paper made of MXene/ CNF. The developed sample had a lamellar structure, similar to nacre, and was fabricated by a self-assembly process persuaded by vacuum filtration. The strong linkage between 2D MXene and 1D CNF caused the nacre-like composite paper to toughen and strengthen, which subsequently improved the tensile strength to around 135.4 MPa. The sample exhibited excellent folding endurance (around 14,300 times), large electrical conductivity (nearly 740 S/m), and an exceptional specific SE equivalent to 2647 dBcm^2/g at a thickness of 47 micrometers. The fabricated sample has applications in different fields, like robot joints, weapon equipment, and flexible wearable devices.

17.3.7 Laminates

Despite it being increasingly easier for us to make use of EM waves as a means to convert signals, energy, and information, EMI pollution has adverse effects on the working of electronic equipments as well as our well-being. As a coping material, we make use of composites with the ability to shield. EM shielding materials can profitably shield the signals by absorbing and reflecting them. MXenes, along with insulating polymers in the form of laminated structures, can bring about incredible efficiency in shielding and can have several applications, including medical, aerospace, military, and many more fields.

Lately, Zhang et al. [75] have processed MXene-based, absorption-dominant EMI shielding fabrics that are lightweight and nonwoven. The material displayed hydrophilicity as well as high conductivity. MXene underwent coating with distinct non-woven laminated fabrics (cotton, calcium alginate and polyester). MXene and the coated fibers formed conductive layers over the sample, which eventually led to heterogeneous interfaces. This, in turn, boosted the reflections of EM waves. MXene/calcium alginate composite with a thickness of 3.17 mm exhibited a reasonable SE of 25.26 dB at the X-band frequency range (precisely at 12.4 GHz). MXene/cotton composite, which was 1.36 mm thick, attained a fair specific shielding effectiveness of 2301.95 dB cm^2/g. Thickening the sample with conductive interfaces can enhance SE. However, it is not the desired strategy since it increases the weight of the sample, and thereupon, it is preferred to increase the laminated nonwoven fabric thickness. Based on the number of layers, the samples were classified into C/M-2, C/M-3, C/M-4, CA/M-2, CA/M-3, CA/M-4 and P/M-2, P/M-3, P/M-4 where C/M represents cotton/MXene composites, CA/M corresponds to calcium alginate/MXene composites and P/M denotes polyester /MXene composites. Out of the synthesized samples, CA/M-4 showed an allowable shielding efficiency of around 25.26 dB. The maximum value for specific shielding effectiveness was obtained from the C/M-2 sample. Their work, being highly innovative, facilitates more research to develop materials with shielding properties.

Huang et al. [76] explored a strategy to fabricate a robust, high-performance MXene composite for EMI shielding. The synthesized sample was an MXene/hydroxyethyl cellulose (HEC)/silicon resin composite, which exhibited high stability, mechanical properties and hydrophobicity. The silicon resin was responsible for the hydrophobic nature of the material, which could be validated by the obtained contact angle range of 151.5°–155°. In fact, such a high contact angle implied that the developed material was superhydrophobic. The sample had a multilayer, stacked structure with MXene, HEC and silicon resin, forming an aerogel out of the components. MXene/HEC mixed aerogels (MHAs) showed exceptional values for electrical conductivity around 3166.4 S/m and SE of 74.5 dB in the X-band. Depending on the stacking layers of MXene, the samples were named MHA-1, MHA-2, MHA-3, MHA-4, and MHA-5, out of which the optimum result was obtained from the MHA-5 sample. A practical way to fabricate high-performance shielding materials was introduced in their work.

In one of the most recent investigations, Gao et al. [77] synthesized flexible, multilayered MXene/polyurethane composites through a L-b-L spraying technique. The multiple-layered composites had 28.6 wt% MXene content in them, were 52 micrometers thick, and possessed a high electrical conductivity of 160 S/m and a notable SE of 50.7 dB in the X-band. Additionally, the composite had shown an excellent joule heating performance at 113 °C when the applied voltage was 5 V. The best result was obtained when the composite had six layers of thermoplastic polyurethane, five layers of MXene and 28.6 wt%. MXene in it, denoted as M5T6-28.6. Multilayers stacked together display incredible thermal management performance, due to their outstanding thermal and electrical conductivity.

Xu et al. [78] demonstrated an effective strategy to fabricate an MXene/PP nanocomposite with a meagre percolation threshold, aiming at efficient shielding of EM waves. The nanocomposite was synthesized by dip-coating MXene sheets over a polypropylene fabric, which was previously grafted with polyethyleneimine and subsequent vacuum-assisted compression molding. The synthesized multilayered nanocomposite displayed a high electrical conductivity of 437.5 S/m at a minimal MXene loading (2.12 vol%). A surprising shielding output of more than 60 dB over the X-band region could be observed. Thus, the synthesized multilayered nanocomposite is one step ahead of most of the similar polymer nanocomposites.

Zhou and group [79] fabricated a multilayer CNF and MXene using an easy and efficient approach of alternating vacuum filtration. The mechanical frame effect by the CNF layers typically restricts the propagation of cracks in MXene layers, and the alternate layered structure enhanced the mechanical strength (around 112.5 MPa) compared to pristine MXene or a CNF/MXene blend film. The synthesized composite demonstrated a decent SE of 40 dB in the X-band and K-band, and the SSE was found to be 7029 $dBcm^2/g$ for a 0.035 mm sample. It also exhibited excellent electrical conductivity and, therefore, an outstanding joule heating performance, which had the scope of reaching an equilibrium temperature of around 100 °C at an applied voltage of 6 V.

17.4 Intrinsically Conducting Polymer-based

Since its discovery in 2011, MXene has been used for EMI shielding applications. $Ti_3C_2T_x$ film, 45 μm thick, exhibits a shielding effectiveness of 92 dB. But, in comparison with other materials with similar thickness, $Ti_3C_2T_x$ has a high dielectric constant, which in turn leads to impedance mismatch. This makes further improvement of EMI shielding difficult. The hybrids of MXene with intrinsically conducting polymers reduce this impedance mismatch and improve the mechanical performance of the material. Some of the works, including intrinsically conducting polymers and MXene hybrids, are discussed.

17.4.1 Aerogels, Hydrogels, and Foams

PANI is a conducting polymer that is also well-established as an EMI shielding material. The synergetic effects of PANI/MXene have been studied by many researchers. Several researchers have created three-composite or four-composite MXene/PANI-based composites to improve shielding performance. Jia et al. [80], for example, developed compressible MXene@PANI/mPP composite foams. After encapsulating with PDMS, the foams demonstrated an absorption-dominant shielding performance with an effective EMI SE of around 23.5–39.8 dB and a low R coefficient of 0.20–0.31.

17.4.2 Polymer Thin Films

Zhang et al. [46] fabricated flexible and ultra-thin $Ti_3C_2T_x$/c-PANI composite films with an EMI SE of 36 dB at a thickness of 40 m and discovered that the shielding mechanism shifts from equal shielding to reflection dominant shielding as $Ti_3C_2T_x$ concentration increases in $Ti_3C_2T_x$/c-PANI composites. Kumar et al. [81] found a lightweight Ti_3C_2/PANI composite with an EMI SE shielding performance of 23 dB and a sample thickness of 1.5 mm. His study elaborated on the role of specific surface area in the shielding performance of the composite prepared. Laminated MXene with a superior specific surface area was found to give the best shielding performance.

Yin et al. [82] developed a light, flexible and multifunctional PANI/MXene-based composite textile with an incredible electrical conductivity (325 S/m) and a remarkable shielding performance (SE of ≈35.3 dB) with a thickness of 0.376 mm. They used optimized organic/inorganic (1D/2D) continuous conductive networks to achieve this excellent shielding performance. Raagulan et al. [83] reported an MXene-PAT-PANI-PpAP composite, which showed a shielding performance of SE of 45.18 dB at 8.2 GHz. They stated that the deducted form of MXene tends to increase total shielding performance, and the increased absorption was attributed to the ant-farm-like structures. Wang et al. [45] $Ti_3C_2T_x/Fe_3O_4$@PANI composite film by vacuum filtration, which was flexible and lightweight. They were able to achieve a shielding value of 62 dB with a film thickness of 16.7 μm. This multifunctional film shows potential for applications in wearable sensors, packaging and in the military.

Making use of a vacuum-assisted filtration process, Liu et al. [84] developed an ultrathin, freestanding and flexible $Ti_3C_2T_x$/PEDOT: PSS composite film with a "brick-and-mortar" structure. The ultrathin polymeric composite film has a high EMI shielding effectiveness (SE) value of 42.10 dB despite its thickness of 11.1 m. The tensile strength was observed to expand with the addition of PEDOT: PSS from 5.62 to 13.71 MPa and the. The film also showed an exceptional conductivity of 340.5 S/cm.

17.4.3 Paper

Yin et al. [85] prepared a multilayered aligned PANI/MXene/CF paper. They used an L-b-L assembly method including 2D transition metal nitrides/carbides (MXenes) sheets and PANI polymer alternately arranged on carbon fiber (CF) fabric via strong interfacial linkage. The fabricated paper exhibited a shielding effectiveness of 26.0 dB. This fabric also retained its flexibility, air permeability, and washability while providing good absorption-dominated shielding performance and reducing secondary pollution caused by wave reflection.

17.5 Graphene-based

Graphene and MXene are two 2D nanomaterials which have been a center of attraction for researchers around the world working on EMI shielding. Several researchers have studied the synergetic effects of these materials, and they tend to form 3D cross-linked graphene/MXene-based materials, where the 2D layers are interconnected with a bulk material [86]. This leads to the enhancement of electromagnetic shielding as well as mechanical properties. Here, we are discussing some relevant works done in graphene/MXene hybrid materials for EMI shielding applications.

17.5.1 Foam/Aerogels

Fan et al. [87] used freeze-drying and reduction heat treatment to introduce $Ti_3C_2T_x$ MXene nanosheets into graphene oxide and fabricate highly conductive lightweight MXene/graphene hybrid foam. The interconnected GO and MXene sheets with the high porosity of the aerogel boost its shielding performance up to 50.7 dB. The material also exhibited the highest ever reported specific EMI shielding value of 6217 $cm^3 g^{-1}$ at very low thickness. This material has the potential to be used in next-generation smart devices.

Zhao et al. [88] constructed 3D $Ti_3C_2T_x$ porous architectures that are largely conductive by graphene oxide-assisted hydrothermal assembly followed by directional freezing and freeze-drying. Graphene oxide was used to elevate the gelation probability of MXene sheets. The prepared hybrid aerogels showed an aligned microstructure in which graphene sheets carried out the role of the inner skeleton. The porous material was used as a reinforcement in an epoxy matrix, which produced an astounding shielding efficiency of 50 dB in the X-Band region. The successful methodology for the alignment of GO/MXnene sheets in a 3D porous architecture is a potential area for further research.

17.5.2 Films

Nguyen et al. [89] developed a diverse EMI shielding skin that was fabricated from graphene foam (GF) decorated by Fe_3O_4 nanoparticles (NPs)-intercalated $Ti_3C_2T_x$ nanosheets. The prepared material has excellent shielding properties up to 80 dB in the X-band region and 77 dB in the Ka-band region with an absorption-dominating shielding mechanism. The prepared film can also be used as a pressure sensor, which widens the application. The flexible, lightweight and highly conductive GF-based composites hold a favorable performance that clears the way for the possibility to utilize them as a multifunctional EMI shielding for robotics and wearable electronics.

Ragulaan et al. [82] generated a cost-effective shielding material using facile techniques such as solvent casting and spray coating. The prepared films were thin, flexible and of low density. The hydrophobic nature of the prepared composites was demonstrated by a higher contact angle of 126°, 43 $mN\,m^{-1}$ wetting energy, 116 $mN\,m^{-1}$ spreading Coefficient, and 30 $mN\,m^{-1}$ last work of adhesion. The material exhibited a shielding capacity of 53.8 dB (99.999%) with a reflection of 13.10 dB and absorption of 43.38 dB in 8–12.4 GHz. A one-time coated carbon fabric showed a SE of 35,369.82 dB $cm^2 g^{-1}$. This low-cost and highly efficient shielding material can find application in the fields of electronics and the military.

17.5.3 Laminates

Using a scalable fabrication method, Minh CanhVu et al. [90] created multifunctional nanocomposites with tunable thermal conductivity and exceptional EMI shielding performance. A hybrid shell of MXene (MX) and reduced graphene oxide (rGO) was arranged on a PMMA bead surface to prepare a 3D structure of the MXene/rGO in PMMA nanocomposite. The hot compression technique was used for the fabrication of this material. The fabricated composites exhibited a tunable thermal conductivity. The EMI shielding performance of the material was between 28 and 61 dB. The shielding effectiveness was found to be elevated with the increase of MXene/rGO. The prepared material can act as a multifunctional material with good EMI shielding properties along with heat-dissipating properties.

17.6 Conclusion

In summary, the current review analysis revealed that research on microwave and EMI shielding materials has brought a great breakthrough in the outline and development of MXenes and their hybrids. MXene has also demonstrated a variety of applications in a multitude of areas, ranging from microwave absorption to sensor-based applications. Exceptional conductivity and 2D structure facilitated MXenes and their hybrids in microwave and EMI shielding applications. Numerous approaches for enhancing their electromagnetic wave attenuation by combining organic/inorganic particles with polymers have also been discussed widely. However, gaining the global utility of MXene materials in the microwave and EMI shielding application still stays a protracted and upsloping journey.

References

1 Ezika, A.C., Sadiku, E.R., Idumah, C.I. et al. (2022). On energy storage capacity of conductive MXene hybrid nanoarchitectures. *J. Energy Storage* 45: 103686.

2 Niu, R., Han, R., Huang, Y. et al. (2022). Hydrothermal ion exchange synthesis of CoM (M= Fe or Mn)/MXene 2D/2D hierarchal architectures for enhanced energy storage. *J. Alloys Compd.* 894: 162385.

3 Baig, M.M., Gul, I.H., Baig, S.M., and Shahzad, F. (2022). 2D MXenes: synthesis, properties, and electrochemical energy storage for supercapacitors—a review. *J. Electroanal. Chem.* 904: 115920.

4 Zhu, Z., Wang, Z., Ba, Z. et al. (2022). 3D MXene-holey graphene hydrogel for supercapacitor with superior energy storage. *J. Energy Storage* 47: 103911.

5 Zhang, B., Wong, P.W., and An, A.K. (2022). Photothermally enabled MXene hydrogel membrane with integrated solar-driven evaporation and photodegradation for efficient water purification. *Chem. Eng. J.* 430: 133054.

6 Sun, Y., Yi, F., Li, R.H. et al. (2022). Inorganic–organic hybrid membrane based on pillararene-intercalated MXene nanosheets for efficient water purification. *Angew. Chem. Int. Ed.* 61 (14): e202200482.

7 Wang, Y., Nie, J., He, Z. et al. (2022). $Ti_3C_2T_x$ MXene nanoflakes embedded with copper indium selenide nanoparticles for desalination and water purification through high-efficiency solar-driven membrane evaporation. *ACS Appl. Mater. Interfaces* 14 (4): 5876–5886.

8 Ranjith, K.S., Vilian, A.E., Ghoreishian, S.M. et al. (2022). Hybridized 1D–2D $MnMoO_4$–MXene nanocomposites as high-performing electrochemical sensing platform for the sensitive detection of dihydroxybenzene isomers in wastewater samples. *J. Hazard. Mater.* 421: 126775.

9 Khazaei, M., Arai, M., Sasaki, T. et al. (2013). Novel electronic and magnetic properties of two-dimensional transition metal carbides and nitrides. *Adv. Funct. Mater.* 23 (17): 2185–2192.

10 Khazaei, M., Arai, M., Sasaki, T. et al. (2014). Two-dimensional molybdenum carbides: potential thermoelectric materials of the MXene family. *Phys. Chem. Chem. Phys.* 16 (17): 7841–7849.

11 Tang, Q., Zhou, Z., and Shen, P. (2012). Are MXenes promising anode materials for Li ion batteries? Computational studies on electronic properties and Li storage capability of Ti_3C_2 and $Ti_3C_2X_2$ (X= F, OH) monolayer. *JACS* 134 (40): 16909–16916.

12 Enyashin, A.N. and Ivanovskii, A.L. (2012). Atomic structure, comparative stability and electronic properties of hydroxylated Ti_2C and Ti_3C_2 nanotubes. *Comput. Theor. Chem.* 989: 27–32.

13 Guo, J., Sun, Y., Liu, B. et al. (2017). Two-dimensional scandium-based carbides (MXene): band gap modulation and optical properties. *J. Alloys Compd.* 712: 752–759.

14 Pang, J., Mendes, R.G., Bachmatiuk, A. et al. (2019). Applications of 2D MXenes in energy conversion and storage systems. *Chem. Soc. Rev.* 48 (1): 72–133.

15 Li, L. (2016). Lattice dynamics and electronic structures of $Ti_3C_2O_2$ and $Mo_2TiC_2O_2$ (MXenes): the effect of Mo substitution. *Comput. Mater. Sci* 124: 8–14.

16 Khazaei, M., Ranjbar, A., Arai, M., and Yunoki, S. (2016). Topological insulators in ordered double transition metals M′2M″C2 (M′=Mo, W; M″=Ti, Zr, Hf) MXenes. *Phys. Rev. B* 94.

17 Hope, M.A., Forse, A.C., Griffith, K.J. et al. (2016). NMR reveals the surface functionalisation of Ti_3C_2 MXene. *Phys. Chem. Chem. Phys.* 18: 5099–5102.

18 Persson, I., Näslund, L.-Å., Halim, J. et al. (2018). On the organization and thermal behavior of functional groups on Ti_3C_2 MXene surfaces in vacuum. *2D Mater.* 5.

19 Wang, X., Shen, X., Gao, Y. et al. (2015). Atomic-scale recognition of surface structure and intercalation mechanism of Ti_3C_2X. *JACS* 137: 2715–2721.

20 Wang, H.W., Naguib, M., Page, K. et al. (2016). Resolving the structure of $Ti_3C_2T_x$ MXenes through multilevel structural modeling of the atomic pair distribution function. *Chem. Mater.* 28: 349–359.

21 Harris, K.J., Bugnet, M., Naguib, M. et al. (2015). Direct measurement of surface termination groups and their connectivity in the 2D MXene V_2CT_x using NMR spectroscopy. *J. Phys. Chem. C* 119: 13713–13720.

22 Srivastava, P., Mishra, A., Mizuseki, H. et al. (2016). Mechanistic insight into the chemical exfoliation and functionalization of Ti_3C_2 MXene. *ACS Appl. Mater. Interfaces* 8: 24256–24264.

23 Sharma, G., Naguib, M., Feng, D. et al. (2016). Calorimetric determination of thermodynamic stability of MAX and MXene phases. *J. Phys. Chem. C* 120 (49): 28131–28137.

24 Magnuson, M., Halim, J., and Näslund, L.Å. (2018). Chemical bonding in carbide MXene nanosheets. *J. Electron. Spectrosc. Relat. Phenom.* 224: 27–32.

25 Sang, X., Xie, Y., Lin, M.W. et al. (2016). Atomic defects in monolayer titanium carbide ($Ti_3C_2T_x$) MXene. *ACS Nano* 10 (10): 9193–9200.

26 Wang, X., Garnero, C., Rochard, G. et al. (2017). A new etching environment (FeF_3/HCl) for the synthesis of two-dimensional titanium carbide MXenes: a route towards selective reactivity vs water. *J. Mater. Chem. A* 5 (41): 22012–22023.

27 Palisaitis, J., Persson, I., Halim, J. et al. (2018). On the structural stability of MXene and the role of transition metal adatoms. *Nanoscale* 10 (23): 10850–10855.

28 Anjum, D.H., Ahmed, B., and Alshaeef, H.N. (2018). Transmission electron microscopy of two-dimensional MXene sheets. *Microsc. Microanal.* 24 (S1): 1608–1609.

29 Bakardjieva, S. and Klie, R. (2018). The morphology of TiiAlC (M2AX) and TiiC (MXene) sheets revealed by HAADF STEM analysis. *Microsc. Microanal.* 24 (S1): 156–157.

30 Hart, J.L., Hantanasirisakul, K., Lang, A.C. et al. (2018). Direct correlation of mxene surface chemistry and electronic properties. *Microsc. Microanal.* 24 (S1): 1606–1607.

31 Persson, I., Halim, J., Lind, H. et al. (2019). 2D transition metal carbides (MXenes) for carbon capture. *Adv. Mater.* 31 (2): 1805472.

32 Hantanasirisakul, K. and Gogotsi, Y. (2018). Electronic and optical properties of 2D transition metal carbides and nitrides (MXenes). *Adv. Mater.* 30 (52): 1804779.

33 Lipatov, A., Alhabeb, M., Lukatskaya, M.R. et al. (2016). Effect of synthesis on quality, electronic properties and environmental stability of individual monolayer Ti_3C_2 MXene flakes. *Adv. Electron. Mater.* 2 (12): 1600255.

34 Zhang, C.J., Pinilla, S., McEvoy, N. et al. (2017). Oxidation stability of colloidal two-dimensional titanium carbides (MXenes). *Chem. Mater.* 29 (11): 4848–4856.

35 Li, R., Zhang, L., Shi, L., and Wang, P. (2017). MXene Ti_3C_2: an effective 2D light-to-heat conversion material. *ACS Nano* 11 (4): 3752–3759.

36 Mashtalir, O., Lukatskaya, M.R., Zhao, M.Q. et al. (2015). Amine-assisted delamination of Nb2C MXene for Li-ion energy storage devices. *Adv. Mater.* 27 (23): 3501–3506.

37 Wang, K., Zhou, Y., Xu, W. et al. (2016). Fabrication and thermal stability of two-dimensional carbide Ti_3C_2 nanosheets. *Ceram. Int.* 42 (7): 8419–8424.

38 Zha, X.H., Yin, J., Zhou, Y. et al. (2016). Intrinsic structural, electrical, thermal, and mechanical properties of the promising conductor Mo_2C MXene. *J. Phys. Chem. C* 120 (28): 15082–15088.

39 Muckley, E.S., Naguib, M., and Ivanov, I.N. (2018). Multi-modal, ultrasensitive, wide-range humidity sensing with Ti_3C_2 film. *Nanoscale* 10 (46): 21689–21695.

40 Muckley, E.S., Naguib, M., Wang, H.W. et al. (2017). Multimodality of structural, electrical, and gravimetric responses of intercalated MXenes to water. *ACS Nano* 11 (11): 11118–11126.

41 Gao, G., Ding, G., Li, J. et al. (2016). Monolayer MXenes: promising half-metals and spin gapless semiconductors. *Nanoscale* 8 (16): 8986–8994.

42 Khazaei, M., Ranjbar, A., Arai, M. et al. (2017). Electronic properties and applications of MXenes: a theoretical review. *J. Mater. Chem. C* 5: 2488–2503.

43 Dayananda, D., Reddy, P.L., Deshmukh, K. et al. (2022). MXene-based flexible polymer composites as high dielectric constant materials. *Mxenes Compos.* 725–758.

44 Zheng, X., Shen, J., Hu, Q. et al. (2021). Vapor phase polymerized conducting polymer/MXene textiles for wearable electronics. *Nanoscale* 13 (3): 1832–1841.

45 Yin, G., Wang, Y., Wang, W., and Yu, D. (2020). Multilayer structured PANI/MXene/CF fabric for electromagnetic interference shielding constructed by layer-by-layer strategy. *Colloids Surf., A* 601: 125047.

46 Yin, G., Wang, Y., Wang, W. et al. (2021). A flexible electromagnetic interference shielding fabric prepared by construction of PANI/MXene conductive network via layer-by-layer assembly. *Adv. Mater. Interfaces* 8 (6): 2001893.

47 Iqbal, A., Sambyal, P., Kwon, J. et al. (2021). Enhanced absorption of electromagnetic waves in $Ti_3C_2T_x$ MXene films with segregated polymer inclusions. *Compos. Sci. Technol.* 213: 108878.

48 Cheng, W., Zhang, Y., Tian, W. et al. (2020). Highly efficient MXene-coated flame retardant cotton fabric for electromagnetic interference shielding. *Ind. Eng. Chem. Res.* 59 (31): 14025–14036.

49 Zheng, X., Hu, Q., Wang, Z. et al. (2021). Roll-to-roll layer-by-layer assembly bark-shaped carbon nanotube/$Ti_3C_2T_x$ MXene textiles for wearable electronics. *J. Colloid Interface Sci.* 602: 680–688.

50 Jia, X., Shen, B., Zhang, L., and Zheng, W. (2020). Waterproof MXene-decorated wood pulp fabrics for high-efficiency electromagnetic interference shielding and Joule heating. *Composites, Part B* 198: 108250.

51 Chen, Y., Yang, Y., Xiong, Y. et al. (2021). Porous aerogel and sponge composites: assisted by novel nanomaterials for electromagnetic interference shielding. *Nano Today* 38: 101204.

52 Shahzad, F., Alhabeb, M., Hatter, C.B. et al. (2016). Electromagnetic interference shielding with 2D transition metal carbides (MXenes). *Science* 353 (6304): 1137–1140.

53 Zhang, H., Chen, J., Ji, H. et al. (2021). Electromagnetic interference shielding with absorption-dominant performance of Ti_3C_2TX MXene/non-woven laminated fabrics. *Text. Res. J.* 91 (21-22): 2448–2458.

54 Guo, Z., Li, Y., Jin, P. et al. (2021). Poly(vinyl alcohol)/MXene biomimetic aerogels with tunable mechanical properties and electromagnetic interference shielding performance controlled by pore structure. *Polymer* 230: 124101.

55 Wang, W., Yuen, A.C., Long, H. et al. (2021). Random nano-structuring of PVA/MXene membranes for outstanding flammability resistance and electromagnetic interference shielding performances. *Composites, Part B* 224: 109174.

56 Han, M., Yin, X., Hantanasirisakul, K. et al. (2019). Anisotropic MXene aerogels with a mechanically tunable ratio of electromagnetic wave reflection to absorption. *Adv. Opt. Mater.* 7 (10): 1900267.

57 Zhou, Q., Lyu, J., Wang, G. et al. (2021). Mechanically strong and multifunctional hybrid hydrogels with ultrahigh electrical conductivity. *Adv. Funct. Mater.* 31 (40): 2104536.

58 Kim, E., Zhang, H., Lee, J.H. et al. (2021). MXene/polyurethane auxetic composite foam for electromagnetic interference shielding and impact attenuation. *Composites, Part A* 147: 106430.

59 Gao, L., Li, C., Huang, W. et al. (2020). MXene/polymer membranes: synthesis, properties, and emerging applications. *Chem. Mater.* 32 (5): 1703–1747.

60 Lipton, J., Weng, G.M., Alhabeb, M. et al. (2019). Mechanically strong and electrically conductive multilayer MXene nanocomposites. *Nanoscale* 11 (42): 20295–20300.

61 Wang, J., Yang, K., Wang, H., and Li, H. (2021). A new strategy for high-performance electromagnetic interference shielding by designing a layered double-percolated structure in PS/PVDF/MXene composites. *Eur. Polym. J.* 151: 110450.

62 Ma, W., Cai, W., Chen, W. et al. (2021). Microwave-induced segregated composite network with MXene as interfacial solder for ultra-efficient electromagnetic interference shielding and anti-dripping. *Chem. Eng. J.* 425: 131699.

63 Wan, Y.J., Li, X.M., Zhu, P.L. et al. (2020). Lightweight, flexible MXene/polymer film with simultaneously excellent mechanical property and high-performance electromagnetic interference shielding. *Composites, Part A* 130: 105764.

64 Ji, H., Zhao, R., Zhang, N. et al. (2018). Lightweight and flexible electrospun polymer nanofiber/metal nanoparticle hybrid membrane for high-performance electromagnetic interference shielding. *NPG Asia Mater.* 10 (8): 749–760.

65 Du, Z., Chen, K., Zhang, Y. et al. (2021). Engineering multilayered MXene/electrospun poly(lactic acid) membrane with increscent electromagnetic interference (EMI) shielding for integrated Joule heating and energy generating. *Compos. Commun.* 26: 100770.

66 Wang, Z., Han, X., Han, X. et al. (2021). MXene/wood-derived hierarchical cellulose scaffold composite with superior electromagnetic shielding. *Carbohydr. Polym.* 254: 117033.

67 Cui, C., Xiang, C., Geng, L. et al. (2019). Flexible and ultrathin electrospun regenerate cellulose nanofibers and d-$Ti_3C_2T_x$ (MXene) composite film for electromagnetic interference shielding. *J. Alloys Compd.* 788: 1246–1255.

68 Yang, S., Yan, D.X., Li, Y. et al. (2021). Flexible poly (vinylidene fluoride)-MXene/silver nanowire electromagnetic shielding films with joule heating performance. *Ind. Eng. Chem. Res.* 60 (27): 9824–9832.

69 Ma, Z., Kang, S., Ma, J. et al. (2020). Ultraflexible and mechanically strong double-layered aramid nanofiber–$Ti_3C_2T_x$ mxene/silver nanowire nanocomposite papers for high-performance electromagnetic interference shielding. *ACS Nano* 14 (7): 8368–8382.

70 Hu, D., Huang, X., Li, S., and Jiang, P. (2020). Flexible and durable cellulose/MXene nanocomposite paper for efficient electromagnetic interference shielding. *Compos. Sci. Technol.* 188: 107995.

71 Zhan, Z., Song, Q., Zhou, Z., and Lu, C. (2019). Ultrastrong and conductive MXene/cellulose nanofiber films enhanced by hierarchical nano-architecture and interfacial interaction for flexible electromagnetic interference shielding. *J. Mater. Chem. C* 7 (32): 9820–9829.

72 Zhan, Y., Meng, Y., and Xie, Q. (2021). Simple approach to fabricate MXene/cellulose paper for electromagnetic interference shielding applications. *J. Appl. Polym. Sci.* 138 (25): 50597.

73 Xie, F., Jia, F., Zhuo, L. et al. (2019). Ultrathin MXene/aramid nanofiber composite paper with excellent mechanical properties for efficient electromagnetic interference shielding. *Nanoscale* 11 (48): 23382–23391.

74 Cao, W.T., Chen, F.F., Zhu, Y.J. et al. (2018). Binary strengthening and toughening of MXene/cellulose nanofiber composite paper with nacre-inspired structure and superior electromagnetic interference shielding properties. *ACS Nano* 12 (5): 4583–4593.

75 Huang, J.J., Wang, T., Su, Y.M. et al. (2021). Hydrophobic MXene/hydroxyethyl cellulose/silicone resin composites with electromagnetic interference shielding. *Adv. Mater. Interfaces* 8 (11): 2100186.

76 Gao, Q., Pan, Y., Zheng, G. et al. (2021). Flexible multilayered MXene/thermoplastic polyurethane films with excellent electromagnetic interference shielding, thermal conductivity, and management performances. *Adv. Compos. Hybrid Mater.* 4 (2): 274–285.

77 Xu, M.K., Liu, J., Zhang, H.B. et al. (2021). Electrically conductive $Ti_3C_2T_x$ MXene/polypropylene nanocomposites with an ultralow percolation threshold for efficient electromagnetic interference shielding. *Ind. Eng. Chem. Res.* 60 (11): 4342–4350.
78 Zhou, B., Zhang, Z., Li, Y. et al. (2020). Flexible, robust, and multifunctional electromagnetic interference shielding film with alternating cellulose nanofiber and MXene layers. *ACS Appl. Mater. Interfaces* 12 (4): 4895–4905.
79 Jia, X., Shen, B., Zhang, L., and Zheng, W. (2021). Construction of compressible polymer/MXene composite foams for high-performance absorption-dominated electromagnetic shielding with ultra-low reflectivity. *Carbon* 173: 932–940.
80 Zhang, Y., Wang, L., Zhang, J. et al. (2019). Fabrication and investigation on the ultra-thin and flexible $Ti_3C_2T_x$/co-doped polyaniline electromagnetic interference shielding composite films. *Compos. Sci. Technol.* 183: 107833.
81 Kumar, S., Kumar, P., Singh, N., and Verma, V. (2019). Steady microwave absorption behavior of two-dimensional metal carbide MXene and polyaniline composite in X-band. *J. Magn. Magn. Mater.* 488: 165364.
82 Raagulan, K., Braveenth, R., Kim, B.M. et al. (2020). An effective utilization of MXene and its effect on electromagnetic interference shielding: flexible, free-standing and thermally conductive composite from MXene–PAT–poly (*p*-aminophenol)–polyaniline co-polymer. *RSC Adv.* 10 (3): 1613–1633.
83 Wang, Y., Gao, X., Zhang, L. et al. (2019). Synthesis of Ti_3C_2/Fe_3O_4/PANI hierarchical architecture composite as an efficient wide-band electromagnetic absorber. *Appl. Surf. Sci.* 480: 830–838.
84 Liu, R., Miao, M., Li, Y. et al. (2018). Ultrathin biomimetic polymeric $Ti_3C_2T_x$ MXene composite films for electromagnetic interference shielding. *ACS Appl. Mater. Interfaces* 10 (51): 44787–44795.
85 Lu, Y., Zhang, S., He, M. et al. (2021). 3D cross-linked graphene or/and MXene based nanomaterials for electromagnetic wave absorbing and shielding. *Carbon* 178: 413–435.
86 Zhou, B., Su, M., Yang, D. et al. (2020). Flexible MXene/silver nanowire-based transparent conductive film with electromagnetic interference shielding and electro-photo-thermal performance. *ACS Appl. Mater. Interfaces* 12 (36): 40859–40869.
87 Fan, Z., Wang, D., Yuan, Y. et al. (2020). A lightweight and conductive MXene/graphene hybrid foam for superior electromagnetic interference shielding. *Chem. Eng. J.* 381: 122696.
88 Zhao, S., Zhang, H.B., Luo, J.Q. et al. (2018). Highly electrically conductive three-dimensional $Ti_3C_2T_x$ MXene/reduced graphene oxide hybrid aerogels with excellent electromagnetic interference shielding performances. *ACS Nano* 12 (11): 11193–11202.
89 Nguyen, V.T., Min, B.K., Yi, Y. et al. (2020). MXene ($Ti_3C_2T_x$)/graphene/PDMS composites for multifunctional broadband electromagnetic interference shielding skins. *Chem. Eng. J.* 393: 124608.
90 Vu, M.C., Mani, D., Kim, J.B. et al. (2021). Hybrid shell of MXene and reduced graphene oxide assembled on PMMA bead core towards tunable thermoconductive and EMI shielding nanocomposites. *Composites, Part A* 149: 106574.

18

Technological Aspects in the Development of MXenes and Its Hybrid Nanocomposites: Current Challenges and Prospects

Sujeet K. Mishra[1] *and Devendra Kumar*[2]

[1] *CCRAS-Central Ayurveda Research Institute, Jhansi, UP, India*
[2] *Department of Applied Chemistry, Delhi Technological University, New Delhi, India*

18.1 Introduction

The development of Graphene, an ultrathin two-dimensional (2D) material, in 2004 has resulted in an unprecedented growth of similar 2D layered materials with applications in various scientific fields, particularly material science and nanotechnology [1]. Subsequently, many 2D nanomaterials have been developed by researchers, such as nanosheets of hexagonal boron nitride [2], antimonene [3], transition metals dichalcogenides (MoS_2, WS_2, etc.) [4], and phosphorene [5] from their bulk layered structures. These 2D nanomaterials possess distinctive physicochemical properties in comparison to their higher dimension counterparts due to the presence of quantum confinement effects in them [6]. In 2011, a new 2D member was added to this group, having excellent electrical conductivity, hydrophilicity, hardness, flexibility, and high chemical stability. These nanomaterials are popularly known as MXenes and are synthesized by researchers from Drexel University. MXenes are basically a family of nitrides, carbides or carbonitrides of transition metals with a wide range of applications in various fields [7, 8]. These are synthesized by selectively etching the A layers (IIIA or IVA elements such as Al, Ga, Si, Pb, In, P, Sn, S, Cd, Ge, As and Ti) from the corresponding 3D transition metal-based carbides and nitrides denoted as $M_{n+1}AX_n$ or MAX phase, where M is an early transition metal, $n = 1–3$, and X represents Carbon and/or Nitrogen [9, 10]. The delamination process results in the generation of terminal surface functional groups such as —OH, —O and/or F on MXene [11]. After synthesis, the molecular formula of MXenes becomes $M_{n+1}X_nT_x$, where T_x is the terminal surface functional group. The amount of different surface functional groups varies with modification in the etching conditions. The 2D morphology of MXenes, along with their good flexibility and lamellar structure, makes them a suitable candidate for forming composites/hybrids with other materials. This combination can result in the formation of composites/hybrids with outstanding properties that are synergistically complement each other. So, both MXenes and MXene-based composites with plenty of mesmerizing electronic, mechanical, magnetic and electrochemical properties have fascinated substantial research interest in various applications such as energy storage

MXenes: Next-Generation 2D Materials: Fundamentals and Applications, First Edition.
Edited by Jay Singh, Kshitij RB Singh, Ravindra Pratap Singh, and Charles Oluwaseun Adetunji.

Table 18.1 Timeline of MXene advancement and breakthroughs thereof.

S. no.	Year	Major breakthroughs
1	2011	Ti_3C_2 discovery using HF etching method
2.	2012	Development of a family of MXenes including Ti_2C, Ta_4C_3, Ti_3CN
3.	2013	Intercalation/delamination of MXenes
4.	2014	(a) Bifluoride (NH_4HF_2) etching method (b) Formation of clay (small flakes) via LiF/HCl etching method and (c) Formation of MXene-polymer composite
5.	2015	(a) Tetrabutylammonium hydroxide (TBAOH) and amine-assisted delamination of layered structures (b) Formation of Mo_2C and Mo_2N MXene via urea glass route and (c) Development of chemical vapour deposition (CVD) method and theoretical estimation of 2D-ordered MXenes
6.	2016	(a) Formation of nitride-based MXene (Ti_4N_3) via molten salt etching method (b) Synthesis of new MXene Zr_3C_2 at room temperature
7.	2017	Development of ordered divacancies in MXene
8.	2018	(a) Development of hydrothermal production technique using NaOH (b) Synthesis of Fluoride MXenes by means of an electrochemical binary water system
9.	2019	(a) Development in synthesis technique of MXene-composite system (b) Synthesis of MXenes using Lewis acid molten salts as etchants
10.	2020	Formation of MXene-polymer composite

systems [12-14], catalysts [15-16], gas sensors [17], biosensors [18] and removal of toxic heavy metal ions from water/wastewater [19].

The 3D layered structure of MAX phases is not easy to convert into 2D and 1D nanomaterials as strong organometallic bonding is present between the early transition metal and C/N atoms, preventing the reduction in their dimensionality [20]. Various strategies have been devised from time to time to extract MXenes from their MAX phases, as shown in Table 18.1. In 1996, Barsoum and El-Raghy did pioneer work by synthesizing a single-phase pristine Ti_3SiC_2 [21] with the MAX phase. In 2011, Naguib et al. synthesized the first MXene, a 2D Ti_3C_2, by immersing Ti_3AlC_2 in aqueous HF at ambient temperature to selectively etch Al layers from the MAX phase structure [22]. Later on, with the discovery of more precursor MAX phases, the family of MXenes also extended with further addition of out-of-plane and in-plane ordered double-transition metal MXenes in 2015 and 2017, respectively.

18.2 Progressive Approach Towards MXene Composites and Hybrids

MXenes possess exceptional properties such as layered structure, large negative zeta-potential, electrical conductivity, hydrophilicity and good flexibility like their counterpart graphene. These properties can further be enhanced by integrating MXene with

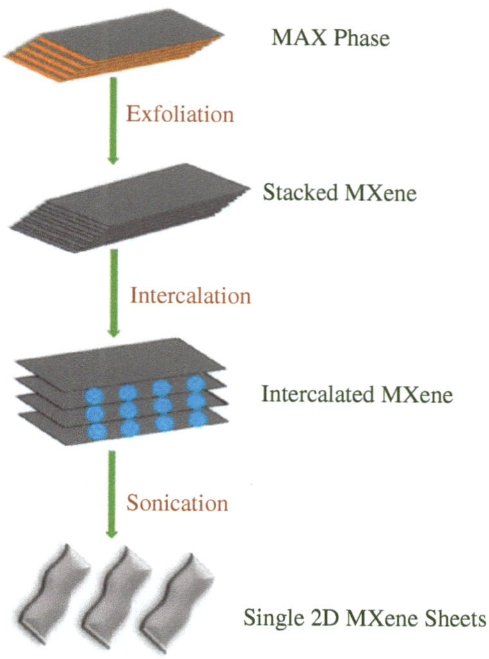

Figure 18.1 Schematic representation of formation of 2D MXene sheets from MAX phase.

polymers and numerous other nanomaterials, such as carbon-based nanomaterials, metal nanoparticles, quantum dots, oxides of metal and metal–organic frameworks. For example, one- or two-layer MXene flakes obtained after delamination of multilayered MXene can be combined with carbon nanotubes or conducting polymers to generate hybrids/nanocomposites with higher conductivities and mechanical strength. The general methodology for the synthesis of MXene from the MAX phase is depicted in Figure 18.1.

18.2.1 MXene-Metal Composite

The addition of 2D MXenes with notable mechanical stiffness [23], high in-plane electrical conductivity and catalytic potential [24] can impart significant enhancement in the properties of metal. The presence of hydrophilic surfaces makes MXene an ideal filler for metal and ceramic matrix-based composites. The integration of MXene with certain metal oxides and sulfides can deter the restacking of MXene layers and, at the same time, provide mechanical stability to the structure of metal oxides/sulphides. This symbiotic relationship has led to the development of a vast number of MXene-metal-based composites [25]. Recently, $Ti_3C_2T_x$ MXenes have been incorporated as a reinforcing material for aluminium using the powder metallurgy technique [26-27]. In this study, first of all, multilayer $Ti_3C_2T_x$ powder was mixed with aluminium using the ball milling process. This was followed by pressureless frottage at 650 °C in an argon environment and then hot rolling [26]. In another study, Fei et al. integrated $Ti_3C_2T_x$ onto Al_2O_3 and studied the effects on the microstructures and mechanical properties of alumina [28]. It was observed that 2 wt% $Ti_3C_2T_x$ led to the increment in the hardness, fracture toughness and bending strength by approximately 300%, 300% and 150%, respectively. It has been observed that the incorporation of a small wt% of a few layers of delaminated MXene into the metal matrix increases the tensile strength more in comparison to non-delaminated multilayer MXene powder. This is mainly due to the presence of a large surface area in the case of single-layer MXene, which results in the transfer of stress from the metal matrix towards the delaminated MXene flake. Also, in the case of single-flake MXenes, only in-plane metal-MXene bonding is present, which is quite stronger in comparison to weaker secondary bonding (van der Waals) present in multilayer MXene flakes.

Ti_3C_2-Ag nanocomposite was formed *in situ* by the direct reduction of $AgNO_3$ solution by $Ti_3C_2(OH)_{0.8}F_{1.2}$ for application as anodes in Li-ion batteries [29]. MXene-Ag composite electrodes exhibited large specific capacities of 310, 260 and 150 $mAh\,g^{-1}$ at current densities of 1, 10 and 50 °C, respectively, and a lengthy cycle lifetime of more than 5000 cycles. The main reason behind the outstanding electrochemical performances is the decrease in interface resistance and conversion of Ti^{2+} to Ti^{3+} during this cycle process. Tang et al. reported the synthesis of Ti_3C_2/Au nanohybrids with applications in nanomedicine [30]. The nanohybrid comprising Au nanoparticles uniformly deposited on the surface and edges of Ti_3C_2 nanosheets exhibited strong near-infrared optical absorption and mild photothermal effect. The nanocomposite has potential applications in radiotherapy with no long-term toxicity. Lorencova et al. utilized the Pt nanoparticles decorated $Ti_3C_2T_x$ nanosheets on a glassy carbon electrode (GCE) as a platform for the electrochemical sensing of H_2O_2 [31]. $Ti_3C_2T_x$-Pt nanocomposite depicted a more steady redox behaviour than the pristine $Ti_3C_2T_x$. The nanocomposite-modified GCE showed a limit of detection value (LOD) of 448 nM towards the quantification of H_2O_2 and has further applications in the analysis of various redox molecules such as acetaminophen, uric acid, ascorbic acid and dopamine.

18.2.2 MXene-Carbon Composites

Carbon-based materials comprise CNTs, mesoporous carbon, graphene and carbon nanofibers with superior electronic and optical properties and high mechanical strength. The integration of these materials into MXene can lead to the formation of composites with enhanced properties. Ghidiu et al. [32] reported the generation of carbon nanofibers (CNF)-MXene composite using the CVD technique with Co as the catalyst. The composite so formed has an open framework like with CNF, forming a 3D bridge-like structure amid layers of MXene, which also act as a pathway for electron transport.

In another method, free-standing $Ti_3C_2T_x$–CNF mats were synthesized via an electrospinning technique for applications in supercapacitors [33]. Lamellar $Ti_3C_2T_x$ nanosheets were uniformly spread inside the CNFs, and nanofiber mats were generated without applying any binder or additive. The composite-based electrode depicted areal capacitance approximately three times that of the pure carbonized polyacrylonitrile nanofiber mats under the same conditions.

CVD technique was also utilized to form vertical heterostructures of graphene-Mo_2C composite on a Cu/Mo substrate by Xu et al. [34]. It was observed that the incorporated MXene in the composite prevented any stacking of 2D graphene layers without altering the high surface area of graphene leading to the enhancement of properties of a composite.

Cai et al. [35] prepared a strain sensor by using a sandwich-like $Ti_3C_2T_x$MXene/CNT composite with thickness 1–2 μm formed by combining delaminated $Ti_3C_2T_x$ flakes with hydrophilic SWCNTs via layer-by-layer spray-coating method. The multipurpose sensor so formed showed high sensitivity with a gauge factor of up to 772.60, low LOD (0.1%), and a tuneable range of sensing from 30% to 130% strain.

Zhao et al. reported the formation of $Ti_3C_2T_x$/CNT composites for application in electrochemical capacitors [36]. In this composite, $Ti_3C_2T_x$ and CNT layers were consecutively placed on top of each other in a sandwich-like manner until 6–10 layers were formed. In comparison to the isolated MXene, the composite exhibited a better volumetric capacitance and rate performance on being used as an electrode of the supercapacitor. The enhancement in interlayer spacing of $Ti_3C_2T_x$ flakes due to the incorporation of CNT increases the cation intercalation and the diffusion path of electrolyte ions, leading to exceptional electrochemical performance of the composite.

CNTs@Ti_3C_2 hybrid structures were synthesized via microwave irradiation technique by first manually mixing ferrocene and MXene powders and subsequently subjecting them to microwave irradiation at 900 W for a designated interval of time [37]. The MWCNTs were cultivated on the top of MXene sheets, leading to the formation of a porous-like structure of the composite. The CNTs@MXene composite was further utilized as electrodes of Li-ion batteries and exhibited the specific capacities of 430 and 175 $mAh\,g^{-1}$ at 1 and 10 $A g^{-1}$, respectively.

Xie et al. described the formation of flexible, free-standing nanocomposite film of Mo_2CT_x-MWCNTs and $Ti_3C_2T_x$-MWCNTs using an electrostatic self-assembly technique for applications in supercapacitors [38]. In this technique, the negatively charged Mo_2CT_x and $Ti_3C_2T_x$ nanosheets were electrostatically grown along with positively charged MWCNTs (imbedded with cetyltrimethylammonium bromide) on filter papers. These paper-based electrodes exhibited high volumetric capacities when used in supercapacitors.

Yang et al. incorporated Ti_3C_2-CNTs nanocomposite films on graphite papers using electrophoretic deposition method [39]. In this technique, two graphite paper electrodes separated by 1 cm were dipped in H_2O:acetone (1:1) suspension of Ti_3C_2 powder and MWCNTs. A constant 8 V DC voltage was passed through the electrodes for 10 min, resulting in the establishment of the composite film on the electrodes. The prepared graphite paper-MXene composite electrodes depicted specific capacitance 2.6 times those of pristine CNT and 1.5 times that of Ti_3C_2.

18.2.3 MXene-Polymer Composites

Polymers are versatile long-chain molecules having surface functional groups that can be integrated with MXenes to modify their physico-chemical properties. The composite can be easily manufactured on a production scale at a low cost with exceptional properties such as anti-corrosion, high mechanical strength, good flexibility, resistance to fracture and impact resistance. 2D MXenes can easily be implanted on various polymers to form composites with unique and interesting applications that can be helpful in the development of unexplored avenues of science and technology. The schematic pathway for the formation of MXene polymer composite is shown in Figure 18.2.

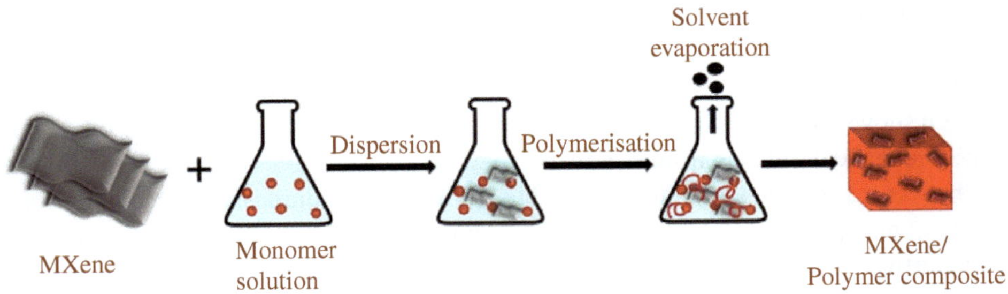

Figure 18.2 Schematic representation of formation of MXene-polymer composites.

Zhang et al. reported a Ti_3C_2/ultra-high molecular weight polyethylene (UHMWPE) composite by homogeneously mixing Ti_3C_2 and UHMWPE with the aid of a high-speed mixer and subsequently moulding by a press vulcanizer [40]. The composite was found to have enhanced mechanical properties due to the dissipation of stress from the UHMWPE matrix to Ti_3C_2, which is physically adsorbed on polymer chains. Another MXene-polymer composite was developed by Wu et al. through the *in situ* polymerization of pyrrole monomer along with Ti_3C_2 to form organ-like Ti_3C_2-PPy nanocomposite [41]. The morphological study of the composite exhibited the presence of spherical agglomerated particles, which enhanced the electrolyte ion diffusion by increasing the interlamellar spacing. Hence, Ti_3C_2-PPy composite can act as a promising material to be used as a supercapacitor electrode. Another high-conducting polymer such as polyaniline (PANI) has been incorporated with $Ti_3C_2T_x$MXene in a sandwich intercalated manner to study its microwave absorption properties by Wei et al. [42]. The MXene-PANI composite was formed using *in situ* polymerization technique where aniline monomer along with MXene was kept in an ice-water bath for hours with ammonium peroxydisulfate as an initiator. The presence of multilayer acted as reflected surfaces in the composite, which resulted in electromagnetic attenuation.

Naguib et al. fabricated a $Ti_3C_2T_x$/polyacrylamide (PAM) composite by first delaminating the individual $Ti_3C_2T_x$ layers using DMSO and then uniformly mixing and drying the $Ti_3C_2T_x$-PAM solution at ambient temperature for 4–5 days [43]. The mass ratio of $Ti_3C_2T_x$ to PAM was varied to prepare a composite of optimum mechanical strength. It was found that the composite with 6 wt% of $Ti_3C_2T_x$ possesses the finest mechanical properties and the peak electrical conductivity of 3.3×10^{-2} S m^{-1}. Apart from the polymers discussed above, polyvinylpyrrolidone [44], poly (acrylic acid) [44] and poly (ethylene oxide) [45] can also combine with MXenes to form composites with excellent mechanical and electrical properties.

18.2.4 MXene-Oxide Composites

MXene–oxide composites have versatile applications in the electrochemical industry as well as in supercapacitors owing to their extra stability and exceptional electrical conductivity. Zhang et al. [46] synthesized Nb_2O_5–$Nb_4C_3T_x$ hierarchical composite by oxidizing $Nb_4C_3T_x$ powders at 850 °C for 30 min in a continuous flow of CO_2. Nb_2O_5 is formed on the surface and in the interlayers of $Nb_4C_3T_x$. The composite exhibited good electrochemical and cycling performances attributed to the shortening of the ion diffusion paths and high electronic conductivity of unoxidized $Nb_4C_3T_x$.

Huang et al. described the formation of $Ti_3C_2T_x$–$Na_{0.23}TiO_2$ nanocomposite by incorporating interwoven amorphous metal oxide nanobelts ($Na_{0.23}TiO_2$)in between the layered structure of MXene using a hot aqueous method [47]. The nanocomposite so formed is tested for a lithium-ion batteries anode and showed a remarkable cycling performance and rate capability. The long cycling test performed on the nanocomposite electrode at an energy density of $5\,A\,g^{-1}$ exhibited an intensification of discharge capacity with increase in the number of cycles, with final stabilization obtained after 4000 cycles. Also, on varying current density in the range from 0.1 to $5\,A\,g^{-1}$, the reversible capacity of the nanocomposite electrode was found to reach a high value of $288\,mAh\,g^{-1}$ throughout the cycle. The exceptional cycling stability and rate capability are attributed to the reduction in transfer pathways of Li^+ and Na^+ ions and the release of the strain of the electrode upon cycling via the sandwich-like structure of the nanocomposite. Apart from the above-mentioned MXene-oxide composites, many other transition metal oxides such as TiO_2 [48], SnO_2 [49], Nb_2O_5 [46], Co_3O_4 [50], Sb_2O_3 [51], Cu_2O [52] and $NiCo_2O_4$ [50] have been used to form composites with MXene.

18.3 Technological Progress of MXene and Its Hybrids for Biomedical Sector

Recent developments have shown that MXene and its hybrids have huge potential in the biomedical industry. MXenes have poor water dispersibility, which somehow limits their application in healthcare sector. However, the development of surface functional groups on MXenes and their composites have tremendously increased their prospective in biomedical sector, such as biosensing, drug delivery, bone regeneration, bioimaging and antimicrobial applications [53]. Some of the recent applications in biomedical sector are discussed:

18.3.1 Bioimaging

The attachment of certain fluorescent molecules on the surface of MXenes enhances their luminescence property, making them a viable option for fluorescent bioimaging. Wang et al. synthesized ultrasmall fluorescent sheets of Ti_3C_2 having a thickness (0.5–1.5 nm) and lateral size (2–5 nm) via simultaneous intra-layer cutting and inter-layer delamination using aqueous tetramethylammoniumhydroxide (TMAOH) [54]. The Ti_3C_2 nanoflakes exhibited showed strong photoluminescence and excitation-dependent emission behaviour.

Liu et al. synthesized ultrathin Ti_3C_2 MXene nanosheets with an excellent photothermal conversion efficiency of approximately 58.3%, remarkable mass extinction coefficient value of $28.6\,L\,g^{-1}\,cm^{-1}$, and efficient production of singlet oxygen on being irradiated with 808 nm laser [55]. The as-obtained Ti_3C_2 nanosheets were subjected to surface modification of each layer with Doxorubicin (DOX), a chemotherapeutic drug to specifically target CD44 antigen cancerous cells and coated with Hyaluronic acid to increase its biocompatibility. Further, both *in vitro* and *in vivo* experiments revealed that Ti_3C_2-DOX exhibit cumulative effect of targeted drug delivery in response to stimuli and killing of tumour tissue and cancer cell via synergistic photodynamic (PDT)/photothermal (PTT)/chemotherapy.

Xue et al. pioneered the synthesis of water-soluble photoluminescent Ti_3C_2 MXene-based Quantum dots (MQDs) using facile hydrothermal technique. A range of MQDs with different diameters (2.9–6.2 nm) was prepared by changing the reaction temperature (100–150 °C). The as-prepared biocompatible Ti_3C_2 MQDs with quantum yields of up to 10% were utilized as multicolour cellular imaging reagents and for the detection of Zn^{2+} ions [56]. Another vital bioimaging tool that has a tendency to generate high-resolution 3D images is an X-ray CT scan. The traditional CT scan uses iodine-containing compounds as contrasting agents that are highly toxic and can cause kidney damage and a short blood circulation rate. This necessitated the advancement of more superior, eco-friendly and biocompatible elements for CT bioimaging. Dai et al. developed biocompatible Ta_4C_3 MXene-based composite nanosheets that can be utilized as high-performance contrasting agents for CT imaging [57].

18.3.2 Antimicrobial Activity

2D nanosheets, including MXenes, have been widely explored for their antibacterial properties to be probably utilized in wastewater treatment and biomedical applications. The bactericidal property of MXenes is assumed to arise from their easy absorption on the surface of bacteria due to pointed edges [58] and then rupturing of cell microstructure by reacting with the biomolecules present in the cytoplasm and cell wall.

Rasool et al. exhibited the antibacterial properties of $Ti_3C_2T_x$-modified membranes by testing against *Bacillus* subtilis (*B. subtilis*) and *Escherichia* coli (*E. coli*) [59]. It was proved using techniques such as flow cytometry, colony forming counts, SEM and AFM that $Ti_3C_2T_x$ membranes impaired the growth of both the bacterial strains. The fresh $Ti_3C_2T_x$ modified membrane exhibited bactericidal rates of 73% and 67% against *B. subtilis* and *E. coli*, respectively, while more than 99% antibacterial rate against both bacteria species was observed in the case of aged $Ti_3C_2T_x$ membrane.

18.3.3 Biosensor

Biosensors make use of biomolecules such as enzymes, DNA, antibodies, etc., immobilized on an electrode, which in certain reactions with an analyte, generates biochemical signals. These signals are then converted into electrical signals by transducer for measurement of analyte concentration. MXene possesses good electrical conductivity, biocompatibility and high surface area, which makes it an ideal contender for protein immobilization matrix in biosensing applications. MXene was first utilized as sensing material by Xu et al., who fabricated ultrathin Ti_3C_2 MXene micropatterns based field-effect transistor (FET) biosensor for label-free detection of dopamine and fast discovery of spiking activity in primary hippocampal neurons [60]. This work opened the avenue for MXenes to be used as a protein immobilizing matrix in biosensing applications. Wang et al. developed an amperometric biosensor using TiO_2-modified $Ti_3C_2T_x$ as haemoglobin immobilization matrix for the detection of H_2O_2 [61]. The bioelectrode (Nafion/Hb/$Ti_3C_2T_x$–TiO_2/GCE) exhibited a linear range from 0.1 to 260 μM for H_2O_2 with 20 nM as the LOD value. The system exhibited a quasi–reversible interfacial electron transfer mechanism, leading to a fast detection of H_2O_2 in less than 3 s.

A novel MXene-$Ti_3C_2T_x$-based bio-electrochemical microfluidic chip sensor was fabricated by Liu et al. for multicomponent (uric acid, urea and creatinine) analysis of whole blood [62]. $Ti_3C_2T_x$ was screen printed as an electrode on a microfluidic chip along with a dialysis membrane for continuous analysis of multiple analytes in raw blood samples. This miniaturized and automated novel biosensor can be utilized to identify the patients requiring hemodialysis treatment. The work demonstrated the use of MXene in point-of-care (POC) devices with the benefits of biocompatibility, stability, economy and automation. The MXene-based POC devices can prove to be a valuable asset for clinicians in understanding and analysing patient's health in a real-time manner. For its full-fledged applications of MXenes in the biomedical sectors, there are some shortcomings that need to be addressed. For example, the researchers have to fabricate new miniaturized designs with specific applications, develop more new and effective functional groups for surface functionalization of MXenes and comprehensively optimize the working conditions for obtaining reliable and repeatable data.

18.4 Technological Progress of MXene and Its Hybrids for Energy Sector

The depletion of conventional sources of energy and increasing demand for energy consumption has shifted the focus towards nonconventional resources such as hydro, solar, wind and tidal energies. These renewable resources are non-reliable as they are largely affected by environmental factors. A more viable option is to develop electrochemical energy storage devices like supercapacitors and batteries with high energy and power density output.

18.4.1 Ion Batteries

Various commercially developed batteries in recent times include Li-ion batteries (LiB), Na-ion batteries (NaB) and Zn-ion batteries (ZnB). LiB possesses good cycle performance and high energy density and is commonly used in portable electronic equipment. To date, LiB utilizes graphite and its composites as anode with a storage capacity of 330 mAh g^{-1}. However, it is not considered sufficient for future energy requirements. Therefore, novel materials with high storage capacity that can substitute graphene in LiB are required. MXenes and its composites, with their 2D layer structure have the capacity to accommodate various ions such as Mg^{2+}, K^+, Li^+ and Na^+ [63]. Also, MXenes are potentially good candidates for LiB anode material as they possess high electron conductivity with low diffusion barrier and high theoretical value of Li storage capacity [64].

Er et al. studied the topographical adsorption of various ions (Ca, Li, K and Na) on Ti_3C_2 MXene surface by means of the first principles of density functional calculations. The group observed a decrease in adsorption energy with an increase in coverage area for Na, K and Ca ions, while for Li it was independent of coverage [65]. This correlation between adsorption energies and surface coverage was explained on the basis of the difference in effective ionic radii of the ions. A larger effective ionic radius results in stronger interaction between the ions, resulting in lesser coverage. Further, the first principles density functional calculations show specific capacity values of 191.8, 319.8, 351.8 and 447.8 mAh g^{-1} for K, Ca, Na and Li, respectively, on the Ti_3C_2 surface. Through theoretical and experimental studies, it was noticed that bare MXene and those containing terminated oxygen groups have a significant influence on increasing the performance of the battery, while fluorine and hydroxyl functionalized MXene derivatives result in a lowering of Li diffusion and storage capacity [66, 67].

The application of LiB is somehow limited by inadequate natural deposits of Li, inflated cost of production, and safety concerns [68]. So, novel non-Li battery systems are explored for the fulfilment of energy requirements. In this regard, Xie et al. thoroughly examined the storage of Al, K, Mg, Ca and Na ions on both O-terminated and bare MXene nanosheets using combined first principles experimental and simulation techniques [67]. The experimental data suggested that both K^+ and Na^+ exhibit good cyclable performance with O-terminated Ti_3C_2nanosheet. Also, adsorption of Mg^{2+} and Al^{3+} was observed in the multilayer MXene nanosheet, leading to remarkable theoretical capacity.

Recently, Meng et al. constructed a nanocomposite of black Phosphorus quantum dots (BPQDs) and Ti_3C_2 nanosheet for the development of a highly efficient battery-capacitive dual-model energy storage system [69]. BPQDs anchored on Ti_3C_2 nanosheet when utilized as anode in LiB instigated superior battery performance with improved conductivity, high capacity (910 mAh g^{-1} at 100 mA g^{-1} current density) and long cyclic stability (2400 cycles with 100% capacity retention). The covalent linkage between BPQDs and MXene nanosheet endows high interfacial charge transfer and mechanical stability, resulting in a remarkable dual battery-capacitive model.

18.4.2 Supercapacitors

Supercapacitors are next-generation energy accumulation devices with exceptional properties such as fast rate of charge-discharge, long cycle stability and high power density [70]. To date, graphene and activated carbon have been utilized as supercapacitor electrodes. However, experimental studies during the last decade have suggested that 2D MXenes with large surface area, hydrophilicity and good conductivity can potentially be used as electrodes in supercapacitors. $Ti_3C_2T_x$ MXene has exhibited spontaneous intercalation with certain monovalent/multivalent cations in their salt solutions, resulting in high electrochemical capacitance behaviour [71].

Hu et al. studied the capacitive mechanism of $Ti_3C_2T_x$ MXene in sulphuric acid electrolyte by means of the Raman spectroscopy technique [72]. It was observed that during the discharging process, H_3O^+ gets bonded with —O functional group of MXene electrode, while during charging, de-bonding of the same took place. This reversible process of bonding and de-bonding results in a reversible change in the oxidation state of Ti in $Ti_3C_2T_x$. However, in other electrolytes, an electric double layer is formed, which results in counter-ion adsorption on MXene-based electrodes. The main shortcoming of using MXene in supercapacitors is their tendency to get agglomerate. The energy density gets diminished due to the redeposition of ions within the 2D layers of MXenes during supercapacitor charge–discharge cycle, which can be rectified by use of a separator in the MXene layers. Recently, Shao et al. synthesized MXene-based 1-D supercapacitors having wire-like structure and properties exhibiting high strength and flexibility [73]. The free-standing electrodes were fabricated using the electrospinning method, where the nanofibres of MXene were self-winded on polyester substrate yarn. Laser Machine technology is used to develop 2D micro-supercapacitors by loading MXene on paper [74]. Electroactive inks generated from hydrophilic MXenes can be easily printed [75], stamped [76] or dip-coated [77] on various substrates for the production of flexible 2D electrodes.

An intrinsically stretchable micro-supercapacitor with macroporous interdigitated electrodes was fabricated using 3D printing technology by Li et al. [78]. $Ti_3C_2T_x/MnO_2/Ag/$ fullerene nanocomposite electrode exhibited an exceptional areal capacitance of $216.2\,mF\,cm^{-2}$ at a scanning rate of $10\,mV\,s^{-1}$ and unprecedented areal energy and power density of $19.2\,\mu Wh\,cm^{-2}$ and $58.3\,mW\,cm^{-2}$, respectively. Although significant improvement is being made in the production of MXene-based 1D, 2D and 3D supercapacitors, glitches remain for the full-fledged applications of MXenes as electrodes in supercapacitors. One such requirement is the development of nonaqueous electrolytes-based MXene electrodes for supercapacitors, which may result in greater energy density due to their wide window of decomposition voltage.

18.5 Technological Progress of MXene and Its Hybrids for Environment Sector

The rapid and vast expansion of global industrialization has resulted in destructive impact on human health and the environment. The growing concern about environmental pollution has drawn the attention of researchers from across the world. The presence of toxic pollutants such as aromatic compounds, heavy metal ions, bio-toxins and textile dyes in environment, particularly in water, poses a serious life-threatening environmental concern. Apart from certain conventional methods such as membrane filtration, aerobic/anaerobic digestion and adsorption, various advanced nanomaterials, including metal–organic frameworks, carbon nanomaterials, ceramics and zeolites, have been utilized for environmental remediation. MXenes with inherent properties such as activated functional group sites, high surface area, high hydrophilicity, conductivity and non-toxicity have stimulated deep interest as promising material for environmental challenges.

18.5.1 Water Treatment

Highly toxic heavy metal ions such as lead, chromium, mercury, cadmium and arsenic are high-risk pollutants, deadly to all living organisms. These ions are nondegradable and are widely present in water bodies. Conventional methods used for the treatment of toxic ions are either chemical (precipitation, adsorption, filtration and electro-dialysis) or biological. These methods are not precise at low concentrations of ions (ppm or ppb) and are also costly. Several reports suggest that MXenes and their composites with high surface area, surface functional groups and interlayer spacing have the tendency to adsorb toxic metal ions, dyes and radionuclides from water efficiently. MXene nanosheets have interlayer spacing of less than 2 Å leading to strong trapping of ions. Also, the adsorption ability of MXenes can be greatly enhanced by surface functionalization of active sites.

Peng et al. functionalized $Ti_3C_2T_x$-MXene with —OH groups by using exfoliation and alkalization with 5% NaOH solution [19]. The functionalized MXene was further used for selective removal of Pb^{2+} in the presence of other rival cations like Mg^{2+} and Ca^{2+} in higher concentrations. The active sites of Ti-OH, along with the layered structure of MXene, endow high adsorption of Pb^{2+}in a short duration of time (120 s). Ying et al. sequestered Cr^{6+} ions from water using delaminated $Ti_3C_2T_x$ nanosheets, which reduced the toxic Cr^{6+} to Cr^{3+} and then adsorbed it with maximum adsorption capacity of 250 mg g^{-1} [79]. The study showed that adsorption of Cr^{6+} ions depends directly on the concentration of MXene and the pH of the water system. These MXene nanosheets can also successfully reduce and remove $NaAuCl_4$, $KMnO_4$ and $K_3[Fe(CN)_6]$ from wastewater. The water redemption can further be enhanced by using MXene-based nanocomposites. For example, a nanocomposite of $Ti_3C_2T_x$-MXene and Fe_2O_3 was utilized for the adsorption of Hg^{2+} ions in wide pH range from industrial wastewater with highest adsorption capacity of 1128.41 mg g^{-1} [80]. The metal ions adsorption on MXene followed pseudo-second-order kinetics in agreement with Redlich–Peterson adsorption isotherm. The synergistic effect of $Ti_3C_2T_x/Fe_2O_3$, along with the active Fe-OH sites, led to high adsorption capacity.

18.5.2 Dye Degradation

Dyes are another nonbiodegradable major water pollutant whose urgent removal is quite necessary for environmental remediation. Mashtalir et al. studied the comparative adsorption of anionic dye (acid blue 80) and cationic dye (methylene blue) by multilayered $Ti_3C_2T_x$-MXene [81]. It was observed that $Ti_3C_2T_x$-MXene being negatively charged electrostatically adsorbed only the positively charged methylene blue (adsorption capacity ~39 mg g^{-1}). The adsorption capacity of MXene can further be increased by treatment with a hot alkaline solution. This treatment can lead to expansion of interlayer spacing in MXene along with incorporation of —OH groups on its surface. Wei et al. treated $Ti_3C_2T_x$-MXene with several alkalis (NaOH, KOH and LiOH) and found that the MXene treated with NaOH showed the highest adsorption of methylene blue dye with an adsorption capacity of 189 mg g^{-1} [82]. Cai et al. changed the morphology of 2D MXene from sheet-like structure to a rod-like architecture by treatment of $Ti_3C_2T_x$ with phytic acid via a single-step hydrothermal technique [83]. The modified phytic acid-doped MXene composite showed an increased adsorption capacity of 42 and 22 mg g^{-1} for methylene blue and Rhodamine B dyes, respectively. The adsorption process for methylene blue was able to sustain 85% adsorption capacity even after 12 cycles and followed Langmuir model. The adsorbed organic dyes can further deteriorate by photocatalytic degradation triggered by generation of $TiO_2/Ti(OH)_4$ on MXene surface using UV illumination [84]. Pristine MXene shows limited photocatalytic effect due to low range of light absorption, less stability in water and rapid recoupling of electron–hole pairs [85]. Gao et al. overcame this deficiency by synthesizing nanocomposite of 0D TiO_2/2D $Ti_3C_2T_x$ and utilized it for photocatalytic degradation of methyl orange. The nanocomposite exhibited 98% degradation of the dye in 30 min, which was higher in comparison to either $Ti_3C_2T_x$ (42%) or TiO_2 (77%) nanoparticles under similar experimental conditions. The better performance of the nanocomposite is due to greater electron–hole separation and the creation of TiO_2 and $Ti_3C_2T_x$ heterojunction on UV light exposure.

18.6 Technological Challenges and Outlook

Since their discovery a decade ago, MXene and MXene-based hybrids/composites have been developed as versatile materials with many promising applications in different sectors of science and technology. In the last decade, many research papers have been published and patents filed on MXenes. The numerous applications of MXene and its composites/hybrids range from medicine to electronics, optoelectronics, sensing, catalysis and so on, as depicted in Figure 18.3.

Although much technological progress has been made in their synthesis, several challenges still exist that need to be tackled for their scalable production. Some of the significant challenges are discussed:

(a) The main technological challenge is the extension of the MXene family by exfoliating new MAX phases. Most of the MXene reported to date are carbides, and only a few are nitrides and carbonitrides [86]. Also, new phases such as MAB are still unexplored and attempts to produce them have not born any fruit yet. Many 2D borides, nitrides and carbides with novel properties have been predicted computationally but are still not synthesized. New methods with better control over surface group functionalization and morphology of MXenes need to be continuously devised for their scalable synthesis and specific applications. The stability of new MXene compositions in etchants is also an issue that needs to be tackled. For example, the attempt to etch Al from Ti_2AlN and Cr_2AlC using HF as an etchant has resulted in the formation of metal oxides and fluorides [87]. The interlayer bonding present in the MAX phase is also a determining

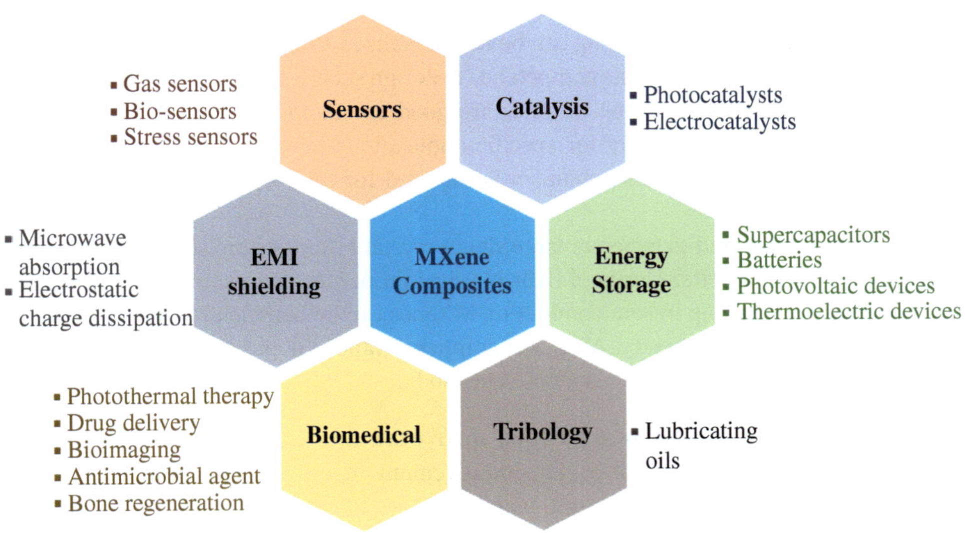

Figure 18.3 Different applications of MXene composites.

factor in the generation of particular MXene. For example, Meshkian et al. delineate the formation of Mo_2CT_x from Mo_2Ga_2C thin films instead of Mo_2GaC as Ga—Ga chemical bond is easier to break than Mo—Ga bonds [88].

(b) Although many theoretical studies have been done in MXene synthesis, their experimental validation still needs to be addressed [89]. Also, the accurate modelling of experimental techniques, including surface terminations of different MXenes compounds and the use of classical molecular dynamics for simulation studies, is still scarce. To date, most of the studies regarding the optimum conditions for generation and processing of MXenes and their effect on structure and properties have been done on $Ti_3C_2T_x$, so for better understanding, these studies should also be carried out on different MXenes.

(c) MXenes are susceptible to oxidations under severe oxidizing environments and at high temperatures which are largely unheeded by many researchers. These oxidized products can be utilized for the synthesis of novel nanohybrid structures for energy storage and conversion [90, 91]. Also, the impurities generated during the synthesis of MXenes, such as unwashed salts or etching products, unreacted MAX phases, etc., must be eliminated to ensure reproducibility and precise reports.

(d) For making MXenes ready for industrial applications, it is necessary to ponder on two main aspects, i.e. their elevated cost and structural/property instability. The high cost can be tackled by devising novel, economical synthesis procedures such as using 3D printing technology and MXene-based aqueous inks [92]. Also, the highly priced MAX phase precursor makes the manufacturing of MXene less economical.

(e) Although MXenes and MXene-based hybrids/nanocomposites have been utilized in many applications, their systematic understanding of the working mechanism is still not very explicit in most cases. For example, intercalation is a vital phenomenon in the synthesis, physicochemical properties and efficiency of MXenes and their composites/hybrid. However, it is unclear how the intercalants interact with MXene nanosheets and solvents during their confinement. MXene-based lamellar structures having large interlayer spacing can be functionalized to adsorb heavy metal ions from water samples. However, the competence to design such structures is still unintelligible or at a trial-and-error level. MXenes are good conductors of electricity but how to customize the conductivity for specific applications requires further investigations. Similarly, a perspicuous mechanism is required for their applications in microwave absorption.

(f) Efforts are required to generate large single-crystal MXenes from single crystals of the MAX phase for better clarity of their physical properties. For example, proper knowledge of the bandgap in semiconducting MXenes can be beneficial in their application in the electronics industry. Also, like graphene, vapour deposition methods have not been able to generate single-layer MXenes so far.

Many researchers have already started their work on the above problems and have shown promising results such as the improvement in the stability of $Ti_3C_2T_x$ by the use of antioxidants [93] or by using capping agents to cover the edges of MXene [94]. Kamysbayev et al. performed several new surface functionalizations of MXenes via substitution/elimination reactions in molten salt [95]. Apart from this, high-entropy

MXenes [96] and hydrophobic MXenes have also been reported recently by modifying the surface groups. With its rich structural and compositional diversities, MXene has lots of possibilities that can open new avenues in fundamental and multidisciplinary research.

References

1 Novoselov, K.S., Geim, A.K., Morozov, S.V. et al. (2004). Electric field in atomically thin carbon films. *Science* 306: 666–669.

2 Chen, H., Yang, Z., Zhang, Z. et al. (2019). Construction of a nanoporous highly crystalline hexagonal boron nitride from an amorphous precursor for catalytic dehydrogenation. *Angew Chem. Int. Ed.* 58: 10626–10630.

3 Peng, L., Ye, S., Song, J., and Qu, J. (2019). Solution-phase synthesis of few-layer hexagonal antimonene nanosheets via anisotropic growth. *Angew Chem. Int. Ed.* 58: 9891–9896.

4 Manzeli, S., Ovchinnikov, D., Pasquier, D. et al. (2017). 2D transition metal dichalcogenides. *Nat. Rev. Mater.* 2: 17033.

5 Ren, X., Zhou, J., Qi, X. et al. (2017). Few-layer black phosphorus nanosheets as electrocatalysts for highly efficient oxygen evolution reaction. *Adv. Energy Mater.* 7: 1–8.

6 Zhao, S., Kang, W., and Xue, J. (2015). MXene nanoribbons. *J. Mater. Chem. C* 3 (4): 879–888.

7 Naguib, M., Mochalin, V.N., Barsoum, M.W., and Gogotsi, Y. (2014). 25th anniversary article: MXenes: a new family of two-dimensional materials. *Adv. Mater.* 26 (7): 992–1005.

8 Sobolčiak, P., Tanvir, A., Sadasivuni, K.K., and Krupa, I. (2019). Piezoresistive sensors based on electrospun mats modified by 2D $Ti_3C_2T_x$ MXene. *Sensors (Switzerland)* 19 (20): 1–8.

9 Tang, J., Mathis, T.S., Kurra, N. et al. (2019). Tuning the electrochemical performance of titanium carbide MXene by controllable in situ anodic oxidation. *Angew Chem. Int. Ed.* 58 (49): 17849–17855.

10 Vahid, M.A., Mojtabavi, M., Caffrey, N.M. et al. (2019). Assembling 2D MXenes into highly stable pseudocapacitive electrodes with high power and energy densities. *Adv. Mater.* 31 (8): 1–9.

11 Li, S., Tuo, P., Xie, J. et al. (2018). Ultrathin MXene nanosheets with rich fluorine termination groups realizing efficient electrocatalytic hydrogen evolution. *Nano Energy* 47: 512–518.

12 Liang, X., Garsuch, A., and Nazar, L.F. (2015). Sulfur cathodes based on conductive MXene nanosheets for high-performance lithium-sulfur batteries. *Angew. Chem. Int. Ed.* 54 (13): 3907–3911.

13 Mashtalir, O., Lukatskaya, M.R., Zhao, M.Q. et al. (2015). Amine-assisted delamination of Nb_2C MXene for Li-ion energy storage devices. *Adv. Mater.* 27 (23): 3501–3506.

14 Wang, X., Kajiyama, S., Iinuma, H. et al. (2015). Pseudo capacitance of MXene nanosheets for high-power sodium-ion hybrid capacitors. *Nat. Commun.* 6: 1–6.

15 Guo, Z., Zhou, J., Zhu, L., and Sun, Z. (2016). MXene: a promising photocatalyst for water splitting. *J. Mater. Chem. A* 4 (29): 11446–11452.

16 Ye, M., Wang, X., Liu, E. et al. (2018). Boosting the photocatalytic activity of P 25 for carbon dioxide reduction by using a surface-alkalinized titanium carbide MXene as cocatalyst. *Chem. Sus. Chem.* 11 (10): 1606–1611.

17 Yu, X.F., Li, Y.C., Cheng, J.B. et al. (2015). Monolayer Ti_2CO_2: a promising candidate for NH_3 sensor or capturer with high sensitivity and selectivity. *ACS Appl. Mater. Interfaces* 7 (24): 13707–13713.

18 Wang, F., Yang, C.H., Duan, M. et al. (2015). TiO_2 nanoparticle modified organ-like Ti_3C_2MXene nanocomposite encapsulating hemoglobin for a mediator-free biosensor with excellent performances. *Biosens. Bioelectron.* 74: 1022–1028.

19 Peng, Q., Guo, J., Zhang, Q. et al. (2014). Unique lead adsorption behavior of activated hydroxyl group in two-dimensional titanium carbide. *JACS* 136: 4113.

20 Schwarz, K. (1987). Band structure and chemical bonding in transition metal carbides and nitrides. *Crit. Rev. Solid State Mater. Sci.* 13 (3): 211–257.

21 Barsoum, M.W. and El-Raghy, T. (1996). Synthesis and characterization of a remarkable ceramic: Ti_3SiC_2. *J. Am. Ceram. Soc.* 79: 1953–1956.

22 Naguib, M., Kurtoglu, M., Presser, V. et al. (2011). Two-dimensional nanocrystals produced by exfoliation of Ti_3AlC_2. 23: 4248–4253.

23 Lipatov, A., Alhabeb, M., Lu, H. et al. (2020). Electrical and elastic properties of individual single-layer $Nb_4C_3T_x$MXene flakes. *Adv. Electron. Mater.* 6 (4): 1–10.

24 Lim, K.R.G., Handoko, A.D., Nemani, S.K. et al. (2020). Rational design of two-dimensional transition metal carbide/nitride (MXene) hybrids and nanocomposites for catalytic energy storage and conversion. *ACS Nano* 14 (9): 10834–10864.

25 Yang, J., Bao, W., Jaumaux, P. et al. (2019). MXene-based composites: synthesis and applications in rechargeable batteries and supercapacitors. *Adv. Mater. Interfaces* 6 (8): 1–32.

26 Zhang, J., Li, S., Hu, S., and Zhou, Y. (2018). Chemical stability of Ti_3C_2MXene with Al in the temperature range 500–700 °C. *Materials (Basel)* 11 (10): 1–9.

27 Zhou, W., Zhou, Z., Fan, Y., and Nomura, N. (2021). Significant strengthening effect in few-layered MXene-reinforced Al matrix composites. *Mater. Res. Lett.* 9 (3): 148–154.

28 Fei, M., Lin, R., Lu, Y. et al. (2017). MXene-reinforced alumina ceramic composites. *Ceram. Int.* 43 (18): 17206–17210.

29 Zou, G., Zhang, Z., Guo, J. et al. (2016). Synthesis of MXene/Ag composites for extraordinary long cycle lifetime lithium storage at high rates. *ACS Appl. Mater. Interfaces* 8 (34): 22280–22286.

30 Tang, W., Dong, Z., Zhang, R. et al. (2019). Multifunctional two-dimensional core-shell MXene@gold nanocomposites for enhanced photo-radio combined therapy in the second biological window. *ACS Nano* 13 (1): 284–294.

31 Lorencova, L., Bertok, T., Filip, J. et al. (2018). Highly stable $Ti_3C_2T_x$ (MXene)/Pt nanoparticles-modified glassy carbon electrode for H_2O_2 and small molecules sensing applications. *Sens. Actuators, B* 263: 360–368.

32 Ghidiu, M., Halim, J., Kota, S. et al. (2016). Ion-exchange and cation solvation reactions in Ti_3C_2MXene. *Chem. Mater.* 28 (10): 3507–3514.

33 Levitt, A.S., Alhabeb, M., Hatter, C.B. et al. (2019). Electrospun MXene/carbon nanofibers as supercapacitor electrodes. *J. Mater. Chem. A* 7 (1): 269–277.

34 Xu, C., Song, S., Liu, Z. et al. (2017). Strongly coupled high-quality graphene/2D superconducting Mo_2C vertical heterostructures with aligned orientation. *ACS Nano* 11 (6): 5906–5914.

35 Cai, Y., Shen, J., Ge, G. et al. (2018). Stretchable $Ti_3C_2T_x$MXene/carbon nanotube composite based strain sensor with ultrahigh sensitivity and tunable sensing range. *ACS Nano* 12 (1): 56–62.

36 Zhao, M.Q., Ren, C.E., Ling, Z. et al. (2015). Flexible MXene/carbon nanotube composite paper with high volumetric capacitance. *Adv. Mater.* 27 (2): 339–345.

37 Zheng, W., Zhang, P., Chen, J. et al. (2018). In situ synthesis of CNTs@Ti_3C_2 hybrid structures by microwave irradiation for high-performance anodes in lithium-ion batteries. *J. Mater. Chem. A* 6 (8): 3543–3551.

38 Xie, X., Zhao, M.Q., Anasori, B. et al. (2016). Porous heterostructured MXene/carbon nanotube composite paper with high volumetric capacity for sodium-based energy storage devices. *Nano Energy* 26: 513–523.

39 Yang, L., Zheng, W., Zhang, P. et al. (2018). MXene/CNTs films prepared by electrophoretic deposition for supercapacitor electrodes. *J. Electroanal. Chem.* 830–831: 1–6.

40 Zhang, H., Wang, L., Chen, Q. et al. (2016). Preparation, mechanical and anti-friction performance of MXene/polymer composites. *Mater. Des.* 92: 682–689.

41 Wu, W., Wei, D., Zhu, J. et al. (2019). Enhanced electrochemical performances of organ-like Ti_3C_2MXenes/polypyrrole composites as supercapacitors electrode materials. *Ceram. Int.* 45 (6): 7328–7337.

42 Wei, H., Dong, J., Fang, X. et al. (2019). $Ti_3C_2T_x$ MXene/polyaniline (PANI) sandwich intercalation structure composites constructed for microwave absorption. *Compos. Sci. Technol.* 169: 52–59.

43 Naguib, M., Saito, T., Lai, S. et al. (2016). $Ti_3C_2T_X$ (MXene)-polyacrylamide nanocomposite films. *RSC Adv.* 6 (76): 72069–72073.

44 Luo, J., Tao, X., Zhang, J. et al. (2016). Sn^{4+} ion decorated highly conductive Ti_3C_2MXene: promising lithium-ion anodes with enhanced volumetric capacity and cyclic performance. *ACS Nano* 10 (2): 2491–2499.

45 Mayerberger, E.A., Urbanek, O., McDaniel, R.M. et al. (2017). Preparation and characterization of polymer-$Ti_3C_2T_x$ (MXene) composite nanofibers produced via electrospinning. *J. Appl. Polym. Sci.* 134 (37): 1–7.

46 Zhang (John), C., Kim, S.J., Ghidiu, M. et al. (2016). Layered orthorhombic Nb_2O_5@$Nb_4C_3T_x$ and TiO_2@$Ti_3C_2T_x$ hierarchical composites for high performance Li-ion batteries. *Adv. Funct. Mater.* 26 (23): 4143–4151.

47 Huang, J., Meng, R., Zu, L. et al. (2018). Sandwich-like $Na_{0.23}TiO_2$ nanobelt/Ti_3C_2MXene composites from a scalable in situ transformation reaction for long-life high-rate lithium/sodium-ion batteries. *Nano Energy* 46: 20–28.

48 Zhu, J., Tang, Y., Yang, C. et al. (2016). Composites of TiO_2 nanoparticles deposited on Ti_3C_2MXene nanosheets with enhanced electrochemical performance. *J. Electrochem. Soc.* 163 (5): A785–A791.

49 Ahmed, B., Anjum, D.H., Gogotsi, Y., and Alshareef, H.N. (2017). Atomic layer deposition of SnO_2 on MXene for Li-ion battery anodes. *Nano Energy* 34: 249–256.

50 Zhao, M.Q., Torelli, M., Ren, C.E. et al. (2016). 2D titanium carbide and transition metal oxides hybrid electrodes for Li-ion storage. *Nano Energy* 30: 603–613.

51 Guo, X., Xie, X., Choi, S. et al. (2017). Sb_2O_3/MXene ($Ti_3C_2T_x$) hybrid anode materials with enhanced performance for sodium-ion batteries. *J. Mater. Chem. A* 5 (24): 12445–12452.

52 Zhang, H., Dong, H., Zhang, X. et al. (2016). Cu_2O hybridized titanium carbide with open conductive frameworks for lithium-ion batteries. *Electrochim. Acta* 202: 24–31.

53 Huang, H., Jiang, R., Feng, Y. et al. (2020). Recent development and prospects of surface modification and biomedical applications of MXenes. *Nanoscale* 12: 1325–1338.

54 Wang, Z., Xuan, J., Zhao, Z. et al. (2017). Versatile cutting method forproducing fluorescent ultrasmall MXene sheets. *ACS Nano* 11: 11559–11565.

55 Liu, G., Zou, J., Tang, Q. et al. (2017). Surface modified Ti_3C_2 MXene nanosheets for tumor targeting photothermal/photodynamic/chemo synergistic therapy. *ACS Appl. Mater. Interfaces* 9: 40077–40086.

56 Xue, Q., Zhang, H., Zhu, M. et al. (2017). Photoluminescent Ti_3C_2 MXene quantum dots for multicolor cellular imaging. *Adv. Mater.* 29: 1604847.

57 Dai, C., Chen, Y., Jing, X. et al. (2017). Two- dimensional tantalum carbide (MXenes) composite nanosheets for multiple imaging-guided photo-thermal tumor ablation. *ACS Nano* 11: 12696–12712.

58 Rasool, K., Helal, M., Ali, A. et al. (2016). Antibacterial activity of $Ti_3C_2T_x$ MXene. *ACS Nano* 10: 3674–3684.

59 Rasool, K., Mahmoud, K.A., Johnson, D.J. et al. (2017). Efficient antibacterial membrane based on two-dimensional $Ti_3C_2T_x$ (MXene) nanosheets. *Sci. Rep.* 7: 1598.

60 Xu, B., Zhu, M., Zhang, W. et al. (2016). Ultrathin MXene-micropattern-based field-effect transistor for probing neural activity. *Adv. Mater.* 28: 3333–3339.

61 Wang, F., Yang, C.H., Duan, C.Y. et al. (2015). An organ-like titanium carbide material (MXene) with multilayer structure encapsulating hemoglobin for a mediator-free biosensor. *J. Electrochem. Soc.* 162: B16–B21.

62 Liu, J., Jiang, X., Zhang, R. et al. (2019). MXene-enabled electrochemical microfluidic biosensor: applications toward multicomponent continuous monitoring in whole blood. *Adv. Funct. Mater.* 29: 1807326.

63 Anasori, B., Shi, C., Moon, E.J. et al. (2016). Control of electronic properties of 2D carbides (MXenes) by manipulating their transition metal layers. *Nanoscale Horiz.* 1: 227–234.

64 Ronchi, R.M., Arantes, J.T., and Santos, S.F. (2019). Synthesis, structure, properties and applications of MXenes: current status and perspectives. *Ceram. Int.* 45: 18167–18188.

65 Er, D.Q., Li, J.W., Naguib, M. et al. (2014). Ti_3C_2 MXene as a high-capacity electrode material for metal (Li, Na, K, Ca) ion batteries. *ACS Appl. Mater. Interfaces* 6: 11173–11179.

66 Anasori, B., Lukatskaya, M.R., and Gogotsi, Y. (2017). 2D metal carbides and nitrides (MXenes) for energy storage. *Nat. Rev. Mater.* 2: 16098.

67 Xie, Y., Dall'Agnese, Y., Naguib, M. et al. (2014). Prediction and characterization of MXene nanosheet anodes for non-lithium-ion batteries. *ACS Nano* 8: 9606–9615.

68 Goodenough, J.B. and Park, K.S. (2013). The Li-ion rechargeable battery: a perspective. *JACS* 135: 1167–1176.

69 Meng, R., Huang, J., Feng, Y. et al. (2018). Black phosphorus quantum dot/Ti_3C_2 MXene nanosheet composites for efficient electrochemical lithium/sodium-ion storage. *Adv. Energy Mater.* 8: 1801514.

70 Mathis, T.S., Kurra, N., Wang, X.H. et al. (2019). Energy storage data reporting in perspective-guidelines for interpreting the performance of electrochemical energy storage systems. *Adv. Energy Mater.* 9: 1902007.

71 Lukatskaya, M.R., Bak, S.M., Yu, X.Q. et al. (2015). Probing the mechanism of high capacitance in 2D titanium carbide using in situ X-ray absorption spectroscopy. *Adv. Energy Mater.* 5: 1500589.

72 Hu, M.M., Li, Z.J., Hu, T. et al. (2016). High-capacitance mechanism for $Ti_3C_2T_X$ MXene by in situ electrochemical Raman spectroscopy investigation. *ACS Nano* 10: 11344–11350.

73 Shao, W., Tebyetekerwa, M., Marriam, I. et al. (2018). Polyester@MXene nanofibers-based yarn electrodes. *J. Power Sources* 396: 683–690.

74 Kurra, N., Ahmed, B., Gogotsi, Y., and Alshareef, H.N. (2016). MXene-on-paper coplanar micro supercapacitors. *Adv. Energy Mater.* 6: 1601372.

75 Xu, S., Dall'Agnese, Y., Wei, G. et al. (2018). Screen-printable microscale hybrid device based on MXene and layered double hydroxide electrodes for powering force sensors. *Nano Energy* 50: 479–488.

76 Quain, E., Mathis, T.S., Kurra, N. et al. (2018). Direct writing of additive-free MXene-in-water ink for electronics and energy storage. *Adv. Mater. Technol.* 4: 1800256.

77 Hu, M., Hu, T., Cheng, R. et al. (2018). MXene-coated silk-derived carbon cloth toward flexible electrode for supercapacitor application. *J. Energy Chem.* 27: 161–166.

78 Li, X., Li, H., Fan, X. et al. (2020). 3D-printed stretchable micro-supercapacitor with remarkable areal performance. *Adv. Energy Mater.* 10: 1903794.

79 Ying, Y., Liu, Y., Wang, X. et al. (2015). Two-dimensional titanium carbide for efficiently reductive removal of highly toxic chromium(VI) from water. *ACS Appl. Mater. Interfaces* 7: 1795–1803.

80 Shahzad, A., Rasool, K., Miran, W. et al. (2018). Mercuric ion capturing by recoverable titanium carbide magnetic nanocomposite. *J. Hazard Mater.* 344: 811–818.

81 Mashtalir, O., Cook, K.M., Mochalin, V.N. et al. (2014). Dye adsorption and decomposition on two-dimensional titanium carbide in aqueous media. *J. Mater. Chem. A* 2: 14334.

82 Wei, Z., Peigen, Z., Wubian, T. et al. (2018). Alkali treated $Ti_3C_2T_x$ MXenes and their dye adsorption performance. *Mater. Chem. Phys.* 206: 270.

83 Cai, C., Wang, R., Liu, S. et al. (2020). Synthesis of self-assembled phytic acid-MXene nanocomposites via a facile hydrothermal approach with elevated dye adsorption capacities. *Colloids Surf. A* 589: 124468.

84 Rasool, K., Pandey, R.P., Rasheed, P.A. et al. (2019). Water treatment and environmental remediation applications of two-dimensional metal carbides (MXenes). *Mater. Today* 30: 80–102.

85 Wang, H., Wu, Y., Yuan, X. et al. (2018). Clay-inspired MXene-based electrochemical devices and photo-electrocatalyst: state-of-the-art progresses and challenges. *Adv. Mater.* 30: 1704561.

86 Khazaei, M., Ranjbar, A., Esfarjani, K. et al. (2018). Insights into exfoliation possibility of MAX phases to MXenes. *Phys. Chem. Chem. Phys.* 20: 8579–8592.

87 Abdelmalak, M.N. (2014). *MXenes: A New Family of Two-Dimensional Materials and Its Application as Electrodes for Li-Ion Batteries.* Philadelphia, PA, USA: Drexel University.

88 Meshkian, R., Näslund, L.Å., Halim, J. et al. (2015). Synthesis of two-dimensional molybdenum carbide, Mo_2C, from the gallium based atomic laminate Mo_2Ga_2C. *Scr. Mater.* 108: 147–150.

89 Lei, J.C., Zhang, X., and Zhou, Z. (2015). Recent advances in MXene: preparation, properties, and applications. *Front. Phys.* 10 (3): 276–286.

90 Naguib, M., Mashtalir, O., Lukatskaya, M.R. et al. (2014). One-step synthesis of nanocrystalline transition metal oxides on thin sheets of disordered graphitic carbon by oxidation of MXenes. *Chem. Commun.* 50 (56): 7420–7423.

91 Zhang, C., Beidaghi, M., Naguib, M. et al. (2016). Synthesis and charge storage properties of hierarchical niobium pentoxide/carbon/niobium carbide (MXene) hybrid materials. *Chem. Mater.* 28 (11): 3937–3943.

92 Zhang, C.J., Kremer, M.P., Seral-Ascaso, A. et al. (2018). Stamping of flexible, coplanar micro-supercapacitors using MXene inks. *Adv. Funct. Mater.* 28 (9): 1–10.

93 Zhao, X., Vashisth, A., Prehn, E. et al. (2019). Antioxidants unlock shelf-stable $Ti_3C_2T_x$ (MXene) nanosheet dispersions. *Matter* 1 (2): 513–526.

94 Natu, V., Hart, J.L., Sokol, M. et al. (2019). Edge capping of 2D-MXene sheets with polyanionic salts to mitigate oxidation in aqueous colloidal suspensions. *Angew. Chem. Int. Ed.* 58 (36): 12655–12660.

95 Kamysbayev, V., Filatov, A.S., Hu, H. et al. (2020). Covalent surface modifications and superconductivity of two-dimensional metal carbide MXenes. *Science* 369 (6506): 979–983.

96 Du, Z.G., Wu, C., Chen, Y.C. et al. (2021). High-entropy atomic layers of transition-metal carbides (MXenes). *Adv. Mater.* 33 (39): 1–9.

Index

MXenes: Next-Generation 2D Materials: Fundamentals and Applications, First Edition.
Edited by Jay Singh, Kshitij RB Singh, Ravindra Pratap Singh, and Charles Oluwaseun Adetunji.